Fascinating Life Sciences

This interdisciplinary series brings together the most essential and captivating topics in the life sciences. They range from the plant sciences to zoology, from the microbiome to macrobiome, and from basic biology to biotechnology. The series not only highlights fascinating research; it also discusses major challenges associated with the life sciences and related disciplines and outlines future research directions. Individual volumes provide in-depth information, are richly illustrated with photographs, illustrations, and maps, and feature suggestions for further reading or glossaries where appropriate.

Interested researchers in all areas of the life sciences, as well as biology enthusiasts, will find the series' interdisciplinary focus and highly readable volumes especially appealing.

Friederike Range • Sarah Marshall-Pescini

Wolves and Dogs

between Myth and Science

Springer

Friederike Range
Domestication Lab, Konrad Lorenz
Institute of Ethology
University of Veterinary Medicine
Vienna
Vienna, Austria

Sarah Marshall-Pescini
Domestication Lab, Konrad Lorenz
Institute of Ethology
University of Veterinary Medicine
Vienna
Vienna, Austria

ISSN 2509-6745 ISSN 2509-6753 (electronic)
Fascinating Life Sciences
ISBN 978-3-030-98410-6 ISBN 978-3-030-98411-3 (eBook)
https://doi.org/10.1007/978-3-030-98411-3

This Springer imprint is published by the registered company Springer Nature Switzerland AG
The registered company address is: Gewerbestrasse 11, 6330 Cham, Switzerland

Dedicated to our families (the two- and four-legged ones)

Foreword

Wolves and dogs—what a pair of species! Everybody knows them—at least most people have an opinion about them. And they are described and referred to in so many different ways. One is the wild, the brave, the fearless, the clever, the selfish, or the greedy, while the other is rather tender, withdrawn, obedient, sharing, or kind, perhaps a friend of us. So something like yin and yang or black and white?

All this may sound familiar to people, especially those living in Europe, who grew up hearing the tales by the Grimm brothers, such as Little Red Riding Hood. But there is also another story to tell in which the same wolves are admired by people for their spirit and pure nature, and are regarded by many historic tribes as their ancestors. This culture of wolves was also resurrected by the Romans claiming that a she-wolf raised the two founders, Romulus and Remus, of the city Rome.

So, what could be a greater challenge than to write a book about wolves, probably one of the most mythical animals in human culture. Not to mention that there have been only a handful of scientists who dared to step on this road. Friederike Range and Sarah Marshall-Pescini are just at the right place at the right time to follow this tradition to update what one may call the "ethology and ecology of wolves". There is no place here to list all the books on wolves that have dealt with their life history, behaviour, and evolution. Many earlier works are mentioned in this book. However, thanks to the advancements of the last 20–30 years, and our growing knowledge about wolves, the content of this book is a very significant addition to the earlier published monographs, for example, by biologists Erik Zimen (1971), David Mech (1981), and Mech and Boitani (2010).

Wolves are always a topic for discussion. In a not too distant past, there was the worry that wolves would be pushed to the brink of extinction, at least in Europe and the Americas, but their populations were also decreasing elsewhere. But thanks to conservationists, their fate has changed during the last 20–30 years, and they are now re-colonising many previously abandoned habitats both in Europe and in North America. Although the re-introduction of wolves to Yellowstone was heatedly debated, their showing up in many regions had a positive effect on the ecosystem. Wolf populations recovered even in the absence of active human involvement; a ban on hunting and providing an area with little disturbance appears to be largely sufficient. International students who visited our department were always asked to give a short presentation about their work. Many years ago, a young Dutch student talked about the measures that may be useful to introduce wolves in the Netherlands. We all listened to her very politely, but after the talk we could not resist commenting: "But there are no wolves in the Netherlands!" She was prepared for such a remark and replied very firmly: "Yes, but they will come." And she was right! In 2019 the media reported several sightings of wolves in this country.

So now we have achieved what we had wanted. Are we satisfied and happy? Well, the respected and sacred wolves can end up quickly on the negative end of the news, if we start to believe or actually find out that they are also killing our livestock. So now we all have to learn how to share the land. We should allow the wolves to establish stable populations which help to balance the ecosystem, but we also have to ensure that humans and their livestock are protected from attacks from this top predator.

Friederike Range and Sarah Marshall-Pescini are not afraid of describing and analysing at least 9 different domestication hypotheses. There is the rumour that at the end of the nineteenth century the French academy "banned" the publication of new hypotheses on the emergence of human language. It seems that many people in the community of wolf and dog research enjoy coming up with "THE" domestication hypothesis that explains "everything" about the history of our animal friends. However, if one appreciates the complexity of evolutionary processes, mechanisms of selection, and the interwoven nature of genetic and environmental factors, then it turns out that most such proposals can explain only a small part of the changes in the variation of traits that gave raise to modern dogs. So, although in most of those hypotheses there is likely some "truth", the problem is that we do not necessarily know where it actually is. The main issue is that we do not have the necessary tools to provide experimental evidence for a specific and unique series of events that occurred in the distant past. This situation can be traced in the scientific adventure towards the understanding of where and when dogs were domesticated. Since the first paper using genetic data to establish the time frame and the location of dog domestication (Vila et al 1997), the evidence for a specific time interval and location changed from key paper to key paper. And one important limitation of this approach (probably unlikely to be solved in the near future) is that it is impossible to save very ancient DNA. Moreover, most of these hypotheses are rather complementary than exclusive; they look at different levels of biological organisation. For example, possible selection for genes that may make dogs look so young (see Kinderschema, Lorenz) in the eyes of their human friends may not exclude earlier/parallel/later selection for more flexible agonistic behaviour at older age. Pleitropy and polygenetic effects are also likely to make the unveiling of an evolutionary process very difficult.

I, personally, totally agree with the authors' preference to regard wolves and dogs as two species, but talking about "wolves" and "dogs" in an evolutionary scenario, as it often happens, is a major oversimplification. "Putting things into boxes" is a typical human way of thinking, probably based on our bias to think in linguistic terms. But in reality there are (were?) at least 10–14 subspecies of wolves ranging from Saudi Arabia to Alaska, and there are more than 400 breeds of dogs dispersed over the entire globe, not to mention the dingoes and the millions of stray and feral dogs. So when scientists seek to explain the emergence of differences in certain phenotypic traits, they often imagine the two species as extremes, corresponding to the adjectives above that typically characterise “wolves” and “dogs”, but in reality there are many wolf-like dogs and also dog-like wolf populations. I think we should take both within and between phenotypic variation in these species seriously.

On the more positive side there is also some good news, when it comes to our much improved knowledge about the social life of wolves. Extensive field observations of wolf populations and individual groups are now complemented with observations on social structure, migration of individuals, and intra- and intergroup competition for resources. And the major discovery, attributed to Jane Packard, is also quite recent: typical wolves live in families; that is a father, a mother, and their offspring make up a group. So the old idea of the wolf pack, in which all individuals fight for being the leader, can be thrown into the garbage heap of outdated scientific ideas. Nevertheless, this discovery finds its place only slowly in the mind of people as many still refer to wolf packs instead of families. Somebody may ask, what is the difference? Well, a lot. If the family is the social unit in wolves, their cooperation at various tasks and challenges is less of a surprise. The parents can realise their fitness only if they are able to raise as many young animals as possible, and the offspring can also gain a lot from cooperating with their parents. The collaboration among siblings can be explained on similar grounds. If collaboration in wolves is based on kin selection, then actually the transition to a collaborative dog is actually not a straightforward evolutionary issue because those kin-related genetic factors do not operate when humans and dogs work together.

This book is packed with much knowledge on wolves most of which was established in the last 20 years. I think we all, who are fascinated by the evolution, ecology, and behaviour of wolves and dogs, have to thank those few enthusiastic scientists who dedicated a part of their careers to follow the lives of these amazing creatures. Friederike Range and Sarah Marshall-Pescini deserve a fine place among these brave researchers together with Zsófia Virányi, Enikő Kubinyi, and Márta Gácsi, to name only a few others. The science that is presented with great care in this book, taking into account different views, perspectives, and lack of knowledge, should invite any readers interested in biology to get a closer look at this species without actually getting cold or dirty, and they also do not need to stand the look of a wolf from a short distance.

Congratulations, Friederike and Sarah!

Family Dog Project, Department of Ethology, Eötvös Loránd University
Budapest, Hungary
31 October 2021

Ádám Miklósi

Foreword

Investigating the relationships between humans, wolves, and dogs provides a more important mirror to understand ourselves and our relationship with the world, than most people would appreciate. *Homo sapiens* is the most invasive species on earth, permanently inhabiting even the most unlikely habitats such as deserts and the high Arctic. Dogs arguably contributed a lot to this roaring success. Unplanned and unintended, our ancestors bumped into wolves on their way into the Eurasian continent. It is still a miracle that this happened—and why it happened remains enigmatic. Even more astonishing is that some of these palaeolithic human–wolf associations turned into permanent companionship with domesticated wolves we call dogs. This amazing story raises many questions, for example, how did humans and wolves get into contact in the first place? How come that such a great variety of dogs originated from a single wolf type? And in what way do modern dogs differ from wolves, i.e. what happened during adapting to live with, or close to, humans during the process of domestication?

One would assume that science has long provided most of the relevant answers, but this is not the case. First of all, the topic is so close (too close) to many people, including scientists, that even research was—and still is—an ideological minefield. Second, and most importantly, only the most recent developments in palaeogenetics, archaeology, experimental behavioural biology, and other disciplines provided the kind of firm ground needed to increasingly allow knowledge to compete with opinionation. This book is a milestone of knowledge, integrating more than a decade of experimental research at the Wolf Science Center (WSC) with the relevant results of colleagues worldwide. As an anchor of facts it will provide the ground for further thinking, discussing, and developing the field.

The contents of this book even radiate into experimental archaeology, enlightening educated guesses as to why wolves and humans may have gotten into contact. The most promising candidate factor is the close match between the ecology, lifestyles, and social organisation of hunter-gatherer humans and wolves. Within-clan and within-pack cooperative orientation of humans and wolves was probably key in both the initial match making and our closer-than-ever companionships with dogs. Our present knowledge even contributes arguments towards the early scenarios, making hand-rearing and a respectful non-dominating human interaction style much more likely than the still prominent idea of "self-domestication", holding that some wolves actively chose the company of humans. But who knows? The last witnesses of these remarkable initial events lived some 100 000 generations ago.

The three founders of the Wolf Science Center—Friederike Range, Zsofia Viranyi, and myself were brought together by their keen common interest in wolf and dog behaviour, in the principles of cooperation, cognitive mechanisms, domestication, and human–dog relationships. Given the complexity and novelty of our goals, as well as the effort it would require working long-term with a considerable number of wolves and dogs, we knew that we had to cooperate, not just over developing the concepts, but also in raising funds and in building the necessary research infrastructure. We hand-reared our first four wolf pups in 2008—with a lot of enthusiasm and excitement, but with no financial plan and pretty naïve about what it would mean to run and fund a private enterprise, with 20 employees on the pay list just a few years later. The WSC not only survived, but quickly became scientifically productive. We were extremely grateful when the University of Veterinary Medicine, Vienna, took over the WSC a few years ago; after all, we are scientists, not business administrators. It was certainly worth all the effort. The WSC not just grew into the biggest and one of the most productive basic experimental canid research units worldwide, it is unique in allowing "fair" comparisons between equally raised and kept wolves and dogs, key to fathom the behavioural consequences of dogification.

This book summarises the major WSC research results over more than a decade in perspective with what other workgroups found out and hypothesised. Often enough, the outcomes contradicted mainstream wisdom and surprised ourselves. Less surprising was that the "one, big, fundamental" difference between wolves and dogs does not exist. Rather, a mosaic of differences (and even more matches) between wolves and dogs emerged. Such patterns make sense in the light of domestication, i.e. the adaptation of dogs to live with people, which resulted in an altered socio-ecology as compared to their wolf ancestors—either as an epiphenomenon of the general selection for tameness, or as adaptations of dogs to their new lifestyle, or both. This includes that dogs mentally and physically live in much steeper dominance hierarchies than wolves, affecting the way they relate to, and cooperate with, each other and their human partners, their neophobia and exploration, and how they act out their cognitive skills and solve problems (or not), just to mention a few aspects of the emerging, pretty dog-specific expression of the domestication syndrome.

Through this book, an amazingly coherent story emerges. Many of the mosaic patterns of mentalities and behaviours, as well as their ultimate level explanations, are novel, or at least offer new perspectives and insights. Still, at least as important as the new knowledge featured in this book is its overarching message that even seemingly simple facts may have much more complex mechanistic causalities, than one would wish. Thereby, this book generates a wealth of novel research questions, many of them as unexpected as they are tantalising. I suggest that at present, this is the closest approximation to wolf–dog–human behavioural reality ever, but as typical for science in general, it is certainly not the final word.

I hope—no—I am sure that this book will be well received by scientists as well as by the interested public, as it has considerable potential to reduce the misconceptions, myths, and outright fake news about wolves and dogs, even science is still burdened with. Provided that the messages hit target, they have the potential to make the world a better place for humans and dogs, and not the least, for wolves. Because this is a sober and (self)critical science text, never claiming "the truth" and avoiding any guru positions, it will spark vivid, controversial, and productive discussions and will motivate substantially more research. As every piece of good science, this book provides answers, but raises even more questions.

Department of Behavioural Biology
University of Vienna
Vienna, Austria

Kurt Kotrschal

Acknowledgements

Writing this book has been quite a journey, giving us the precious opportunity to take time to sit and read, walk, and talk, discussing ideas—basically enjoying the best part of science—and trying to critically put all our thoughts on paper. Just as precious is that we have the incredibly good fortune of working with a fantastically collaborative team of researchers, as well as many colleagues around the world who generously gave us great feedback on different parts of this book.

We particularly thank Kurt Kotrschal for taking the time to carefully read through the entire manuscript, chapter by chapter, giving us suggestions on how to improve the readability of the book and, importantly, many wonderful comments to the scientific content. Given we do not always agree, his comments were the most thought-provoking and challenging and therefore immensely useful!

Several other people also provided very valuable feedback to one or several chapters including Andreas Berghänel, Giulia Cimarelli, Rachel Dale, Martina Lazzaroni, Malgorzata Pilot, Sabine Tebbich, Gwen Wirobski, and Clive Wynne. The lively questions and sometimes challenging comments from many different perspectives enriched our thinking, and we hope this comes through in the book. Renate Hengsberger helped us sorting out the references in the various chapters.

We would also like to acknowledge all the researchers who have worked with us over the years, observing and conducting experiments with the animals, but also contributing to our thinking, in particular Zsófia Virányi and Simona Cafazzo. In a memorable trip to Rome the 4 of us laid the ground for much of the work that is summarised in this book.

We would also like to thank the entire trainer team at the Wolf Science Center that made the work at the Wolf Science Center possible over the last decade: Bea Belenyi, Lars Burkert, Charles Gent, Anna Griebler, Nadine Gross, Marianne Heberlein, Marleen Hentrup, Corinna Kratz, Katharina Kriegler, Christine Mayer, Lina Oberliessen, Caroline Ritter, Rita Takacs, Laura Stott, and Cindy Voigt. Their dedication to the animals and support to the students greatly contribute to the unique environment, which makes our science possible. Equally essential is the support of all the administration staff at the Vetmed University: Brigitte Pavlik, Sandra Sperlinger, and Christina Szaga-Doktor and at the Wolf Science Center Marlene Fusek, Tina Gunhold-de Oliveira, Aleksander Orlic, Esther Rudolph, Claudia

Wittman, and Lena Schaidl. A thriving research group produces an enormous amount of bureaucracy, and without their help, we would be lost.

We are immensely grateful for all the people and institutions who supported us with the financial means to set up the Wolf Science Center and conduct our research. We would like to thank the University of Veterinary Medicine, Vienna, and its employees for the generous support—both at the financial and personal level—appreciating our work and always having an open ear for our needs. Thanks also to the University of Vienna, which supported us especially in the early years. A number of funding agency have supported our past and present scientific work.

We would like to thank the FWF (Austrian Science Fund), WWTF (Vienna Science and Technology Fund), ERC (European Research Council), ÖAW (Austrian Academy of Science), and OeAD (Austria's Agency for Education and Internationalisation). We would also like to thank all the companies and private sponsors that have supported the Wolf Science Center and thus our work over many years. Last but certainly not least, we would like to thank our families. Especially the two "amore" in our lives, who put up with our endless talk of wolves and dogs and fueled us with good food, glorious wine, and happy holiday times all together. Finally, a gentle thanks to the hairy ones in our lives, past and present.

Contents

1 Introduction 1
1.1 Why Study the Cognitive Abilities of Wolves? 1
1.2 How Wolves Became Dogs . . . Maybe 3
1.2.1 The 'Human-Selection' Scenario 4
1.2.2 The Self-Domestication Scenario 6
1.3 Wolves and Dogs Today 9
1.4 Book Overview 11
References 13

2 Comparing the Behaviour and Cognition of Wolves and Dogs 19
2.1 Wolf Park Indiana, Battle Ground, IN (1972–Present) 22
2.2 Kiel 23
2.3 Erik Zimen 24
2.4 University of Michigan Canine Information-Processing Project (1979–1984) 25
2.5 Family Dog Project of the Department of Ethology, Budapest (2002–2004, Partially Ongoing) 26
2.6 The Dog Cognition Programme at Max Planck Institute in Leipzig 27
2.7 Tovetorp Zoological Research Station of Stockholm University (2014–2016) 28
2.8 Wolf Science Center (2008–Present) 28
2.8.1 Decisions Taken for the Set-up of the Wolf Science Center 30
2.8.2 The Reality 36
2.8.3 Socialization of Our Animals 36
2.8.4 Other Research Projects Connected to the Wolf Science Center 40
References 43

3 The Socio-Ecology of Wolves 47
3.1 A Few General Facts 48
3.2 Environment and Diet: The Wolf as the Epitome of the Hunter? 51
3.2.1 Factors Influencing Hunting Success 51
3.2.2 The Diet of Wolves 54
3.3 Social Organization 55
3.3.1 Pack Composition and Pack Size 55
3.3.2 Mating and Reproduction 57
3.3.3 Parental Care: Cooperative Breeding 60
3.3.4 Pup Mortality and Dispersal 61
3.3.5 Social Dynamics 65
3.3.6 Territory Defence 66
3.4 Recolonization of Wolves 69
3.5 Summary 74
References 74

4 The Socio-Ecology of Free-Ranging Dogs 83
4.1 A Few General Facts 87
4.2 Environment and Diet: Scavenger or Hunter? 88
4.2.1 Dog: The Scavenger 88
4.2.2 Dog: The Hunter 90
4.3 Social Organization 92
4.3.1 Group Size: When Are Dogs Solitary and When Social? 92
4.3.2 Group Composition 95
4.3.3 Mating and Reproduction 95
4.3.4 Parental Care 98
4.3.5 Social Dynamics 102
4.4 Summary 104
References 104

5 Social Dynamics Between Conspecifics 111
5.1 Dominance and Aggression 112
5.1.1 Formal Signals of Dominance in Wolves and Dogs . . . 114
5.1.2 Dominance, Hierarchy, and Leadership in Captive and Free-Living Wolves and Dogs 116
5.1.3 Dominance and Aggression: Wolves and Dogs Compared 120
5.2 Dominance and Access to Resources 125
5.2.1 Dominance, Aggression, and Access to Resources: Wolves and Dogs Compared 127
5.3 Affiliative Relationships 131
5.3.1 Affiliation: Wolves and Dogs Compared 133
5.4 Play 135

5.4.1 Summary Play 140
5.5 Reconciliation 140
5.5.1 Summary Conflict Management 143
5.6 Do the Affiliative and Dominance Dimensions Affect Wolves' and Dogs' Behaviour in Different Contexts? 144
5.7 Summary 147
References 147

6 Social Behaviour with Humans 159
6.1 Attachment and Sociability 159
6.2 Wolves and Dogs Compared 162
6.2.1 Puppy Tests 162
6.2.2 Summary Studies with Puppies 165
6.2.3 Adult Tests 165
6.3 Physiological Correlates of the Human-Animal Social Bond . . . 178
6.3.1 Wolves and Dogs Compared 179
6.3.2 Summary Bonding and Physiology 185
6.4 Overall Summary 187
References 189

7 Social Cognition Part I: Communication with Humans 195
7.1 Understanding (Human) Communicative Cues 197
7.1.1 Gaze Following 198
7.1.2 Following Pointing 201
7.1.3 Do Dogs and Wolves Differ in Regard to Understanding Communicative Cues? 201
7.1.4 Sensitivity to Ostensive Cues 211
7.1.5 Summary 213
7.2 Communicating with Humans 215
7.2.1 Sensitivity to the Attentional States of Others 215
7.2.2 Showing Behaviour 217
7.2.3 'Looking Back' Behaviour in the Impossible Task 218
7.2.4 Summary Communication with Humans 224
7.3 Summary Communication 224
References 226

8 Social Cognition Part II: Social Learning and Cooperation 233
8.1 Social Learning 233
8.1.1 Local Enhancement 235
8.1.2 Imitation 239
8.1.3 Summary Social Learning 241
8.2 Cooperation 242
8.2.1 Prosocial Behaviour 243
8.2.2 Inequity Aversion 253
8.2.3 Coordination in a Cooperative Problem-Solving Task . . . 258
8.2.4 Summary Cooperation 270
References 271

9 Physical Cognition Part I: Problem Solving and Its Correlates 277
9.1 Motivational Factors 279
9.1.1 Neophobia and Exploration 279
9.1.2 Persistence and a Little Bit of Motor Diversity 284
9.1.3 Summary: Motivational Factors 288
9.2 Attention and Working Memory 288
9.3 Inhibitory Control 290
9.3.1 Cognitive Inhibition 291
9.3.2 Motor Inhibition 291
9.3.3 Self-Control 298
9.3.4 Summary Inhibition 301
9.4 General Summary 302
References 303

10 Physical Cognition Part II: Learning and Reasoning 311
10.1 Learning 312
10.1.1 Summary: Learning 315
10.2 Inferential Reasoning 315
10.2.1 Reasoning by Exclusion 317
10.2.2 Diagnostic Inference 320
10.2.3 Predictive Inference 321
10.2.4 Summary: Inference 323
10.3 Numerical Competence 324
10.3.1 Summary: Numerical Competence 330
10.4 General Summary 330
References 331

11 Domestication Hypotheses Relating to Behaviour and Cognition: Which Are Supported by the Current Data? 335
11.1 A Comprehensive Theory for All Domesticated Animals: 'The Domestication Syndrome' 337
11.1.1 Evidence For and Against the Domestication Syndrome 338
11.2 What Brought About the Behavioural Wolf-Dog Differences We See Today and Which Hypotheses Are Supported by the Current Data? 341
11.2.1 Interactions with Humans 341
11.2.2 Social Behaviour with Conspecifics 357
11.2.3 Interaction with the Environment 362
11.3 Summary and Overall Conclusions 366
References 367

WSC (Wolf Science Center) Ethogram for Pocket Observer 375

Introduction 1

The most erroneous stories are those we think we know best—and therefore never scrutinize or question. Stephen Jay Gould

1.1 Why Study the Cognitive Abilities of Wolves?

Most importantly, they are very charismatic animals, and they are just fantastic to watch and work with. But of course, there are also good scientific reasons why it is exciting to study wolves and dogs from a theoretical but also practical perspective.

First, the quest to understand the evolution of the large primate brain—especially the human brain—has governed a lot of research in cognitive biology over the past decades. Based on the assumption that, similar to morphological features, cognitive abilities are shaped by natural selection (Byrne 1997), researchers have focused on studying key traits thought to be evolutionarily relevant in this selection process. For example, it has been proposed that the large neocortex of primates evolved as an adaptation to cope with social complexity such as kinship networks, dominance hierarchies, and alliance formation ('the social brain hypothesis'; Byrne and Whiten 1988; Dunbar 1998). Others have suggested that the development of tools to extract food (Foraging or Technical Intelligence Hypothesis; Clutton-Brock and Harvey 1980; Rosati 2017) played a major role in the evolution of the large primate brain. Finally, it has been argued that the motivational and cognitive consequences of cooperative breeding played a major role in the evolution of human cognition (Burkart et al. 2009).

To test the general validity of hypotheses regarding the evolution of intelligence, it is important to also investigate the cognitive abilities of non-primate species. Wolves are actually a fantastic model species, because in contrast to chimpanzees, wolves live in family packs that largely depend on cooperative hunting. Thus, the socio-ecological niche of wolves is rather similar to the one humans have lived deep

F. Range, S. Marshall-Pescini, *Wolves and Dogs*, Fascinating Life Sciences,
https://doi.org/10.1007/978-3-030-98411-3_1

in pre-history: small-scale societies of hunter–gatherers (Hall and Sharp 1978; Marlowe 2005). Like wolves, traditional human groups rely much on meat. Meat is a high-value resource, but it requires a lot of effort to obtain when hunting is involved and success rates are rather low (Kaplan et al. 2000). To acquire food, humans, like wolves, often need to travel relatively far on a daily basis since these resources are widely dispersed (Marlowe 2005). Moreover, humans usually bring the food back to a central place, and the males provide females and their offspring with food (Kaplan et al. 2000). Similarly, during the puppy-raising period, the breeding male and other non-reproductive members of the pack will bring back food to a den or rendezvous site to share both with the breeding female and the puppies (see Chap. 3). Also, in terms of their social organization and social skills, wolves resemble traditional human groups: they live in family groups and cooperate during territory defence, raising of the young, hunting, and sharing of the food afterwards. This lifestyle requires not just high mutual tolerance but likely also skills like social learning (to learn how to hunt cooperatively), attentiveness, and coordination—all aspects that characterize our own species. Thus, overall, the 'similar' social organization and social skills of wolves and traditional human hunter–gatherer societies might reflect convergent adaptations to similar ecological niches (Hall and Sharp 1978), suggesting wolves may be a good model to investigate the evolution of human social skills.

Second, according to genetic studies, wolves are the closest wild-living relatives of dogs (Savolainen et al. 2002; Scott and Fuller 1965). The latter have shared the human environment for the last thousands of years (Frantz et al. 2016; Thalmann et al. 2013; Wang et al. 2014) and in some ways have become human's 'best' friends, i.e. social and working companions. It has been argued that since they are living in such close proximity to us, it is likely that they will enjoy advantages if they are able to communicate and cooperate with people (Miklósi et al. 2004). As such, dogs are thought to have developed cognitive–emotional responses, analogous to the ones that differentiate humans from other primates, in comparison to other animal species (Lakatos et al. 2009; Soproni et al. 2001), including chimpanzees (Bräuer et al. 2006; Hare et al. 2002, but see Mulcahy and Hedge 2012). Assuming that today wolves still strongly resemble the common Paleolithic ancestor of dogs and wolves (Lindblad-Toh et al. 2005), comparing their behaviours and cognitive skills gives us the possibility to explore the changes brought about by the evolutionary process of domestication,[1] i.e. the adaptation to the human environment. To date, many different hypotheses have been proposed to explain how exactly dog cognition might have changed during this process (Hare et al. 2002; Hare and Tomasello 2005a, b; Miklósi et al. 2003). We will discuss these hypotheses in detail in the light of the data on wolf–dog comparisons in the last chapter of this book.

[1]Domestication can be generally considered a selection process for adaptation to human agro-ecological niches and, at some point in the process, human preferences (Larson et al. 2014).

Finally, understanding wolf cognition and behaviour in itself and in comparison to dogs has a high practical relevance. First, a better understanding of wolves may actually pave the way for their return into our daily lives in light of the fact that wolves are spreading all over Europe regaining their past distribution. The return of the wolf into populated areas, however, does posit many challenges in terms of the protection of livestock, and it will be important to find the means for an acceptable coexistence. Understanding the behaviour and motivation of wolves might help in that regard.

Moreover, understanding differences and similarities between wolves and dogs may contribute to the development of socialization and training techniques that can facilitate the life of dogs and on the other hand help debunk some problematic but still highly popular notions that owners should behave 'like alpha wolves' in order to naturally educate their dogs. A closer look at this notion reveals that it is often a supposed justification for the use of force, inhibition, and aggression to control dogs' behaviour. The assumption that owners should behave like 'alpha' wolves is wrong on so many levels. To start with, as we shall see, wolf packs are largely based on affiliative and cooperative relationships with high levels of tolerance; hence, using wolf social behaviour to justify any form of aggression is incorrect (as well as entirely unethical). Second, while wolves and dogs are closely related, their social interactions with each other and humans are not the same but an outcome of the domestication process. So yet again, using a wolf model—especially the wrong one—to inform training aspects is inaccurate. Finally, we humans are primates, and the challenge, joy, and uniqueness of a fruitful and positive relationship with dogs is to find a common language, which can engender a strong positive bond between us, since what is certain from behavioural research is that for both canids and primates, social bonds are massively important.

1.2 How Wolves Became Dogs . . . Maybe

The domestication process is associated with rapid and extensive changes in phenotypic traits, either under the direct pressure of artificial selection by humans or as a result of adaptation to new habitats created by humans (Larson and Fuller 2014). Actually, the first animal to turn domestic was the wolf. This is one of the most intriguing domestication processes as wolves are highly successful group-hunting predators and are potentially dangerous to man (Linnell et al. 2002); the more as Paleolithic wolves may have been much less shy of humans than present day wolves, due to the lack of any severe hunting pressure. Nevertheless, of all domesticated species, dogs are probably those that form the closest social relationship with us. Based on many recent publications of genetic studies, the domestication process started about 15,000 to max. 30,000 years ago (Frantz et al. 2016; Larson and Bradley 2014; Skoglund et al. 2015; Thalmann et al. 2013; Wang et al. 2015) and happened somewhere in the Old World, however, where exactly is less clear (Freedman et al. 2014; Pang et al. 2009; Shannon et al. 2015; Thalmann et al. 2013; Wang et al. 2013; Wang et al. 2015). It is also not quite clear how many independent wolf populations were involved in how many founding events (Botigué et al. 2017; Thalmann and Perri 2018). There is, for example, growing evidence that the wolf ancestral branch of dogs is now extinct (Bergström et al. 2020; Freedman et al. 2014), which is not surprising as many species are dynamically changing over time.

Despite these debates, most researchers agree that dogs were well established across Eurasia by the late Pleistocene. While during the Mesolithic period, humans lived in nomadic hunter–gatherer societies with a geographical overlap with wolf populations, with the climate change from the glacial times to more agreeable weather, humans started to settle in many areas, slowly engaging more and more into agriculture rather than living as hunter–gatherers about 10,000 BP (Richerson et al. 2001). A recent study suggests that five lineages of dogs already existed in the late Pleistocene, prior to the transition to the Holocene, and that very little wolf introgression is evident in these early dogs (Bergström et al. 2020). This suggests that, by this time, dogs and wolves were already genetically quite distinct.

There are two major scenarios relating to dog domestication: the human-selection and the self-domestication scenario.

1.2.1 The 'Human-Selection' Scenario

The 'human-selection' scenario (or 'directed pathway'; see Zeder 2012; Larson and Fuller 2014) suggests that people actively adopted wolf pups, raised them, and then selected the most docile animals (likely killing the others), which then reproduced with each other (Clutton-Brock 1992; Clutton-Brock 1995; Galton 1865). Based on pioneering studies by Belyaev and colleagues on silver foxes (see Box 1.1 and Chap. 11 for a discussion of these studies hypothesis), direct selection for a reduced fear response towards humans could result in approximately 50% of the population showing dog-like friendly responses after around 30 generations (Trut et al. 2009; Trut et al. 2004). However, the sustained effort to maintain strict control over the animal's breeding choices for so many generations appears, intuitively, somewhat at odds with the capacity and lifestyle of early humans (or in fact even modern humans considering the little control we have of free-ranging dogs' mating behaviours).

Box 1.1 The Tame Foxes and the Pioneering Ideas of Belyaev

The distinguished Russian geneticist, Dmitry K. Belyaev, was the first researcher proposing that the key factor of domestication is selection for *tameness* producing strikingly similar results in many species (Belyaev 1979; Trut et al. 2009). He hypothesized that by selecting against aggression and fright to facilitate human–animal symbiosis, significant behavioural responses (i.e. tame animals) appeared through destabilizing selection (*Destabilizing Selection for Tameness Hypothesis*; Belyaev 1979). The basis

(continued)

Box 1.1 (continued)
for this idea is the close relationship between the nervous and endocrine systems. By selecting for behaviour, the hormonal status of an animal can be changed, which in turn may have consequences for the ontogenetic development of the animals—after all, hormones regulate gene functions. Belyaev thought that this process was brought about by destabilizing selection, which results in 'breaking up previously integrated ontogenetic systems and thus leads to multiple phenotypic effects that seem genetically unrelated to the selected character, namely tame behavior'. This destabilization of the neuro-endocrine control of ontogenesis seems to occur when stress factors in the environment appear or increase in strength (cited from Belyaev 1979). Since mammals share similar regulatory hormonal and neurochemical mechanisms, similar effects are expected across domesticated mammals.

To test his ideas, in the 1960s, Belyaev (1969) together with several colleagues, in particular with his student Lyudmila Trut, started a now 'text-book famous' experiment to investigate the effect of domestication. He used silver foxes bred for the fur industry and always selected the top 3–5% of animals that readily approached the hands of humans standing in front of their cages for breeding. This selection for presumably reduced fear and aggression towards humans led to animals that showed no fear and aggression but rather showed tail-wagging, whining, and licking of the experimenters' hand and face when approached as well as following the experimenter when allowed out of the cage (Trut et al. 2004). Moreover, the domesticated foxes had lower stress hormone levels than the unselected foxes, suggesting that the mechanistic basis of the tame behaviour is a change in the hypothalamic–pituitary–adrenal system (Trut et al. 2009).

The programme is still running today at the Cytology and Genetics Department of the Russian Academy of Sciences in Novosibirsk (for a lovely popular science book narrating the story of this project, see *How to Tame a Fox*; Dugatkin et al. 2017). Thanks to a chance meeting with Anna Kukekova, a Russian geneticist based in America who has been working with the foxes for many years, we were lucky enough to be able to visit the fox farm in Novosibirsk (in December!, Fig. 1.1). We were blown away by the dedication, persistence, and commitment of the team of scientists and keepers taking care of the animals against so many odds. It is an extraordinary living testament to the passion that science and animals can inspire! From a scientific perspective, it is also the best available model to better understand whether 'selection for tameness' in domestic dogs is indeed the trait which is (alone?) responsible for the wolf–dog differences we observe.

Fig. 1.1 Two enthusiastic faces, paying homage to Belyaev and his foxes at the Cytology and Genetics Department of the Russian Academy of Sciences in Novosibirsk

1.2.2 The Self-Domestication Scenario

The other scenario of dog domestication is the 'self-domestication' or 'commensal pathway', which proposes a two-stage process. In the first stage, dogs' ancestors started exploiting a new feeding niche provided by humans (Coppinger and Coppinger 2001, 2016; Hulme-Beaman et al. 2016; Larson and Fuller 2014; Larson et al. 2012; Zeder 2012). The beginning of dogs' domestication appears to coincide with the late Pleistocene megafauna extinction (Cooper et al. 2015) and is associated with a dramatically reduced prey availability, which resulted in a drastic fall in wolves' population size (Pilot et al. 2014). This also coincided with the disappearance of a subspecies, which exhibited a more robust craniodental morphology (Baryshnikov et al. 2009; Kuzmina and Sablin 1994; Leonard et al. 2007) likely specialized in hunting of and scavenging on megafauna (Fox-Dobbs et al. 2008; Germonpré et al. 2017). This subspecies was replaced by a less robust wolf likely originating from the Beringia region, where more reduced climatic fluctuation may have reaped less havoc with their environment and predator–prey equilibrium (Loog et al. 2020).

In this rapidly changing environment, some wolves might have discovered and slowly started to rely upon what could be found around human encampments that, at this time, were largely nomadic. While these were certainly not food waste dumps (see Jung and Pörtl 2018), which emerged only thousands of years after the start of domestication, it might have been human faeces or other leftovers that humans could not digest, which attracted the animals. A new 'excess protein' hypothesis suggests that humans might have shared certain meat types with wolves (Lahtinen et al. 2021). Since humans are not fully adapted to a carnivorous diet due to the limited capacity of the liver to metabolize protein, they might have shared surplus of animal-derived protein with wolves/proto dogs especially during harsh winters when game is lean. While the authors suggest that humans actively shared their excess meat, no good argumentation is provided for it. Instead, they might as well have just left it, attracting the animal's presence. In any case, from the wolves' perspective, a new source of food linked to humans might have become available at a time of food scarcity.

Wolves being a highly adaptable species might have exploited this new niche, which likely leads to new adaptations selected for to increase fitness in the new environment (Zohary 2004). Indeed, a change in diet is linked to rapid adaptions. One study has found that differences in habitat, specifically precipitation, temperature, vegetation, and prey specialization, strongly affect wolves' craniodental morphology (Flower and Schreve 2014; Geffen et al. 2004; O'Keefe et al. 2013; Pilot et al. 2006). Despite living side by side and without physical barriers to prevent gene flow between them, North American wolf populations (Larson and Fuller 2014; Musiani et al. 2007; Stronen et al. 2014), as well as European wolf populations that have specialized in different prey and habitat use, show distinct genetic and morphological differentiation. This suggests that even without direct human control on the breeding choices of wolves, adaptation to a novel ecological niche could have resulted in morphological and genetic changes in direction towards dogs. Importantly, this natural selection process continues during the entire domestication process and is independent from the behaviour of the domesticator (Alter 2007; Heiser 1988; Zohary 2004).

A further catalyst for changes in dogs' temperament and behaviour may have been the rise in refuse availability, which accompanied the development of sedentary human communities associated with the agricultural revolution (Newsome and van Eeden 2017). While humans might have actively shared some resources (Lahtinen et al. 2021), the huge majority of dog food likely consisted of human faeces, which are surprisingly nutritious (Butler et al. 2018; see also Chap. 4), and animal or plant parts, which humans do not usually eat. Indeed, numerous studies have suggested that the availability of anthropogenic foods might alter a series of behaviours including predator–prey relationships, hunting behaviour, trophic interactions, and bottom-up and top-down processes as well as sociality, pack size, breeding rates, dispersal, and group density (Butler et al. 2018).

The importance of starch, and evidence of a shift in feeding ecology from hunting to scavenging, emerges from dogs showing genetic adaptations for starch digestion, which are absent in wolves (Axelsson et al. 2013) but already present in the ancient Newgrange dog (4800 BP, Botigué et al. 2017; Frantz et al. 2016). Analyses of DNA from ancient canines suggest that the adaptation was present already in some dogs

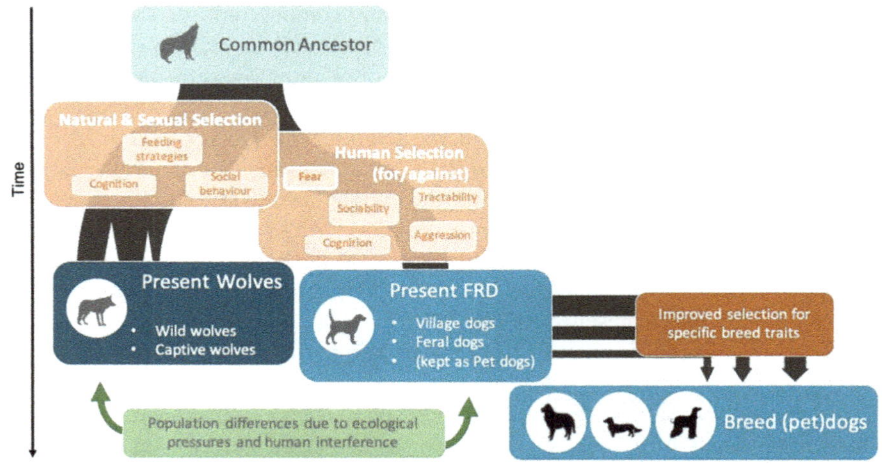

Fig. 1.2 Diagram of the dog domestication process. Figure reprinted from Trends in Cognitive Sciences, Range and Marshall-Pescini, Comparing wolves and dogs: current status and implications for human 'self-domestication, 2022, with permission from Elsevier (license number: 5297040356361)

living with hunter–gather human societies, but it emerged in full strength only after several thousands of years of living with humans, who had adopted an agricultural lifestyle (Bergström et al. 2020).

Overall, whereas it is likely that during the domestication process, from the very start, humans did put some selective pressure on the populations living in close proximity to them by removing unwanted animals (e.g. those that were too aggressive; see MacLean et al. 2019), it was only at a later stage of this domestication process, when the animals were likely already more amenable to being controlled, that humans directly selected for specific, desired traits (e.g. more attentive, responsive, cooperative, etc.) (Fig. 1.2).

However, this 'self-domestication' hypothesis that identifies a shift in wolves' niche as the initial stages of the domestication process is not without its critics. For example, as pointed out above, food waste dumps did not exist during the Pleistocene, and, if indeed food waste attracted wolves, it is unclear why other animals like foxes and jackals were not domesticated (Jung and Pörtl 2018). Moreover, wolves and humans likely encountered each other regularly since they were hunting the same prey, and there are some hints that they respected each other (Jung and Pörtl 2018). Ultimately, we will never know what happened, and most likely it was a combination of several factors leading wolves to being domesticated including that they are very social animals relying on cooperation with each other. Independent of the exact process, it is important to keep in mind that the domestication process involved initial stages, where humans lived as hunter–gatherers, then in small settlements, and finally in larger villages. Therefore, the change in the dogs' environment needs to be seriously considered, as well as the direct selection for certain traits by humans, when trying to understand the domestication process and the differences between wolves and dogs today.

Interestingly, a version of these two potential paths of domestication was already proposed by Darwin. He suggested that observed changes in domestic animals during domestication could occur either as methodical or unconscious selection (Darwin 1859). As implied by the names, Darwin saw the difference between the two selection processes as whether there was a purpose to improve the breed behind any selection or not. For example, if humans preserved the most valued and prevented the less valued individuals from reproducing, without any thought of altering the breed, he would term it unconscious selection. However, at the same time, humans would hope that the preserved animals would produce offspring with similar traits (see Bidau 2009 for a nice review on Darwin's ideas).

Building on Darwin's idea, we would argue that there are actually three types of selective pressures that formed and are still forming dogs. First, we often select animals for specific phenotypic traits that we would like to maintain and promote in the population (improved selection)—this is probably mostly, but not exclusively, conscious. Then there is selection 'against' a specific trait, which can be both conscious and unconscious. Unconscious selection likely defined many 100 s of years of cohabitation and still does in the free-ranging populations—e.g. too bold/aggressive/pushy animals get killed. And then, finally, there is likely also a selection process, which is not directly linked to human decisions (conscious and unconscious) but is rather determined by 'environmental/conspecific' selection. This was probably important during early domestication but still applies to FRDs today. It is what we usually refer to as natural and sexual selection in non-domesticated species and refers to the adaptation to the environment and the selection processes deriving from mate choice. This raises the question of how much the process of domestication is an intentional process based on human's deliberate decisions (Morey 1994). Or, at least in part, a 'self-domestication' process, where wild animals adapted to new, man-made ecological niches.

To sum up, while the self-domestication hypothesis suggests that dog behaviour evolved as a result of the combined pressures of natural and sexual selection that have enabled dog populations to adapt to a new niche that differs from those of grey wolves (see Fig. 1.2; Chaps. 3 and 4), as well as artificial selection imposed by humans, the human-selection hypothesis, as the name implies, focuses on the artificial selection.

1.3 Wolves and Dogs Today

Having outlined how the domestication *event* may have started, it is important to understand that domestication is in fact an *ongoing process*. Whereas it is perhaps easier to see how dogs continue to evolve today because it happens in front of our eyes, it is important to remember that also wolves did not 'stop evolving'. This is also why we should not refer to today's wolves as dogs' ancestors, but rather as their 'closest living relatives'.

The wolf population(s) giving rise to dogs during the early phases of domestication are likely extinct (Freedman et al. 2014; Thalmann et al. 2013). The wolves living during the Pleistocene in Northern Eurasia and America represented a continuous population genetically and morphologically (e.g. a skull shape allowing for much greater bite force) distinct from the wolves living in this area today (Hofreiter 2007; Leonard et al. 2007). They were likely adapted to hunting and scavenging large ungulates such as bison (*Bison bison*) and horses (*Equus lambei*) and vanished with the extinction of the megafauna in the late Pleistocene (Leonard et al. 2007). Today's wolf populations all derive from a single wolf population from the Beringia areas, which successfully survived the megafauna extinction (Loog et al. 2020). In Europe, the haplogroup associated with the Beringian wolves also substantially decreased in frequency likely due to the same ecological processes (Pilot et al. 2010). The extinction of the megafauna and the resulting bottleneck in wolves soon after their divergence from dogs implies that at the onset of domestication, there was substantially more genetic diversity for selection to act upon than is present in our wolves today.

In the course of their history, wolves have undergone severe 'genetic bottlenecks', i.e. periods of their existence in which as a population they underwent drastic and significant reductions in numbers (Fan et al. 2016; Pilot et al. 2014). In recent centuries, these bottlenecks were caused by humans; however, for earlier cases, it is not as clear. In general, how severely humans affected wolf populations depends on location, which resulted in different degrees of genetic variability in populations across Eurasia as well as in America. For example, populations from Siberia and Mongolia show high genetic variability, whereas Italian and Iberian populations show low variability (Pilot et al. 2014), which probably reflects hunting pressure by humans. In North America, wolves originated (most likely) from a single expansion from Beringia, and as a result, their genetic variability has been low to start with, due to a process known as 'founder effect', which is similar to a population bottleneck.

Severe hunting pressures by humans on wolves may have effectively selected for 'shyer' animals, in that more elusive, fearful, maybe nocturnal animals may have succeeded in evading hunters more successfully than bolder ones. Interestingly, circumstantial support for the possibility that different hunting pressure between American and European wolves may have affected wolves on the shyness–boldness axis comes from the general perception of zookeepers that European wolves, compared to American ones, are more skittish, reactive, and shy, leading most centres to prefer either American or Arctic wolves that do not mind the presence of visitors as much as European wolves do. However, it is still possible that the inadvertent selection for shyer wolves might have also resulted in the selection for correlated traits such as higher aggression (Fadel et al. 2016; Wright et al. 2012).

The majority of the dogs we meet today at least in the Western world are modern breeds that originated from recent, stringent breeding efforts during the last 200 years (Lindblad-Toh et al. 2005; von Holdt et al. 2010). This strong artificial selection to breed dogs with specific characteristics to fulfil certain functions (e.g. herding, guarding, hunting) (Careau et al. 2010; Coppinger and Coppinger 2001; Svartberg 2006; Zapata et al. 2016) is responsible for the striking morphological differences among dog breeds (Ostrander et al. 2000; Spady and Ostrander

2008; Zapata et al. 2016) but also for a reduction in their genetic variability at least in pure-bred animals (Lindblad-Toh et al. 2005).

Another important aspect is that while wolves and dogs have clearly diverged in morphology, behaviour, physiology, and genetics, they can still produce fertile hybrid offspring. Thus, in terms of the biological species definition, they cannot be considered two different species, and, accordingly, dogs are often referred to as *Canis lupus familiaris*, i.e. as a subspecies of wolves. However, this is somewhat problematic, since there is also evidence that wolves mate with coyotes and produce fertile offspring (e.g. Bohling et al. 2016; Hinton et al. 2018; Mech et al. 2017) suggesting that maybe all *Canis* species can successfully interbreed and thus could be considered as a single species. On the other hand, it is clear that wolves and dogs differ in many aspects and are adapted to different ecological niches. Accordingly, and to facilitate terminology in this book, we will refer to wolves and dogs here as two species: *Canis lupus* and *Canis familiaris*.

However, the potential level of hybridization with dogs in any wolf population is an important point to keep in mind when comparing the two species, since 'hybrids' may well behave differently than both their original species. Studies so far have shown that dog introgression in North American wolf populations is much less of an issue compared with Eurasian wolves (Pilot et al. 2018, 2019).

In summary, while we can assume that the change from a common ancestor to today's wolves might not have been as profound as the change observed from Pleistocene wolf to modern dogs, it is also unlikely that wolves, as a species, have remained unchanged since the start of the domestication process. Indeed, at least old-world wolves and dogs might actually have been selected for opposite sides of a continuum: shyness towards humans versus affiliation and closeness to humans. So, in the end, what we can investigate is how wolves and dogs differ from each other at this current stage of the domestication process. We are unable to go back in time and understand every part of this journey, but based on the differences and similarities in the behaviour of wolves and dogs, we can propose likely scenarios about the possible selective pressures that shaped the domestic dog.

1.4 Book Overview

Having outlined the important steps in the domestication process (see Fig. 1.2; Boyko 2011), in the following chapters, we will first introduce the main (past and present) research programmes which have engaged in this comparison of wolves and dogs and then discuss results from studies investigating different topics. The aim of this book is not to review all literature on dog and wolf behaviour and cognition but rather to focus on studies where direct comparisons between the two species were conducted and then discuss what we can learn from such results about the domestication process.

We start the book by introducing the important aspects that need to be kept in mind when comparing wolf and dog behaviour and the main research groups (past and present) that focused on the wolf–dog comparison (Chap. 2). We then summarize the social ecology of wolves (Chap. 3) and dogs (Chap. 4) to give the reader an idea of the environment and the potential challenges that the two species have adapted to. In the wolf chapter, we will review studies from many different researchers studying wolves in the wild and in captive settings, while in the dog chapter we highlight the distinction between pet dogs and free-ranging dogs and largely focus on studies of the latter.

In Chap. 5, we review studies comparing the social interactions of wolves and dogs with conspecifics focusing on dominance, affiliation, play, and conflict behaviours. In Chap. 6, we outline the social relationship both species can form with humans if they have received intensive socialization. These two chapters provide a good overview of the differences in the animals' social relationships and form the basis for the next two chapters, which focus on social cognition. In Chap. 7, after outlining what social cognition refers to, we review studies that investigate the abilities of wolves and dogs to use human communicative cues and how they themselves might use communicative behaviours to reach certain goals. In Chap. 8, we analyse the cooperative skills of wolves and dogs both with conspecifics and humans and try to understand the mechanism underlying such skills.

In Chap. 9, we will explore how motivational factors and low-level mechanisms such as persistence and inhibitory control might influence how wolves and dogs solve the problems that we reviewed in the previous chapters. Chapter 10 is dedicated to the few studies investigating how wolves and dogs perceive their non-social environment paying special attention to their ability to learn and make inferences. Finally, in the last chapter, we will consider the various domestication hypotheses that have been proposed and critically evaluate which are supported by the available data and where future researchers may need to focus in order to further our understanding of the two species.

While reviewing all the different literature on wolf and dog behaviour, we came across the following note from Erik Zimen (1981, page 5), which we wholeheartedly embrace both in reference to our own research and also to all the other studies reviewed here:

'Naturally I ***(we)*** will try to paint as true as possible a picture. But who can claim to paint a picture free of error? The acquisition of knowledge, even when carried out with scientific methods, is an ambiguous process, often the result of constructive misinterpretations, sometimes of chance, and always dependent on the inquirer himself (***themselves***) as well as on his (***their***) social and political consciousness and environmental conditions. So, I (***we***) invite the reader to keep an open mind about what I (***we***) have to say'.

We would also like to add that our results, both observational and experimental, are, strictly speaking, only valid for the specific contexts in which we raise, keep, and experiment with our animals. Consequently, the results can only be generalized to a certain degree and remain a matter of informed interpretation. While we can somewhat tackle the question of the context specifically in regard to experiments by using diverse paradigms to ask the same question, in the end, we will need verification of our results in other wolf and dog populations (which we are working on ☺) to pinpoint if and how results are specific to the dog and wolf populations studied so far.

References

Alter SG (2007) Separated at birth: the interlinked origins of Darwin's unconscious selection concept and the application of sexual selection to race. J Hist Biol 40:231–258. https://doi.org/10.1007/s10739-006-9113-5

Axelsson E et al (2013) The genomic signature of dog domestication reveals adaptation to a starch-rich diet. Nature 495:360–364. https://doi.org/10.1038/nature11837

Baryshnikov GF, Mol D, Tikhonov AN (2009) Finding of the Late Pleistocene carnivores in Taimyr Peninsula (Russia, Siberia) with paleoecological context. Russ J Theriol 8:107–113. https://doi.org/10.15298/RUSJTHERIOL.08.2.04

Belyaev DK (1969) Domestication of animals. Sci J 5:47–52

Belyaev DK (1979) Destabilizing selection as a factor in domestication. J Hered 70:301–308. https://doi.org/10.1093/oxfordjournals.jhered.a109263

Bergström A et al (2020) Origins and genetic legacy of prehistoric dogs. Science 370:557–564. https://doi.org/10.1126/science.aba9572

Bidau CJ (2009) Domestication through the Centuries: Darwin's Ideas and Dmitry Belyaev's long-term experiment in silver foxes. Gayana 73:55–72. https://doi.org/10.4067/S0717-65382009000300006

Bohling JH, Dellinger J, McVey JM, Cobb DT, Moorman CE, Waits LP (2016) Describing a developing hybrid zone between red wolves and coyotes in eastern North Carolina, USA. Evol Appl 9:791–804. https://doi.org/10.1111/eva.12388

Botigué LR et al (2017) Ancient European dog genomes reveal continuity since the Early Neolithic. Nat Commun 8:16082. https://doi.org/10.1038/ncomms16082

Boyko AR (2011) The domestic dog: man's best friend in the genomic era. Genome Biol 12:216. https://doi.org/10.1186/gb-2011-12-2-216

Bräuer J, Kaminski J, Riedel J, Call J, Tomasello M (2006) Making inferences about the location of hidden food: social dog, causal ape. J Comp Psychol 120:38–47. https://doi.org/10.1037/0735-7036.120.1.38

Burkart JM, Hrdy SB, van Schaik CP (2009) Cooperative breeding and human cognitive evolution. Evol Anthropol 18:175–186. https://doi.org/10.1002/evan.20222

Butler JRA, Brown WY, du Toit JT (2018) Anthropogenic food subsidy to a commensal carnivore: the value and supply of human faeces in the diet of free-ranging dogs. Animals 8:67. https://doi.org/10.3390/ani8050067

Byrne RW (1997) The technical intelligence hypothesis: an additional evolutionary stimulus to intelligence? In: Whiten A, Byrne RW (eds) Machiavellian intelligence II: extensions and evaluations. Cambridge University Press, Cambridge, pp 289–211. https://doi.org/10.1017/CBO9780511525636.012

Byrne RW, Whiten A (1988) Machiavellian intelligence. Social expertise and the evolution of intellect in monkeys, apes, and humans. Clarendon Press, Oxford

Careau V, Réale D, Humphries MM, Thomas DW (2010) The pace of life under artificial selection: personality, energy expenditure, and longevity are correlated in domestic dogs. Am Nat 175: 753–758. https://doi.org/10.1086/652435

Clutton-Brock J (1992) The process of domestication. Mammal Rev 22:79–85 https://doi.org/10.1111/j.1365-2907.1992.tb00122.x

Clutton-Brock J (1995) Origins of the dogs: domestication and early history. In: Serpell JA (ed) The domestic dog: its evolution, behaviour and interactions with people. Cambridge University Press, Cambridge, pp 199–216

Clutton-Brock TH, Harvey PH (1980) Primates, brains and ecology. J Zool 190:309–323. https://doi.org/10.1111/j.1469-7998.1980.tb01430.x

Cooper A, Turney C, Hughen KA, Brook BW, McDonald HG, Bradshaw CJ (2015) Abrupt warming events drove Late Pleistocene Holarctic megafaunal turnover. Science 349:602–606. https://doi.org/10.1126/science.aac4315

Coppinger R, Coppinger L (2001) Dogs. A new understanding of canine origin, behavior and evolution. The University of Chicago Press, Chicago
Coppinger R, Coppinger L (2016) What is a dog? The University of Chicago Press, Chicago
Darwin C (1859) On the origin of species by means of natural selection, or the preservation of favoured races in the struggle for life. John Murray, London
Dugatkin LA, Trut L, LudN T (2017) How to tame a fox (and build a dog). Visionary Scientists and a Siberian Tale of Jump-Started Evolution. University of Chicago Press. https://press.uchicago.edu/ucp/books/book/chicago/H/bo25568406.html
Dunbar RIM (1998) The social brain hypothesis. Evol Anthropol 6:178–190. https://doi.org/10.1002/(SICI)1520-6505(1998)6:5<178::AID-EVAN5>3.0.CO;2-8
Fadel FR, Driscoll P, Pilot M, Wright H, Zulch H, Mills D (2016) Differences in trait impulsivity indicate diversification of dog breeds into working and show lines. Sci Rep 6:1–10. https://doi.org/10.1038/srep22162
Fan Z et al (2016) Worldwide patterns of genomic variation and admixture in gray wolves. Genome Res 26:163–173. https://doi.org/10.1101/gr.197517.115
Flower LOH, Schreve DC (2014) An investigation of palaeodietary variability in European Pleistocene canids. Quat Sci Rev 96:188–203. https://doi.org/10.1016/j.quascirev.2014.04.015
Fox-Dobbs K, Leonard JA, Koch PL (2008) Pleistocene megafauna from eastern Beringia: Paleoecological and paleoenvironmental interpretations of stable carbon and nitrogen isotope and radiocarbon records. Palaeogeogr Palaeoclimatol Palaeoecol 261:30–46. https://doi.org/10.1016/j.palaeo.2007.12.011
Frantz LAF et al (2016) Genomic and archaeological evidence suggest a dual origin of domestic dogs. Science 352:1228–1231. https://doi.org/10.1126/science.aaf3161
Freedman AH et al (2014) Genome sequencing highlights the dynamic early history of dogs. PLoS Genet 10:e1004016. https://doi.org/10.1371/journal.pgen.1004016
Galton F (1865) The first steps towards the domestication of animals. Trans Ethno Soc Lond 3:122–138. https://doi.org/10.2307/3014161
Geffen E, Anderson MJ, Wayne RK (2004) Climate and habitat barriers to dispersal in the highly mobile grey wolf. Mol Ecol 13:2481–2490. https://doi.org/10.1111/j.1365-294X.2004.02244.x
Germonpré M, Fedorov S, Danilov P, Galeta P, Jimenez E-L, Sablin M, Losey RJ (2017) Palaeolithic and prehistoric dogs and Pleistocene wolves from Yakutia: identification of isolated skulls. J Archaeol Sci 78:1–19. https://doi.org/10.1016/j.jas.2016.11.008
Hall RL, Sharp HS (1978) Wolf and man, evolution in parallel. Academic Press, New York
Hare B, Brown M, Williamson C, Tomasello M (2002) The domestication of social cognition in dogs. Science 298:1634–1636. https://doi.org/10.1126/science.1072702
Hare B, Tomasello M (2005a) The emotional reactivity hypothesis and cognitive evolution. Trends Cogn Sci 9:464–465. https://doi.org/10.1016/j.tics.2005.08.010
Hare B, Tomasello M (2005b) Human-like social skills in dogs? Trends Cogn Sci 9:439–444. https://doi.org/10.1016/j.tics.2005.07.003
Heiser CB (1988) Aspects of unconscious selection and the evolution of domesticated plants. Euphytica 37:77–81. https://doi.org/10.1007/BF00037227
Hinton JW, Gittleman JL, van Manen FT, Chamberlain MJ (2018) Size-assortative choice and mate availability influences hybridization between red wolves (*Canis rufus*) and coyotes (*Canis latrans*). Ecol Evol 8:3927–3940. https://doi.org/10.1002/ece3.3950
Hofreiter M (2007) Pleistocene extinctions: haunting the survivors. Curr Biol 17:R609–R611. https://doi.org/10.1016/j.cub.2007.06.031
Hulme-Beaman A, Dobney K, Cucchi T, Searle JB (2016) An ecological and evolutionary framework for commensalism in anthropogenic environments. Trends Ecol Evol 31:633–645. https://doi.org/10.1016/j.tree.2016.05.001
Jung C, Pörtl D (2018) Scavenging hypothesis: lack of evidence for dog domestication on the waste dump. Dog Behavior 4:41–56. https://doi.org/10.4454/db.v4i2.73

Kaplan H, Hill K, Lancaster J, Hurtado A (2000) A theory of human life history evolution: diet, intelligence, and longevity. Evol Anthropol 9:156–185. https://doi.org/10.1002/1520-6505(2000)9:4<156::AID-EVAN5>3.0.CO;2-7

Kuzmina IE, Sablin MV (1994) Wolf, *Canis lupus* L. from the late paleolithic sites Kostenki on the Don River. Proceedings of the Zoological Institute of the Russian Academy of Sciences 256:44–58

Lahtinen M, Clinnick D, Mannermaa K, Salonen JS, Viranta S (2021) Excess protein enabled dog domestication during severe ice age winters. Sci Rep 11:1–5. https://doi.org/10.1038/s41598-020-78214-4

Lakatos G, Soproni K, Dóka A, Miklósi Á (2009) A comparative approach to dogs' (*Canis familiaris*) and human infants' comprehension of various forms of pointing gestures. Anim Cogn 12:621–631. https://doi.org/10.1007/s10071-009-0221-4

Larson G, Bradley DG (2014) How much is that in dog years? The advent of canine population genomics. PLoS Genet 10:e1004093. https://doi.org/10.1371/journal.pgen.1004093

Larson G, Fuller DQ (2014) The evolution of animal domestication. Ann Rev Ecol Evol Syst 45: 115–136. https://doi.org/10.1146/annurev-ecolsys-110512-135813

Larson G et al (2012) Rethinking dog domestication by integrating genetics, archeology, and biogeography. Proc Natl Acad Sci U S A 109:8878–8883. https://doi.org/10.1073/pnas.1203005109

Larson G et al (2014) Current perspectives and the future of domestication studies. Proc Natl Acad Sci U S A 111:6139–6146. https://doi.org/10.1073/pnas.1323964111

Leonard JA, Vilà C, Fox-Dobbs K, Koch PL, Wayne RK, Van Valkenburgh B (2007) Megafaunal extinctions and the disappearance of a specialized Wolf Ecomorph. Curr Biol 17:1146–1150. https://doi.org/10.1016/j.cub.2007.05.072

Lindblad-Toh K et al (2005) Genome sequence, comparative analysis and haplotype structure of the domestic dog. Nature 438:803–819. https://doi.org/10.1038/nature04338

Linnell JDC et al (2002) The fear of wolves: a review of wolf attacks on humans, vol 731. NINA, Trondheim

Loog L et al (2020) Ancient DNA suggests modern wolves trace their origin to a Late Pleistocene expansion from Beringia. Mol Ecol 29:1596–1610. https://doi.org/10.1111/mec.15329

MacLean EL, Snyder-Mackler N, VonHoldt BM, Serpell JA (2019) Highly heritable and functionally relevant breed differences in dog behaviour. Proc R Soc B 286:20190716. https://doi.org/10.1098/rspb.2019.0716

Marlowe FW (2005) Hunter-gatherers and human evolution. Evol Anthropol 14:54–67. https://doi.org/10.1002/evan.20046

Mech LD, Asa CS, Callahan M, Christensen BW, Smith F, Young JK (2017) Studies of wolf x coyote hybridization via artificial insemination. PLoS One 12:e0184342. https://doi.org/10.1371/journal.pone.0184342

Miklósi Á, Kubinyi E, Topál J, Gácsi M, Virányi Z, Csányi V (2003) A simple reason for a big difference: wolves do not look back at humans, but dogs do. Curr Biol 13:763–766. https://doi.org/10.1016/s0960-9822(03)00263-x

Miklósi Á, Topál J, Csányi V (2004) Comparative social cognition: what can dogs teach us? Anim Behav 67:995–1004. https://doi.org/10.1016/j.anbehav.2003.10.008

Morey DF (1994) The early evolution of the domestic dog. Am Sci 82:336–347. http://www.jstor.org/stable/29775234

Mulcahy NJ, Hedge V (2012) Are great apes tested with an abject object-choice task? Anim Behav 83:313–321. https://doi.org/10.1016/j.anbehav.2011.11.019

Musiani M et al (2007) Differentiation of tundra/taiga and boreal coniferous forest wolves: genetics, coat colour and association with migratory caribou. Mol Ecol 16:4149–4170. https://doi.org/10.1111/j.1365-294X.2007.03458.x

Newsome T, van Eeden L (2017) The effects of food waste on wildlife and humans. Sustainability 9:1269. https://doi.org/10.3390/su9071269

O'Keefe FR, Meachen J, Fet EV, Brannick A (2013) Ecological determinants of clinal morphological variation in the cranium of the North American gray wolf. J Mammal 94:1223–1236

Ostrander EA, Galibert F, Patterson DF (2000) Canine genetics comes of age. Trends Genet 16: 117–124. https://doi.org/10.1016/S0168-9525(99)01958-7

Pang J-F et al (2009) mtDNA data indicate a single origin for dogs south of Yangtze river, less than 16,300 years ago, from numerous wolves. Mol Biol Evol 26:2849–2864. https://doi.org/10.1093/molbev/msp195

Pilot M et al (2010) Phylogeographic history of grey wolves in Europe. BMC Evol Biol 10:104. https://doi.org/10.1186/1471-2148-10-104

Pilot M et al (2014) Genome-wide signatures of population bottlenecks and diversifying selection in European wolves. Heredity 112:428–442. https://doi.org/10.1038/hdy.2013.122

Pilot M et al (2018) Widespread, long-term admixture between grey wolves and domestic dogs across Eurasia and its implications for the conservation status of hybrids. Evol Appl 11:662–680. https://doi.org/10.1111/eva.12595

Pilot M, Jędrzejewski W, Branicki W, Sidorovich VE, Jędrzejewska B, Stachura K, Funk SM (2006) Ecological factors influence population genetic structure of European grey wolves. Mol Ecol 15:4533–4553. https://doi.org/10.1111/j.1365-294X.2006.03110.x

Pilot M et al (2019) Global Phylogeographic and admixture patterns in grey wolves and genetic legacy of an ancient Siberian lineage. Sci Rep 9:17328. https://doi.org/10.1038/s41598-019-53492-9

Range F, Marshall-Pescini S (2022) Comparing wolves and dogs: current status and implications for human 'self-domestication'. Trends Cogn Sci 26(4):337–349

Richerson PJ, Boyd R, Bettinger RL (2001) Was agriculture impossible during the Pleistocene but mandatory during the Holocene? A climate change hypothesis. Am Antiq 66:387–411. https://doi.org/10.2307/2694241

Rosati AG (2017) Foraging cognition: reviving the ecological intelligence hypothesis. Trends Cogn Sci 21:691–702. https://doi.org/10.1016/j.tics.2017.05.011

Savolainen P, Zhang Y, Luo J, Lundeberg J, Leitner T (2002) Genetic evidence for an east asian origin of domestic dogs. Science 298:1610–1613. https://doi.org/10.1126/science.1073906

Scott JP, Fuller JL (1965) Genetics and the social behavior of the dog. University of Chicago Press, Chicago

Shannon LM et al (2015) Genetic structure in village dogs reveals a Central Asian domestication origin. Proc Natl Acad Sci U S A 112:13639–13644. https://doi.org/10.1073/pnas.1516215112

Skoglund P, Ersmark E, Palkopoulou E, Dalén L (2015) Ancient wolf genome reveals an early divergence of domestic dog ancestors and admixture into high-latitude breeds. Curr Biol 25: 1515–1519. https://doi.org/10.1016/j.cub.2015.04.019

Soproni K, Miklósi Á, Topál J, Csányi V (2001) Comprehension of human communicative signs in pet dogs (*Canis familiaris*). J Comp Psychol 115:122–126. https://doi.org/10.1037//0735-7036.115.2.122

Spady TC, Ostrander EA (2008) Canine behavioral genetics: pointing out the phenotypes and herding up the genes. Am J Hum Genet 82:10–18. https://doi.org/10.1016/j.ajhg.2007.12.001

Stronen AV, Navid EL, Quinn MS, Paquet PC, Bryan HM, Darimont CT (2014) Population genetic structure of gray wolves (*Canis lupus*) in a marine archipelago suggests island-mainland differentiation consistent with dietary niche. BMC Ecol 14:11. https://doi.org/10.1186/1472-6785-14-11

Svartberg K (2006) Breed-typical behaviour in dogs–Historical remnants or recent constructs? Appl Anim Behav Sci 96:293–313. https://doi.org/10.1016/j.applanim.2005.06.014

Thalmann O, Perri AR (2018) Paleogenomic inferences of dog domestication. In: Lindqvist C, Rajora O (eds) Paleogenomics. Population genomics. Springer, Cham, pp 273–306. https://doi.org/10.1007/13836_2018_27

Thalmann O et al (2013) Complete mitochondrial genomes of ancient Canids suggest a european origin of domestic dogs. Science 342:871–874. https://doi.org/10.1126/science.1243650

Trut L, Oskina I, Kharlamova A (2009) Animal evolution during domestication: the domesticated fox as a model. BioEssays 31:349–360. https://doi.org/10.1002/bies.200800070

Trut LN, Plyusnina IZ, Oskina IN (2004) An experiment on fox domestication and debatable issues of evolution of the dog. Russ J Genet 40:644–655. https://doi.org/10.1023/B:RUGE.0000033312.92773.c1

von Holdt BM et al (2010) Genome-wide SNP and haplotype analyses reveal a rich history underlying dog domestication. Nature 464:898–902. https://doi.org/10.1038/nature08837

Wang G-D et al (2014) Genetic convergence in the adaptation of dogs and humans to the high-altitude environment of the Tibetan plateau. Genome Biol Evol 6:2122–2128. https://doi.org/10.1093/gbe/evu162

Wang G-d et al (2013) The genomics of selection in dogs and the parallel evolution between dogs and humans. Nat Commun 4:1860. https://doi.org/10.1038/ncomms2814

Wang G-D et al (2015) Out of southern East Asia: the natural history of domestic dogs across the world. Cell Res 26:21. https://doi.org/10.1038/cr.2015.147

Wright HF, Mills DS, Pollux PMJ (2012) Behavioural and physiological correlates of impulsivity in the domestic dog (*Canis familiaris*). Physiol Behav 105:676–682. https://doi.org/10.1016/j.physbeh.2011.09.019

Zapata I, Serpell JA, Alvarez CE (2016) Genetic mapping of canine fear and aggression. BMC Genomics 17:572. https://doi.org/10.1186/s12864-016-2936-3

Zeder MA (2012) Pathways to animal domestication. In: Gepts P, Famula TR, Bettinger RL, Brush SB, Damania AB, McGuire PE, Qualset CO (eds) Biodiversity in agriculture. Cambridge University Press, Cambridge, pp 227–259

Zimen E (1981) The wolf, a species in danger. Delacorte Press

Zohary D (2004) Unconscious selection and the evolution of domesticated plants. Econ Bot 58: 5–10

2 Comparing the Behaviour and Cognition of Wolves and Dogs

When comparing the behaviour between wolves and dogs that live today, several aspects have to be considered if the aim is to investigate the effects of domestication on behaviour and cognition. Of course, dependent on the question asked, some aspects might be more important than others, and researchers have to decide which of the following points are relevant for them. However, for most of the research reviewed in this book, probably all aspects should be taken into account.

First, we have to ensure that the experiences of the wolves and dogs that are investigated are comparable. Domestic dogs in the Western world usually grow up as pets—often as a family member—and thus have ample opportunities for learning, especially from humans. On the one hand, this experience can affect their cognitive skills due to living in a much more complex (human) social environment than wolves usually do. This ontogenetic process has been named 'enculturation' when referring to great apes living in close human contact (Call and Tomasello 1996). For example, people often take their pet dogs into the cities and the woods or to visit friends, where they are exposed to lots of new objects and challenges that they have to learn to cope with. However, such experiences also give them the opportunity to learn things about their social and physical environment. On the other hand, just having to interact and communicate with humans on a daily basis might greatly influence how attentive they are towards humans (even without specific training), which in turn may affect their performance in social tasks (see also Miklósi et al. 2004; Topál et al. 2009; Udell et al. 2008). Moreover, many people, especially those that usually participate in our scientific studies, invest much time and effort training their animals and teaching them various different useful (and not so useful) things. All these interactions likely promote the development of the social and cognitive abilities of pet dogs. Accordingly, it would be rather inconclusive to compare these pet dogs with wolves that have been kept in enclosures without comparable experiences and interactions with humans. If such comparisons are done anyway, we cannot automatically attribute all dog–wolf differences to the domestication process, since the different experiences may have also played a (potentially large) part in causing such differences (Pang et al. 2009; Udell et al. 2010). Thus, to be sure

F. Range, S. Marshall-Pescini, *Wolves and Dogs*, Fascinating Life Sciences,
https://doi.org/10.1007/978-3-030-98411-3_2

that dog–wolf differences reflect evolutionary changes and do not result from different individual experiences, we need to raise wolves and dogs identically and keep them in the same environment.

Second, wolves have a strongly ingrained fear of people and the human environment—for good reasons. They have been hunted almost to extinction, and avoiding humans has been a survival strategy for hundreds of years. Hence, the shyness of wolves is due to the genetic footprint humans left by creating a landscape of fear over thousands of years. Accordingly, it is likely that only the shyest animals survived and produced offspring, which selected for wolves that are, on average, rather shy of humans and their artefacts. Exceptions to this are populations of wolves that live in remote areas, where they have not been hunted extensively. One such population is on Ellesmere Island, where Dave Mech observed wolves over many years. While Dave mainly reports that the wolves were not very interested in him, he reported a few instances when animals did approach to investigate his quad (personal communication). This apparent lack of shyness in Arctic wolves is met also in other animals from part of the world, where humans did not create a 'landscape of fear' (such as the Pacific islands). In any case, in order to investigate the animals' behaviour and cognitive abilities in experimental settings and especially in interactions with humans, it is necessary that they are not afraid of the people they work with; otherwise, we measure their fear reactions but not much else. To prevent this kind of reaction, wolves need to be socialized with humans and the testing environment, to make sure they are thoroughly comfortable with both. Unfortunately, wolves and dogs and also individual wolves differ considerably in regard to when and for how long socialization to humans has to occur to avoid fear reactions.

The socialization of wolf pups must start before they open their eyes, usually well before 3 weeks of age (otherwise it becomes difficult), and continuous human contact must be provided for 24 h a day for a considerable amount of time thereafter (Klinghammer and Goodmann 1987). Moreover, Klinghammer and Goodmann (1987) and Frank et al. (1989) show that wolf puppies, which were exposed to adult conspecifics for extended periods of time before the age of 8–10 weeks, later developed wariness and avoidance in the presence of humans. Hence, it is not sufficient to have a tame wolf female raise their pups in a human environment; wolf pups need to be thoroughly hand-raised to become well socialized at least with their raisers. Importantly though, they usually do not readily generalize the build-up trust to other non-familiar people (see Chaps. 6 and 11). Finally, the socialization of the adult wolves needs to be kept up continuously, for example, by daily interaction routines, even if properly socialized as puppies; otherwise, at least some animals may start to become wary of humans.

In contrast, dogs can be properly socialized with as little as half an hour daily interaction for 1 week between the age of 2 and 9 weeks (Scott and Marston 1950), and, in contrast to wolves, they generalize this to non-familiar people (see Chaps. 6 and 11). However, without any socialization with humans during the first 14 weeks of a pup's life, dogs also develop fear and avoidance of humans, and a 'normal' social relationship with humans is no longer possible (Freedman et al. 1961). It is likely that this need for at least moderate early socialization with humans explains the highly diverse reaction to humans observed in free-ranging dogs (Ortolani et al. 2009).

Thus, socializing captive wolves with humans at an early age is often done in zoos and is achieved by hand rearing the animals (Klinghammer and Goodmann 1987). While this procedure ensures that the animals become quite trusting (some more, some less, depending on their personality) to some humans, it also has a positive effect on their welfare, since hand rearing reduces their stress responses when exposed to unfamiliar people (as in zoo settings) and considerably enhances their manageability (Frank et al. 1986; Rushen et al. 1999). For example, all our animals at the Wolf Science Center can be checked by a local veterinarian with whom they are socialized; they can be vaccinated, and we can draw blood from several animals without sedating them. Moreover, instead of hiding in the back of the enclosure during opening hours of the game park, they will come to the fence to inspect visitors (which also enhances the experience for the visitors).

Third, we have to keep in mind that even if the wolves and dogs are raised and kept in a comparable way, they might not necessarily experience the environment in the same way and learn the same things but instead adapt their behaviour to different aspects of the same environment (Virányi and Range 2014). Thus, especially when comparing older animals, we will detect genetically based differences potentially enhanced by differential learning processes (epigenetic differences). However, given that they do live in the same environment, these differences stem from species-specific preferences and sensitivities and thus will allow us to deduce that the domestication process has influenced the behaviour of the dogs, though how much of this difference is driven by actual genetic differences or by the subsequent learning or experiences we cannot say. One way to avoid these learning effects is testing very young animals, which has been done in many wolf–dog comparisons (see various chapters). On the other hand, cognitive abilities change over the lifespan, increasing rapidly from infancy to young adulthood with the canine brain only fully developed at about 18 months. Thus, testing young animals may carry the risk of testing animals that are not cognitively fully developed as well as detecting 'false' differences since dogs and wolves develop at a different speed.

Finally, while the first two points are important when comparing wolves and dogs, we also need to keep in mind that the socialization and manner of keeping the animals might not necessarily reflect their species' usual environment and thus alter the animals' behaviour and how they try to solve certain problems. And because wolves and dogs are different, the same raising and keeping procedures may affect the two species differently. Thus, ideally, the behaviour of dogs and wolves kept under these special conditions should be compared also to other populations of the same species. In the case of dogs, it could be compared with dogs raised as pet dogs or with free-ranging dogs. In the case of wolves, we should compare behaviour of the socialized animals with non-socialized animals in captivity as well as with wild-living wolves. Obviously, not all studies can be replicated in free-ranging populations or non-socialized animals, but wherever possible an effort should be made.

Research Groups that Have Compared Wolves and Dogs

Raising and keeping a sufficient sample of dogs and wolves socialized with humans under identical conditions and satisfying the animals' needs at the same time requires a lot of effort, rendering it understandable that this has been done only rarely. Some research groups have opted to compare wolves with different populations of pet dogs instead. These research groups have invested a considerable effort to ensure that wolves are comfortable with the testing procedures (i.e. they are hand-raised and well socialized to humans), and thus their results are important since they enhance our understanding of how the species reacts to specific stimuli and allows for an analysis of the generalizability of results across labs. One of the places that have raised many wolves and compared them with different populations of dogs is Wolf Park in Indiana.

2.1 Wolf Park Indiana, Battle Ground, IN (1972–Present)

Wolf Park was founded by Erich Klinghammer in 1972. He and his team raised many generations of wolves to the present day and thus have accumulated enormous experience and knowledge about socialization and wolf behaviour. Klinghammer's primary goal was to learn about wolves' behaviour and communication by observing their interactions.

Importantly, Klinghammer and his colleague Patricia Goodmann were interested in the socialization process and established the best protocol to raise wolves so that they are fully comfortable around people (Klinghammer and Goodmann 1987). For example, they realized that humans can only socialize wolf pups successfully if the puppies are separated from their mother before they open their eyes, suggesting that they either learn early in life to prefer their own species or have a strong genetic preference for conspecifics (Klinghammer and Goodmann 1987). Accordingly, at Wolf Park, wolves are hand-raised in peer groups (mainly animals from the same litter) by staff starting at the age of 10–14 days with continued interactions with humans on a daily basis. In the first 6 weeks, they have 24-h contact with human caretakers followed by contact during daylight hours until 4 months of age and then daily human interaction with caretakers and other humans thereafter.

While strongly recommending human contact early on during the socialization process, Klinghammer and Goodmann (1987) also state that it might be important to raise them with conspecifics (i.e. littermates) to ensure normal intraspecific behaviour. They start bringing in adult wolves into the nursery once the pups are about 8 weeks of age and then start integrating the puppies with adult packs at the age of 3 months—on weekends. At 4.5 months of age, they completely move in with the adult animals.

The people working at Wolf Park are special in many ways, having been around for a very long time and gathering an immense experience in how to hand-raise and handle wolves. The relationships certain people have established with some of the animals are amazing—my 'aha' moment was when I saw Monty Sloan alone in an enclosure and lying down amid four to five adult wolves playing with them in what seemed to be a mutual enjoyment for both parties.

Today, the main purpose of Wolf Park has shifted a bit to being a public education facility. However, research is still one of their primary goals, and Clive Wynne and Monica Udell have conducted several studies on wolf social interactions with humans and cognition in recent years (Udell et al. 2008, 2011, 2012; Udell and Wynne 2011) that are reviewed in the respective chapters. However, while the animals at Wolf Park are trained for safe interactions with visitors, they have less experience with cooperation with humans via commands that would be important for at least some research projects. For example, most of their animals are not used to being separated from other pack members or to being held on the leash while watching a demonstration. These limitations are sometimes reflected in the methods of the specific studies (e.g. Udell et al. 2008).

Recognizing the importance of keeping the animal's raising and living experience the same, a few research projects have endeavoured to do so in different ways. Here we will introduce the main groups and the different approaches taken. These differences in raising wolves and dogs and their living experiences are important background information which will then help in the interpretation of results from studies conducted at the different facilities. In the later chapters, we will then review the results of the different research groups according to the various topics being addressed.

2.2 Kiel

In the 1960s, Wolfgang 'Wolf' Herre and later Dorit Feddersen-Petersen crossed wolves and standard poodles (as well as coyotes and miniature poodles and golden jackals and toy poodles) to study the genetics of the relative brain size and other morphological features (fur structure and colour) at the 'Institute for Pet Science' (Institut für Haustierkunde) of the University of Kiel (Germany). The poodles were chosen because they looked very different than non-domesticated canines and, importantly, they come in three different sizes (Standard, Miniature, and Toy), which allowed the researchers to breed them with various wild canines. The poodle–wolf hybrids were called Puwos or Wopus depending on whether the mother or the father was a wolf, respectively.

The wolves and poodles were kept in separate packs, but some were also kept in mixed packs in enclosures. The pups were raised by their natural mother. Interestingly, it was easy to socialize the wolves and the poodles with each other even at adult age, whereas this was not possible with jackals and poodles; they avoided each other as adults, and in peer groups the jackals played only with other jackals.

Feddersen-Petersen continued the behavioural observations after the genetic studies stopped in 1975 (Feddersen-Petersen 1991, 1994, 2000) and raised additional animals. To allow for valid behavioural observations, the wolves were either hand-raised or semi-hand-raised (if the parents were human raised, they allowed early human contact to the puppies) to avoid the animals becoming wary of the human observers. However, while the animals seemed to like the experiments that were conducted by the researchers, they ignored them when engaged in their own interactions. Together with her group, she collected observational data of social interactions on 189 canids (of whom 74 were hand-raised) including not only the poodles and wolves of the genetic studies but also dogs of several different breeds that were kept also in packs (Feddersen-Petersen 1994; see also Feddersen-Petersen 1991, 2000). Many of her results will be reviewed in the following chapters since they are the only systematic comparative studies of conspecific social behaviours.

2.3 Erik Zimen

In the mid-1960s, Erik Zimen started to observe the behaviour of the animals in Kiel to do his PhD thesis (Zimen 1971). One of his and his supervisor Wolf Herre's primary goals was to decide on the base of comparing behaviour, whether dogs originated from wolves or whether golden jackals also had a share in their ancestry, as Konrad Lorenz suggested. Moreover, from June 1967 to December 1969, he also hand-raised several wolves ($N = 10$, 4 different litters (1, 1, 3, 5 animals)), poodles ($N = 8$, 3 litters (3, 3, 2), with litter 2 and 3 being born by mothers from the first litter), and their crosses ($N = 8$), comparing their human-directed behaviour with the mother-raised animals (Zimen 1987). All animals were raised on a farm in Northern Germany; after the socialization period, they were kept together in one enclosure, though it seemed that the poodles also spent some time in the courtyard. All animals were raised as similar as possible. However, four wolves and one poodle were already over 21 days old when they arrived. All wolves and one litter of poodles ($N = 2$) were artificially fed by inserting a tube into their stomach. Starting at the age of 21 days, they were also fed with milk from a bowl, and from 4 weeks of age they also received solid food. Due to intensive sucking on the genitals of the other puppies, they were mainly kept in isolation except periods around the feeding. By interacting intensively with the puppies, Zimen managed to tame the six wolves that were under 2 weeks of age when they arrived. These animals were taken for leash walks almost daily. The non-tamed animals that remained in the enclosure started to show stereotypic behaviours at 1 year of age and were apathetic in comparison to the tamed animals at the age of 3 years. The crosses were returned to Kiel at the age of 1 year.

As far as we can tell based on his descriptions living with these animals, he treated them with respect but also using harsher and more physical training methods than have been employed at the other facilities that hand-raised wolves. He described his own role as that of the super-alpha, which nicely portrays the relationships he established with the animals. However, he also gave them more freedom compared to the more recent projects. For example, he took them on walks through the fields and forest without putting them on a leash until they were about 6–9 months old. In his diary entries, he also describes instances when the wolves (and dogs) went off and tried to hunt cows and a horse—the wolves (or dogs) were never successful at this age. Even after his thesis, Zimen continued to hand-raise wolves, and, in his books, there are many funny stories about the relationship he built up with these animals. For example, one of the wolves really liked to sleep in his bed! Interestingly, he made clear that he never had any issues with the wolves (puppies and adults) that were growing up in his house and with his children.

During his thesis, Zimen observed the animals over a period of 2 years conducting 2-min scans of their behaviours almost on a daily basis. Although no statistics were performed on the data he presents in his thesis, he reached a good understanding of the animals' behaviours and their differences that we will review in the different chapters.

2.4 University of Michigan Canine Information-Processing Project (1979–1984)

Martha and Harry Frank (Frank and Frank 1987) ran this project starting in 1979 to test their theoretical model of information processing in wolves and dogs. In total they raised 11 wolves and four malamute pups. The main aim of this project was comparing wolves' and dogs' behavioural and cognitive performance. First, the Franks raised four wolves and then four malamute pups in two different years, each cohort consisting of one male and one female from each of two litters, in the same environment. From the age of 10 to 13 days, all pups were taken care of by the same lactating wolf female. Until 8 weeks of age, the wolves spent 12 h per day with the foster mother and 12 h per day with the two experimenters. Sleeping was rotated so that each pup spent one night with each experimenter and then two nights with the foster mother. From 8 to 20 weeks of age, the wolves spent approximately 4 h per day with the experimenter. From week 20 onwards, they were confined to the barn.

Since the malamutes were easier to socialize, they spent less time with the experimenter. While the foster mother lactated a little bit, the pups received milk from bottles until about 3 weeks of age and thereafter were fed in the house until 20 weeks of age.

Despite this reduced socialization of the dog puppies and otherwise virtually identical conditions during hand-raising, the dog pups were highly socialized to humans, but the wolf pups remained essentially unsocialized to humans. To remedy this difference and to explore how a better socialization with humans would influence their cognitive skills, the Franks raised eight additional wolf pups (5 males and 3 females, not known if they were from one or two litters) in 1983. These pups were raised exclusively by humans starting at the age of 6 or 8 days. Until week 6, they had no contact with other canids and they spent 20 h a day in contact with the human experimenters. Interestingly, their performance differed from previously raised wolf pups, which was attributed to the mother reared animals being a lot more fearful, not concentrating on the task at hand (Frank 2011; Frank et al. 1989; Chap. 10). Since the Franks compared the performance of the differently raised pups by subjecting them to the same experimental tasks, these studies highlight two important facts: (1) that the early hand-raising of the wolves is necessary to build up a trustful and relaxed relationship between humans and wolves and (2) that it is paramount to hand-raise wolves if one aims at investigating the cognitive skills of these animals rather than how they behave in a stressful situation.

Overall, the research by the Franks was the first to directly compare wolves' and dogs' cognitive abilities and as such provide the foundations for many subsequent studies on these topics.

2.5 Family Dog Project of the Department of Ethology, Budapest (2002–2004, Partially Ongoing)

Between 2000 and 2003, a group of researchers led by Ádám Miklósi of the Department of Ethology, Eötvös Loránd University, Budapest, Hungary, raised 13 European wolves (six males and seven females, from five different litters but only two genetic lines) and 11 dog mongrels from three shelters (six males and five females from five different litters). The aim of this project was to compare dogs and wolves in their communication, interactions, and relationship with humans. Accordingly, the raising regime was set up to best fulfil this aim.

The wolves and dogs were raised individually by researchers and students in separate years. Five researchers raised both wolves and dogs, which controlled for individual differences in interaction styles of the caretakers. The animals were separated from their mothers and littermates on days 4–6 after birth (before their eyes opened) and then raised like well-socialized pet dogs. Each caregiver spent 24 hr in close contact with 'their' animal. They lived in the caregivers' homes and slept together at night. For the first 4–5 weeks, they were bottle-fed and afterwards also fed with solid food. The caregivers devised a pouch that allowed them to carry the pups around during their everyday activities including traveling by public transport, attending classes at the university, visiting friends, etc. (Kubinyi et al. 2007 for a review). Once old enough, the puppies were also regularly walked on leash and most attended a dog school where they could meet and interact with unfamiliar dogs and humans. At least twice a week, the pups also met conspecifics of about the same age. Accordingly, the puppies were regularly exposed to novel stimuli and situations as well as to familiar and unfamiliar individuals. Importantly, the researchers tried to avoid any competitive situations and aggressive interactions with the animals and rather tried to behave like a mother than a dominant conspecific (Ujfalussy et al. 2003).

However, at the age of 2 to 4 months, the wolf puppies were returned to a private wolf farm (Horatius Wolf Park, Gödöllö). Here they lived around the owner's house with his dogs and could interact with humans and other wolves daily. Once or twice per week, they were visited by their hand-raisers for half a day (Virányi et al. 2008) until the animals were approximately 1.5–2 years old (Kubinyi et al. 2007). During such visits, the animals were taken out from their group for training, testing, and free social interactions with their caregivers. Moreover, there were many regular and occasional visitors at the wolf farm, most of whom made short contacts with the wolves. The wolves were gradually integrated into packs of older animals until 1 year of age. They continued to have daily contact with several familiar humans (who regularly entered their enclosure and interacted with them) and regularly met unfamiliar humans outside the enclosure individually either during visits, film shootings, shows, or behavioural tests (Kubinyi et al. 2007).

The dogs were rehomed to volunteers or stayed at their hand-raiser's home at the age of 2 months and lived as pet dogs for the rest of their lives.

Thus, while this socialization regime ensured that the wolves displayed no sign of wariness or avoidance in the presence of humans (even strangers)—just the opposite—(see Gácsi et al. 2005; Topál et al. 2005) and also showed no fear of the various testing apparatuses at any time, from the age of 2–4 months on, the individual experiences of the wolves differed from the ones of the mongrels.

The Budapest project introduced two particularly novel aspects into the wolf–dog comparison. Firstly, the focus on the potential changes between wolves and dogs in how they interact and relate to humans. This aspect, until then, had not really been explored, and the pioneering studies by this group on this topic opened a highly fruitful area of research. The second aspect is the emphasis on raising the animals with intense human contact but also with the kind of socialization (environment, conspecific, training etc.), which may have a combined effect on the animals' behaviour. The fact that the wolf hand-raisers relatively quickly had to 'give up' the wolf pups indicates quite clearly the difficulties related to living with a wild animal and the long process which has led dogs to display the necessary adaptability to the human environment.

2.6 The Dog Cognition Programme at Max Planck Institute in Leipzig

Approximately at the same time as the Family Dog Project in Budapest was on its way, another group of researchers working at the Department of Comparative and Developmental Psychology of the Max Planck Institute in Leipzig became interested in dog cognition, particularly in comparison to human infants and apes. Brian Hare was particularly instrumental in developing the research on dog cognition and suggested, similarly to Belyaev (see Chap. 11), that the parallel process of selection for a tamer, friendly temperament drove the evolution of dogs and humans. This hypothesis was derived from a number of studies he conducted with colleagues comparing pet dogs with zoo-kept, non-socialized wolves and silver foxes in Russia, particularly focusing on these animals' abilities to follow the human pointing gesture in a two-choice task (see Chap. 7). However, due to the differences in experience and socialization, conclusion that can be drawn from the studies with wolves is somewhat problematic.

2.7 Tovetorp Zoological Research Station of Stockholm University (2014–2016)

Recently, Christina Hansen Wheat, together with several helpers, hand-raised two litters of Alaskan huskies ($N = 12$, eight males, four females) and three litters of European grey wolves for her PhD to investigate the development of social behaviour over time and how social dynamics, individuality, and hormonal profiles might interact. The first two wolf litters were from the same breeding pair so that the pups were full siblings ($N = 7$, four males, three females), whereas the third litter was unrelated ($N = 6$, four males, two females). The wolves and dogs were extensively socialized starting at the age of 10 days and kept under similar conditions. The pups were socialized within litters, and human caregivers were present for 24 h in the first 2 months. From 2 months of age onwards, the human presence was stepwise reduced so that at the age of 4 months the human caregivers spent only every other night with the pups. Further reduction led to humans only spending 4–6 h a day with the animals at the age of 6 months. Until 5 weeks of age, the animals were kept indoors; afterwards, they had access to an outdoor enclosure during the day and to the indoor and a small roofed outdoor enclosure during the night. At the age of 3 months, the pups from the respective litter were moved to a 2000 m^2 outdoor enclosure, where they stayed until the end of the study period. Testing procedures were standardized across the 3-year period of the research project. Importantly, puppies were never disciplined, trained, or forced to have contact with their caregivers (Hansen Wheat et al. 2019, 2020). Contact with strangers was gradually started through the fence with the support of one or more human caregivers at the age of 8 weeks. The main research question that was studied with these animals is the development of the human–animal bond, and these studies are reviewed in Chap. 6.

2.8 Wolf Science Center (2008–Present)

The Wolf Science Center (WSC) is a facility in Austria that was established in 2008 by Kurt Kotrschal, Friederike Range, Zsofia Viranyi (in alphabetical order) (see Box 2.2 for the story of the idea and the early years). While we started as private entrepreneurs with some support by the government and by the University of Veterinary Medicine Vienna (Vetmed) and the University of Vienna, in 2017 the Vetmed took over the WSC in acknowledgement of our scientific achievements and to provide long-term stability. Taking in the results and experiences from other research groups, our idea was to focus on all aspects of the dog–wolf comparison, the social behaviour between conspecifics (Feddersen-Petersen), the relationship with humans (Budapest Group), and the cognitive differences between the animals (Franks). And, importantly, we intended to keep the animals, provide optimal living conditions, and explore their abilities until they were grown-up and old (Box 2.1).

Box 2.1 The Beginning...
by Friederike Range

Once upon a time, there were three researchers interested in wolves' social behaviour and cognition. Zsofia Viranyi had participated in the Budapest wolf/dog project as a hand-raiser and scientist (at that time PhD student). However, the Budapest project had ended, and all the wolves had been released into pack life, while the dogs were given away to owners or the hand-raisers kept them (Todor, whom Zsofi hand-raised, remained with her and would play an important role also at the WSC). But while Zsofi had learned a lot about wolves, she still had many more questions about their social skills. Kurt Kotrschal, head of the Konrad Lorenz Forschungsstelle in Grünau, had a picture posted on the webpage proclaiming that he was interested in working with wolves. And, actually, this picture was likely responsible for us three coming together. I, Friederike Range, had just finished my PhD in the USA on non-human primates running through the forest in Africa and wanted to return to Europe and work on the social behaviour of wolves, and so, while searching for potential research labs, I stumbled across this picture. Together with Kurt, we set out to get money for the research we had in mind. Kurt's idea was to get running belts, and I was interested in cooperation, so we could easily combine these two ideas to write fantastic grant applications. Unfortunately, or actually fortunately in hindsight, the reviewers did not think the applications were that fantastic and especially criticized that we had no animals to do our studies with and no experience in raising them—both valid points. Anyhow, we did not give up! I got a position as a postdoc in Ludwig Huber's lab, so I could stay in Austria, and since Ludwig was a really good boss and was always open for new ideas, we laid the foundation to what later became the Clever Dog Lab together after I realized that my young Border Collie Guinness might be easier to train as a demonstrator for social learning experiments than the marmosets. However, because marmosets were still an important part of our research, the Konrad Lorenz Institute in Altenberg put out a call for a researcher to work with these animals. And fate brought Zsofi along! The two of us started to work together, and while we did not do much research with the marmosets, we were quite productive with the dogs. But, when you work with dogs and you realized the things they can do, you really start to wonder where these abilities come from—socialization/experience or domestication. So here we were again; we just needed the wolves! With Zsofi on board, we now could rule out at least one of the reviewer's concerns, since obviously Zsofi had previous wolf-raising experience! But no, this was not enough. The reviewers were still concerned that we still had no wolves available for testing.

So, the three of us sat together and decided that, well, we really want to do this, so let's do it—even without money. We were very optimistic that money would come once we actually had some animals! Luckily Kurt was in the

(continued)

Box 2.1 (continued)

position that he could convince the game park in Grünau to give us an enclosure and an old goat stable, where we could hand-raise wolves once they were about 4 weeks old. Next step: we went to Herberstein and got four wolf puppies (Kaspar, Tayanita, Shima, and Aragorn). The first few weeks, we raised the animals in our private apartments until we moved them to Grünau at the age of 4 weeks. With the help of a few journalists, our project got better known, and we could recruit the first god parents for our animals rather quickly. These god parents did not just believe in us (one is still one of our best supporters), but they also provided the financial means to hire our first animal trainer, Bea Belenyi. Bea had been part of the Budapest wolf project and thus also had a lot of experience with wolves and was also a really good dog trainer. She stayed with us for the next 8 years and was helping us immensely not just with our first wolf generation but also the following generations.

The three founders with the first wolf generation (picture taken by Rosmarie Kotrschal)

2.8.1 Decisions Taken for the Set-up of the Wolf Science Center

Our aims as outlined above had certain consequences for the set-up of the WSC:

First, we needed to make sure that we could continue working with the wolves, once they were older than 1 year of age. Thus, before the start of the project, Zsófia and Friederike first went on an information hunt to the USA and visited some of the wolf facilities (Wolf Park, Wildlife Science Center, International Wolf Center) that have long-time experiences with raising wolves. We received valuable insights and recommendations from Peggy Callahan, Prof. Erich Klinghammer, Pat Goodmann, Monty Sloan, and many more people working at these places. To all these people, we will always be very grateful for their time and help, which allowed us to run the project without repeating all the mistakes that were made early on in some of the other projects (we still made enough on our own ☺). Of course, we also profited from Zsófia's experience from Hungary, where she was involved in the family dog project and had already hand-raised two wolves.

Based on the advice we received and our own experiences, we planned our own facility and how to work with the animals:

A. We set up the facilities in a way that we have the possibility to work hands-off with animals if the need arises.
B. We implemented a lot of routines for the wolves early on, since this makes handling them easier. For example, we run a long-term learning set and object permanence study with each animal starting when they were only a few weeks old. In these experiments, the animals could choose between 2 and 3 objects, and only one was baited. Importantly, during the time of the experiment (1 year at least), they learn that if they make a wrong choice, we get to take the object with the food, even if it is right in front of their nose. This routine then allows us to test even the adult animals in full contact and take the food away if they make incorrect choices without them trying to fight us for the rewards.
C. When interacting with the animals, we always give them a choice. For example, we have designated areas in the enclosures where we take visitors, and if the wolves (or dogs) do not want to join, they can always withdraw, and nobody will follow them and force them to get into contact. This gives even shyer animals the confidence to come and check out strangers.
D. We try to set up a relationship based on trust, not hierarchical power or force, with the animals. Accordingly, we treat them with respect and use only positive reinforcement for training. Having said that however, we also take care that the animals respect us as much as we respect them. We achieve the latter with consistent behaviour, much continuing routine and interactions, and a lot of patience. For example, when Kaspar, one of our first wolf pups, was just 6 months old, we wanted to run a pointing study and asked Kaspar to come into the testing house. Once he was in, however, he refused to participate in the test. We just sat down and waited, giving him opportunities every 10 min to work with us. He refused for 2 h until he was ready to work (it was a Sunday evening in October, and it was around 10°C in the test room). The next morning, he played the same game for 10 min and then never again.

In general, among lots of rules, we have two main unspoken ones: if we go and visit the animals in their enclosures, we are the guests, while, if they decide to come out of the enclosures and work with us, we set the rules (in a way that we never breach the trust of the animals) and we never do anything during puppyhood that we would not do with adult animals, so no forceful domination or punishment. So far, we have done pretty well with our animals, though we do have a few animals that can only be handled by the primary caregivers that were present during the hand-raising.

Second, to have the possibility of working with the animals over the course of their lives, we established a team of people which would be responsible, together with the founders, for the development of a strong, lasting relationship with the animals. The first two wolf generations were raised to a large degree by the three founders together with two trainers as well as several students and our partners. Some of the students stayed on and became part of our trainer team. After the first generations, while we were still involved in the hand-raising of our later dog and wolf generations (especially Kurt; Zsofi and Friederike spent more time (hand-) raising their human children), other duties took more and more time (grant writing, publications, teaching, planning, and conducting experiments), and we relied more and more on our fantastic long-term trainers and caretakers. They help us shoulder the responsibilities for the animals, taking care that they are adequately trained to participate in the scientific tests and that they receive the best medical care possible. They have trained the animals on a number of routines that allows them to vaccinate, draw blood, and collect urine and saliva (the latter two mainly for research) of the animals without sedating them.

Third, because we were interested in both the social behaviour between conspecifics and the human–animal relationship, we chose to raise our animals in peer groups with intensive human socialization giving them a well-rounded healthy lifestyle. Early contact with our adult animals further contributed to the puppies and adult animals building up relationships with each other and allowed us to integrate the puppies into conspecific packs once old enough (Fig. 2.1).

Fig. 2.1 Our conspecific peer groups. The dogs (left) (picture taken by Peter Kaut) and the wolves (right) (picture taken by Walter Vorbeck)

Fourth, another big decision that we needed to take at the beginning of the project was which wolves (European or American) and which dogs (a specific breed or mongrels) we would raise. To recap, our main research interest lay (1) in the proximate decision-making processes governing cooperation, (2) in the social relationships these animals form with each other and with humans, and (3) if or how the domestication process changed these processes and interactions. The rationale to choose American grey wolves instead of European wolves was threefold: first, American wolves live in larger packs than European wolves and mostly hunt larger prey, which requires more cooperation. Second, probably because of the high hunting pressure and mainly shy animals surviving this prosecution, European wolves seem—according to zookeepers' experience—more difficult to handle; they are more reactive and less predictable. Accordingly, American wolves seemed better suited for what we had in mind for hands-on work. Finally, we also argued that for the same reasons American wolves might be more similar to the ancestral wolf population in terms of their behaviour towards humans (e.g. being less scared of humans) than European wolves. Whether this was the best choice is a point of discussion of course, given that genetic analyses indicate that dogs were domesticated in the Old World. On the other hand, recent analyses suggest that wolf–dog long-term admixture might be higher in Europe compared to America (Pilot et al. 2018, 2019), supporting our choice of working with American wolves (Fig. 2.2).

In terms of dogs, we discussed whether to use a so-called ancient breed such as the husky that is more closely related to wolves or to use mongrels. The husky might be favoured because this breed is generally considered fit to be kept in outdoor enclosures in groups. In the end, however, we decided to use mongrels, since we thought they are more representative of dogs in general and not a specific breed (i.e. containing a much wider array of alleles than any single breed). Additionally, we wanted to have pups from parents that themselves have lived in packs to increase our chances that such a life would be feasible for our animals. Accordingly, we (Zsofi mainly) reached out to Hungarian Shelters, which often collected mongrels from the streets and kept them in packs in the shelters. In hindsight, this was a good decision. As outlined above, a large part of the dogs we meet today, at least in the Western world, are modern breeds that originated from recent, stringent breeding efforts during the last 200 years (Lindblad-Toh et al. 2005; von Holdt et al. 2010; but see New Jr. et al. 2004), who reported that about 50% of dog litters in the USA were unplanned). This strong artificial selection to breed dogs with specific characteristics to fulfil certain functions (e.g. herding, guarding, hunting; Careau et al. 2010; Coppinger and Coppinger 2001; Svartberg 2006; Zapata et al. 2015) is responsible for the striking morphological differences among dog breeds (Ostrander et al. 2000; Spady and Ostrander 2008; Zapata et al. 2016) but also for a reduction in the genetic variability of dogs (at least within breeds) (Lindblad-Toh et al. 2005). In contrast, free-ranging dogs are genetically more variable than breed dogs (Pilot et al. 2015). A genetic analysis of our dogs revealed that they are genetically more similar to free-

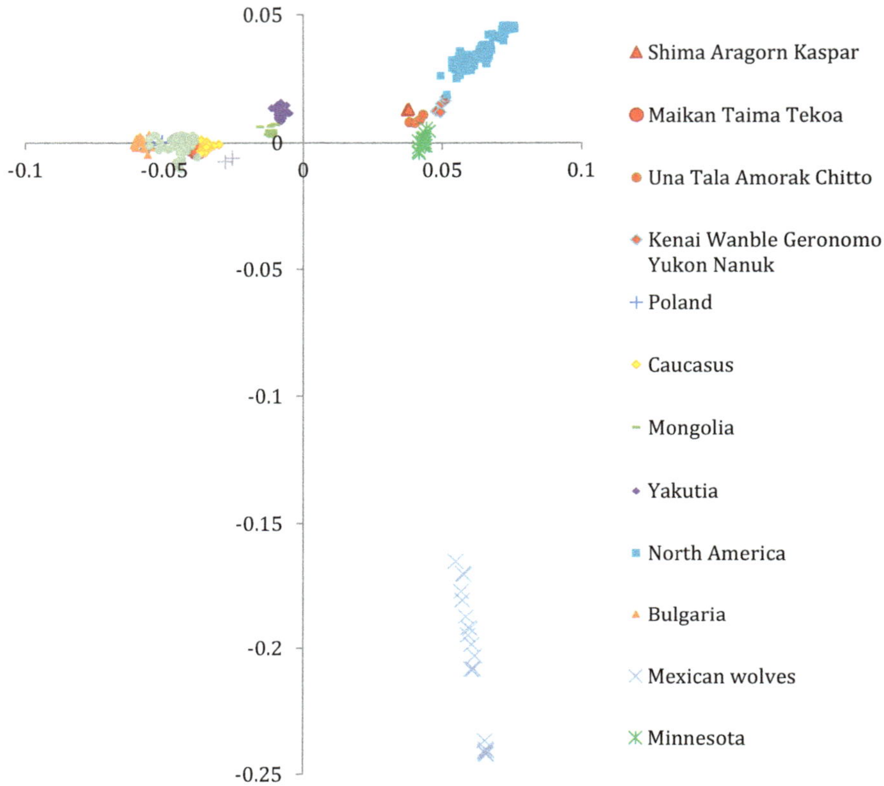

Fig. 2.2 Genetic differentiation between the WSC wolves (marked in red) and wolf populations from across the Holarctic. The WSC samples were genotyped at genome-wide SNP loci using Axiom Canine Genotyping Array Set A (Thermo Fisher Scientific) and compared with the samples genotyped using either the same array or CanineHD Whole-Genome Genotyping BeadChip (Illumina). Data on some wolf populations were obtained from Pilot et al. (2019). The figure shows that most of the WSC wolves originated from North America and Maikan, Tekoa, and Taima from East Eurasia

ranging dogs and thus show a higher genetic diversity than breed dogs (unpublished data, Fig. 2.3).

As explained above, most of our dogs came from animal shelters in Hungary. The last two litters of dog puppies, however, were born by two of our females from the third dog generation. The fathers were Alaskan huskies, a type of dog specifically bred for dog sledding, consisting of a blend of registered dog breeds including Greenland dog, Siberian husky, Alaskan malamute, and various pointer breeds. Alaskan huskies are bred for running performance but are not a registered dog breed.

Last but not least, we wanted to have an adequate sample size to be able to say something about wolves and dogs in general—in other words, we wanted a good

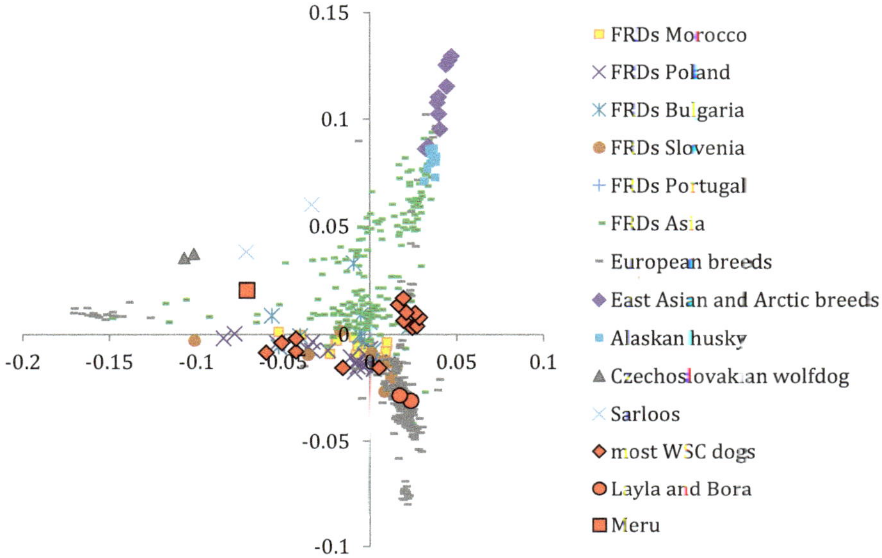

Fig. 2.3 Genetic differentiation between the WSC dogs (marked in red) and free-ranging dog populations as well as pure-bred dogs, reconstructed using the principal component analysis. The WSC samples were genotyped at genome-wide SNP loci using Axiom Canine Genotyping Array Set A (Thermo Fisher Scientific) and compared with the samples genotyped using either the same array or CanineHD Whole-Genome Genotyping BeadChip (Illumina). Data on some dog populations and breeds were obtained from Vaysse et al. (2011) and Pilot et al. (2015). The figure shows that most WSC dogs originated from free-ranging populations, while two individuals (Layla and Bora) are likely mixed-breed dogs

genetic variability. Since in a family the animals (except the breeding pair) are closely related, from a scientific point of view, a family pack would really only represent one data point. Accordingly, we decided to take only two puppies, one female and one male, from any given litter so as to have a good genetic variability in our sample but at the same time also have a female and a male from the same litter allowing us to look at sex differences while controlling for genetics. While this approach is of course a bit better for science in terms of the sample size, it also means that we do not have the same pack composition as wolves in the wild, though for the dogs this criticism probably does not apply.

For the same reason, we decided to vasectomize our males at the age of 1 year so that they are hormonally intact but cannot father any offspring. Vasectomizing and not castrating the males or sterilizing the females is the procedure with the least effect on circulating hormonal levels and thus behaviours.

2.8.2 The Reality

Although we made all these nice and well-thought-through decisions at the beginning of our project, life did catch up with us, and we had to make a few compromises along the way. These include that we sometimes could only take one animal from a litter or there were only males in a litter, leading to a higher number of males than females in our wolves (for the dogs we did not have this issue). Also, in the last generation of wolves, we had to resort to wolves (Taima, Tekoa, and Maikan) which later turned out to be cross-breeds between American and European wolves since we could not find any suitable American wolf puppies in any zoos worldwide. These three animals were also from the same litter.

Also, in regard to the dogs, we made some adjustments: since we had quite some aggression within the original packs, when the animals became of age. We, opted to raise two family packs hoping it would give us the opportunity to keep the animals in larger packs rather than pairs. For the same reason, we raised these family packs with more contact with their mothers (see below). However, both these measures did not attenuate the levels of aggression once sexual maturity was reached and our dog packs today consist in most cases of a male female dyad. This latter aspect also applies to the wolves: it seems as they get older, they are not so open-minded to new animals joining their packs anymore, and we have several packs consisting of only two animals. Part of this intra-group aggression might stem from the fact that the animals are kept in enclosures which prevents them leaving their natal packs and emigrating as they would likely do under more natural conditions. Also, we have to note here that our animals are hormonally intact, which also leads to higher levels of aggression during the breeding seasons due to competition for access to the females.

Altogether, since the start of the project in 2008, we have raised 23 wolves (from 13 litters, although some of the parents of the different litters were related) and 25 dogs (from 11 litters) for our research at the WSC (Fig. 2.4, Table 2.1) (Box 2.2).

2.8.3 Socialization of Our Animals

According to the experiences of researchers that have worked with wolves, we carefully socialized all our animals. Importantly, we used mostly the same procedure for the wolves and the dogs so that their experiences are similar. All our animals are hand-raised starting at the age of 10–12 days (Fig. 2.5). We raise the animals in small groups of peers (four to six animals, wolves and dogs separately) to guarantee that each pup has ample opportunity to socialize with the humans but also has a nice cohort of conspecifics—ideally not just relatives—to facilitate later pack life. Only Wapi and Kenai were raised in a group of two since we could not find any other puppies in that year.

The animals have close contact (24/7) with six to ten human raisers for 4–-5 months in addition to their conspecifics. Moreover, four to six private pet dogs of the hand-raisers also have regular contact with the puppies. The reason to involve the pet dogs was twofold. First, since we planned to conduct social learning

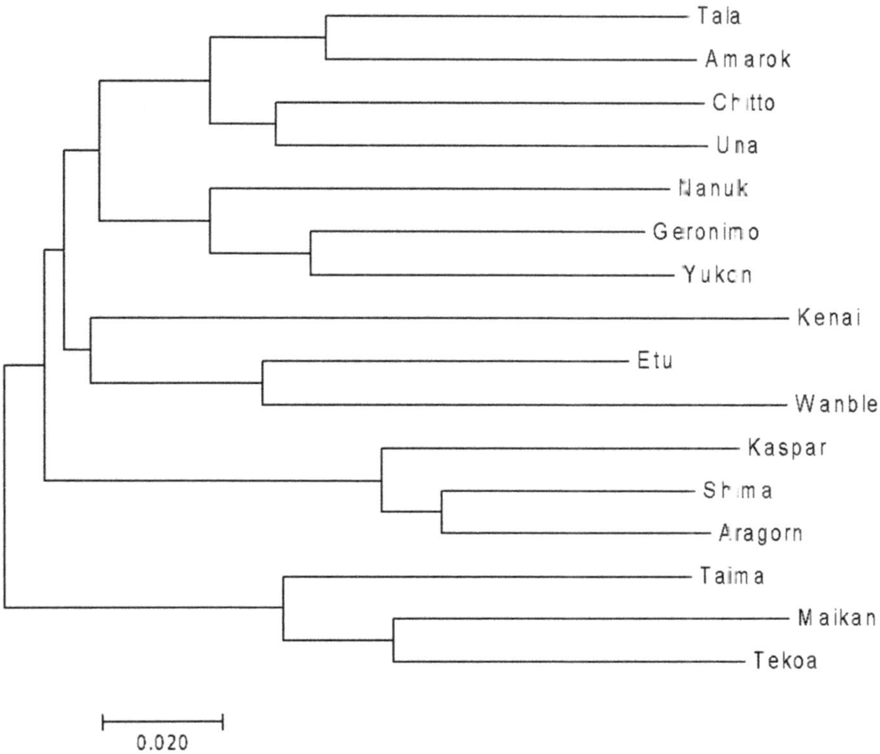

Fig. 2.4 Genetic similarity among our wolves that were still around in 2018 (Malgorzata Pilot, unpublished data)

experiments, we needed demonstrators of the actions. Of course, it would have been possible to train pack members for this role, but then we would have also lost potential subjects for the experiment, since the demonstrator cannot be also a subject at the same time. Having pet dogs well socialized with our dogs and wolves allowed us to use these animals in social learning studies as demonstrators. Moreover, since the pet dogs were all well trained and socialized, our shyer animals often used them as social support in stressful situations, such as when meeting unfamiliar people on a walk. These pet dogs behave somewhat like parents to the puppies, and the puppies readily submit to them for several years (this can look very funny if the pet dogs are rather small and an adult wolf tries to actively submit to the three times smaller dog). During these first months, the human care keepers sleep on mattresses together with the animals, bottle- and later hand-feeding them every few hours. In the first few weeks only during the day, but later also at night, the puppies have ad libitum access to an outside enclosure. When the animals are about 3–4 weeks old, we start with training them to sit, lay, and walk on a leash using positive reinforcement.

Table 2.1 All animals that were raised at the Wolf Science Center

Species	Animal	Sex	Litter	Born	Died/ left	Origin
Wolf	Aragorn	M	1	2008	2019	Herberstein, Austria
Wolf	Shima	F	1	2008	2019	
Wolf	Kaspar	M	2	2008		
Wolf	Tayanita*	F	2	2008	2009[a]	
Wolf	Tatonga	F	4	2009	2013[a]	Triple D Farm, USA
Wolf	Nanuk	M	5	2009		
Wolf	Geronimo	M	6	2009		
Wolf	Yukon	F	6	2009		
Wolf	Cherokee	M	7	2009	2012[a]	Zoo Basel, Switzerland
Wolf	Apache	M	7	2009	2013[a]	
Wolf	Kenai	M	8	2010		Parc Safari, Canada
Wolf	Wapi	M	8	2010	2012[a]	
Wolf	Tala	F	9	2012		Minnesota Wildlife Connection, USA
Wolf	Amarok	M	9	2012		
Wolf	Una	F	10	2012		
Wolf	Chitto	M	10	2012		
Wolf	Wamblee	M	11	2012		Haliburton Forest, Canada
Wolf	Kay	F	11	2012	2014[a]	
Wolf	Etu	M	12	2016		
Wolf	Ela	F	12	2016	2018[a]	
Wolf	Taima	F	13	2016		Moscow, Russia
Wolf	Tekoa	M	13	2016		
Wolf	Maikan	M	13	2016		
Dog	Rafiki	M	1	2009	2014[b]	Tengelic, Hungary
Dog	Alika	F	1	2009	2011[b]	Tengelic, Hungary
Dog	Kilio	M	2	2009	2014[b]	Paks, Hungary
Dog	Maisha	M	2	2009	2018[b]	Paks, Hungary
Dog	Asali	M	3	2010	2019[c]	Siofok, Hungary
Dog	Binti	F	3	2010	2018[c]	Siofok, Hungary
Dog	Bashira	F	4	2010	2013[c]	Paks, Hungary
Dog	Hakima	M	4	2010	2013[b]	Paks, Hungary
Dog	Meru	M	5	2010	2019	Velence, Hungary
Dog	Tana	M	5	2010	2011	Velence, Hungary
Dog	Nuru	M	6	2011	2017[c]	Paks, Hungary
Dog	Zuri	F	6	2011		Paks, Hungary
Dog	Layla	F	7	2011		Györ, Hungary
Dog	Bora	F	7	2011	2019[c]	Györ, Hungary
Dog	Nia	F	8	2011	2018[c]	Paks, Hungary
Dog	Yera	M	9	2011	2012[b]	Hungary
Dog	Kali	M	9	2011	2011	Hungary
Dog	Gombo	M	10	2014	2019*	WSC (Nia)
Dog	Hiari	M	10	2014		WSC (Nia)

(continued)

Table 2.1 (continued)

Species	Animal	Sex	Litter	Born	Died/ left	Origin
Dog	Imara	F	10	2014		WSC (Nia)
Dog	Sahibu	M	10	2014	2018[b]	WSC (Nia)
Dog	Panya	F	11	2014		WSC (Layla)
Dog	Pepeo	M	11	2014	2017[b]	WSC (Layla)
Dog	Banzai	M	11	2014	2018[b]	WSC (Layla)
Dog	Enzi	M	11	2014		WSC (Layla)

[a] died due to health issues (e.g. epileptic seizures due to inbreeding, heart problems)
[b] given into private home for various reasons (social partner died or had to be given away, did not feel comfortable in the pack)
[c] given to private home due to health problems (e.g. infection of the uterus leading to castration, hip dysplasia, injury)

Fig. 2.5 Hand-raising of the pups by the three founders (from left to right: Kurt Kotrschal, Zsófia Virányi, and Friederike Range) (pictures taken by Daniel Zupanc)

Importantly, already at this early age, we try to avoid any conflict situations with the animals, and all interactions are based on voluntary participation.

We also start to socialize the puppies with our adult animals at the age of 4–6 weeks and integrate them in the packs of adult animals between 4 and 5 months of age in accordance with the individual preferences the animals expressed during the prior socialization sessions (Fig. 2.6). Once integrated, the animals live in large enclosures (2000–8000 m^2) in packs of two to six animals but still have daily interactions with their raisers and other trainers: they receive basic training to sit, lay down, walk on a leash, etc., strictly on the base of positive reinforcement and have a firm routine of being separated from the packs in order to participate in regular cognitive and behavioural testing. This method guarantees that both dogs and wolves readily interact and work with human partners and at the same time develop relationships with each other allowing not just interspecific comparisons but importantly also intraspecific interactions.

Fig. 2.6 Socialization of the wolf pups with the adult animals (picture taken by Friederike Range)

While we have tried to standardize the raising and keeping conditions as well as possible, we made one exception in 2014, when we raised one group of dogs as family packs. Two of our females were inseminated by mixed-breed dogs that were not part of the centre, and we reduced daily contact to humans to 6 h in the hope that this would lead to more stable social dog groups. Despite the increased socialization with the rest of the dogs of the respective dog packs, the dogs once again became aggressive towards each other as they became sexually mature, so that we had to separate them into smaller groups again. Moreover, several of these animals were rehomed as pet dogs since the life in the conspecific groups did not agree with them. Thus, the next generation will be raised again in the same way as the wolves.

Unfortunately, over the years, we have lost animals due to old age, health issues (especially ultimate consequences of inbreeding in the wolves ($N = 7$) and poor health status of dog puppies that came from the shelters ($N = 11$)), and rehousing dogs with private owners (in case they needed to be castrated for health reasons or needed operations requiring long recovery times afterwards ($N = 11$)). This was the case despite the optimal vet care our animals always received. Our oldest animals are currently (2022) 13 years old.

2.8.4 Other Research Projects Connected to the Wolf Science Center

Due to the initiative of Sarah Marshall-Pescini, who joined the WSC in 2013 as a postdoc and now is a senior scientist strongly involved in conducting research at the WSC, we started to investigate (1) whether the data we collect on our captive populations is representative of wolves and dogs in general and (2) to what degree the socio-ecology of the dogs and wolves may drive the differences we observe in their behaviours. To this end, we have carried out a number of studies with other captive wolf packs in the Czech Republic (Cafazzo et al. 2016; Lazzaroni et al. 2017) and in Germany (not published yet), and we have an ongoing study with wild wolves in Italy (also not published yet). Moreover, together with Martina Lazzaroni, we have established a field site in Morocco, where we study the social interactions and the cognitive skills of free-ranging dogs (Lazzaroni et al. 2019, 2020). We hope that by studying these other populations of dogs and wolves using the same methods as at the WSC, we will ultimately reach a better understanding of the domestication process and the changes it brought about.

But if you think that from then on everything went smoothly, you are mistaken (see Box 2.2).

Box 2.2 The First Year and the Present

by Friederike Range

Before we moved with the pups to the game park in the Almtal, we first had to clean and renovate the old goat stable and build a wolf-proof fence. We had four small rooms in the stable: one was the designated kitchen (sink with running water, albeit only cold water, and a table with a microwave), one room was kept empty to run small experiments and to put a mattress at night to sleep with the animals, in one room we put an apparatus with a touchscreen and a feeder to run automatic learning studies, and one was a 'proper' bed room with a small bed. To go to the toilet, we had to leave the house and the enclosure and go to the public toilets of the game park. This was also the only place where we could find hot water and thus running water in the winter—something we were very much unaware of when we first arrived in spring! So, in other words, it was rustic, but it had flair and was the ideal place to start our wolf adventure for real—it reminded us a bit of a movie (in retrospect). Anyhow, we moved in and we were really happy! We enjoyed raising the puppies and learning how they react and what would work with them and what not. One of the lessons we learned is that it is important to give wolves the opportunity to withdraw if strangers come and visit. So we created an imaginary line in the wolf enclosure, a boundary which marked where visitors were allowed to go and where wolves could retreat to if they so choose. This made a huge difference: the animals that had been rather shy and withdrawn when visitors came, started to come forward to greet newcomers—they had quickly learned that if they did not want contact, they could just move away a few meters, and this seemed to give them the sense of security and confidence to explore newcomers in a more relaxed manner.

Another aspect which was quite surprising to us was that the wolf pups really did not care where they landed their poop and pee—as we learned later dog puppies are a lot more considerate. But one of the cutest things is that they really like to sleep next to your neck—it seems to be very cosy there.

Another thing we learned is that wolves really like to play in the middle of the night—unfortunately, the best place to play is on the hand-raisers' stomach! This is great as long as the animals are small, but they keep this up also when they weigh 15 kg and when its muddy, cold, and wet outside, and the

(continued)

Box 2.2 (continued)

best game is to romp through the mud playing chase and then catch your littermate, landing directly on the humans' stomach.

We also learned a lot about the personalities of these animals. Kaspar wanted to be the boss from the beginning on; Aragorn was very relaxed as long as he had access to food (Kaspar realized early on that as long as Aragorn can have priority of access to the food, he will not complain about Kaspar being the leader and thus insisted on the privilege of first access to the food only rarely); Shima was always the nerdy one, more interested in working with us than anything else; and Taya (Tayanita) was as obsessed with food as Aragorn, but due to her blindness she was usually second at the food. However, due to her ferocity, Aragorn always willingly shared with her.

(**a**) A walk with a wolf in the fall in the Almtal (picture taken by Florian Schwärzler), (**b**) the WSC facility in Grünau during winter (picture taken by Friederike Range)

There are of course many stories and very fond memories, since it was really a time where we learned a lot about the animals—how they behave, how best to react, how to build up a trustful relationship, and also how to rely on this trust.

The shock came then in the spring, when we learned that we were no longer welcome in the game park for various reasons and had to find a new home. Luckily, through different channels, Sebastian Schmid, responsible for the game park in Ernstbrunn, heard of our plight and contacted us. After our first visit, the Duke Reuss—owner of the park—offered us a new home. It was a real saviour especially since we had our next batch of wolves coming in May.

Since the game park in Ernstbrunn did not have any facilities for us, we first moved into the old kitchen garden and the orangery across from the castle. Again, we did a lot of renovations to prepare for our animals; however, in contrast to the goat stable, we had the luxury of real walls now, minus the wind blowing through the holes. We moved the first generation in April, and then the six puppies from our second generation (four from the USA and two from

(continued)

Box 2.2 (continued)

Basel) joined us in May. We spent a fantastic time up in the kitchen garden, while we built our new facilities in the game park.

In Summer 2010, we finally moved into our present test house and enclosures. The next big change came in 2017, when the University of Veterinary Medicine Vienna offered to properly integrate the WSC into the university structure. This change was completed in 2021 with the establishment of the Core Facility WSC. For us this was a huge relief since it meant some financial stability and a lot less responsibility resting on our three shoulders.

And well, in a fairy tale one would say, 'And they lived happily ever after...', but after navigating university politics for the last years, we rather would put it as 'and the adventure continues'. But so far it works well!

Marianne Heberlein with our six wolf pups from the second generation

References

Cafazzo S, Lazzaroni M, Marshall-Pescini S (2016) Dominance relationships in a family pack of captive arctic wolves (*Canis lupus arctos*): the influence of competition for food, age and sex. PeerJ 4:e2707. https://doi.org/10.7717/peerj.2707

Call J, Tomasello M (1996) The effect of humans on the cognitive development of apes. In: Russon AE, Bard KA, Parker ST (eds) Reaching into thought. Cambridge University Press, New York, pp 371–403

Careau V, Réale D, Humphries MM, Thomas DW (2010) The pace of life under artificial selection: personality, energy expenditure, and longevity are correlated in domestic dogs. Am Nat 175: 753–758. https://doi.org/10.1086/652435

Coppinger R, Coppinger L (2001) Dogs. A new understanding of canine origin, behavior and evolution. The university of Chicago Press, Chicago

Feddersen-Petersen D (1991) The ontogeny of social play and agonistic behaviour in selected canid species. Bonn Zool Beitr 42:97–114. https://www.zobodat.at/pdf/Bonner-Zoologische-Beitraege_42_0097-0114.pdf

Feddersen-Petersen D (1994) Social behavior of wolves and dogs. Vet Q 16:51–52. https://doi.org/10.1080/01652176.1994.9694490

Feddersen-Petersen DU (2000) Vocalization of European wolves (*Canis lupus lupus* L.) and various dog breeds (*Canis lupus* f. fam.). Arch Tierz 43:387–397. https://doi.org/10.5194/aab-43-387-2000

Frank H (2011) Wolves, dogs, rearing and reinforcement: complex interactions underlying species differences in training and problem-solving performance. Behav Genet 41:830–839. https://doi.org/10.1007/s10519-011-9454-5

Frank H, Frank M (1987) The University of Michigan canine information-processing project (1979-1981). In: Frank H (ed) Man and wolf. Advances, issues, and problems in captive wolf research. Dr. W. Junk Publishers, NL, Dordrecht, pp 143–167

Frank H, Frank MG, Hasselbach LM, Littleton DM (1989) Motivation and insight in wolf (*Canis lupus*) and Alaskan malamute (*Canis familiaris*): Visual discrimination learning. Bull Psychon Soc 27:455–458. https://doi.org/10.3758/BF03334654

Frank H, Hasselbach LM, Littleton DM (1986) Socialized vs. unsocialized wolves (*Canis lupus*) in experimental research. In: Fox MW, Mickley LD (eds) Advances in animal welfare science 1986/87, vol 3. Springer, Dordrecht, pp 33–49. https://doi.org/10.1007/978-94-009-3331-6_4

Freedman DG, King JA, Elliot O (1961) Critical period in the social development of dogs. Science 133:1016–1017. https://doi.org/10.1126/science.133.3457.1016

Gácsi M, Győri B, Miklósi Á, Virányi Z, Kubinyi E, Topál J, Csányi V (2005) Species-specific differences and similarities in the behavior of hand-raised dog and wolf pups in social situations with humans. Dev Psychobiol 47:111–122. https://doi.org/10.1002/dev.20082

Hansen Wheat C, Larsson L, Temrin H (2020) Hand-reared wolves show similar, or stronger, attachment toward human caregivers compared to hand-reared dogs. bioRxiv. https://doi.org/10.1101/2020.02.17.952663

Hansen Wheat C, van der Bijl W, Temrin H (2019) Dogs, but not wolves, lose their sensitivity toward novelty with age. Front Psych 10:2001. https://doi.org/10.3389/fpsyg.2019.02001

Klinghammer E, Goodmann PA (1987) Socialization and management of wolves in captivity. In: Frank H (ed) Man and wolf. Advances, issues, and problems in captive wolf research. Dr. W. Junk Publishers, NL, Dordrecht, pp 31–59

Kubinyi E, Virányi Z, Miklósi A (2007) Comparative social cognition: from wolf and dog to humans. Comp Cogn Behav Rev 2:26–46. https://doi.org/10.3819/ccbr.2008.20002

Lazzaroni M, Marshall-Pescini S, Cafazzo S (2017) Post-conflict opponent affiliation reduces victim re-aggression in a family group of captive arctic wolves (*Canis lupus arctos*). PLoS One 12:e0187450. https://doi.org/10.1371/journal.pone.0187450

Lazzaroni M et al (2020) Why do dogs look back at the human in an impossible task? Looking back behaviour may be over-interpreted. Anim Cogn. https://doi.org/10.1007/s10071-020-01345-8

Lazzaroni M et al (2019) The role of life experience in affecting persistence: a comparative study between free-ranging dogs, pet dogs and captive pack dogs. PLoS One 14:e0214806. https://doi.org/10.1371/journal.pone.0214806

Lindblad-Toh K et al (2005) Genome sequence, comparative analysis and haplotype structure of the domestic dog. Nature 438:803–819. https://doi.org/10.1038/nature04338

Miklósi Á, Topál J, Csányi V (2004) Comparative social cognition: what can dogs teach us? Anim Behav 67:995–1004. https://doi.org/10.1016/j.anbehav.2003.10.008

New JC Jr, Kelch WJ, Hutchison JM, Salman MD, King M, Scarlett JM, Kass PH (2004) Birth and death rate estimates of cats and dogs in U.S. households and related factors. J Appl Anim Welf Sci 7:229–241. https://doi.org/10.1207/s15327604jaws0704_1

Ortolani A, Vernooij H, Coppinger R (2009) Ethiopian village dogs: behavioural responses to a stranger's approach. Appl Anim Behav Sci 119:210–218. https://doi.org/10.1016/j.applanim.2009.03.011

Ostrander EA, Galibert F, Patterson DF (2000) Canine genetics comes of age. Trends Genet 16: 117–124. https://doi.org/10.1016/S0168-9525(99)01958-7

Pang J-F et al (2009) mtDNA data indicate a single origin for dogs south of yangtze river, less than 16,300 years ago, from numerous wolves. Mol Biol Evol 26:2849–2864. https://doi.org/10.1093/molbev/msp195

Pilot M et al (2018) Widespread, long-term admixture between grey wolves and domestic dogs across Eurasia and its implications for the conservation status of hybrids. Evol Appl 11:662–680. https://doi.org/10.1111/eva.12595

Pilot M et al (2015) On the origin of mongrels: evolutionary history of free-breeding dogs in Eurasia. Proc R Soc B 282:20152189. https://doi.org/10.1098/rspb.2015.2189

Pilot M et al (2019) Global Phylogeographic and admixture patterns in grey wolves and genetic legacy of an ancient Siberian lineage. Sci Rep 9:17328. https://doi.org/10.1038/s41598-019-53492-9

Rushen J, Taylor AA, de Passillé AM (1999) Domestic animals' fear of humans and its effect on their welfare. Appl Anim Behav Sci 65:285–303. https://doi.org/10.1016/S0168-1591(99)00089-1

Scott JP, Marston M-V (1950) Critical periods affecting the development of normal and mal-adjustive social behavior of puppies. J Genet Psychol 77:25–60. https://doi.org/10.1080/08856559.1950.10533536

Spady TC, Ostrander EA (2008) Canine behavioral genetics: pointing out the phenotypes and herding up the genes. Am J Hum Genet 82:10–18. https://doi.org/10.1016/j.ajhg.2007.12.001

Svartberg K (2006) Breed-typical behaviour in dogs–Historical remnants or recent constructs? Appl Anim Behav Sci 96:293–313. https://doi.org/10.1016/j.applanim.2005.06.014

Topál J, Gácsi M, Miklósi Á, Virányi Z, Kubinyi E, Csányi V (2005) Attachment to humans: a comparative study on hand-reared wolves and differently socialized dog puppies. Anim Behav 70:1367–1375. https://doi.org/10.1016/j.anbehav.2005.03.025

Topál J et al (2009) The dog as a model for understanding human social behavior. In: Advances in the study of behavior, vol 39. Academic Press, pp 71–116. https://doi.org/10 1016/S0065-3454(09)39003-8

Udell MAR, Dorey NR, Wynne CDL (2008) Wolves outperform dogs in following human social cues. Anim Behav 76:1767–1773. https://doi.org/10.1016/j.anbehav.2008.07.028

Udell MAR, Dorey NR, Wynne CDL (2010) What did domestication do to dogs? A new account of dogs' sensitivity to human actions. Biol Rev 85:327–345. https://doi.org/10.1111/j.1469-185X.2009.00104.x

Udell MAR, Dorey NR, Wynne CDL (2011) Can your dog read your mind? Understanding the causes of canine perspective taking. Learn Behav 39:289–302. https://doi.org/10.3758/s13420-011-0034-6

Udell MAR, Spencer JM, Dorey NR, Wynne CDL (2012) Human-socialized wolves follow diverse human gestures… and they may not be alone. Int J Comp Psychol 25:97–117. https://escholarship.org/uc/item/9rb9d1zk

Udell MAR, Wynne CDL (2011) Reevaluating canine perspective-taking behavior. Learn Behav 39:318–323. https://doi.org/10.3758/s13420-011-0043-5

Ujfalussy D, Virányi Z, Kubinyi E (2003) Individual and intensive hand-rearing of wolves: the method and its results in handling the wolves. In: ASAB 2003 Summer Meeting. Grünau, Austria

Vaysse A et al (2011) Identification of genomic regions associated with phenotypic variation between dog breeds using selection mapping. PLoS Genet 7:e1002316. https://doi.org/10.1371/journal.pgen.1002316

Virányi Z, Gácsi M, Kubinyi E, Topál J, Belényi B, Ujfalussy D, Miklósi Á (2008) Comprehension of human pointing gestures in young human-reared wolves (*Canis lupus*) and dogs (*Canis familiaris*). Anim Cogn 11:373–387. https://doi.org/10.1007/s10071-007-0127-y

Virányi Z, Range F (2014) On the way to a better understanding of dog domestication: aggression and cooperativeness in dogs and wolves. In: Kaminski J, Marshall-Pescini S (eds) The social dog. Behaviour and Cognition, Elsevier, Amsterdam, pp 35–62

von Holdt BM et al (2010) Genome-wide SNP and haplotype analyses reveal a rich history underlying dog domestication. Nature 464:898–902. https://doi.org/10.1038/nature08837

Zapata I, Serpell JA, Alvarez CE (2016) Genetic mapping of canine fear and aggression. BMC Genomics 17:572. https://doi.org/10.1186/s12864-016-2936-3

Zimen E (1971) Wölfe und Königspudel. Vergleichende Verhaltensbeobachtungen. Ethologische Studien. Piper, München

Zimen E (1987) Ontogeny of approach and flight behavior towards humans in wolves, poodles and wolf-poodle hybrids. In: Frank H (ed) Man and Wolf. Advances, Issues, and Problems in Captive Wolf Research. Dr. W. Junk Publishers, NL, Dordrecht, pp 275–292

The Socio-Ecology of Wolves

3

The grey wolf (*Canis lupus*) is quite a special animal that has become interwoven with human evolutionary and cultural history (see also Perri et al. 2021). Wolves had one of the most extensive historical geographic distributions of any mammal, occurring throughout the Northern Hemisphere (north of 15–20°N) (Musiani et al. 2007; Paquet and Carbyn 2003) and up to an altitude of 5000 m in the Himalayas, showing an extreme adaptability in terms of the environments they can live in, foods they can live off, and flexibility of their social organization. This adaptability is also quite obvious at the moment, with the wolf returning all over Europe despite the fact that humans have altered the natural environment quite a bit: wolves adjust the use of their home range according to where we are and what we do (Carricondo-Sanchez et al. 2020); they change their diet, eating human discarded food on rubbish dumps if necessary (Newsome et al. 2016); and they breed with pet dogs when opportunities to mate with their own species are absent or reduced (e.g. Hindrikson et al. 2012; Khosravi et al. 2013; Pacheco et al. 2017; Vilà and Wayne 1999), hence supporting the idea of them being two separate species.

Similar to early human hunter–gatherer societies, wolves live in family units that extensively cooperate with their clans (packs) to acquire resources and raise their young (see below) (Hall and Sharp 1978), suggesting that wolves may be a good model for understanding parts of the human evolutionary history by investigating convergent evolution of cooperation as an adaptation to specific environments (see Chap. 11).

Likely because of these 'social' characteristics matching human social orientation as well as their flexibility illustrated through their extraordinary adaptability, wolves were also the first animals to be domesticated (Larson and Fuller 2014), despite they are highly successful group-hunting predators and potentially dangerous to humans (Linnell et al. 2002). Wolves—as dogs—have formed probably the closest relationship with us of all domesticated species.

While there is quite a lot of information about wolves—especially in regard to prey-predator relationships—this review will concentrate on those aspects relevant to the wolf–dog comparison that this book focuses on—the feeding ecology,

F. Range, S. Marshall-Pescini, *Wolves and Dogs*, Fascinating Life Sciences,
https://doi.org/10.1007/978-3-030-98411-3_3

territoriality, social organization and social relationships. It is also important to keep in mind that many of the studies on free-ranging wolf populations have been conducted in the Yellowstone National Park and to some extent Isle Royale in North America.

3.1 A Few General Facts

Wolves are large carnivores belonging to the genus *Canis*. They are genetically sufficiently closely related to the other species of the genus *Canis* (coyotes, golden jackals, 'domestic dogs') that they can interbreed and produce fertile hybrids with all of them. According to the classical biological definition of speciation proposed by Mayr 1942 (Mayr 1999), this does not qualify any of the *Canis* species as separate species. However, there are different concepts relating to the definition of 'species' including a revised, population-based definition suggesting that a species constitutes interbreeding natural populations that are separated from other similar groups (Mayr 1963) and an ecological concept where species are defined as being adapted to a specific environment as a consequence of evolutionary processes (Ridley 2004). Based on these definitions, the various *Canis* species, including the domestic dog, can be referred to as separate species. Further supporting this notion, vonHoldt et al. (2011), using a principal component analysis of individual genome-wide single-nucleotide polymorphisms (SNPs), showed that domestic dogs, the various wolf populations across the world, and coyotes are largely genetically distinct. Virtually all members of the genus *Canis*, however, are incredibly adaptable and opportunistic. This is particularly evident when considering coyotes' or red foxes' successful colonization of urban environments (Bateman and Fleming 2012; Gehrt et al. 2011), but it is also evident in cases of wolves living off human-generated resources (Newsome et al. 2016).

Compared to the other species of the genus *Canis*, wolves display an enormous size range, with Arabian wolves being rather small (64–66 cm at shoulder height; 15–20 kg), whereas the Northwestern wolf can measure up to 80–90 cm at the shoulder and weight up to 72 kg. Females are on average 20% smaller than males, ranging in weight from 25 to 39 kg, but can grow up to 54 kg (Mech 1970). These differences in body size between sexes are thought to be the result of selection for their respective reproductive strategies and roles in hunting and territoriality (see below).

Moreover, wolves have relatively long legs and huge paws, which enables them to travel fast also during winter over snow. Their mean rate of travel is 8.7 km/h, but they can run several kilometres at the speed of 56–64 km/h (Mech 1970). Mech and Cluff (2011) recorded the movement of a wolf and presumably his pack during the winter months on Ellesmere Island. During a 9-day foray, they covered a 263 km straight-line distance and returned within 4 days travelling an average of 41 km/day.

The wolf's fur is also perfectly adapted to the sometimes harsh environmental conditions they live in. Their undercoat is fantastically dense and fluffy during the winter time, and even after rain, if one strokes the animals, they are never cold or wet close to the skin. Their winter fur allows wolves to rest comfortably in open areas at −40°C. They shed their undercoat in spring, leaving only the guard hairs. In the early summer, our animals at the Wolf Science Center look suddenly different: the compact, furry animals become long-legged thin racing machines (Fig. 3.1).

The colour of their coat ranges from white in the Arctic to all black (Musiani et al. 2007). Interestingly, black-coloured wolves are found with relatively high frequency in several North American populations (Anderson et al. 2009; Mech et al. 2003; Musiani et al. 2007). The black coat colour of North American wolves was derived from a single adaptive introgression (via interbreeding) from early Native American dogs between 1598 and 7248 years ago in the Yukon (Schweizer et al. 2018). However, the original mutation for the black coat colour is estimated to be at least 46,886 years old (95% confidence limit: 12,779 to 121,182 years) (Anderson et al. 2009); thus, it is unclear whether it occurred before or after the onset of domestication. In case it indeed occurred before domestication, it must have been lost from the gene pool of wolves and reacquired in North America.

Originally, researchers thought that the black coat colour yields adaptive advantages in forested habitats and thus led to the selective sweep across northern wolf populations (Anderson et al. 2009). However, more recent studies suggest immunological benefits connected to the K_B (referred to as K) locus. The dominant allele K at the beta-defensin protein locus is responsible for the black (melanistic) coat colour in wolves—as it is in dogs (Anderson et al. 2009)—while animals that are homozygotes for the recessive, wild-type allele k_y (referred to as k) are grey. In the Yellowstone National Park wolf population, grey homozygous females had the

Fig. 3.1 (**a**) Kasper in summer 2009 (picture taken by Friederike Range); (**b**) Amarok in summer (**c**) and winter (pictures taken by Rooobert Bayer)

highest annual reproductive success (Stahler et al. 2013); black heterozygotes had the highest fitness in terms of annual survival, lifespan, and lifetime reproductive success; and black homozygotes had the lowest fitness (Coulson et al. 2011). Since black heterozygotes and homozygotes have the same black colour, these fitness differences cannot be explained by the pigmentation. Beta-defensin genes have been previously identified to play a role in innate and adaptive immunity (Yang et al. 1999), and black heterozygote wolves of Yellowstone National Park had higher survival rates after distemper outbreaks than grey homozygous wolves, showing that selection for enhanced immunity in response to distemper may be responsible for the spread of the black coat colour in wolves (Schweizer et al. 2018). In other words, the gene for black colour is linked to a better immune response resulting in black animals surviving distemper outbreaks better and thus producing more offspring in the long run. Since the black gene is dominant over the gene coding for grey, the black colour will spread through the population. The problem with this hypothesis is that it does not explain why there are no black wolves in migrating populations (Musiani et al. 2007). Thus, since this mutation is also linked to the serotonin metabolism, it likely also affects the behaviour of the black animals. For example, it has been shown over multiple wild vertebrate species that darker individuals are more aggressive, sexually active, and more resistant to stress than lighter individuals (Ducrest et al. 2008). Accordingly, the more aggressive behaviour of black individuals might have a higher selective advantage in territorial packs but less so in wolves that do not defend their territories but instead migrate with caribou herds. In these migrating wolves, 93% are grey, whereas in territorial wolves, only 38% are grey (Musiani et al. 2007).

Thus, while the black colour in wolves is indicative of a hybridization event with dogs, this can have happened a long time ago and does not suggest that currently hybridization is an issue in a given population. Rather, there must have been adaptive advantages of the underlying genetics of the black coat colour, which have fixed this mutation in Northern American territorial wolf populations.

Compared to other animals of the same size, wolves have a fast life history. Their generation time is relatively short with animals in the wild living 5 years on average—50% of female breeders die by age 5 (Stahler et al. 2013). They start reproducing at the age of 2 years, and females reach their highest reproductive success between 2 and 4 years of age, with typically one litter per year. They have a high fecundity with five to six pups per litter, and the pups reach 80% of their adult body size at their first birthday (Fuller et al. 2003; MacNulty et al. 2009a; Peterson et al. 1998).

Finally, especially in the light of the many experiments relying on visual information that we will summarize in the later chapters of this books, it should be noted that while wolves have a well-developed sense of smell, vision is also a critical sense for these animals (Gable and Gable 2019). As coursing hunters, wolves need to position their attacks on kicking, fleeing predators that could injure them (Mech et al. 2015). Moreover, wolves rely on different ear and tail positions as well as facial features when communicating with each other, which are detected by vision (Goodmann et al. 2002).

3.2 Environment and Diet: The Wolf as the Epitome of the Hunter?

Wolves are primarily cursorial hunters that often hunt smaller prey alone but coordinate with other pack members in a differentiated way when hunting larger prey. However, despite the fact that wolves are primarily hunters and often hunt cooperatively, hunting success ranges between 10% and 50% based on the number of prey groups attacked in North America (Mech et al. 2015). For moose, hunting success in Scandinavia was estimated at 45% or 64% depending on the method used for a 5-year observation period (Sand et al. 2006). This hunting success is somewhat comparable to other socially hunting mammals: wild dogs Kenya, mean 51% (Fuller and Kat 1993); Tai chimpanzees mean, 52–56% (Boesch 1994; Samuni et al. 2018); and hyenas, mean 33% (Holekamp et al. 1997). Accordingly, researchers have described wolves' feeding ecology as a 'fast-or-feast' affair, i.e. food is unpredictable and infrequent, but when successful, food is of high quality and potentially abundant (depending on size of prey) (Mech et al. 2015). In line with this feeding ecology, wolves can gorge as much as 10 kg at a time, but then they can also fast for several days if they do not find food (Mech and Boitani 2003).

3.2.1 Factors Influencing Hunting Success

Wolves have a specialized skeleton for running down prey but not so much for killing (MacNulty et al. 2016). They need to rely on their incisors and canines to grab and subdue prey, and these wear out with age (Gipson et al. 2000). Actually, at the Wolf Science Center, several of our older animals, with 'old' starting at 6 years of age, had dental operations, which resulted with two of our animals sporting two steel canines. However, the luxury of a dentist is of course not available to wild wolves. Moreover, their long snout reduces the force of jaw-closing muscles, preventing them from delivering killing bites as strong as those of felids and hyenas (wolf, 593 N/cm^2; lion, 1768 N/cm^2; hyena, 9000 N/cm^2), and they cannot lock or stabilize their jaw during the bite (Wang and Telford 2008).

In addition, most of the wolves' prey is well equipped with formidable defensive weapons and displays different anti-predator behaviours including standing their ground (bison, moose, or elk) and engaging in group defence (musk oxen or bison), requiring the wolves to be careful in order to avoid injuries or being killed. This applies not just to their larger prey but also smaller hoofed prey, such as the white-tailed deer which has been observed to kill wolves (Mech et al. 2015). Thus, every attack needs to be carefully assessed, and cost–benefit ratios need to be judged, explaining why wolves are very apt at targeting young, old, and disabled animals during their hunt (Gazzola et al. 2005; Mattioli et al. 1995; Mech et al. 2015; Wagner et al. 2012). Moreover, running their prey to exhaustion before attacking and selecting weak animals will be less energy consuming. Therefore, it is a mandatory part of their hunting tactic to get big prey moving, which is easy in case of caribou and elk, more difficult in case of bison, and really difficult with musk oxen.

This careful selection of prey is in contrast with human behaviour. For example, a study by Wright and colleagues in the northern Yellowstone area (Wright et al. 2006) showed that while human hunters mainly killed prime-aged female elk with the greatest reproductive values, wolves selected a large proportion of elk calves and older females with low reproductive values during the same time showing that hunters exerted a greater total reproductive impact on the herd than wolf predation. In line with this implication, another study found that while elk responded similarly to wolf and human predation risk by changing grouping patterns and increasing movement rates, they responded more strongly to the human predation risk (Proffitt et al. 2009).

While wolves have been shown to be able to take down all prey alone, including bison, musk oxen, and moose, they typically hunt the larger animals in packs. This is not really surprising given the average body weight of 50 kg in wolves and of 500–1000 kg in these large prey species. According to the old rule of field ecology, total predator mass should equal prey mass, which is coarsely true for wolves—hence, they may hunt prey less than 50 kg alone. Thus, cooperative hunting of the larger prey might increase hunting success but also might offset the risk of injury to some degree (Mech 1970). One other very interesting aspect, especially in the light of wolves returning all over Europe, is that one study, examining 1457 scats between 2008 to 2013 in Italy, calculated a model that showed that the presence of stable packs, instead of dispersing wolves, reduced predation on livestock (Imbert et al. 2016).

Some support for why wolves live at least in pairs or small packs comes from a detailed analysis on elk hunting in Yellowstone National Park, which revealed that packs were most successful if there were four animals. While fewer animals decreased the success, more animals in the pack did not show a measurable effect on the outcome of wolf–elk interactions (MacNulty et al. 2012). This latter result was interpreted as being due to the fact that while the breeding pair actively hunted, additional wolves—pups as well as adults—were 'free-riders', e.g. they run along to reap the benefits of a kill, but did not actively help to bring down the prey, thus reducing their own risk of injuries and their energy expenditure. In a subsequent study, MacNulty et al. (2014) investigated how the success of wolves increased with pack size when hunting bison—a much larger, more formidable opponent. They found that here wolves were more successful the larger the pack size was and success rate started to level off only at 9–13 animals. In contrast to the hunting of elk, all wolves in the packs were more cooperative and did not tend to free-ride. This more cooperative behaviour of the wolves when hunting bison compared to when hunting elk is suggested to be due to the fact that a lone wolf has a low probability of killing an adult bison, whereas it has about a three times greater chance of killing an adult elk by itself. Overall, given that a pack usually consists of kin, being flexible in when to cooperate with each other and when not is of course beneficial in terms of inclusive fitness. More precisely, for the parents, the benefits of youngsters not participating in a hunt, where they are not needed to bring down the prey, and thus not risking injury and potential death might be higher than the potential decrease in energy expenditure when hunting cooperatively.

One other aspect that is interesting in this regard is a story from Smith and Ferguson (2012), describing how a wolf pack that had been primarily focused on hunting bison—and was very successful—basically fell apart after their breeding female was killed during a hunt, suggesting that social aspects and preferences of the leading animals also contribute to the packs' hunting preferences and success.

Wolves' age—probably representing experience and skills—and body size influence hunting success at the individual level and also, depending on the pack composition, at the pack level. While the capacity to hunt and kill is innate in canids (Fox 1969), it is likely that youngsters need to learn to fine-tune these behaviours (e.g. see Mech 2014) for an observation of an adult wild wolf providing an opportunity for pups to learn to kill live prey. The general idea is that wolf pups start to accompany the pack during hunting at the age of about 4–6 months, where they have the opportunity to learn from the adult hunters to kill prey (Packard 2003), which may take up to 2 or 3 years. A study in Scandinavia found that pups did not kill any deer in their first winter but only fed of kills made by their parents or other carcasses they found (Zimmermann et al. 2015). Interestingly, in Italy, where we have observed free-ranging wolf puppies in the last few years at their rendezvous sites, the pups did not go hunting with their parents until at least 8 months old but rather remained at the rendezvous site and waited for their parents' return. This could be due to several reasons. First, in general, wolves in Italy hunt for relatively (to the USA) smaller prey, thereby not requiring larger packs. Second, it may also be an adaptive behaviour to avoid being spotted by humans since our study area is in a densely populated area and, traditionally, tolerance towards wolves is low. And, finally, competition between packs might not be as severe as in North America making it relatively safer for pups to stay alone. It could of course also be just the relaxed Italian way of life shining through!

According to long-term data from Yellowstone, the optimal performance of hunting elk is reached between 2 and 3 years of age, after which the hunting ability decreases again due to physiological senescence (MacNulty et al. 2009b). Larger body size has been shown to improve the ability of wolves to attack and kill elk, even if the animals were of the same age (MacNulty et al. 2009a). The improved attack ability is hypothesised to be due to bigger animals being less intimidated by their large prey, while the better killing ability is attributed to the bigger mass, since they need their weight to topple the prey. Body size also explains why males outperform females in attacking and killing prey given that, normally, females are about 20% smaller than males. If, however, females are as big as males, sex does not play a role. It is important to note here that the influence of age and size is likely dependent on the size and speed of the prey. Especially in the case of fast-footed prey, lighter animals might have an advantage due to their higher mobility.

Similar results were shown in a study in Scandinavia that used snow tracking to investigate the hunting behaviour of wolves on moose in 17 wolf territories. The authors found that the age of breeding males, but not hunting group size, snow depth, or moose density, predicted hunting success in 142 observations. The prime age for hunting moose was at 4.5–5.5 years. Interestingly, one pack that lost the breeding male switched to hunting roe deer, whereas another pack, where the breeding female was lost, continued hunting moose. The authors suggest that the effect of male age on the hunting success reflects the experience and greater size of adult male wolves compared to the females (Sand et al. 2006).

3.2.2 The Diet of Wolves

Due to the importance of understanding predator–prey dynamics for wildlife management strategies and likely the competitive relationship between wolves and man when it comes to hunting, there is a suite of studies on wolves' diet composition ranging from Asia, over Europe, to America. The percentage of prey species in the wolf's diet depends (1) on their diversity; (2) sometimes, but not always, on their relative density (Martin et al. 2018; Mattioli et al. 2004; Mori et al. 2016; Nowak et al. 2011; Tallian et al. 2017); and (3) on the time of the year (summer vs. winter) (e.g. Gazzola et al. 2005; Kojola et al. 2004; Metz et al. 2012).

In a recent review, Newsome et al. (2016) compiled dietary data from 177 studies incorporating 94,607 scat and stomach samples. They report that overall in North America ($N = 77$ studies), the diet of grey wolves was dominated by large- and medium-sized ungulates including different deer species (mule, white-tailed, and black-tailed deer), elk, moose, and caribou (see also Peterson and Ciucci 2003 for similar results). In a few studies, bison (four studies, average 44%) and rubbish (three studies, average 21%) also constituted a high percentage of the diet. Beavers were also part of the wolf's diet in 54 studies, while domestic animals were only featured in ten studies, comprising no more than 8% of wolves' diet.

In Europe ($N = 85$ studies), medium-sized wild ungulates (especially wild boar, roe deer, and chamois) and, to a lesser degree, large wild ungulates (moose and red deer) constituted the highest proportions of the diets (see also Mori et al. 2016 for similar results). In contrast to North America, domestic species contributed much more to the wolf diet (average of 33% across 73 studies) with the primary candidates being pigs, goats, and horses. However, also garbage and fruit were consumed in many more studies ($N = 39$ compared to only $N = 13$ in America).

In Asia ($N = 15$ studies), wolves' main food resources were domestic species, which contributed to 50% of the diet in 14 studies and medium-sized wild ungulates. Domestic animals included in most studies were sheep and also quite often cattle and goats, whereas yaks and poultry were the dominant domestic animals eaten in two studies. Also, in two studies, a relatively high percentage of fruits (10%) were reported.

The differences in diet contributions between continents were significant in that large- and medium-sized ungulates were more often consumed in North America than Europe and Asia, medium-sized wild ungulates constituted the main prey in Europe, and domestic animals were more often part of the diet in Asia than the other two continents. The reliance on domestic animals in Asia is likely due to the lower abundance of wild ungulates compared to North America and Europe (Ripple et al. 2015). Further differences were reported for different bioregions within the continents.

While this review shows that wolves can survive on almost any food resources, it also clearly indicates that grey wolves are carnivores relying mainly on wild ungulates (Paquet and Carbyn 2003). Only if wild ungulate species and medium-sized wild mammals are not available or their abundance is low, wolves seem to shift their diet to waste and domestic animal species, as seen in Asia and specific regions such as the Yazd province in central Iran (Tourani et al. 2014), central Greece (Migli et al. 2005), Western Galicia in Spain (López-Bao et al. 2013), and Belarus (Sidorovich et al. 2003). There are also some studies indicating that wolves might prefer wildlife over livestock consumption. For example, one study showed that while the presence of domestic ungulates on high-altitude pastures during summer (May–October) influenced wolf diet (summer 19.0%, winter 0.3%) in the Italian Alps, wolves still preferred wild ungulates despite the higher density of domestic livestock (Gazzola et al. 2005; Imbert et al. 2016; see also Meriggi and Lovari 1996). Other studies have found that consumption of livestock by grey wolves decreased over time, coincident with an increase in relying on wild ungulates in southern Italy (Newsome et al. 2016). However, these data need to be taken with caution since it is likely dependent on the protection measures taken by the livestock owners and maybe herd size.

Also, it should be noted that wolves, like all mammals except cats, need to eat some plant component, since they cannot produce essential unsaturated fatty acids themselves (Bauer 2006) and hence have to take plants or gut contents of herbivores (e.g. in the Arctic).

3.3 Social Organization

3.3.1 Pack Composition and Pack Size

Wolves live in family packs, led by a breeding male and female, who often mate for life (Mech and Boitani 2003). One breeding season per year, in which both males and females become sexually active in late winter, results in a single litter being born in a pack in early spring (April–May). The offspring emigrate at the age of 1–3 years. Thus, most packs are family packs composed of the breeding pair with the offspring from the past several years. Apart from this normal pack composition, many other combinations have been observed, where usually siblings of one or both of the mated pair join the pack as well. However, some packs may also contain unrelated, non-reproducing animals (Lehman et al. 1992). In line with this apparently high flexibility, pack sizes in America vary between 2 and 42 animals with an average of eight animals. In Europe, average pack size seems to be a bit smaller with five to six animals (Mech and Boitani 2003), though, in Germany, the packs seem to be larger based on the number of pups reported per year and pack (https://www.dbb-wolf.de/Wolfsvorkommen/territorien/entwicklung-der-rudel).

Pack size seems to primarily be dependent on the abundance of food. For example, the largest pack sizes, sometimes comprising multiple breeding females, were observed during the re-introduction of wolves to Yellowstone, which might have been due to the high abundance of prey at the beginning of the project (Smith and Ferguson 2012). In contrast, if the number of ungulates is small, packs often consist of only the breeding pair (Mech and Boitani 2003). However, see Ausband et al. (2016) for a study where pack size was not dependent on prey abundance. A number of other factors seem to affect packs size on top of prey availability.

One factor that has been discussed to influence pack size, not just in wolves but in group hunting carnivores in general, is prey size. The optimal number of animals in the pack should be a function of the costs of energy expenditure to kill the prey and the benefits in terms of energy return from the meat of the prey (though it may also be dependent on other factors as described below). A review by Schmidt and Mech (1997) of 11 studies including different prey species across different areas, however, showed that the larger the pack, the less food each pack member receives (with the possible exception of packs hunting primarily moose). These data suggest that facilitation of predation on large prey is not the main reason why wolves live in packs, with the possible exception of when the packs specialize on difficult prey (MacNulty et al. 2014).

One other factor that needs to be considered is that wolf packs also need to defend carcasses against scavengers (e.g. coyotes, ravens). Ravens can remove (eating and caching) up to 43 kg of meat per day from a kill (Heinrich 1994) cited in Kaczensky et al. (2005)), which may lead to wolves losing an estimated 2–20 kg of food per day to scavenging ravens (Hayes et al. 2000; Kaczensky et al. 2005; Vucetich et al. 2004). By defending the carcass together, scavenging can be drastically reduced, e.g. ravens can usurp 66% of a lone wolf's kill but only about 10% of the kill of a pack of ten wolves. Accordingly, it has been proposed that the greater food-sharing costs in larger packs might be compensated by smaller losses to scavengers, suggesting overall benefits of increased pack sizes in wolves (Kaczensky et al. 2005; Vucetich et al. 2004).

Other studies have shown that pack size is an important driver of reproductive success in wolves, probably due to the additional help in pup-rearing provided by auxiliaries (see below) (Stahler et al. 2013). But again, results differ, with at least one study, analysing an 11-year data set from several populations across America, showing that the helpers guarded the pups more in smaller than larger packs (Ausband et al. 2016). Interestingly, whether pack members help with raising the young seems to be sometimes dependent on the food availability (Harrington et al. 1983) and sometimes not (Ausband et al. 2016). See below for factors influencing litter size.

Though people often think first about prey size and cooperative hunting when considering pack size, one of the more important factors might be inter-pack competition for territories. Being the owner of a good territory has the benefit of providing access to uncontested prey and the protection of pups. Since winning fights with other packs is dependent on the composition of the pack and also pack size (see below; Cassidy et al. 2015), larger pack size might have benefits for holding onto prime real estate, which in turn may increase, or stabilize, group size. Since a good territory likely increases inclusive fitness for the owner and might be inheritable, it also increases the benefits for the young adults to stay around, 'being helpers' rather than leaving to find their own territory. And, as long as enough prey is available to feed all, the parents might tolerate their older offspring to stay around, since it is likely to increase their own inclusive fitness by better protection of the territory and thus higher survival of the pups. This hypothesis could explain the increase of multiple litters per pack as well as the increase in bite wounds in Germany with the stabilization and saturation of the population in the past 10 years (personal communication Ilka Reinhardt) and also the development in Yellowstone.

However, Vucetich et al. (2004) found no support for this argument for the wolves on Isle Royale in the USA.

To conclude, while the core of a wolf pack is the family unit, pack size seems to be primarily limited by prey abundance. Beyond this, territory defence in saturated populations, the type of prey being hunted, competition at carcasses, and reproductive success are important factors determining pack size.

3.3.2 Mating and Reproduction

3.3.2.1 Mating

Wolves mate sometime between January and April (Schmidt et al. 2008). About 1–2 months before the onset of the mating season, the hormone levels of the males (testosterone) and females (oestrogen) increase, which actually leads to males having visible testes (the rest of the year, they are bean-sized and hardly visible). The breeding male starts guarding 'his' female, and, at least at the Wolf Science Center, the males have not much time anymore for anything else (even feeding is a problem for some of them). The increasing hormone levels are also reflected in their sleeping distances: Knick and Mech (1980) reported an average sleeping distance of 4.48 m (range = 0.61 to 8.62 m) in 24 packs in Minnesota, but during the mating season the breeding pairs often slept less than 1 m apart (Mech and Knick 1978). Similar observations were made in Yellowstone National Park over two winters (Peterson et al. 2002) and in captive packs (Zimen 1976).

Also the stress hormone levels of wolves increased during the mating season in Yellowstone National Park leading to more intrasex agonistic interactions within the packs (Sands and Creel 2004). This is probably due to intense breeding competition since, contrary to earlier ideas, the breeding pair does not hormonally suppress other pack members from breeding (Sands and Creel 2004). For example, in a study of two captive wolf packs at Burger's Zoo in Arnhem from 1977 to 1985, authors found that while males intervened and separated contacts between other males and females, especially if the interaction involved their own preferred female, and showed high frequencies of intrasexual aggression during the mating season, females, especially the breeding female, engaged in a high level of intrasexual dominance display in and outside the mating season (Derix et al. 1993). Accordingly, it seems that monogamy in wolf packs mainly depends on the behaviour of the breeding pair with the male ensuring his paternity by defending this female, whereas the breeding female ensures the necessary resources for her pups. The latter argument would also explain the observation that breeding females sometimes commit infanticide of pups from subdominant females (McLeod 1990). At the Wolf Science Center, Kaspar and also Geronimo were very intolerant towards other males in their pack approaching and interacting with Tala or Yukon during their receptive days, and they seemed to be very good at pinpointing those days.

There are different options wolves can pursue in order to breed:

1. They leave their natal pack and find their own territory. This is likely the most frequent strategy.
2. Alternatively, they wait for an open position in their own pack or in another one (i.e. the death of an established breeder).
3. They become an extra breeder within their own pack (usually females closely related to the breeding female) (VonHoldt et al. 2008); this might be linked to high food availability (Resource Abundance Hypothesis; Macdonald 1983) and/or intense between-group competition.
4. They actively displace the current breeder of their pack (Mech and Boitani 2003; VonHoldt et al. 2008).

One problem associated with some of these strategies is inbreeding, which may lead to the expression of recessive deleterious alleles and loss of heterozygosity and thus to reduced fitness (see Robinson et al. 2019; but see Kokko and Ots 2006 for possible advantages of inbreeding). There is some evidence of vertebral malformations in inbred wolf populations both at Isle Royale and in the Scandinavian wolf population (Räikkönen et al. 2006; Räikkönen et al. 2009), and we have had some severe problems with epileptic seizures, likely due to inbreeding, claiming four of our animals since 2008. Not surprisingly, wolves usually do not mate with individuals from their natal pack (Geffen et al. 2011; Smith et al. 1997; VonHoldt et al. 2008). Interestingly, outside their natal pack, wolves mate indiscriminately in regard to relatedness with individuals and independent of how often they encounter relatives (Geffen et al. 2011). This raises the questions whether wolves actually recognize their kin or not, as kin recognition has been proposed as one mechanism to avoid inbreeding. Since wolves live in family packs and later disperse over large distances, making it rather unlikely that they encounter relatives and mate with them, kin recognition might not be an important ability for wolves. In fact, it has been proposed that mechanisms to avoid inbreeding are more likely to develop in species where kin encounter each other as adults (Pusey and Wolf 1996). Although we have never investigated whether the wolves at the Wolf Science Center recognize their kin, the relationship between the siblings Yukon and Geronimo suggests that they may. While we combined pups from three different litters when they were just 10 days old, raising them all together, the sister–brother pair Geronimo and Yukon (Fig. 3.2) always got along very well. Even after they were separated for several years living in different packs, it was no problem to put them back together, and they were happy to see each other, which we have not observed in unrelated pairs that were raised together. Since, differently than in wild populations, they were raised with other unrelated puppies, this observation suggests that experience with each

Fig. 3.2 Geronimo and Yukon—siblings—that really enjoy each other's company (picture taken by Rooobert Bayer)

other is not sufficient to explain this preference. One mechanism that has been suggested to function as a kinship marker across vertebrates is odours associated with the highly polymorphic major histocompatibility complex (MHC) (Brown and Eklund 1994; Penn 2002; but see Green et al. 2015); however, whether this applies to wolves will be a future avenue of research to pursue.

However, even if Geronimo and Yukon recognized each other as kin, Geronimo really likes to mate with her (which we have also observed in other packs where related male/female pairs were kept), while Yukon might actually prefer Wamblee—the lower-ranking male in that pack of three. Thus, kin recognition in this case did not lead to inbreeding avoidance (which might of course be due to the fact that there was no other option for Geronimo to mate and/or males being less selective than females in terms of whom to mate with according to the different reproductive strategies of males and females) (Trivers 1972). Interestingly, this is in line with a study from the Isle Royale population, where wolves even preferred to mate with related individuals of other packs, suggesting (1) that wolves recognize kin and (2) that, under certain circumstance, they do not actively avoid inbreeding (Geffen et al. 2011). Unfortunately, as mentioned already above, this inbreeding on Isle Royale had severe consequences for the population (Robinson et al. 2019).

Another aspect that seems to influence mating patterns is the coat colour. In a recent study, a strong (63%) negative-assortative mating for grey and black coat colour was recorded in Yellowstone National Park (e.g. grey and black wolves prefer mating with animals *not* of the same coat colour), which likely maintains the colour polymorphism (Hedrick et al. 2016). One mechanism of this assortative mating might be the more aggressive and sexually active behaviour of darker compared to lighter individuals (Ducrest et al. 2008).

3.3.2.2 Reproduction

Female wolves in the wild usually do not ovulate until their second, third, or even forth winter and accordingly have their first pups between 2 and 5 years of age (Lehman et al. 1992; Mech et al. 1998). However, as mentioned above, whether they have pups at a certain age strongly depends on their social status. In most cases, unless food abundance is very high, only one female per pack, the breeding female, will reproduce successfully. The gestation period is 61–64 days, and thus the birth of the pups coincides with the birth season of herbivores and the abundance of weak prey at the end of the winter months. Not surprisingly, wolves breed earlier at lower latitudes (between 12 and 80°N), probably because of differences in seasonality of their prey species (Mech 2002).

Litter size can vary between 1 and 11 pups. In the Denali National Park, one to five pups/pack per year were observed over a 7-year period (Mech and Boitani 2003). In Yellowstone National Park, researchers observed a range of 0–11 puppies per litter (average of 4.74 puppies per year and pack) over 14 years (Stahler et al. 2013). In the latter study, it was found that litter size at emergence from the den at about 49–64 days improved with the mother's body mass (estimates of individual, age-specific body mass—although not significant—and age). Female breeders had the largest litters at the age of 4 years. On average, mothers gave birth for the first time at 2.7 years, and 50% of the breeding females died—probably were killed by other wolves—not older than 5 years of age. Some exceptions were observed though with some females still having pups at 9 years of age. Litter size was also influenced by pack size with the largest litters observed in packs of eight wolves and was negatively correlated with population density and distemper outbreaks (Stahler et al. 2013) (see below).

3.3.3 Parental Care: Cooperative Breeding

All pack members are usually involved in raising the young by providing either direct care such as regurgitating food to pups or indirect care including digging the dens, providing food to the lactating female, and defending the pups. The first indirect investment in the pups is digging dens. Females will dig dens even if not pregnant, ovariectomized in the case of captive wolves, or if they do not produce pups in the whelping season, suggesting that den digging is a function of increasing prolactin in spring rather than an increase in oestrogen or progesterone (Mech et al. 1996).

In human-populated areas in Eurasia, the main factors influencing the den location seem to be the distance to villages, forest edges, and intensively used roads (Theuerkauf et al. 2003 and studies referenced therein). In areas less populated by humans, the dens are usually located towards the centre of the territory (Mech

1994). The denning period lasts for 49–64 days with some wolf packs using several different dens during that time (Packard 2003; Schmidt et al. 2008). While some females never reuse dens, but return to the same parts of the territories in subsequent years as breeding sites (Schmidt et al. 2008), some females may reuse dens and, sometimes, excellent dens might be used over generations (reviewed in Packard 2003). The denning period ends about 2 months after the birth, when the pups start to stay in so-called rendezvous sites, which can include 'nests', playgrounds, and trails, until they are about 5 months old and join the pack during hunting and travelling. In Italy, we have observed some litters remaining at the rendezvous site until they were about 7–8 months old (Lazzaroni, Marshall-Pescini, personal observations).

During the first 1–4 weeks of their life, direct care is mainly provided by the female in terms of milk, warmth to the pups, belly licking to provoke poop elimination, and swallowing pup faeces and urine, while the father hunts and provides the nursing female with food (Ballard et al. 1991; Harrington and Mech 1982). In Poland, the mothers spent 65–98% of their time with the pups, decreasing the time until they resume full mobility with the weaning of the pups at the age of 50–70 days (Schmidt et al. 2008). In a study of three packs in the Denali National Park, pups were attended for 27–60% of the time (Chapman 1977). Likely, these different times spent at the den are also dependent on the female's access to food, which is likely dependent on the number of pack members that bring back food to the den and how abundant food is close to the den if the female goes hunting herself.

Starting at the age of about 5 weeks, pups can also digest solid food. Now, (all) pack members contribute to feeding the puppies by regurgitating food, including the mother once she joins the hunts again. A 6-year study on regurgitation in a pack of wolves on Ellesmere Island showed that all adult wolves, including yearlings and a post-reproductive female, regurgitated food—mostly to the pups (81%), but also in some cases to the breeding female (14%)—however, parents contributed about twice as much as other pack members. The more effort the breeding female invested, the larger the litter (Mech et al. 1999).

When the mother is absent during the pup-raising time, anecdotal observations from a study in Poland suggested that the pups were guarded by other pack members (Schmidt et al. 2008) (Fig. 3.3). A recent study (Ausband et al. 2016) analysing an 11-year data set from different wolf populations from North America showed that before, but not after, weaning non-breeding female helpers guarded more than non-breeding males, while age of the helper had no influence on the helping behaviour. Moreover, while wolf density did not predict helping behaviour, individuals spent more time guarding the pups in smaller rather than in larger packs.

3.3.4 Pup Mortality and Dispersal

3.3.4.1 Factors Influencing Pup Mortality

Mortality rates of wolf pups are highly dependent on ecological and social conditions. For example, during recolonization of an area in British Columbia, puppy survival rates were high with 85% pups surviving at least until December

Fig. 3.3 Being a cooperative helper can be a lot of fun (pictures taken by Friederike Range)

(Pletscher et al. 1997). In contrast, in Yellowstone, survival can be rather low with 30% in certain years if ecological conditions are bad (Mech et al. 2008). In Minnesota, over a 6-year period of the mean 6.6 pups per litter at birth, 3.2 survived until mid-November (Fuller 1989). The survival of the pups is dependent on several factors. Not surprisingly, body weight of the pups in their first months of life is rather important (Van Ballenberghe and Mech 1975). In captivity, wolf pups weigh about 450 g at birth and then gain about 1200–1500 g (females–males) per week (International Wolf Center; see also Fig. 3.4). One study based on live-trapped wolves in three different areas in Minnesota showed that pups' weights (relative to standard weight curves of captive pups) varied greatly within and between litters and across different years with extremes of 31–144% of the standard weight. The data also suggested that pups with a body weight below 65% had in general low chances of survival, whereas pups with at least 80% of the standard weight had high survival rates (Van Ballenberghe and Mech 1975).

In addition to pup weight, body mass and age of the mother played a role in the Yellowstone National Park (13-year data set) with fewer pups surviving the first year if the mother was older than 5 years and a 39% higher chance of survival for each 10 kg increase in the mother's body mass (MacNulty et al. 2007). The body mass

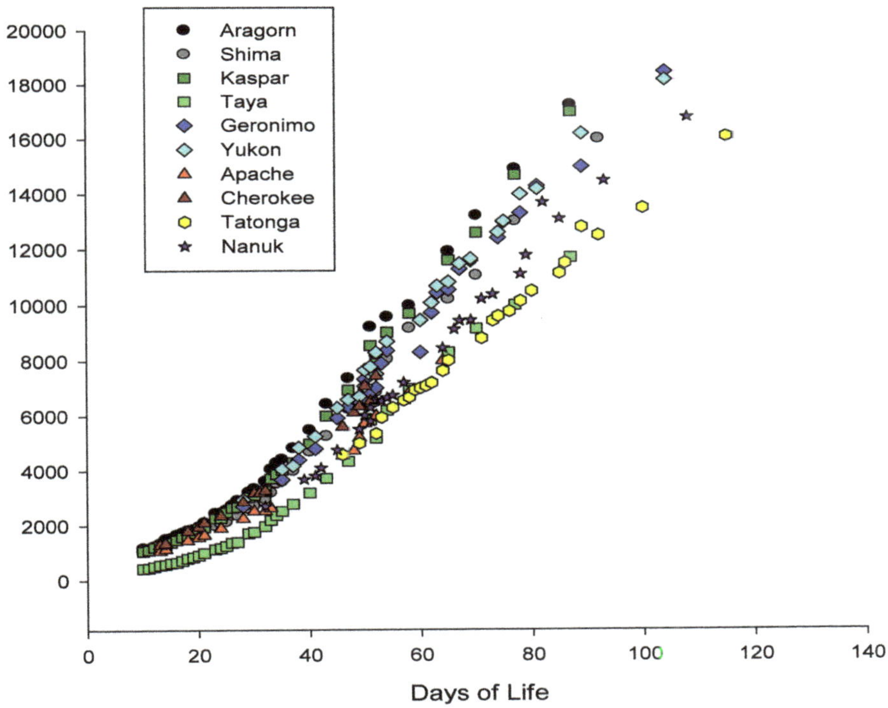

Fig. 3.4 Growth curve of the first two wolf puppy generations of the Wolf Science Center (unpublished data)

effect might be related to the pups' body mass, which is likely higher if the female is better at foraging, e.g. hunting (see above for the link between body mass and hunting success), and thus can provide the pups with more milk, or later food. Interestingly, coat colour also influences survival with grey-coloured females having 25% more surviving pups than black females (Stahler et al. 2013), which, in the long run, seems to be reversed, since black heterozygote females seem to have a higher lifetime reproductive success (Coulson et al. 2011) possible due to longer lifespan linked to the enhanced immunity to distemper (see above; coat colour).

Moreover, pup survival increased with the number of wolves in a pack in the Yellowstone National Park (Stahler et al. 2013). If the breeding female had only one helper, there was a profound effect (55% increased survival), and more additional helpers resulted only in a gradual increase in survival rates. Another study investigated pups' survival until the age of 7–8 months and pup weights at 5–6 months for two wolf populations in northern Minnesota (Harrington et al. 1983). No link between the pups' weights and pack size was found for either population. In the Superior National Forest population, pairs produced more pups than larger packs probably due to the scarcity of prey, while in the Beltrami Island State Forest population, with high prey abundance and the wolf population on the rise, pack size and litter size were positively correlated. Finally, Brainerd et al. (2008) investigated the impact of the loss of one or both territorial, breeding wolves (134 cases, 75 M and 73 F) on wolf pup survival until their first birthday using their own data from Scandinavia and published records. In 58 of 71 known cases (84%), one or more pups survived independently of who was lost and the age of the pups (older or younger than 6 months of age) with the exception of older pups surviving slightly better if the female rather than the male was around. Moreover, pup survival was higher in packs with more than six wolves and for larger litter sizes. The most important factor influencing whether pups survived after a breeder was lost was the number of adult-sized wolves remaining in the pack. Since, in combination with the relatively low life expectancy, the chances for losing a parent are really high per reproductive event, compensating the loss may be an important driver of inclusive fitness and thus pack formation.

Both population size and distemper outbreaks had a negative effect on litter size in the Yellowstone National Park (Stahler et al. 2013). Furthermore, a 15-year study in Belarus found that pup survival was higher under better environmental conditions (less rain, fewer mosquitoes) and with less human presence (Sidorovich et al. 2017). The latter effect was suggested to be due to wolves being stressed and moving their pups more often to different dens, reducing their hunting and protection efficiency. Also, survival of puppies was lower if the litter was larger compared to smaller. However, from 2004 to 2005, a study of 51 pups (age 3.5–31 weeks) in Algonquin Provincial Park, Canada, found no influence of pack or litter size on pup survival (Mills et al. 2008).

And, finally, as mentioned above, it is likely that infanticide by the dominant female affects mortality rates of pups from subdominant females (McLeod 1990).

Accordingly, several factors seem to influence pup survival including the body condition of both the pups and the mother, in some populations the pack and population size (which both might be linked to prey abundance and thus can have positive or negative effects), distemper outbreaks, weather conditions, and disturbance by humans.

3.3.4.2 Dispersal

Wolves can disperse over distances up to 1092 km (Boyd and Pletscher 1999; Wabakken et al. 2007) and may actually travel up to 10,000 km in their search of a new territory (Kirilyuk et al. 2020). The known world record was achieved by the female wolf OR-54, who travelled more than 14,000 km in 2 years from Oregon to California where she was found dead (https://wildlife.ca.gov/Search-Results?q=OR-54). Moreover, wolves are good at overcoming geographical obstacles. Hence, potential of wolves for invasiveness and gene flow is high. Usually both males and females disperse, though some biases have been reported in some locations with females sometimes dispersing further than males and sometimes males dispersing further than females (see Mech and Boitani 2003 for a review). Wolves dispersed as early as 15 weeks of age, and monthly dispersal rates were high for young pups in Algonquin Provincial Park, Canada, from 2004 to 2005 (Mills et al. 2008). In Minnesota, about 80% of the animals left before 2 years of age. However, it also seems like animals were more successful in finding new mates and territories when they are older than 1 year at dispersal (67% success rate). Similarly, during 1986–1992, in the Gates of the Arctic National Park and Preserve in northern Alaska, USA, 1- and 2-year-old youngsters emigrated at high annual rates (47% and 27% of pups born, respectively, in the two study years) (Adams et al. 2008).

In Spain, in an agricultural area, of 14 radio-collared wolves, 9 dispersed during the study period from 1997 to 2004 at the mean age of 24.8 months (Blanco and Cortés 2007). In Finland, of 60 wolves, at least 30 wolves (50%) departed their home territory during 1998–2004 as pups and yearlings. The survival and success rate (e.g. reproduction) was dependent on the direction in which the animals dispersed, since all animals that dispersed to reindeer management areas were shot, while 10 of 16 wolves travelling into the other direction reproduced (Kojola et al. 2006).

3.3.5 Social Dynamics

Group living species tend to have strong mechanisms regulating both their *within-* and *between*-group social relationships. Here we focus on the between-group relationships in wolves, since these can be studied only in free-ranging animals. Within-group dynamics, such as the affiliative and dominance bonds, which characterize the relationship between individuals and how they manage conflicts, will be the focus of Chap. 5, where we will summarize studies conducted both in free-living and captive animals.

Most wolf populations seem to be territorial (Kittle et al. 2015), defending the entire area they inhabit against other wolves and packs (Yellowstone National Park; vonHoldt et al. 2020). However, also in this regard, wolves show a high flexibility (similar to orcas) with some populations migrating with their prey. For example, the tundra wolf migrates with the caribou herds annually and differs from other nearby territorial wolves in terms of genetics and their lighter coat colour (Musiani et al. 2007). However, we could not find any studies that investigated the behaviour of the migrating wolves, e.g. whether or not they defend the caribou herds against other wolves or, in other words, are still territorial just not in regard to land but rather food.

Territory size is somewhat dependent on prey biomass/density, and this is reflected in the increase in territory size at higher latitudes in America (Minnesota 33 km^2 vs. Alaska 4335 km^2) (Jędrzejewski et al. 2007; Mech and Boitani 2003). In Ontario, the quality of the habitat (projected wolf use, probability of moose occupancy, and proportion of preferred land cover) was inversely related to territory size but unrelated to pack size (Kittle et al. 2015). In Europe, in Germany, territories are around 200 km^2 (Bundesministerium für Umwelt 2015); in the Bialowieza Primeval Forest, pack territories ranged from 154 to 343 km^2 ($N = 15$) (Jędrzejewski et al. 2007; Jedrzejewski et al. 2001); and in Dalmatia territories are between 141 and 160 km^2 (Kusak et al. 2005). Also here, territory sizes were negatively correlated to ungulate encounter rates. Interestingly, while in Poland the pack size (three to eight wolves) had no influence on territory size, in North America pack sizes were larger at higher latitudes; however, the increase was not in relation to the enlargement of territories (Jędrzejewski et al. 2007). The authors suggested that the larger pack size in the northern parts of Canada and Alaska may be due to the lower level of prosecution from humans and, likely more important, due to the size of the prey. The size of the most dominant prey must allow all pack members to satisfy their daily food requirement at a single kill, thus determining the pack size (Fuller et al. 2003).

Based on different studies, it has been suggested that there are core areas of territories and so-called buffer zones between pack territories that are avoided by wolves (Mech and Harper 2002). In Poland, the season, breeding status, and the activity of the animals determine territory use. During the reproductive period, breeding females and sub-adults had the shortest daily movement distances in May, while the longest distance was observed in an adult male during the mating season. Moreover, wolves constantly altered their hunting ground, revisiting the same part of their territory only every 6 days—probably in order not to keep prey in a constant state of alertness and thereby generating a landscape of fear, prompting potential prey to alter habitat use, which would in turn affect their huntability. Not surprisingly, the distance travelled became shorter with increased prey abundance (Jedrzejewski et al. 2001).

3.3.6 Territory Defence

Since fitness is determined by reproducing offspring and making sure it survives to reproduction, being able to defend territories is important for wolves since it secures both the survival of the pups and the necessary food resources. Wolves defend their territories in different ways. Scent marking including urine, scats, and ground scratching is used by wolves to indicate the borders of their territories. Usually the breeding pair marks around every 240 meters and about twice as often at the edges than in the core of the territory (Mech and Boitani 2003). However, in Poland, due to spending more time in the core area of the territories, the density of markings was higher in the centre of the territories as well as in hot spots where skirmishes with neighbouring packs occurred (Zub et al. 2003). Urine marking and ground scratching also occurred more often during the winter season than summer months with a peak during the mating season (Zub et al. 2003). Scent marks have the advantage that they last up to 2–3 weeks but the disadvantage of being short distance signals (Peters and Mech 1975). Howling, on the other hand, can be heard up to 11 km in the forest and about 16 km in the tundra, and wolves seem to be good at pinpointing the location of the howler—one study showed that the wolves could pinpoint a human howler from a distance of 2.7 km (Harrington and Mech 1983). They also seem to be able to estimate the number of animals of the neighbouring packs based on howling (Harrington and Mech 1983). If other wolves do not respect the boundaries as indicated by scent marking and howling, or a pack decides to engage another, between-pack fights are the result, with wolves often trying to catch and kill adversaries. This often results in the eradication of the smaller pack.

These severe inter-pack attacks constitute one of the highest causes of mortality for wolves. In an early study, Mech (1994) found that of 22 territorial wolves that were killed by conspecifics in Minnesota between 1968 and 1992, 23% were killed at the border of their territories, 41% within 1 km of their border, and 91% within 3.2 km of their border. In the newly created Gates of the Arctic National Park and Preserve in northern Alaska, mortality, primarily due to intraspecific attacks, equalled 11% per year (Adams et al. 2008), while in the Yellowstone Natinal Park it accounted for two-thirds of natural mortality from 1995 to 2012 (Smith et al. 2015). Prime-aged adults—especially breeders and wolves older than 6 years—made up the largest percentage of wolves killed by other wolves (53%). In the northern Yellowstone National Park, a study showed that population size was regulated by an increase in intraspecific aggression leading to high mortality rates (41%) of adult wolves (again older than 6 years) independent of the relative abundance of vulnerable elk. However, this population is characterized by generally high ungulate densities and low human-caused mortality, suggesting that the wolves were competing for space rather than prey (Cubaynes et al. 2014). It has been suggested that wolves need a minimum exclusive territory size to raise puppies (Jędrzejewski et al. 2007), even if the density of prey would be sufficient to sustain the pack in a smaller territory.

In comparison, mortality due to intergroup competition was lower (30%) in the interior of the park, where prey densities are lower and thus should increase food competition (Cubaynes et al. 2014). However, in the interior of the park, wolf densities are also lower, which could be due to reduced recruitment rates (low pup survival, high dispersal rates) because of the lower prey density. Thus, wolf-density-dependent survival rates may only occur when prey density is high and human-caused mortality is low. This also means that in areas that are 'saturated' with wolf packs, wolf densities do not vary much and—before all—do not increase.

In Yellowstone National Park, several studies investigated the behaviour of wolves during between-group conflicts between 1995 and 2010/2011. In 121 inter-group conflicts, 34 (28.1%) escalated to an attack and 11 (9.1%) resulted in a fatality and relative pack size increased the probability of winning the conflicts. However, having more old members (over 6 years of age) or more adult males could offset the advantage of larger pack size. Interestingly, the outcome of the interactions did not depend on the location of the conflicts—suggesting that no matter where in their territory, wolves will always give their best to defend it. Moreover, while these results suggest that pack size matters and thus might contribute to the evolution of group living in wolves, pack composition is also crucial with some individuals being more important than others (Cassidy et al. 2015). In a follow-up study, Kira Cassidy examined exactly this question. During 292 aggressive interactions (including pack–pack but also pack–lone wolf interactions), she, together with her colleagues (Cassidy et al. 2017), found that all wolves ($N = 534$ individuals) were more likely to participate in chasing rivals if they were part of the larger pack (each additional wolf increased the odds of chasing by 1.6%). These results also suggest that wolves are able to accurately assess their opponent's pack size, adjusting their behaviour accordingly (Cassidy et al. 2017), and their motivation to engage in aggression against another pack is modulated by the amount of social support received via the sheer size of their own pack (see also Harrington and Mech 1979; Chap. 9). Males were more active during the chasing than females and also chased more when their pack was outnumbered. Interestingly, some results suggested that grey animals were more aggressive than black ones, though this was not definite and is in contrast to previous ideas (Ducrest et al. 2008) (see beginning of chapter). It was also the case that male and grey-coloured wolves had higher cortisol levels (e.g. being stressed) compared to females and black wolves. Another curious fact was that the females did not really participate in chasing lone male wolves away, suggesting that they might have had some different interests. A recent study investigated the heritability of aggression in 141 wolves of Yellowstone National Park (vonHoldt et al. 2020). The authors found that heritability of aggression lies somewhere between 14% and 37% and an additional 14–16% is explained by the shared environment—e.g. being in the same natal pack. Additionally, breeders were found to be more aggressive than non-breeders, and aggressiveness increased with pack size. In contrast to previous results, sex and coat colour had no significant effect.

Another aspect that Cassidy investigated was whether animals would help each other in potentially dangerous situations (Cassidy and McIntyre 2016). Of the 34 observed inter-pack conflict situations that escalated into an attack (see above), six included interference behaviour where another group member (four times a closely related animal, one time a second-order relative, and one time the breeding partner; overall four of them male) tried to help the victim by attacking the attackers. While three of these supports were successful, three failed and ended in the death of the victim.

Several studies have tried to elucidate why the majority of wolf populations are territorial. One hypothesis has suggested that it is largely linked to protection of food (resource defence) (Mech 1973; Mech and Harper 2002; Packard and Mech 1980), while a more recent study suggests that the main reason is the defence of pups against infanticide (Smith et al. 2015). The latter study reports peaks of wolf–wolf killings during the denning season (March–May). Conversely, aggressive (but not necessarily fatal) between-group interactions were highest during the mating season in February, when packs roam their territories and encounter rates are higher than during the rest of the year. Of the 13 wolves killed during the denning season, 12 were breeders, and in four of six observed cases, pups were likely killed as well. The loss of the pups resulted in the loss of territory in the following years (due to smaller pack sizes), and in one case the pack ceased to exist altogether. While this is a small sample size and restricted to the Yellowstone situation, it suggests that cooperative breeding is the main driver of pack formation. On the other hand, killing competitors during the mating season might not be as effective as killing the pups, since there is the chance that then other individuals from the losing pack might breed instead and thus keep the pack size intact. Thus, intergroup competition might be most effective during the denning period, when killing the pups might lead to a decrease in pack size and thus to the opportunity to take over the territory. In combination with the short (reproductive) lifespan and the 'long' waiting period until the next birth season, the loss of the pups may lead to a collapse of the potential indirect fitness benefits and thus favour leaving vs. staying in order to breed on one's own due to the potential benefits.

In general, one needs to keep in mind that most of the more detailed studies currently come from the Yellowstone National Park, where, overall, wolf densities are probably higher than at most other places. The results may thus create a bias considering a lot of the observed features to be wolf-general, which are in fact just characteristic of high-density (prey and wolves) sites.

3.4 Recolonization of Wolves

Wolf populations are nowadays globally stable or expanding, recolonizing numerous areas despite having faced a long period of intense persecutions in the past, resulting in a drastic range and demographic decline. However, especially in Europe following protection laws mostly initiated in the 1970s and 1980s, wolves have shown their resilience and their numbers have steadily increased (Dorresteijn et al. 2015), expanding even into areas densely populated by humans. In general, wolves try to avoid heavily anthropic environments or use them when humans are not around. For example, a number of studies in both Europe and America have shown that road use by wolves (which likely makes it easier for them to travel around) is largely confined to nighttime (Zimmermann et al. 2014; Nowak et al. 2007), as are wolves' visit to human settlements (Ciucci et al. 1997; Kojola et al. 2016). Road use in Italy was also affected by season (higher use in winter) and the wolves' social status (packs avoided roads more than 'floaters') (Mancinelli et al. 2019). These studies suggest that wolves are adapting to life near humans, and indeed in some cases wolves appear to specialize on exploiting anthropogenic food such as the carcasses of domestic animals, preferring these to hunting (Ciucci et al. 2020).

Despite little being known about wolves living in more urbanized environments (Kuijper et al. 2016), there is emerging recognition that relying on anthropogenic foods can dramatically alter ecology and behaviour of wildlife, including wolves (Newsome et al. 2016, 2017). It has been suggested that the availability of anthropogenic foods could alter aspects such as evolved predator–prey relationships, hunting behaviour, trophic interactions, and bottom-up and top-down processes (Dorresteijn et al. 2015; Newsome et al. 2015) as well as many aspects of sociality (Newsome et al. 2017). To date, studies analysing the effect of wolves' dependence on anthropogenic resources on these aspects are largely missing; however, considering wolves' extraordinary flexibility and adaptability, it is not unlikely that once 'wild' places are saturated, wolves will learn to adapt to more anthropogenic environments (Mech 2017) similarly to how coyotes and foxes have shown a strong capacity to adapt to suburban environments (Contesse et al. 2004; Ellington and Gehrt 2019; Larson et al. 2020). Indeed, Newsome et al. (2017) speculate that in areas where grey wolves are living near humans relying heavily on anthropogenic resources and potentially hybridizing with free-ranging dogs, a 'novel' wolf domestication process may be taking place.

Although it may be perhaps somewhat premature to consider this a 'new' domestication event, there are interesting parallels between changes brought about by 'urbanization' and those linked to 'domestication'. Indeed, in a review of behavioural differences between urban and rural populations of birds, Møller (2010) first notes that many urbanization effects on behaviour appear to mirror changes found between domestic and wild species (e.g. both urban and domestic animals appear to be less neophobic). Some studies in mammals appear to support this idea, with urban populations showing more domestication-like behavioural patterns such as a decreased neophobic response, reduced fear of humans, and increased risk-taking behaviours, compared to more 'wild' or 'rural' populations (but see Breck et al. 2019; Greenberg and Holekamp 2017; Griffin et al. 2017; Miranda et al. 2013).

In general, although specific studies comparing more vs. less urbanized wolves are missing, the growing literature on the effects of 'urbanization' on animal's behaviour and morphology (e.g. see urban foxes showing skull changes and reduced sexual dimorphism; Parsons et al. 2020) points to the considerable selective pressure induced by the 'anthropogenic niche' (diet, human presence, environmental challenges) on animal's adapting to it. This aspect likely had a fundamental role, alongside direct human selection, in dogs' domestication process (Marshall-Pescini et al. 2017).

Box 3.1 The Italian Invasion and 'Going into the Wild'

The first time I (Sarah) heard about the Wolf Science Centre project, I was sitting on a bus with colleagues from all over the world attending a conference on animal behavior. Ludwig Huber was chatting to someone (no recollection who), telling them how Zsofi and Friederike were both somewhere in the middle of the Austrian mountains raising wolf pups. I sat on the edge of my chair and, feigning disinterest, tried to get as close as possible to eavesdrop on the rest of the conversation! I was mesmerized!

Not long after, I made it to Vienna and met Zsofi and Friderike. Together with the excitement of meeting the wolves, there was this awesome feeling you get when you have long passionate conversations and you knew that the people you've just met are your very same brand of 'crazy'! At the time I was working at the University of Milan so we started with just small collaborative research projects at the WSC. Then Simona Cafazzo, a good friend and colleague from Italy, decided to apply for funding to study conflict management in the wolves and dogs at the WSC, and I was super happy to get involved with the project. In an epic trip to Rome, Zsofi, Friederike, Simona, and I talked, wolves, dogs, and the meaning of life until we really thought we had it all solved.

Soon after, Simona won her grant, and Friederike an ERC grant. When, Friederike then offered me a postdoc position at the Wolf Science Center, it was impossible to say no!

But this was not an easy choice, with my partner and his two kids in Milan, the reality of my private life clashing with my research life hit home. It is thanks to the extraordinary flexibility of our research team and my partner's unwavering certainty that 'we will make it work somehow' that 7 (happy) years later I am still here;-).

I missed out on the first stormy years of the Wolf Science Center. But joining a project once it is already on its feet has its own challenges! There is the process of finding out how best to fit in the existing team and then this urge to contribute your own ideas and thoughts to what is already there—and all of this, together with Simona's stories of 'her' Rome dogs really fired up this desire to find field sites in which to study wild wolves and free-living dogs!

Both of them, strangely enough, materialized through chance meetings with non-scientists on Facebook! The Morocco free-ranging dog project started because I contacted two Italian dog trainers (Clara Caspani and Lorenzo Niccolini) who regularly visited the area and posted stunning videos of the dogs' lives on their Facebook page. Their generosity in sharing the information about their experience meant we could seriously consider the possibility of setting up a field site there. Martina Lazzaroni was then able to secure her own PhD grant to start this new adventure with the Morocco dogs (see box in Chap. 4).

(continued)

Box 3.1 (continued)

The wild wolves that we were able to study every year at their rendezvous site from 2014 to 2019 consisted of a few packs living in the Italian Apennines who were being closely monitored by a passionate ex park ranger turned police officer (Massimo Colombari) who enthusiastically offered his support when we came knocking at his door. The situation is ever changing, and we don't know whether we will be able to continue following these guys, but new opportunities have opened up, so we hope Italian wild wolves will stay on our research agenda a bit longer.

The best part of this story, for me, is that it keeps changing—new research opportunities and chance meetings pop up, which often turn into great experiences and new friends! Research is just never a job but more an adventure, and despite being sometimes a little resentful when I get the feeling that our students are doing all the fun bits (gallivanting off to Morocco or watching wild wolf pups grow up) while I write grants to get more money to keep the whole thing going, I am in reality enormously proud and grateful of being part of this team! And I do love writing grants.

A sheep protection dog and Sarah spy on the wolves' rendezvous site in the Italian Apennines. The farmer had both sheep and wolves on his land. He also has three big Maremmani dogs and two wolf researchers camping out in his barn from August to December! (photo by Martina Russignan)

(continued)

Box 3.1 (continued)

Passing ships in the night: same video trap, different times of day. 'Mascellone' (Italian for big jaws) was the breeding male of the pack, Simona Cafazzo and Martina Lazzaroni off to look for wolves! (photo courtesy of Massimo Colombari)

3.5 Summary

This short review shows that wolves usually live in family packs that cooperate in hunting, territory, and carcass defence as well as in the rearing of the pups. However, despite the fact that wolves have been extensively studied, some basic questions, such as the evolution of the incredible flexibility in the sociability of wolves, have not really been resolved. In the end, there are probably two questions that need to be answered in that regard. The first one is: why do wolves live in packs rather than just form pair bonds (like most other canids)? And, second, when are packs larger than just the breeding pair?

Why the breeding pair tolerates additional pack members may be due to the increase in reproductive success with additional pack members due to increased hunting efficiency, success in territory defence, and/or better-quality pup care. Why non-breeding individuals stay though is the bigger mystery and may be partly explained by the ecological constraints like saturated habitats in combination with strong territoriality, kin selection, inclusive fitness benefits, and territory inheritance (also important for unrelated group members).

The 'when they form larger packs' is dependent on various factors that may change continuously including prey abundance, prey size (hunting efficiency of certain larger animals increases with pack size and/or because more surplus food is available; Mech and Boitani 2003*), the number of pups born in the previous year, the dispersal of older animals, and number of fatalities in between-group encounters. All these factors also influence each other, with dispersal, for example, likely being dependent on prey abundance and also population density. In essence, wolves have evolved a great context-dependent flexibility in their social organization and behaviour, making them (as humans) a species with a high adaptability and invasiveness into virtually all Northern Hemisphere habitats, from the Arabian deserts to the high Arctic.*

References

Adams LG, Stephenson RO, Dale BW, Ahgook RT, Demma DJ (2008) Population dynamics and harvest characteristics of wolves in the central brooks range, Alaska. Wildl Monogr 170:1–25. https://doi.org/10.2193/2008-012

Anderson TM et al (2009) Molecular and evolutionary history of Melanism in North American gray wolves. Science 323:1339–1343. https://doi.org/10.1126/science.1165448

Ausband DE, Mitchell MS, Bassing SB, Morehouse A, Smith DW, Stahler D, Struthers J (2016) Individual, Group, and environmental influences on helping behavior in a social carnivore. Ethology 122:963–972. https://doi.org/10.1111/eth.12566

Ballard WB, Ayres LA, Gardner CL, Foster JW (1991) Den site activity patterns of gray wolves, *Canis lupus*, in South central Alaska. Can Field-Nat 105:497–504. https://biodiversitylibrary.org/page/34349015

Bateman PW, Fleming PA (2012) Big city life: carnivores in urban environments. J Zool 287:1–23. https://doi.org/10.1111/j.1469-7998.2011.00887.x

Bauer JE (2006) Metabolic basis for the essential nature of fatty acids and the unique dietary fatty acid requirements of cats. J Am Vet Med Assoc 229:1729–1732. https://doi.org/10.2460/javma.229.11.1729

Blanco JC, Cortés Y (2007) Dispersal patterns, social structure and mortality of wolves living in agricultural habitats in Spain. J Zool 273:114–124. https://doi.org/10.1111/j.1469-7998.2007.00305.x

Boesch C (1994) Cooperative hunting in wild chimpanzees. Anim Behav 48:653–667. https://doi.org/10.1006/anbe.1994.1285

Boyd DK, Pletscher DH (1999) Characteristics of dispersal in a colonizing wolf population in the central Rocky Mountains. J Wildl Manag 63:1094–1108. https://doi.org/10.2307/3802828

Brainerd SM et al (2008) The effects of breeder loss on wolves. J Wildl Manag 72:89–98. https://doi.org/10.2193/2006-305

Breck SW, Poessel SA, Mahoney P, Young JK (2019) The intrepid urban coyote: a comparison of bold and exploratory behavior in coyotes from urban and rural environments. Sci Rep 9:2104. https://doi.org/10.1038/s41598-019-38543-5

Brown JL, Eklund AC (1994) Kin recognition and the major histocompatibility complex: an integrative review. Am Nat 143:435–461. https://doi.org/10.1086/285612

Bundesministerium für Umwelt NunS (2015) Bericht des Bundesministeriums für Umwelt, Naturschutz, Bau und Reaktorsicherheit zur Lebensweise, zum Status und zum Management des Wolfes (*Canis lupus*) in Deutschland. Deutscher Bundes-tag Ausschuss für Umwelt, Naturschutz, Bau und Reaktorsicherheit. https://www.bundestag.de/resource/blob/393542/5e21bfea995e1f0f0f19271d442f365d/bericht-bmub-data.pdf

Carricondo-Sanchez D et al (2020) Wolves at the door? Factors influencing the individual behavior of wolves in relation to anthropogenic features. Biol Conserv 244:108514. https://doi.org/10.1016/j.biocon.2020.108514

Cassidy KA, MacNulty DR, Stahler DR, Smith DW, Mech LD (2015) Group composition effects on aggressive interpack interactions of gray wolves in Yellowstone National Park. Behav Ecol 26:1352–1360. https://doi.org/10.1093/beheco/arv081

Cassidy KA, McIntyre RT (2016) Do gray wolves (*Canis lupus*) support pack mates during aggressive inter-pack interactions? Anim Cogn 19:939–947. https://doi.org/10.1007/s10071-016-0994-1

Cassidy KA, Mech LD, MacNulty DR, Stahler DR, Smith DW (2017) Sexually dimorphic aggression indicates male gray wolves specialize in pack defense against conspecific groups. Behav Proc 136:64–72. https://doi.org/10.1016/j.beproc.2017.01.011

Chapman RC (1977) The effects of human disturbance on wolves (*Canis lupus* L.). Master Thesis. University of Alaska

Ciucci P, Boitani L, Francisci F, Andreoli G (1997) Home range, activity and movements of a wolf pack in central Italy. J Zool 243:803–819. https://doi.org/10.1111/j.1469-7998.1997.tb01977.x

Ciucci P, Mancinelli S, Boitani L, Gallo O, Grottoli L (2020) Anthropogenic food subsidies hinder the ecological role of wolves: Insights for conservation of apex predators in human-modified landscapes. Global Ecol Conserv 21:e00841. https://doi.org/10.1016/j.gecco.2019.e00841

Contesse P, Hegglin D, Gloor S, Bontadina F, Deplazes P (2004) The diet of urban foxes (*Vulpes vulpes*) and the availability of anthropogenic food in the city of Zurich, Switzerland. Mamm Biol 69:81–95. https://doi.org/10.1078/1616-5047-00123

Coulson T, MacNulty DR, Stahler DR, vonHoldt B, Wayne RK, Smith DW (2011) Modeling effects of environmental change on wolf population dynamics, trait evolution, and life history. Science 334:1275–1278. https://doi.org/10.1126/science.1209441

Cubaynes S, MacNulty DR, Stahler DR, Quimby KA, Smith DW, Coulson T (2014) Density-dependent intraspecific aggression regulates survival in northern Yellowstone wolves (*Canis lupus*). J Anim Ecol 83:1344–1356. https://doi.org/10.1111/1365-2656.12238

Derix R, Van Hooff J, De Vries H, Wensing J (1993) Male and female mating competition in wolves: female suppression vs. Male Intervention. Behaviour 127:141–174. https://doi.org/10.1163/156853993X00461

Dorresteijn I et al (2015) Incorporating anthropogenic effects into trophic ecology: predator-prey interactions in a human-dominated landscape. Proc R Soc B 282:20151602. https://doi.org/10.1098/rspb.2015.1602

Ducrest AL, Keller L, Roulin A (2008) Pleiotropy in the melanocortin system, coloration and behavioural syndromes. Trends Ecol Evol 23:502–510. https://doi.org/10.1016/j.tree.2008.06.001

Ellington EH, Gehrt SD (2019) Behavioral responses by an apex predator to urbanization. Behav Ecol 30:821–829. https://doi.org/10.1093/beheco/arz019

Fox MW (1969) Ontogeny of prey-killing behavior in Canidae. Behaviour 35:259–272. https://doi.org/10.1163/156853969X00233

Fuller TK (1989) Population dynamics of wolves in north-central Minnesota. Wildl Monogr 105:1–41. http://www.jstor.org/stable/3830614

Fuller TK, Kat PW (1993) Hunting success of African wild dogs in south western Kenya. J Mammal 74:464–467. https://doi.org/10.2307/1382403

Fuller TK, Mech LD, Cochrane JF (2003) Wolf population dynamics. In: Mech LD, Boitani L (eds) Wolves: behavior, ecology, and conservation. University of Chicago Press, Chicago & London, pp 161–191

Gable TD, Gable DP (2019) Wolf (*Canis* sp.) attacks life-like deer decoy: insight into how wolves hunt deer? Can Field-Nat 133:16–19. https://doi.org/10.22621/cfn.v133i1.2044

Gazzola A, Bertelli I, Avanzinelli E, Tolosano A, Bertotto P, Apollonio M (2005) Predation by wolves (*Canis lupus*) on wild and domestic ungulates of the western Alps, Italy. J Zool 266: 205–213. https://doi.org/10.1017/S0952836905006801

Geffen E et al (2011) Kin encounter rate and inbreeding avoidance in canids. Mol Ecol 20:5348–5358. https://doi.org/10.1111/j.1365-294X.2011.05358.x

Gehrt SD, Brown JL, Anchor C (2011) Is the urban coyote a misanthropic Synanthrope? The case from Chicago. Cities and the Environment (CATE) 4:3. http://digitalcommons.lmu.edu/cate/vol4/iss1/3

Gipson PS, Ballard WB, Nowak RM, Mech DL (2000) Accuracy and precision of estimating age of gray wolves by tooth wear. J Wildl Manag 64:752–758. https://doi.org/10.2307/3802745

Goodmann PA, Klinghammer E, Willard J (2002) Wolf ethogram. In: Ethology Series Battle Ground, vol 3

Green Jonathan P, Holmes Andrew M, Davidson Amanda J, Paterson S, Stockley P, Beynon Robert J, Hurst Jane L (2015) The genetic basis of kin recognition in a cooperatively breeding mammal. Curr Biol 25:2631–2641. https://doi.org/10.1016/j.cub.2015.08.045

Greenberg JR, Holekamp KE (2017) Human disturbance affects personality development in a wild carnivore. Anim Behav 132:303–312. https://doi.org/10.1016/j.anbehav.2017.08.023

Griffin AS, Netto K, Peneaux C (2017) Neophilia, innovation and learning in an urbanized world: a critical evaluation of mixed findings. Curr Opin Behav Sci 16:15–22. https://doi.org/10.1016/j.cobeha.2017.01.004

Hall RL, Sharp HS (1978) Wolf and man, evolution in parallel. Academic Press, New York

Harrington FH, Mech L (1982) Patterns of homesite attendance in two Minnesota wolf packs. In: Harrington FH, Paquet PC (eds) Wolves of the world: perspectives of behavior, ecology, and conservation. Noyes, Park Ridge, New Jersey, USA, pp 81–105. https://pubs.er.usgs.gov/publication/5210299

Harrington FH, Mech LB (1979) Wolf howling and its role in territory maintenance. Behaviour 68: 207–249. https://doi.org/10.1163/156853979X00322

Harrington FH, Mech LD (1983) Wolf pack spacing: howling as a territory-independent spacing mechanism in a territorial population. Behav Ecol Sociobiol 12:161–168. https://doi.org/10.1007/BF00343208

Harrington FH, Mech LD, Fritts SH (1983) Pack size and wolf pup survival–their relationship under varying ecological conditions. Behav Ecol Sociobiol 13:19–26. https://doi.org/10.1007/BF00295072

Hayes R, Baer AM, Wotschikowsky U, Harestad AS (2000) Kill rates by wolves on moose in Yukon. Can J Zool 78:49–59. https://doi.org/10.1139/z99-187

Hedrick PW, Smith DW, Stahler DR (2016) Negative-assortative mating for color in wolves. Evolution 70:757–766. https://doi.org/10.1111/evo.12906

Heinrich B (1994) Does the early common raven get (and Show) the meat? Auk 111:764–769. https://doi.org/10.1093/auk/111.3.764

Hindrikson M, Männil P, Ozolins J, Krzywinski A, Saarma U (2012) Bucking the trend in wolf-dog hybridization: first evidence from Europe of hybridization between female dogs and male wolves. PLoS One 7:e46465. https://doi.org/10.1371/journal.pone.0046465

Holekamp KE, Smale L, Berg R, Cooper SM (1997) Hunting rates and hunting success in the spotted hyaena (*Crocuta crocuta*). J Zool (Lond) 242:1–15. https://doi.org/10.1111/j.1469-7998.1997.tb02925.x

Imbert C et al (2016) Why do wolves eat livestock? Factors influencing wolf diet in northern Italy. Biol Conserv 195:156–168. https://doi.org/10.1016/j.biocon.2016.01.003

Jędrzejewski W, Schmidt K, Theuerkauf J, Jędrzejewska B, Kowalczyk R (2007) Territory size of wolves *Canis lupus*: linking local (Białowieża Primeval Forest, Poland) and Holarctic-scale patterns. Ecography 30:66–76. https://doi.org/10.1111/j.2006.0906-7590.04826.x

Jedrzejewski W, Schmidt K, Theuerkauf J, Jedrzejewska B, Okarma H (2001) Daily movements and territory use by radio-collared wolves (*Canis lupus*) in Bialowieza Primeval Forest in Poland. Can J Zool 79:1993–2004. https://doi.org/10.1139/z01-147

Kaczensky P, Hayes RD, Promberger C (2005) Effect of raven *Corvus corax* scavenging on the kill rates of wolf *Canis lupus* packs. Wildl Biol 11:101–108. https://doi.org/10.2981/0909-6396(2005)11[101:EORCCS]2.0.CO;2

Khosravi R, Rezaei HR, Kaboli M (2013) Detecting hybridization between Iranian wild wolf (*Canis lupus pallipes*) and free-ranging domestic dog (*Canis familiaris*) by analysis of microsatellite markers. Zool Sci 30:27–34. https://doi.org/10.2108/zsj.30.27

Kirilyuk A, Kirilyuk VE, Ke R (2020) Long-distance dispersal of wolves in the Dauria ecoregion. Mamm Res 65:639–646. https://doi.org/10.1007/s13364-020-00515-w

Kittle AM et al (2015) Wolves adapt territory size, not pack size to local habitat quality. J Anim Ecol 84:1177–1186. https://doi.org/10.1111/1365-2656.12366

Knick ST, Mech LD (1980) Sleeping distance in wild wolf packs. Behav Neural Biol 28:507–511. https://doi.org/10.1016/S0163-1047(80)91932-9

Kojola I, Aspi J, Hakala A, Heikkinen S, Ilmoni C, Ronkainen S (2006) Dispersal in an expanding wolf population in Finland. J Mammal 87:281–286. https://doi.org/10.1644/05-Mamm-a-061r2.1

Kojola I et al (2016) Wolf visitations close to human residences in Finland: the role of age, residence density, and time of day. Biol Conserv 198:9–14. https://doi.org/10.1016/j.biocon.2016.03.025

Kojola I, Huitu O, Toppinen K, Heikura K, Heikkinen S, Ronkainen S (2004) Predation on European wild forest reindeer (*Rangifer tarandus*) by wolves (*Canis lupus*) in Finland. J Zool 263:229–235. https://doi.org/10.1017/S0952836904005084

Kokko H, Ots I (2006) When not to avoid inbreeding. Evolution 60:467–475. https://doi.org/10.2307/4095309

Kuijper DPJ, Sahlén E, Elmhagen B, Chamaillé-Jammes S, Sand H, Lone K, Cromsigt JPGM (2016) Paws without claws? Ecological effects of large carnivores in anthropogenic landscapes. Proc R Soc B 283:20161625. https://doi.org/10.1098/rspb.2016.1625

Kusak J, Skrbinšek AM, Huber D (2005) Home ranges, movements, and activity of wolves (*Canis lupus*) in the Dalmatian part of Dinarids, Croatia. Eur J Wildl Res 51:254–262. https://doi.org/10.1007/s10344-005-0111-2

Larson G, Fuller DQ (2014) The evolution of animal domestication. Ann Rev Ecol Evol Syst 45:115–136. https://doi.org/10.1146/annurev-ecolsys-110512-135813

Larson RN, Brown JL, Karels T, Riley SPD (2020) Effects of urbanization on resource use and individual specialization in coyotes (*Canis latrans*) in southern California. PLoS One 15: e0228881. https://doi.org/10.1371/journal.pone.0228881
Lehman N, Clarkson P, Mech LD, Meier TJ, Wayne RK (1992) A study of the genetic-relationships within and among wolf packs using DNA fingerprinting and mitochondrial-DNA. Behav Ecol Sociobiol 30:83–94. https://doi.org/10.1007/BF00173944
Linnell JDC et al (2002) The fear of wolves: a review of wolf attacks on humans, vol 731. NINA, Trondheim
López-Bao JV, Sazatornil V, Llaneza L, Rodríguez A (2013) Indirect effects on heathland conservation and wolf persistence of contradictory policies that threaten traditional free-ranging horse husbandry. Conserv Lett 6:448–455. https://doi.org/10.1111/conl.12014
Macdonald DW (1983) The ecology of carnivore social behaviour. Nature 301:379–384. https://doi.org/10.1038/301379a0
MacNulty DR, Mech LD, Smith DW (2007) A proposed ethogram of large-carnivore predatory behavior, exemplified by the wolf. J Mammal 88:595–605. https://doi.org/10.1644/06-Mamm-a-119r1.1
MacNulty DR, Smith DW, Mech LD, Eberly LE (2009a) Body size and predatory performance in wolves: is bigger better? J Anim Ecol 78:532–539. https://doi.org/10.1111/j.1365-2656.2008.01517.x
MacNulty DR, Smith DW, Mech LD, Vucetich JA, Packer C (2012) Nonlinear effects of group size on the success of wolves hunting elk. Behav Ecol 23:75–82. https://doi.org/10.1093/beheco/arr159
MacNulty DR, Smith DW, Vucetich JA, Mech LD, Stahler DR, Packer C (2009b) Predatory senescence in ageing wolves. Ecol Lett 12:1347–1356. https://doi.org/10.1111/j.1461-0248.2009.01385.x
MacNulty DR, Stahler DR, Smith DW (2016) Understanding the limits to wolf hunting ability. Yellowstone Sci 24:34–36
MacNulty DR, Tallian A, Stahler DR, Smith DW (2014) Influence of group size on the success of wolves hunting bison. PLoS One 9:e112884. https://doi.org/10.1371/journal.pone.0112884
Mancinelli S, Falco M, Boitani L, Ciucci P (2019) Social, behavioural and temporal components of wolf (*Canis lupus*) responses to anthropogenic landscape features in the central Apennines, Italy. J Zool 309:114–124. https://doi.org/10.1111/jzo.12708
Marshall-Pescini S, Cafazzo S, Virányi Z, Range F (2017) Integrating social ecology in explanations of wolf–dog behavioral differences. Curr Opin Behav Sci 16:80–86. https://doi.org/10.1016/j.cobeha.2017.05.002
Martin HW, Mech LD, Fieberg J, Metz MC, MacNulty DR, Stahler DR, Smith DW (2018) Factors affecting gray wolf (*Canis lupus*) encounter rate with elk (*Cervus elaphus*) in Yellowstone National Park. Can J Zool 96:1032–1042. https://doi.org/10.1139/cjz-2017-0220
Mattioli L, Apollonio M, Mazzarone V, Centofanti E (1995) Wolf food habits and wild ungulate availability in the Foreste Casentinesi National Park, Italy. Acta Theriol 40:387–402. https://doi.org/10.4098/AT.arch.95-36
Mattioli L, Capitani C, Avanzinelli E, Bertelli I, Gazzola A, Apollonio M (2004) Predation by wolves (*Canis lupus*) on roe deer (*Capreolus capreolus*) in north-eastern Apennine, Italy. J Zool 264:249–258. https://doi.org/10.1017/s095283690400576x
Mayr E (1963) Animal species and evolution. Cambridge, Belknap
Mayr E (1999) Systematics and the origin of species, from the viewpoint of a zoologist. Harvard University Press, Cambridge, MA and London, UK. https://books.google.at/books?id=mAIjnLp6r_MC
McLeod PJ (1990) Infanticide by female wolves. Can J Zool 68:402–404. https://doi.org/10.1139/z90-058
Mech LD (1970) The wolf. The ecology and behaviour of an endangered species. Natural History Press, Garden City, NY

Mech LD (1973) Wolf numbers in the superior national forest of minnesota. U.S. Dept. of Agriculture, Forest Service, North Central Forest Experiment Station, St. Paul, MN. https://www.fs.usda.gov/treesearch/pubs/10618

Mech LD (1994) Buffer zones of territories of gray wolves as regions of intraspecific strife. J Mammal 75:199–202. https://doi.org/10.2307/1382251

Mech LD (2002) Breeding Season of Wolves, *Canis lupus*, in Relation to Latitude. Can Field-Nat 116:139–140. https://www.biodiversitylibrary.org/page/35151373

Mech LD (2014) A gray wolf (*Canis lupus*) delivers live prey to a pup. Can Field-Nat 128:189–190. https://doi.org/10.22621/cfn.v128i2.1584

Mech LD (2017) Where can wolves live and how can we live with them? Biol Conserv 210:310–317. https://doi.org/10.1016/j.biocon.2017.04.029

Mech LD, Adams LG, Meier TJ, Burch JW, Dale BW (1998) The wolves of Denali. University of Minnesota Press, Minneapolis

Mech LD, Adams LG, Meier TJ, Burch JW, Dale BW (2003) The wolves of Denali. University of Minnesota Press, Minneapolis

Mech LD, Boitani L (2003) Wolf social ecology. In: Mech LD, Boitani L (eds) Wolves: behavior, ecology, and conservation. The University of Chcago Press, Chicago, London, pp 1–35

Mech LD, Cluff HD (2011) Movements of wolves at the northern extreme of the species' range, including during four months of darkness. PLoS One 6:e25328. https://doi.org/10.1371/journal.pone.0025328

Mech LD, Goyal SM, Paul WJ, Newton WE (2008) Demographic effects of canine parvovirus on a free-ranging wolf population over 30 years. J Wildl Dis 44:824–836. https://doi.org/10.7589/0090-3558-44.4.824

Mech LD, Harper EK (2002) Differential use of a Wolf, *Canis lupus*, pack territory edge and core. Can Field-Nat 116:315–316. https://www.biodiversitylibrary.org/page/35151553

Mech LD, Knick ST (1978) Sleeping distance in wolf pairs in relation to breeding season. Behav Biol 23:521–525. https://doi.org/10.1016/S0091-6773(78)91631-0

Mech LD, Phillips MK, Smith DW, Kreeger TJ (1996) Denning behaviour of non-gravid Wolves, *Canis lupus*. Can Field-Nat 110:343–345. https://www.biodiversitylibrary.org/page/34343245

Mech LD, Smith DW, MacNulty DR (2015) Wolves on the hunt. The behavior of wolves hunting wild prey. University of Chicago Press, Chicago, London. https://doi.org/10.7208/chicago/9780226255286.001.0001

Mech LD, Wolf PC, Packard JM (1999) Regurgitative food transfer among wild wolves. Can J Zool 77:1192–1195. https://doi.org/10.1139/cjz-77-8-1192

Meriggi A, Lovari S (1996) A review of wolf predation in southern Europe: does the wolf prefer wild prey to livestock? J Appl Ecol 33:1561–1571. https://doi.org/10.2307/2404794

Metz MC, Smith DW, Vucetich JA, Stahler DR, Peterson RO (2012) Seasonal patterns of predation for gray wolves in the multi-prey system of yellowstone national park. J Anim Ecol 81:553–563. https://doi.org/10.1111/j.1365-2656.2011.01945.x

Migli D, Youlatos D, Iliopoulos Y (2005) Winter food habits of wolves in central Greece. J Biol Res 4:217–220. http://ikee.lib.auth.gr/record/261317/files/WinterFood.pdf

Mills KJ, Patterson BR, Murray DL (2008) Direct estimation of early survival and movements in eastern wolf pups. J Wildl Manag 72:949–954. https://doi.org/10.2193/2006-457

Miranda AC, Schielzeth H, Sonntag T, Partecke J (2013) Urbanization and its effects on personality traits: a result of microevolution or phenotypic plasticity? Glob Chang Biol 19:2634–2644. https://doi.org/10.1111/gcb.12258

Møller AP (2010) Interspecific variation in fear responses predicts urbanization in birds. Behav Ecol 21:365–371. https://doi.org/10.1093/beheco/arp199

Mori E, Benatti L, Lovari S, Ferretti F (2016) What does the wild boar mean to the wolf? Eur J Wildl Res 63. https://doi.org/10.1007/s10344-016-1060-7

Musiani M et al (2007) Differentiation of tundra/taiga and boreal coniferous forest wolves: genetics, coat colour and association with migratory caribou. Mol Ecol 16:4149–4170. https://doi.org/10.1111/j.1365-294X.2007.03458.x

Newsome TM et al (2016) Food habits of the world's grey wolves. Mammal Rev 46:255–269. https://doi.org/10.1111/mam.12067
Newsome TM, Dellinger JA, Pavey CR, Ripple WJ, Shores CR, Wirsing AJ, Dickman CR (2015) The ecological effects of providing resource subsidies to predators: resource subsidies and predators. Glob Ecol Biogeogr 24:1–11. https://doi.org/10.1111/geb.12236
Newsome TM, Fleming PJS, Dickman CR, Doherty TS, Ripple WJ, Ritchie EG, Wirsing AJ (2017) Making a new dog? BioSci 67:374–381. https://doi.org/10.1093/biosci/bix022
Nowak S, Jędrzejewski W, Schmidt K, Theuerkauf J, Mysłajek RW, Jędrzejewska B (2007) Howling activity of free-ranging wolves (*Canis lupus*) in the Białowieża primeval forest and the Western Beskidy Mountains (Poland). J Ethol 25:231–237. https://doi.org/10.1007/s10164-006-0015-y
Nowak S, Mysłajek RW, Kłosińska A, Gabryś G (2011) Diet and prey selection of wolves (*Canis lupus*) recolonising Western and Central Poland. Mamm Biol 76:709–715. https://doi.org/10.1016/j.mambio.2011.06.007
Pacheco C, López-Bao JV, García EJ, Lema FJ, Llaneza L, Palacios V, Godinho R (2017) Spatial assessment of wolf-dog hybridization in a single breeding period. Sci Rep 7:42475. https://doi.org/10.1038/srep42475
Packard JM (2003) Wolf behavior: Reproductive, social and intelligent. In: Mech LD, Boitani L (eds) Wolves: behaviour, ecology and conservation. University of Chicago Press, Chicago, pp 35–65
Packard JM, Mech LD (1980) Population regulation in wolves. In: Cohen MM, Malpass RS, Klein HG (eds) Biosocial mechanisms of population regulation Yale University Press. New Haven, Connecticut, pp 135–150
Paquet PC, Carbyn LN (2003) Gray wolf. *Canis lupus* and allies. In: Feldhamer GA, Thompson BC, Champan JA (eds) Wild mammals of north America: biology, management, and conservation. JHU Press, Baltimore, USA, pp 482–510
Parsons KJ, Rigg A, Conith AJ, Kitchener AC, Harris S, Zhu H (2020) Skull morphology diverges between urban and rural populations of red foxes mirroring patterns of domestication and macroevolution. Proc R Soc Lond B 287:20200763. https://doi.org/10.1098/rspb.2020.0763
Penn DJ (2002) The scent of genetic compatibility: sexual selection and the major histocompatibility complex. Ethology 108:1–21. https://doi.org/10.1046/j.1439-0310.2002.00768.x
Perri AR, Feuerborn TR, Frantz LAF, Larson G, Malhi RS, Meltzer DJ, Witt KE (2021) Dog domestication and the dual dispersal of people and dogs into the Americas. Proc Natl Acad Sci U S A 118:e2010083118. https://doi.org/10.1073/pnas.2010083118
Peters RP, Mech LD (1975) Behavioural and intellectual adaptations of selected mammalian predators to the problem of hunting large animals. In: Tuttle RH (ed) Socioecology and psychology of primates. Mouton Publishers, The Hague Paris, pp 279–300
Peterson RO, Ciucci P (2003) The wolf as a carnivore. In: Mech LD, Boitani L (eds) Wolves: behaviour, ecology and conservation. The University of Chicago Press, Chicago, London, pp 104–130
Peterson RO, Jacobs AK, Drummer TD, Mech LD, Smith DW (2002) Leadership behavior in relation to dominance and reproductive status in gray wolves, *Canis lupus*. Can J Zool 80:1405–1412. https://doi.org/10.1139/z02-124
Peterson RO, Thomas NJ, Thurber JM, Vucetich JA, Waite TA (1998) Population limitation and the wolves of Isle Royale. J Mammal 79:828–841. https://doi.org/10.2307/1383091
Pletscher DH, Ream RR, Boyd DK, Fairchild MW, Kunkel KE (1997) Population dynamics of a recolonizing wolf population. J Wildl Manag 61:459–465. https://doi.org/10.2307/3802604
Proffitt KM, Grigg JL, Hamlin KL, Garrott RA (2009) Contrasting effects of wolves and human hunters on Elk behavioral responses to predation risk. J Wildl Manag 73:345–356. https://doi.org/10.2193/2008-210
Pusey AE, Wolf M (1996) Inbreeding avoidance in animals. Trends Ecol Evol 11:201–206. https://doi.org/10.1016/0169-5347(96)10028-8

Räikkönen J, Bignert A, Mortensen P, Fernholm B (2006) Congenital defects in a highly inbred wild wolf population (*Canis lupus*). Mamm Biol 71:65–73. https://doi.org/10.1016/j.mambio.2005.12.002

Räikkönen J, Vucetich JA, Peterson RO, Nelson MP (2009) Congenital bone deformities and the inbred wolves (*Canis lupus*) of Isle Royale. Biol Conserv 142:1025–1031. https://doi.org/10.1016/j.biocon.2009.01.014

Ridley M (2004) Evolution, 3rd edn. Blackwell Science, Oxford

Ripple WJ et al (2015) Collapse of the world's largest herbivores. Sci Adv 1:e1400103. https://doi.org/10.1126/sciadv.1400103

Robinson JA, Räikkönen J, Vucetich LM, Vucetich JA, Peterson RO, Lohmueller KE, Wayne RK (2019) Genomic signatures of extensive inbreeding in Isle Royale wolves, a population on the threshold of extinction. Sci Adv 5:eaau0757. https://doi.org/10.1126/sciadv.aau0757

Samuni L, Preis A, Deschner T, Crockford C, Wittig RM (2018) Reward of labor coordination and hunting success in wild chimpanzees. Commun Biol 1:138. https://doi.org/10.1038/s42003-018-0142-3

Sand H, Wikenros C, Wabakken P, Liberg O (2006) Effects of hunting group size, snow depth and age on the success of wolves hunting moose. Anim Behav 72:781–789. https://doi.org/10.1016/j.anbehav.2005.11.030

Sands J, Creel S (2004) Social dominance, aggression and faecal glucocorticoid levels in a wild population of wolves, *Canis lupus*. Anim Behav 67:387–396. https://doi.org/10.1016/j.anbehav.2003.03.019

Schmidt K, Jędrzejewski W, Theuerkauf J, Kowalczyk R, Okarma H, Jędrzejewska B (2008) Reproductive behaviour of wild-living wolves in Białowieża Primeval Forest (Poland). J Ethol 26:69–78. https://doi.org/10.1007/s10164-006-0031-y

Schmidt PA, Mech LD (1997) Wolf pack size and food acquisition. Am Nat 150 513–517. https://doi.org/10.1086/286079

Schweizer RM et al (2018) Natural selection and origin of a melanistic allele in North American gray wolves. Mol Biol Evol 35:1190–1209. https://doi.org/10.1093/molbev/msy031

Sidorovich V, Schnitzler A, Schnitzler C, Rotenko I (2017) Wolf denning behaviour in response to external disturbances and implications for pup survival. Mamm Biol 87:89–92. https://doi.org/10.1016/j.mambio.2016.11.011

Sidorovich VE, Tikhomirova LL, Jędrzejewska B (2003) Wolf *Canis lupus* numbers, diet and damage to livestock in relation to hunting and ungulate abundance in northeastern Belarus during 1990-2000. Wildl Biol 9:103–111. https://doi.org/10.2981/wlb.2003.032

Smith D, Meier T, Geffen E, Mech LD, Burch JW, Adams LG, Wayne RK (1997) Is incest common in gray wolf packs? Behav Ecol 8:384–391. https://doi.org/10.1093/beheco/8.4.384

Smith DW, Ferguson G (2012) Decade of the wolf. Revised and updated. Returning the Wild to Yellowstone. Rev. and updated edn. Lyons Press, Guilford, Conn

Smith DW, Metz MC, Cassidy KA, Stahler EE, McIntyre RT, Almberg ES, Stahler DR (2015) Infanticide in wolves: seasonality of mortalities and attacks at dens support evolution of territoriality. J Mammal 96:1174–1183. https://doi.org/10.1093/jmammal/gyv125

Stahler DR, MacNulty DR, Wayne RK, vonHoldt B, Smith DW (2013) The adaptive value of morphological, behavioural and life-history traits in reproductive female wolves. J Anim Ecol 82:222–234. https://doi.org/10.1111/j.1365-2656.2012.02039.x

Tallian A et al (2017) Predator foraging response to a resurgent dangerous prey. Funct Ecol 31: 1418–1429. https://doi.org/10.1111/1365-2435.12866

Theuerkauf J, Rouys S, Jedrzejewski W (2003) Selection of den, rendezvous, and resting sites by wolves in the Bialowieza Forest, Poland. Can J Zool 81:163–167. https://doi.org/10.1139/Z02-190

Tourani M, Moqanaki EM, Boitani L, Ciucci P (2014) Anthropogenic effects on the feeding habits of wolves in an altered arid landscape of central Iran. Mammalia 78:117–121. https://doi.org/10.1515/mammalia-2012-0119

Trivers RL (1972) Parental investment and sexual selection. In: Campbell B (ed) Sexual selection and the descent of man 1871–1971. Aldine, Chicago, pp 136–179
Van Ballenberghe V, Mech LD (1975) Weights, growth, and survival of timber wolf pups in Minnesota. J Mammal 56:44–63. https://doi.org/10.2307/1379605
Vilà C, Wayne RK (1999) Hybridization between wolves and dogs. Conserv Biol 13:195–198. https://doi.org/10.1046/j.1523-1739.1999.97425.x
vonHoldt BM et al (2020) Heritability of interpack aggression in a wild pedigreed population of North American grey wolves. Mol Ecol 29:1764–1775. https://doi.org/10.1111/mec.15349
vonHoldt BM et al (2011) A genome-wide perspective on the evolutionary history of enigmatic wolf-like canids. Genome Res 21:1294–1305. https://doi.org/10.1101/gr.116301.110
VonHoldt BM, Stahler DR, Smith DW, Earl DA, Pollinger JP, Wayne RK (2008) The genealogy and genetic viability of reintroduced Yellowstone grey wolves. Mol Ecol 17:252–274. https://doi.org/10.1111/j.1365-294X.2007.03468.x
Vucetich JA, Peterson RO, Waite TA (2004) Raven scavenging favours group foraging in wolves. Anim Behav 67:1117–1126. https://doi.org/10.1016/j.anbehav.2003.06.018
Wabakken P, Sand H, Kojola I, Zimmermann B, Arnemo JM, Pedersen HC, Liberg O (2007) Multistage, long-range natal dispersal by a global positioning system–collared Scandinavian wolf. J Wildl Manag 71:1631–1634. https://doi.org/10.2193/2006-222
Wagner C, Holzapfel M, Kluth G, Reinhardt I, Ansorge H (2012) Wolf (*Canis lupus*) feeding habits during the first eight years of its occurrence in Germany. Mamm Biol 77:196–203. https://doi.org/10.1016/j.mambio.2011.12.004
Wang X, Telford RH (2008) Dogs: their fossil relatives and evolutionary history. Columbia University Press, New York, NY, USA
Wright GJ, Peterson RO, Smith DW, Lemke TO (2006) Selection of Northern Yellowstone Elk by Gray Wolves and Hunters. J Wildl Manag 70:1070–1078. https://doi.org/10.2193/0022-541x(2006)70[1070:Sonyeb]2.0.Co;2
Yang D et al (1999) β-Defensins: linking innate and adaptive immunity through dendritic and T cell CCR6. Science 286:525–528. https://doi.org/10.1126/science.286.5439.525
Zimen E (1976) On the regulation of pack size in wolves. Z Tierpsychol 40:300–341. https://doi.org/10.1111/j.1439-0310.1976.tb00939.x
Zimmermann B, Nelson L, Wabakken P, Sand H, Liberg O (2014) Behavioral responses of wolves to roads: scale-dependent ambivalence. Behav Ecol 25:1353–1364. https://doi.org/10.1093/beheco/aru134
Zimmermann B, Sand H, Wabakken P, Liberg O, Andreassen HP (2015) Predator-dependent functional response in wolves: from food limitation to surplus killing. J Anim Ecol 84:102–112. https://doi.org/10.1111/1365-2656.12280
Zub K, Theuerkauf J, Jędrzejewski W, Jędrzejewska B, Schmidt K, Kowalczyk R (2003) Wolf pack territory marking in the Białowieża primeval forest (Poland). Behaviour 140:635–648. https://doi.org/10.1163/156853903322149478

The Socio-Ecology of Free-Ranging Dogs 4

Dogs, together with humans, are one of the most successful species on the planet. They can be found in most ecological niches, on all continents, and their global population is thought to range between 700 million and 1 billion individuals (Hughes and Macdonald 2013; Wandeler et al. 1993). Dogs have (and have had also in the past) a valued position in many human societies (Serpell 2016). In the Western world, dogs usually live as companions in our human families sharing our daily lives and quite often also our sofa or bed for cuddle sessions or sleeping. They rely on humans in terms of food, a safe place to stay, and health benefits but also—maybe more to the dislike of the dogs—in regard to whether they can mate and, if so, with whom. This intimate relationship does not just allow a carefree life for the dogs, but it has been shown to be also associated with positive health benefits for their owners (Cutt et al. 2007; Feng et al. 2014), although this association has not emerged in every study (McNicholas et al. 2005; see also Beetz et al. 2012). Apart from these emotional benefits to humans, a number of studies have also demonstrated the functional benefits of dogs, e.g. both in modern (Ruusila and Pesonen 2004) and hunter–gatherer societies (Koster 2008; Lupo 2011, although see Lupo 2017 for a critical review). For example, dogs often increase hunting success (providing a clear material advantage to humans), are a successful tool in the protection of livestock from large predators (such as wolves), and, today, also have an important role as police, military, therapy, and guide dogs for the blind.

Accordingly, in our 'Western' minds, thinking of a dog likely conjures up the image of a labrador (or another such popular breed) cuddling with their owner on a sofa or playing a vigorous game of tug-of-war in a back garden. Indeed, even in the research community, there is a tendency to think of dogs almost exclusively as 'companion dogs'; the vast majority of studies on dog behaviour and cognition have been conducted on this population. However, the concept of 'companion' dogs, although probably not new—ancient Romans did like their lap and hunting dogs (e.g. see *Cynegeticus* by Xenophon)—is a rather narrow perspective of what dogs are. In fact, it is estimated that 75% of the world dog population lives as free-ranging animals (World Health Organization [WHO] 2011), and as we will see later, the

F. Range, S. Marshall-Pescini, *Wolves and Dogs*, Fascinating Life Sciences,
https://doi.org/10.1007/978-3-030-98411-3_4

majority of these dogs are not 'abandoned' companion dogs that have become stray because they are no longer cared for by their owners, but rather these are dogs that were born 'free-ranging' after generations of free-ranging dogs (FRDs). 'FRDs' is an umbrella term which has been loosely used to refer to dogs that are not directly controlled by humans. In fact, with very few exceptions, they live in and around human settlements within the anthropogenic niche and thus feed on food provided either directly or indirectly by humans. The World Health Organization categorizes FRDs according to their dependence on different food types and how restricted they are by humans; however, other authors have also included the type of environment (urban, rural, etc.) as descriptive factors (Ritchie et al. 2013). While the categorization based on the degree of freedom in roaming behaviour, dependence on human resource, and environment allows a certain degree of precision, it is important to note that overlap between categories is still high (Boitani and Ciucci 1995; Boitani et al. 2007; Daniels and Bekoff 1989; Macdonald and Carr 2016; Nesbitt 1975). The following categorization of FRD populations is usually applied:

1. Owned free-roaming dogs: these are dogs that are associated with a specific household, but they are allowed to roam freely. They are normally provided for intentionally by the household, but in many cases they also scavenge around the neighbouring areas. This type of management of the domestic dog has been reported in, e.g. Nigeria (Atuman et al. 2014), Philippines (Beran 1982), Zambia (De Balogh et al. 1993), Zimbabwe (Butler and Bingham 2000), and Kenya (Kitala et al. 2001). Although in principle they can occur in both urban and rural areas, they are mostly associated with the latter.
2. Village/community dogs: dogs that are not linked to a specific household (and as such are not 'owned' by specific individuals) but they depend on human-generated resources. There is a gradient in this category in terms of food provisioning. Dogs may be dependent on a mixture of food intentionally provided for by one or more people (e.g. shops leaving food out for them or people putting their waste in specific places for the animals) and food obtained by scavenging on human refuse including human faeces as well as roadkills, etc. They are also more associated with a rural environment, but they can occur also in towns and larger cities. In urban areas, these may be 'neighbourhood dogs'. In some localities, programs may be in place to assign ownership to the local municipalities that take responsibility for the vaccination and sterilization of these dogs, so in this case they are 'owned' by the municipality (e.g. southern Italy, Cuba).
3. Feral dogs: dogs that live mostly in rural/forested areas, are shy/avoidant of humans, and rely mostly on non-human-generated food; they tend to rather depend on carcass scavenging and hunting (e.g. see studies from Poland: Krauze-Gryz and Gryz 2014 and India: Home et al. 2018). However, shy dogs may also live on the outskirts of towns depending on scavenging of human refuse, limiting their activity to times and places when fewer people are around. In general, contact with humans as puppies increases their socialization towards humans and prevents them from being shy as adults (Scott and Fuller 1965).

As is apparent from the description of 'feral' dogs, the level of sociability towards humans has also been used as a descriptor. Indeed, Boitani et al. (2007) consider the defining factor of feral dogs their shyness towards people, which may be impossible to overcome, if they have had no contact with humans in their first few weeks of life. Sociability towards humans in different populations of FRDs is rarely described (but see Ortolani et al. 2009; Ruiz-Izaguirre et al. 2015); however, it may have important implications from both the management and theoretical perspectives. In the population of FRDs we are studying in Morocco, most dogs are quite social towards humans; however, we also have a number of individuals that are very shy despite living in the same environment. We suspect that they may be the ones that are more likely to roam at night, when the number of people is smaller. This difference in behaviour could lead to (1) social traditions, as shy mothers would give birth and raise their young off people, and (2) a spatiotemporal resource partitioning, resulting in a greater carrying capacity of the environment for these dogs. Further studies have to show how the early life experience of the dogs influences their adult human-directed behaviour.

Finally, geneticists interested in these populations (see Pilot et al. 2015) refer to FRDs as 'free-breeding', emphasizing the fact that their (promiscuous) mating choices are largely their own and not directly determined by humans (although, of course, human intervention as we will see below does affect them). This is indeed the case at the population level, although the many ABC programs (animal birth control, spay and neuter campaigns where animals are re-released in their home ranges after intervention) around the world may reduce this freedom in the sense that it reduces the dogs' interest in mating.

In conclusion, these FRD populations provide us with the possibility of understanding more how dogs 'would live' if humans were not interfering quite as much as we do with pet dogs, and this alone makes them appealing for researchers. However, despite the explosive popularity of dog research in the last 20 years, research on FRDs has been slower. A limited number of studies on their feeding ecology and ranging patterns stem mostly from conservationists interested in assessing the impact of FRD on local wildlife (in terms of competition, disease transmission, and direct predation). Similarly, information on FRD population dynamics comes predominantly from studies aimed at gathering information for rabies-prevention programs. Dogs impose extremely high risks through the transmission of diseases with over seven million people being infected with rabies per year and 55,000 dying in developing countries (Knobel et al. 2005), with 99% of these deaths occurring in countries where rabies is endemic in the domestic dog population (Cleaveland et al. 2006; see also Linnell et al. 2002; Tang et al. 2005). This high risk of dogs transmitting diseases if not vaccinated might in parts also explain the difference in attitudes towards dogs across the world: while in general there seems to be a rather positive attitude towards dogs in most Westernized countries (Lakestani et al. 2011; Miura et al. 2000), in the only larger scale cross-cultural study dogs are considered 'friends' and 'family members' in only 5 of the 60 cultures tested, suggesting that elevating pets to these roles seems to be the exception rather than the rule cross-culturally (Gray and Young 2011).

As regards the social behaviour of FRDs, until just a few years ago only a few studies had been conducted (see studies by Pal (1990), Pal et al. (1998a and 1999), Beck (1975a), Beck et al. (1975), Fox et al. (1975), and Boitani (Boitani and Ciucci 1995; Boitani et al. 1995)). More recently, a group of ethologists in Italy (Cafazzo et al. 2014) carried out a number of studies on a population of FRDs living on the outskirt of Rome, and a lab in India, headed by Prof. Anindita Bhadra, has also focused on behavioural studies of FRDs. Together, these studies provide us with the first insights on how dogs live when humans do not completely control them.

In the following pages, we aim to bring together what is known of the behaviour of FRDs and review studies stemming from the different disciplines. As highlighted above, there is some difficulty in categorizing FRD populations; however, we have kept the five main factors (environment, foraging habits, ownership, sociability, and breeding status) in mind and, where possible, provide information in relation to these aspects.

Box 4.1 A journey into the world of FRDs
By Martina Lazzaroni

Before starting my PhD on FRDs for some years, I studied pet dogs in a lab in Italy and then captive dogs living at the Wolf Science Center.

I had always known about the existence of FRDs, of course, but I had never thought about them too much. I had never interacted with them because in the region I come from in Italy and in Austria where I was studying, they do not exist. So FRDs were something a bit alien to me.

But one day I discovered a remarkably simple fact: FRDs represent 80% of the world dog population! Oh, that changed everything for me! Basically pet dogs that I had been studying until then were just a tiny proportion of 'dogs'. But because they are the most common dogs living in my areas, my vision was totally biased toward them. This simple fact made me consider that perhaps it should be FRDs that ought to be considered as the 'real' representative of the dog species. Yet FRDs have received so little attention by behavioural scientists!

So I started my journey to discover more about them. I won a grant from the Austrian Academy of Science to carry out my PhD, and I went to Morocco, a wonderful country with an abundance of FRDs. We (me and other two students) moved to Taghazout, a small village on the coast close to Agadir, which is quite well known for its surfing and for being dog friendly!

We were quite surprised to discover that the dogs living there, basically village dogs, are well socialized with people and that it is actually pretty easy to engage them in our behavioural tests. After some initial difficulties and defeats (never try to put a leash on a FRD!), we refined our methods, adapting our testing procedures to their free-living condition, and I was able to carry out quite a few experimental studies for my thesis.

(continued)

Box 4.1 (continued)

But aside from the tests, we also carried out daily monitoring of their location and activities, and we identified a few packs where we carried out more in-depth analyses of their social behaviours and interactions; then, we started to collect DNA samples, search for their dens, and figure out their feeding sources, and what was supposed to be a small project turned into the beginning of a more permanent field site. Every day we discovered new aspects about their lives that were worth studying! The project has now been going on for 4 years, and many new students and now also postdocs have started contributing new knowledge and new ideas to discover more about these dogs.

Now that my PhD has ended and I am back in a country deprived of FRDs, I often look back and think of the intensity of that period. Every day something new and unexpected happened: a puppy sneaking into our car and eating all the food we used for testing; finding a den below a stack of beach beds, with squirming pups huddling together; searching for hours for our study pack; and finally finding them on the shore fishing (or trying to anyway!) or being chased and trying to hide from a dog that wanted to be tested again!

In our Western country, FRDs no longer exist; we think dogs' lives are better in our homes. This may be true in some respects; however, the life of the dogs I studied, although admittedly short, was also very rich, and they enriched the lives of the people around them, blending into the social and cultural texture of the society.

4.1 A Few General Facts

The few genetic studies on FRD populations indicate that they typically show a higher genetic diversity than breed dogs, and, in most cases, they form populations distinct from both the local breed and mixed-breed (dogs whose parents were breed dogs) populations (e.g. Africa: Boyko et al. 2009; South and Central Asia: Shannon et al. 2015; and Eurasia: Pilot et al. 2015).

FRDs are generally smaller than wolves; their dentition, and in particular the size of the canines and their mandibular power, is noticeably reduced, and they tend to show more diverse coat colours and specific characteristics such as floppy ears, which are absent in wolves. This aligns with the general 'domestication syndrome' and clearly indicates that FRDs are domesticated. Coppinger and Coppinger (2016) suggested that many of these morphological differences point at how dogs have adapted to their new environmental niche going from a hunting to a human refuse scavenging lifestyle. Indeed, further support for this view comes from specific genetic adaptations for starch digestion (increase in the number of copies of the amylase gene AMY2B), which are absent in wolves but are found in most breed

Fig. 4.1 Pictures of FRDs in Morocco taken by Friederike Range (left) and Larissa Darc (right)

(Pilot et al. 2015) and FRDs (Pendleton et al. 2018). Interestingly, huskies and dingoes show a lack of Amy2B copy number (similar to wolves), likely because of a less pronounced dependence on starch-rich resources in these populations (Arendt et al. 2016; Freedman et al. 2016). These data suggest that dogs showed rapid adaptations to the changes in diet and that 'natural selection' continued to act on domestic dogs alongside the direct selection by humans. This is further evidenced by specific genetic adaptations occurring in some village dogs (but not others). For example, dogs (and humans) from the Tibetan plateau show specific adaptions to life at high altitudes that are absent from their lowland Chinese neighbours (Wang et al. 2014). Although at present few studies have investigated the genetic profile of FRDs, it is likely that they will show many more potential adaptations to their specific environment, similarly to humans (see Reich 2018). Indeed geneticists working with village dogs have suggested that these populations should be considered for conservation efforts, because of their potentially unique genetic (and culturally induced) diversity (Boyko and Boyko 2013) (Fig. 4.1).

4.2 Environment and Diet: Scavenger or Hunter?

As evidenced by the genetic data, a shift in diet likely played a major role in the domestication process. So the first question is: what do FRDs eat today?

4.2.1 Dog: The Scavenger

A series of studies were conducted in various homesteads neighbouring wildlife reserves in Zimbabwe, by using questionnaires, focal sampling with GPS collars of 16 dogs, and censusing techniques. Based on the questionnaire data, of the 236 owned but free-roaming dogs, monitored over a 1-year period, researchers were shown to be primarily scavengers of human waste and animal carcasses: sadza (the local staple maize/millet porridge), human faeces, and cow carrion were the three most important elements in their diet. Of the food dogs obtained independently, scavenged mammalian carrion (55%) and faeces (25%, of which 93% was human) were the most frequent components. Overall, human-derived food (i.e. sadza, cultivated vegetables and fruits, domestic mammals, poultry fish, and faeces) contributed 88% of the diet in these studies (Butler and du Toit 2002). Somewhat surprisingly, in an in-depth analyses of the nutritional values of the various components of the dogs' diet, it emerged that fresh human faeces is highly nutritious, yielding 18.7% crude protein and 18.7 KJ/kg gross energy, which exceeded the results for sadza and was equivalent to the mammal remains (Butler et al. 2018). Moreover, the researchers presented carcasses placed in open areas just on the border between homesteads and the wildlife reserve, as well as within it (Butler and du Toit 2002). They found that regarding carcasses closer to human habitations, dogs were responsible for locating and consuming the majority of these, outcompeting the natural scavengers in the area (vultures and hyenas). Monitoring of spontaneously occurring carcasses showed dogs located and fed on them in 50% of cases, even if they were 1 km within the park boundaries (Butler and du Toit 2002). The latter observations indicate that dogs, due to their reduced fear of humans and due to their acceptance by humans, obtained certain competitive advantages over other wildlife when in proximity to humans.

In another study in Africa, Atickem et al. (2009) surveyed the diet of free-roaming owned dogs in the Bali mountains in Ethiopia, where they live in close proximity to the highly endangered Ethiopian wolves. Focal observations and faecal analyses were carried out, and both methods showed dogs' diet, similarly to the one in Zimbabwe, consisted predominantly of barley husks, human faeces, and animal carcasses.

A number of studies have also been conducted on the diet of FRDs in India. In all such reports, the major food sources were identified as garbage and leftover food dumped outside houses and food shops by people (Oppenheimer and Oppenheimer 1975; Pal 2001, 2015; Paul et al. 2014) and in some cases carcasses of domestic animals and human faeces, particularly from children (Oppenheimer and Oppenheimer 1975).

Although we are able to get some insights in dogs' diet from the studies above, we know a lot less about their actual foraging behaviour. There are a number of descriptions from around the world of dogs visiting specific 'feeding places' where human-generated food is known to be available (Macdonald and Carr 1995; Majumder et al. 2014; Pal 2015), and in some such cases, groups of dogs were observed at the feeding spots (Macdonald and Carr 1995). However, the numerosity of dogs observed foraging together is likely linked to the type of resource and, in other cases, likely due to the inert, itemized, small, and dispersed nature of the food. Thus, scavenging on human refuse appears to be mostly a solitary or dyadic activity, except when juveniles are around which increase the group size of the foraging parties (Majumder et al. 2014). The studies looking at dogs' activity budget suggest that they are on the move (potentially scavenging–foraging) mostly when temperatures are low (early morning and late afternoon) (Majumder et al. 2014). However, despite reports of FRDs being active at night, no studies to our knowledge included night follows to evaluate this aspect in greater depth (Fig. 4.2).

Fig. 4.2 Competition over a carcass. The adult dog is eating first; when it leaves, the puppy/juvenile is approaching and taking over, while the vulture waits and finally leaves (pictures taken by Friederike Range)

4.2.2 Dog: The Hunter

As outlined above and in contrast to wolves that rely mainly on hunting (see Chap. 3), the anthropogenic niche of FRDs seems to be usually characterized by immobile and often reliable and abundant food resources, often making hunting unnecessary. However, that does not mean that dogs are completely inept at hunting and killing live animals; rather, it indicates that dogs, like any other predator species, respond to the availability of food resources. In a highly anthropic environment where 'leftover' food resources are available and live prey might not be very abundant, dogs may rely more heavily on the former; however, different conditions will likely affect what type of foraging behaviour they adopt. In accordance, there are numerous reports from different sources showing that FRD engage in hunting: however, there are hardly any studies directly investigating dogs' hunting behaviour.

In Israel, Manor and Saltz (2004) reported a direct negative relationship between kid/mother ratio of mountain gazelles and the numerosity of free-roaming dogs in the area. Direct evidence of the potential prey–predator relationship between the gazelles and the dogs was also obtained in the form of two incidents of dogs preying on gazelle kids and ten incidents of dogs (in groups) chasing adult gazelles, during the course of the 2-year study. In Ethiopia, rodents (the Ethiopian wolves' natural prey) were also successfully hunted by dogs (see also Sillero-Zubiri and Gottelli 1995) although they made up only the minority (3–4%) of the dogs' diet. In Poland, based on scat analyses, researchers reported dogs feeding on hare and roe deer, and prey abundance in the scat was related to species' abundance in the area, as well as showing a pattern of higher abundance in winter than summer and during harsher years. Results have led authors to suggest that dogs supplement their human-related diet with game when the former is not sufficient (Krauze-Gryz and Gryz 2014).

However, the authors recognize that the problem with scat analyses is that there is no way to determine if the prey was scavenged or hunted. They suggest that it is likely that most of the game observed in the dogs' scats comes from carrion, although they also report direct observation of hunting and killing of both hare and roe deer, the latter by dogs hunting in groups. Other reports of dogs preying on wildlife come from questionnaire and survey data (Hughes and Macdonald 2013; Vanak and Gompper 2009; Young et al. 2011). Based on an online survey and analyses of media reports across India, Home et al. (2018) found that of the 249 responses obtained, 73% reported sightings of dogs attacking wildlife. The sightings (i.e. a total of 460 attacks on wildlife) were shown to be distributed across India but with a high percentage of occurrences on the borders of wildlife reserves. Of the total events, 56% were carried out by packs of dogs (i.e. range 3–25 individuals), and approximately 45% of the dog–wildlife encounter incidences led to the death of the 'prey'. Dogs in packs were much more successful at killing the prey than dogs alone or in pairs (of a total of 205 attacks on wildlife which resulted in the death of an animal, 58% were carried out by packs, 9% by pairs, and 13% by single animals).

In semi-urban areas, potential prey is often rather small or domesticated. For example, in a semi-urban area in Brazil, Campos et al. (2007) identified 42 FRDs (no specification if owned, etc.), and based on the analyses of 137 scats, they found that the dogs' highest consumption was of invertebrates (57%), followed by mammals (25%), birds (15%), and reptiles. Of the mammals, rodents were the most frequent (45%). Occasional group hunting events were also observed in the predominantly human refuse scavenging populations in the semi-urban Kolkata area. In this environment, goats, cows, and pigs graze in open fields, and researcher reported five kills carried out by two to three dogs together on rats and juvenile goats and pigs. The authors describe dogs hunting in a coordinated fashion, i.e. 'surrounding the prey from different directions, barking loudly, and chasing it, closing the circle to finally kill it' (Paul et al. 2014).

In contrast to these studies indicating successful and in some cases even cooperative hunting, the above-mentioned studies on FRDs in Zimbabwe report dogs as 'inefficient predators' that largely failed to kill their prey (Butler et al. 2004, pp. 269: a total of 20 kills, most of which (30%) were represented by domestic goats). On average, killing prey involved 1.7 dogs (range 1–4). Of the 16 radio-collared animals followed closely, none were observed to kill a prey although some chasing behaviour of small prey was observed (Butler et al. 2004). Based on surveys of other communal lands, dogs were also rarely recorded taking live prey (Butler 1998). However, considering dogs can successfully hunt live prey, these failures could of course also be due to the dogs' low motivation considering the otherwise abundant food resources (see above).

As mentioned before, whether dogs engage in hunting or not is likely dependent on the availability of other food resources easier and less costly to obtain (both in terms of energy expenditure and risk of injury) and on the abundance of the local prey species (see Chap. 3 for a discussion of these points). In line with this, Vanak and Gompper (2009) emphasized the potential effect of location and ownership status (feral, free-ranging, owned) on the dogs' feeding habits and highlighted that a gradient was present regarding dependence on human food which largely coincided with the geographical closeness to human habitations/settlements and home range size of dogs. The further away from humans, and the closer to 'wilder' areas, the less dogs depended on human-generated resources, and the more hunting behaviour was observed. This same pattern appears to be consistent also with reports from Poland (Krauze-Gryz and Gryz 2014). This clearly indicates that dogs are indeed able to live in 'wilder' settings by hunting on small- (e.g. rodents, hares, etc.) and medium-sized (e.g. small ungulates) prey (although perhaps the latter less successfully) as well as in more anthropic environments and, like wolves, are quite flexible in their foraging strategies which they can adjust to the local conditions, e.g. whether suitable prey is available. So far, there has been no study (to our knowledge) specifically assessing the dogs' more preferred 'niche', although a study on food preferences of wolves and dogs at the Wolf Science Center revealed no differences between the two species (Rao et al. 2018). Nevertheless, it would appear that the anthropic niche (from more rural to highly urbanized) is their typical choice, but being opportunistic feeders, they can and do supplement their largely scavenging-based diet also with occasionally hunting of other animals, when available.

4.3 Social Organization

4.3.1 Group Size: When Are Dogs Solitary and When Social?

Animals can either be solitary or live in social groups that can range from temporary aggregations to permanent groups. The formation of social groups is facilitated if its benefits (e.g. reduced per capita risk of predation, enhanced access to and defence of food resources, access to safe places; Fig. 4.3) compensate for its costs, which relate largely to increased competition for resources (Ward and Webster 2016). The competition for resources is highly dependent on the distribution of the resources and can be of two general forms: contest or scramble competition (Łomnicki 2009; Nicholson 1954). Under pure contest competition, individuals compete for priority of access (PoA) to a monopolizable resource, and, thus, the higher an individual's dominance rank, the better access to resources it has. In other words, an individual's

Fig. 4.3 Found a safe place? Picture taken by Rachel Dale

access to the resource depends on the number of individuals ranking above it (queuing for access; Alberts et al. 2003; Altmann 1962; Koenig 2002; Łomnicki 2009; van Schaik 1989). Under pure scramble competition, a resource cannot be monopolized, e.g. because it is too widely distributed or too large, and is therefore equally accessible and exploited by all individuals, so individual access to the resource depends only on group size (or, more generally, population density; Koenig 2002; Łomnicki 2009; van Schaik 1989). Thus, the social organization is expected to be more or less dependent on the resource distribution. One other aspect important to consider here is that the level and form of competition can be quite different for females and males even within the same population as they rely on different key resources, resulting in sex-specific socio-ecologies. Thus, while females compete primarily for food and safety, males compete primarily for access to receptive females, suggesting that male socio-ecology is largely determined by female socio-ecology and distribution (Altmann 1962; Clutton-Brock and Parker 1992; Emlen and Oring 1977; Lukas and Clutton-Brock 2013). Not surprisingly, in non-human primates as well as several canid species, the abundance, quality, and distribution of food have been shown to be linked to sociality by shifting the social structure from a solitary or pair-bonded system to multi-male/multi-female groups (foxes: Macdonald and Johnson 2015; dingo: Newsome et al. 2013; coyotes: Hidalgo-Mihart et al. 2004; Ethiopian wolves: Marino et al. 2012).

In regard to FRDs, some studies have reported stable social structures (Cafazzo et al. 2010; Freedman et al. 2016), while others have suggested that FRDs do not form stable social groups, but are mostly solitary, or fluid agglomerations centred around females in heat (Berman and Dunbar 1983; Daniels 1983). Moreover, seasonality has been shown to affect grouping behaviour in at least some FRD populations. Using a censusing method over a 2-year period around Kolkata (West Bengal, India), Majumder et al. (2014) found that dogs were more likely to be observed in groups of 3 in summer, when juveniles are starting to move around outside the den, whereas in autumn (during the breeding season) they were more likely to be observed in pairs (see also Majumder et al. 2014). Finally, in several locations, fission–fusion groups have been observed showing daily variations in grouping patterns. Font (1987) monitored a population of FRDs in Valencia, Spain, and focused on the social behaviours of a pack of four individuals (one adult female, two adult males, one juvenile male). He noted that these individuals travelled alone or in pairs during the day but were nevertheless part of a group of 4 that rested, defended their territory, and showed both affiliative and agonistic behaviours. Similarly, solitary foraging has been observed also in a number of studies of FRDs in India (Majumder et al. 2014). These differences in social organizations of FRDs could potentially be linked to the food distribution and the resulting distribution of the females, since as pointed out above, while in many environments FRDs face rather reliable and abundant food resources, food can also highly vary in patch size, distribution, and quality, ranging from small, widely distributed items through clumped household and restaurant leftovers to large rubbish dumps. This high variability in the characteristics of anthropogenic resources is predicted to result in variable monopolization potentials between individuals and between groups and

Fig. 4.4 FRDs feeding on meat provided by humans and a garbage dump (pictures taken by Giulia Cimarelli)

populations (Marshall-Pescini et al. 2017; Sterck et al. 1997; van Schaik 1989). Although the evidence is at the moment largely circumstantial, since a systematic assessment of the feeding resources and the role of mate competition has not been carried out, there are suggestions that, as for other mammals, also in FRDs food quality and distribution of resources affect sociality (Font 1987; Fox et al. 1975; Macdonald and Carr 2016) (Fig. 4.4).

However, in regard to dogs, humans might have some influence on these patterns as well. For example, it was found that the environment (urban, suburban vs. rural) may affect dogs' sociality. Indeed, in a number of studies, dogs living in urban or village contexts were observed to live more often either solitary (Daniels 1983) or in small groups of up to three individuals (Berman and Dunbar 1983; Font 1987; Macdonald and Carr 2016), whereas in more rural or sparsely inhabited areas, dogs appeared to be more often in larger packs (Krauze-Gryz and Gryz 2014; Macdonald and Carr 2016; Manor and Saltz 2004). In a national park and adjacent area in Bulgaria, using a camera-trap method, researchers directly assessed this hypothesis and found that bigger groups of dogs (3–4 animals) were observed at greater distances from human habitations, whereas solitary dogs were more likely to be observed closer to houses (Doykin et al. 2015). It may be possible that this variation is due to humans' acceptance of dogs living in larger packs or, alternatively, that the numerical variation relating to closeness to human habitation is linked to the dogs' free-ranging category, i.e. whether they are free-roaming owned dogs, village dogs, or feral dogs. If the latter explanation is valid, then it is a methodological difference. However, it might also be due to the food resources available. If the animals close to human proximity feed more on the small chunks of food found on the street, which can be easily monopolized and eaten in a few seconds by a single dog, one would not expect them to forage in groups, whereas animals in more rural areas, which might more often also find larger food resources such as carcasses that cannot be monopolized and devoured quickly, might more often forage in groups, potentially also to defend these larger food resources against other dogs.

What emerges from these studies is a great variability and flexibility in dogs' group composition which is potentially affected by the food distribution and the resulting female distribution but may also be influenced by human interferences and potentially also genetic differences between populations, though the latter point has not been addressed yet. However, the different grouping patterns also highlight the need to be careful in regard to the data collection methods to be sure to capture the variation in the social organizations given that fission–fusion groups seem to be common (e.g. census methods might not be very informative; Font 1987). Moreover, groups should not be defined based only on their temporal association but rather including their behavioural interactions (i.e. 'two or more individuals that were seen to show affiliative interactions like allogrooming, nuzzling, playing, walking together, sharing food, etc., or dogs that were resting peacefully within about 3 ft. of each other') (Font 1987; Majumder et al. 2014; see also Macdonald and Carr 1995 for a similar approach).

4.3.2 Group Composition

Where packs do form, they usually include multiple males and multiple females (see Bonanni and Cafazzo 2014 for a review). However, at present little is known of within-pack relatedness, especially paternity, since no study, to our knowledge, has included genetic analyses of pack structure in FRDs. Nevertheless, in a number of longer-term studies, mother–offspring relationships were known. In India, Pal (2017) reports three of his seven groups studied being composed of multiple related females, and similarly in 15 groups studied, where the pedigree was maintained, Paul and Bhadra (2017) report that often multiple females were related to one another.

4.3.3 Mating and Reproduction

In contrast to wolves and as a consequence of the domestication process, males are sexually active all year round, whereas most females have two breeding seasons per year. Synchronicity in breeding does not appear to occur in FRDs, although there are some reports of a seasonal increase in breeding in some populations. In India, the mating season coincides with the monsoon periods (Chawla and Reece 2002; Pal 2001; Sen Majumder and Bhadra 2015; Totton et al. 2010) with a peak pregnancy prevalence occurring in the late monsoon season (September to November) (Totton et al. 2010) or winter season (Oppenheimer and Oppenheimer 1975; Pal 2001; Pal 2017). Hence, in India at least, although two distinct mating seasons have been observed (Sen Majumder and Bhadra 2015), one litter per year is produced (Pal 2005, 2017; Pal et al. 1998a). In contrast, there was no evidence of seasonality in free-roaming (but owned) dogs in Indonesia and South Africa (Morters et al. 2014) and in a provisioned FRD population in Italy (Cafazzo et al. 2014). Thus, at present it is not clear to what extent and which factors may be affecting seasonality in some FRD populations but not others.

In general, FRDs have been described as having a promiscuous mating system, with multiple males and females mating within a group (America: Daniels 1983; India: Pal 2001; Pal et al. 1999; Italy: Cafazzo et al. 2014; Macdonald and Carr 2016).

Due to the differential costs of reproduction in general, but also specifically in promiscuous species (East et al. 2003; Hoffman et al. 2008; Longpre et al. 2011), females should be choosy and mate preferentially with high-quality males, while males should attempt to mate with as many females as possible. Indeed, it has been shown in several studies that female dogs do not mate with any male and their rejection behaviour can range from simple avoidance of males (or sitting down) to actively attacking them (Beach and LeBoeuf 1967; Cafazzo et al. 2014). Still, mating with more than one (high-quality) male may benefit females in several ways. First, it might confuse paternity and thus discourage infanticide. In wolves, infanticide by neighbouring packs is a high risk (see Chap. 3) and has been reported previously also in dogs (Pal 1990). Especially, the observation of females in Morocco roaming around more widely during the mating season might support this hypothesis. However, even mating with several males within the female's own group may be advantageous by reducing aggression of the males towards the female, as well as increasing the males' social support and willingness to defend the female and her pups against other males or predators. Second, mating with more males increases male–male as well as sperm competition and thus increases the chance for high-quality genes. Finally, mating with several males might also reduce the risk that no fertilization occurs. Given that dogs are only receptive for a few days every 6 months, missing one reproductive period might have high consequences for lifetime reproductive success. However, if you have access to a prime male, like the alpha, it could also be of advantage to be faithful, having the advantage of knowing who the father of your offspring is and ensuring his help to a higher degree. In any case, at the moment we know hardly anything about female choice in dogs.

For males, mating with more than one female has the advantage of increasing fitness, and since mating usually does not come with high costs, this is the best strategy for them. Accordingly, males compete for access to females engaging in aggressive interactions with other males, interrupting matings (Cafazzo et al. 2010; Daniels 1983; Pal 2003a; Pal et al. 1999), building up affiliative relations with certain females (Bonanni et al. 2010), which might buy them mating opportunities later, or even coercing females into mating (Ghosh et al. 1984; Pal 2003a, 2011; Pal et al. 1999). However, mate choice in dogs is still relatively understudied.

In an early study, Daniels (1983) observed urban FRDs in Newark, New Jersey, and found that, during estrous periods, females would be accompanied by groups of between 2 and 12 males (mean 5), which he categorized on the basis of their familiarity with the female (i.e. proximity during non-estrous periods). He found that females tended to show a preference for mating with known rather than stranger dogs. In a series of studies on the mating behaviour of Indian FRDs, Pal found that over the course of the 6 peak days of their receptive period, oestrous females attracted between 6 and 60 males, and in any given observed association, the female may have between 1 and 10 males simultaneously courting her (Pal 2005, 2011; Pal et al. 1999). Most females showed some selectivity in their mating preferences, in that they did not indiscriminately mate with all available mates but rather mated on average with between 38% and 57% of the courting males. However, in all studies, some females mated with all available males, and in two studies (Pal 2005, 2011), a number of females (four and two females, respectively) mated with a single courting male.

Cafazzo et al. (2014) studied a pack of 27 FRDs on the outskirts of Rome in Italy. Ten females (six pack members and four transient females) were followed during their estrous period. Both males belonging to the pack ($N = 16$) and from neighbouring packs/unknown dogs ($N = 7$) were observed courting the estrous females. Females showed clear preferences for specific males by clearly refusing others. Except for the most dominant female in the pack, who mated repeatedly only with the dominant male, all other females copulated with more than one male. Four of these females copulated also with non-pack males (two transient and four pack males). Among pack males, higher-ranking non-aggressive males were preferred by females (the latter seeking to stay in proximity with them), and they enjoyed higher copulatory success (ties). Because of the lack of synchronicity in the females' estrous, males could court all females; however, males showed clear preferences for higher-ranking rather than lower-ranking females. The latter preference is in accordance with the idea that higher-ranking females have access to better resources and thus likely also better reproductive success.

The results indicate that mate competition in FRDs is high and that the population and potentially the genetic makeup (descendants of breed dogs or long established FRD populations) might influence the mating strategies of both males and females. In Italy, rank position is strongly positively correlated with age, suggesting that female preferred higher-ranking/older males (while old here is seldom older than 3 to 4 years). At present such differences need to be taken with caution since not all factors (e.g. pack membership, familiarity, rank position, age) are considered in all studies. For example, in some populations, there might have been a selection for familiar males over strangers due to high prevalence of diseases in that location, whereas in others receptive females may show enhanced roaming behaviour far beyond territory borders, travelling from one male to the next, thereby causing strong and repeated direct competition (personal observation Morocco FRDs; see also Pal 2011) and eventually ending up with the most superior male. There is a clear need for future studies taking all these aspects into account and potentially standardizing data collection methods.

4.3.4 Parental Care

4.3.4.1 Maternal Care

Maternal care starts before parturition in the form of den selection. In a study involving 20 pregnant females of a FRD population in India, researchers found that all females tried out at a least one den (and up to 6) before settling for the final one. Although there was no preference by mothers for sheltered dens, mothers sought out denning sites close to resources and water and were largely unaffected by the level of human disturbance (Paul et al. 2016). In accordance with these observations, Daniels and Bekoff (1989) report mothers raising the pups away from the pack (although mothers themselves maintained contact with the other pack members) for the first 2 months after parturition. However, Boitani and Ciucci (1995) studying feral dogs report most mothers denning within the pack territory and pups being visited by other members of the pack. Both strategies were observed in the FRD pack studied in the outskirts of Rome. Of the ten females observed, four left the pack before giving birth (and had relatively little contact with the pack in the latter stages of pregnancy), and of the six resident females, the high-ranking females denned in the core area, while the others situated their den within the pack territory but not as central. In the latter study, researchers were not able to observe the dens directly; hence, no direct observations of other individuals interacting with the pups were possible (Cafazzo et al. 2014). Similarly, in a study in the Abruzzo region in Italy, Macdonald and Carr (1995) observed that some females stayed close to the other group members and allowed males to interact with the pups, whereas others remained with the other pack members only until they were pregnant and then left the pack to lead a 'semi-detached' existence for the first few months of the pups' existence. During this time, they never observed other pack members approach or interact with the pups until the pups started to visit the feeding sites at around 14 weeks of age (and pups still slept in a different location than the adults until 20 weeks of age).

Whether or not the females allow other members of the pack access to the pups may depend on with whom they mated, since potential fathers would likely be prone to protect the pups, whereas if the potential father was an outsider, infanticide might be a risk. Moreover, for low-ranking females, intrasexual competition with higher-ranking females for resources like food, good denning places, and male support might drive them to avoid the proximity of these females. In general, the behaviour is in stark contrast to the one exhibited by wolf females, who will dig dens preferable in the middle of the territory and avoid areas associated with humans (see Chap. 3). Due to the dependence on other pack members both in terms of food provisioning and protection of the pups, for female wolves, spacing the dens in different areas of the territory would not be feasible, nor would be avoiding the contact between other pack members and the pups.

After a mean gestation period of 63 days, pups are born. Generally, the larger a female in terms of body size, the larger the litter (Fig. 4.5). Given that most FRDs are of medium size, it is not surprising that the mean litter size seems to be slightly smaller than in wolves (Chap. 3) and ranges from 3/4 to 6 pups (3/4: Boitani et al.

Fig. 4.5 Picture of FRD mother with her pups in Morocco (picture taken by Giulia Cimarelli)

1995; Hyeroba et al. 2017; Paul and Bhadra 2018; Paul et al. 2016; 4.6: Totton et al. 2010; 5.03: Pal 2001; Pal et al. 1998a; 5.8: Pal 2005; 6.08: Pal 2017). It is in unclear whether there is a biased sex ratio at birth or not, since two studies in the Indian population report differing results. Based on a 5-year census, Pal (2017) reported a male-biased sex ratio both at birth and as adults; however, no sex bias was found at birth in another study conducted over a 4-year sampling period (Paul et al. 2017).

Pups emerge from the den at around 3 weeks of age (Pal 2008), and studies on parental behaviour have been carried out from this time on. Maternal care has been reported to be the predominant form of care received by the pups and normally continues until the tenth to 15th week of a pups life, by which time pups start foraging alone (Macdonald and Carr 1995; Pal 2008; Paul and Bhadra 2017). The general pattern of maternal care that has been observed in Indian FRDs is that pups are milk-fed until around 10–11 weeks of age, with mothers supplementing this with regurgitated food from around the fifth/sixth week, coinciding with a reduction in suckling (Pal 2005, 2017; Paul and Bhadra 2017). Regurgitation continues until between the tenth (Pal 2005, 2017) and 15th week, and in some cases mothers also bring back scavenged food for the pups during this time (Paul and Bhadra 2017). The intensity of care per pup is affected by litter size: pups with fewer siblings receive more care (Paul et al. 2017).

Based on the changing maternal care behaviours, three phases in the ontogeny of FRDs have been identified (Paul et al. 2017): (1) from birth to 6 weeks, coinciding with ample maternal care (Fig. 4.2); (2) from the seventh to the 13th week, during which weaning gradually occurs; and (3) from the 13th week on, when the pups enter the 'juvenile' stage in which they largely forage independently although potentially still sharing space and externally obtained resources with the mother.

4.3.4.2 Paternal Care

In wolves, providing male care to the breeding female and her pups is adaptive since the breeding male in the pack is very likely the father and thus benefits directly by the survival of his offspring via inclusive fitness. In FRDs, however, females are promiscuous and mate with several different males, resulting in a high paternity confusion that reduces the risk of male infanticide. However, also helping the female and her pups is likely to be adaptive only under special circumstances, where the male can be sure that he is tending his own offspring and not the one of the 'cuckoo', e.g. of another male. Not surprisingly, reports about paternal care in dogs are mixed. In India, in about 50% of cases, a male observed copulating with the female during her estrus then remained with the mother during pup-rearing (Pal 2005, 2017; Paul and Bhadra 2018). Where the 'putative father' was present, paternal behaviour mostly related to guarding/protection (fighting and chasing of strangers) and playing with the pups, although in all studies a few males were also observed regurgitating or provisioning pups. Macdonald and Carr (1995) also observed that in some cases males remained with the pregnant females, and after she gave birth, they remained in close proximity to the bitch and her litter, playing with the pups when they became mobile.

4.3.4.3 Alloparental Care

Alloparental care can include many different types of behaviours, from allo-nursing (nursing by a female which was not the mother) to other forms of alloparental care such as protection, play and grooming. It is important to note here that these behaviours differ in regard to the cost incurred by the donor. In regard to allo-nursing, regurgitation and protection, costs can potentially be very high and might only occur between relatives where inclusive fitness benefits balance these costs, which is the case in wolves (see Chap. 3). In contrast, play and grooming mainly require time and energy and likely are rather cheap especially if food resources are abundant. Early studies of FRDs in the Italian Apennines reported no occurring alloparental care (Boitani and Ciucci 1995; Boitani et al. 2007); however, more recently, a few studies (all conducted in India) have investigated this topic more closely. In an in-depth study of alloparental care in which the categorization of different types of care was carried out, the majority of care behaviours exhibited were guarding, piling up, grooming, and play, although also some regurgitation events by allomothers were observed (Paul and Bhadra 2017). Allogrooming and pup guarding was particularly concentrated during the pups' earlier weeks, continuing until the 15th week of age (Paul and Bhadra 2017). Unfortunately, the authors do not define or describe what they define as pup guarding, and thus it is unclear whether it is indeed a costly behaviour or not.

Overall, when provisioning/regurgitation was observed, these were performed by another female known (or strongly suspected) to be related to the pups' mother (i.e. either the pups' grandmother or aunt) (Pal 2005; Paul and Bhadra 2017; Paul et al. 2014). In one case, following the death of the pups' mother when they were 1-day-old, the grandmother took over the maternal role, showing behaviours indistinguishable in type and frequency from that of other mothers in the study (see Paul et al. 2014). Based on their observations, Paul and Bhadra (2018) suggested that, while FRDs are not cooperative breeders like wolves, they are communal breeders. Communal breeders are defined as multiple females sharing dens or birthing sites, helping each other to rear offspring (e.g. allo-nursing, defending the pups). The difference in regard to cooperative breeding, where older, related offspring help in raising the pups, is that the females do not forfeit their opportunities to reproduce and thus do not need to be related to explain their costly helping behaviour (Paul and Bhadra 2018). This breeding strategy is likely to have played an important role in increasing pup survival in a stochastic environment and helping to adapt to living among humans during the domestication of dogs (cited from Paul and Bhadra 2018). However, at present it is unclear how frequent alloparental care occurs in FRD packs, with some studies reporting fairly low rates (e.g. three out of seven packs: Pal 2017) and others quite high rates (9 out of 11 litters received female alloparental care: Paul and Bhadra 2017; see also Paul and Bhadra 2018). Moreover, it seems that in those populations where the information was available, alloparental care does stem from related females, which would suggest a mix of cooperative and communal breeding in dogs.

4.3.4.4 Dispersal and Mortality

By the age of 4 months, pups start moving around outside their usual range (Pal 2017) and seem to either establish their own groups in vacant territories or likely join other packs peacefully (Macdonald and Carr 2016). However, only two studies by the same author (Pal et al. 1998b) (Pal 2017) investigated the dispersal behaviour of FRDs. Results showed approximately 30% of the monitored juvenile and adult dogs 'dispersed' during late monsoon season (coinciding with the peak estrous season). In contrast to wolves that often disperse over long distances, the mean (±S.D.) dispersal distance of the dogs was 1.702 (±0.425) km. Of these dispersing animals, 67% were under a year old, and specifically, although some individuals started to disperse at 6 months, the majority were between 8 and 12 months of age. The majority of dispersers (adults and juveniles together) were males (67%). This sex-specific dispersal might be linked to the breeding system with related females denning together and helping each other raise their offspring—see discussion above on communal breeding (Paul and Bhadra 2018). However, in this population of dogs studied by Pal, alloparental care was infrequent.

Mortality rates are reported to be high by most studies. The mortality rate of pups below 3 months has been reported to be around 60% across several populations (Pal 2001, 2005, 2017), and approximately 80% of pups die within the first 8–12 months of life (Pal 2017; Paul et al. 2016). In all the studies on this Indian population, the main cause of death was human-related (e.g. removal of pups by children, poisoning, car accidents). However, Pal also reports the infanticide of pups by adult dogs from another group (Pal 2005, 2017). In a study of dogs living in a mountainous area of Italy, a similar mortality rate of 84% (survival until 20 weeks) was reported (Macdonald and Carr 2016). Survival to 1 year of age was reported to be higher in the FRDs studied in the outskirts of Rome, with 50% survival rate (Cafazzo, personal communication). Not surprisingly, high-ranking females enjoyed higher reproductive success (measured as number of pups survived until 8 months of age) than lower-ranking females (Cafazzo et al. 2014), which could be due to their better access to food resources or the protection of the pups by other group members: in Italy, the higher-ranking females have been observed to stay more often in the centre of the territory and benefited more from parental care (see discussion above).

4.3.5 Social Dynamics

As in the chapter relating to wolf social dynamics (Chap. 3), here, we focus on the between-group relationships in dogs, since these can be studied only in free-ranging animals, whereas the within-group dynamics (of both wolves and dogs) will be the focus of the next chapter (Chap. 5).

4.3.5.1 Home Range and Between-Pack Relationships

FRD packs establish home ranges, which have been shown to considerably vary in size. Home ranges of Indian FRDs are reported to vary in size between 0.04 and 0.07 km^2 (Pal 2015), with mean home ranges around 0.06/0.07 km^2 (Pal 2003b). Size of home range in these studies is similar to those described for FRDs living in the USA in Baltimore (Beck 1973, 1975b); Rubin and Beck (1982). However, in a study of a pack of 27 FRDs on the outskirts of Rome, group territory size was considerably larger measuring 0.22 km^2 (Cafazzo et al. 2012). In the Indian population, home range size varied significantly throughout the year, with a maximum size in late monsoon when the females were in estrus and then in winter when the females were lactating (Pal 2003b, 2015; Pal et al. 1998a). Seasonal (summer/winter) variability in home range size was also found in another study in Mexico (Daniels and Bekoff 1989). The seasonal variability as well as the differences between populations in home range size are likely connected to the availability and abundance of food resources. This also explains the much smaller home ranges of dogs compared to wolves (33–4000 km^2). Interestingly, however, the mating season also seems to influence the ranging patterns of dogs at least in one location, which might be explained by females moving around more to look for suitable partners (see discussion above).

FRDs defend their territories from other packs and appear to use scent marking in territory maintenance (Boitani and Ciucci 1995; Boitani et al. 1995; Macdonald and Carr 1995). Indeed, in Indian FRDs, Pal (2003b) reports that 20% of all urine marking behaviours observed were associated with territorial boundaries, which were marked more in the late monsoon season and more by males than females. Furthermore, urine marking was frequently observed in association with territorial defence (Pal 2003b). Urine marking in the pack of dogs studied on the outskirts of Rome was also observed to occur more often on territorial boundaries and was carried out especially by the male members of the pack, and similarly to results above, urine marking was associated with agonistic encounters between packs (Cafazzo et al. 2012).

Fig. 4.6 Between-group aggression in Thailand (pictures taken by Friederike Range)

Between-group aggression has also been observed in FRDs, though it does not seem to be as devastating as in wolves, where it accounts for a high percentage of natural mortality at least in some populations (see Chap. 3). In the Indian population, intergroup aggressive interactions were more frequent during the late monsoon months, when females were in estrous. In accordance with their respective socio-ecological strategies, male dogs were observed to perform more agonistic behaviours in mating contexts and at the boundaries of their territories, whereas female dogs displayed more agonistic behaviours in feeding contexts and in the vicinity of the den. The rate of 'collective' intergroup aggression (i.e. involving at least two pack members) was higher (63%) than the rate of 'individual' (37%) intergroup aggression, suggesting that within-pack members cooperated in the defence of their territory/resources (Pal 2015) (Fig. 4.6).

Intergroup aggression between established groups of FRDs was also reported in the Abruzzo region in central Italy (Macdonald and Carr 2016). In a 3-year study, the researchers witnessed the arrival of new dogs, which in most cases disappeared only days after being sighted. However, in a few cases, dogs were accepted in existing packs/pairs ($N = 4$), and in one confirmed and one suspected case, dogs were killed by resident packs. Similarly, joint intergroup aggression was also shown in two studies involving three packs (comprising 11, 10, and 4 individuals, including adults and juveniles) of FRDs living on the outskirts of Rome (Bonanni et al. 2011; Bonanni et al. 2010). Results showed that the decision whether to engage in or withdraw from an agonistic encounter was based on the number of individuals in ones' own and the opposing pack. The smaller the ratio between own and opponent pack size, the greater the likelihood of at least one pack member engaging in an agonistic interaction. Conversely, the higher the likelihood of a retreat of more than 50% of the pack, the larger the ratio between packs (Bonanni et al. 2011). The decision to participate in the agonistic encounter was also affected by a number of other 'within-group' variables. Overall the proportion of group members cooperating in the inter-pack encounters (i.e. showing agonistic/aggressive behaviours towards members of the other pack) decreased significantly with the increase in the number of group members present. Members of the smallest pack observed showed higher cooperative participation in intergroup conflicts than the larger packs. At the individual level, participation in the conflict encounter was more likely, the higher the number of affiliative partners present (Bonanni et al. 2010).

4.4 Summary

As outlined in the present chapter, there is some debate regarding the sociality of dogs. The discordant reports, going from solitary to structured packs of 27 individuals, likely reflect the flexibility and adaptability of this species as well as possible genetic differences between different populations of FRDs. This high variation in sociality begs the question of which socio-ecological factors, such as food distribution and abundance, interdependence, closeness to humans, etc., may be affecting the dogs' social structure. Where packs do form, the internal structure remains still scarcely understood. It appears that, in most cases, groups are composed of multiple males and females, but mostly due to the absence of genetic data and/or long-term studies, it remains unclear whether the pack members are related with each other or not. Both males and females are largely promiscuous, and parental care seems to be mostly provided by the mother, with varying support from putative fathers and other (related?) females. Dog packs collectively defend established territories, marking and engaging in defensive aggression against other packs.

References

Alberts SC, Watts HE, Altmann J (2003) Queuing and queue-jumping: long-term patterns of reproductive skew in male savannah baboons, *Papio cynocephalus*. Anim Behav 65:821–840. https://doi.org/10.1006/anbe.2003.2106

Altmann SA (1962) A field study of the sociobiology of the rhesus monkey, *Macaca mulatta*. Ann N Y Acad Sci 102:338–435. https://doi.org/10.1111/j.1749-6632.1962.tb13650.x

Arendt M, Cairns KM, Ballard JWO, Savolainen P, Axelsson E (2016) Diet adaptation in dog reflects spread of prehistoric agriculture. Heredity 117:301–306. https://doi.org/10.1038/hdy.2016.48

Atickem A, Bekele A, Williams SD (2009) Competition between domestic dogs and Ethiopian wolf (*Canis simensis*) in the Bale Mountains National Park, Ethiopia: competition between domestic dogs and Ethiopian wolf. Afr J Ecol 48:401–407. https://doi.org/10.1111/j.1365-2028.2009.01126.x

Atuman YJ, Ogunkoya AB, Adawa DAY, Nok AJ, Biallah MB (2014) Dog ecology, dog bites and rabies vaccination rates in Bauchi State, Nigeria. Int J Vet Sci Med 2:41–45. https://doi.org/10.1016/j.ijvsm.2014.04.001

Beach FA, LeBoeuf BJ (1967) Coital behaviour in dogs. I. Preferential mating in the bitch. Anim Behav 15:546–558. https://doi.org/10.1016/0003-3472(67)90057-7

Beck AM (1973) The ecology of stray dogs: a study of free-ranging urban animals. York Press, Baltimore, MD

Beck AM (1975a) Behavior of dogs: canid behavior in a natural setting. In: Price EO, Stokes AW (eds) Animal behavior in laboratory and field. W.H. Freeman and Co., San Francisco, pp 125–127

Beck AM (1975b) The ecology of 'feral' and free-roving dogs in Baltimore. In: Fox MW (ed) The wild canids: Their systematics, behavioural ecology and evolution. Van Nostrand-Reinhold, New York, pp 380–390

Beck AM, Loring H, Lockwood R (1975) The ecology of dog bite injury in St. Louis, Missouri. Public Health Rep 90:262–267

Beetz A, Uvnäs-Moberg K, Julius H, Kotrschal K (2012) Psychosocial and psychophysiological effects of human-animal interactions: the possible role of oxytocin. Front Psych 3:1–15. https://doi.org/10.3389/fpsyg.2012.00234

Beran GW (1982) Ecology of dogs in the Central Philippines in relation to rabies control efforts. Comp Immunol Microbiol Infect Dis 5:265–270. https://doi.org/10.1016/0147-9571(82)90047-9

Berman M, Dunbar I (1983) The social behaviour of free-ranging suburban dogs. Appl Animal Ethol 10:5–17

Boitani L, Ciucci P (1995) Comparative social ecology of feral dogs and wolves. Ethol Ecol Evol 7:49–72. https://doi.org/10.1080/08927014.1995.9522969

Boitani L, Ciucci P, Ortolani A (2007) Behaviour and social ecology of free-ranging dogs. In: Jensen P (ed) The behavioural biology of dogs. CAB International, Wallingford (UK), pp 147–165

Boitani L, Francisci F, Ciucci P (1995) Population biology and ecology of feral dogs in central Italy. In: Serpell J (ed) The domestic dog: its evolution, behaviour and interactions with people. Cambridge University Press, Cambridge, pp 218–245

Bonanni R, Cafazzo S (2014) The social organisation of a population of free-ranging dogs in a suburban area of Rome: a reassessment of the effects of domestication on dogs' behaviour. In: Kaminski J, Marshall-Pescini S (eds) The social dog. Behaviour and cognition. Elsevier, pp 65–104. https://doi.org/10.1016/B978-0-12-407818-5.00003-6

Bonanni R, Natoli E, Cafazzo S, Valsecchi P (2011) Free-ranging dogs assess the quantity of opponents in intergroup conflicts. Anim Cogn 14:103–115. https://doi.org/10.1007/s10071-010-0348-3

Bonanni R, Valsecchi P, Natoli E (2010) Pattern of individual participation and cheating in conflicts between groups of free-ranging dogs. Anim Behav 79:957–968. https://doi.org/10.1016/j.anbehav.2010.01.016

Boyko AR et al (2009) Complex population structure in African village dogs and its implications for inferring dog domestication history. Proc Natl Acad Sci U S A 106:13903–13908. https://doi.org/10.1073/pnas.0902129106

Boyko RH, Boyko AR (2013) Dog conservation and the population genetic structure of dogs. In: Gompper ME (ed) Free-ranging dogs and wildlife conservation. Oxford University Press, pp 185–210

Butler JRA (1998) The ecology of domestic dogs *Canis familiaris* in the communal lands of Zimbabwe. Dissertation. University of Zimbabwe

Butler JRA, Bingham J (2000) Demography and dog-human relationships of the dog population in Zimbabwean communal lands. Vet Rec 147:442–446. https://doi.org/10.1136/vr.147.16.442

Butler JRA, Brown WY, du Toit JT (2018) Anthropogenic food subsidy to a commensal carnivore: the value and supply of human faeces in the diet of free-ranging dogs. Animals 8:67. https://doi.org/10.3390/ani8050067

Butler JRA, du Toit JT (2002) Diet of free-ranging domestic dogs (*Canis familiaris*) in rural Zimbabwe: implications for wild scavengers on the periphery of wildlife reserves. Anim Conserv 5:29–37. https://doi.org/10.1017/S136794300200104X

Butler JRA, du Toit JT, Bingham J (2004) Free-ranging domestic dogs (*Canis familiaris*) as predators and prey in rural Zimbabwe: threats of competition and disease to large wild carnivores. Biol Conserv 115:369–378. https://doi.org/10.1016/S0006-3207(03)00152-6

Cafazzo S, Bonanni R, Valsecchi P, Natoli E (2014) Social variables affecting mate preferences, copulation and reproductive outcome in a pack of free-ranging dogs. PLoS One 9:e98594. https://doi.org/10.1371/journal.pone.0098594

Cafazzo S, Natoli E, Valsecchi P (2012) Scent-marking behaviour in a pack of free-ranging domestic dogs. Ethology 118:955–966. https://doi.org/10.1111/j.1439-0310.2012.02088.x
Cafazzo S, Valsecchi P, Bonanni R, Natoli E (2010) Dominance in relation to age, sex and competitive contexts in a group of free-ranging domestic dogs. Behav Ecol 21:443–455. https://doi.org/10.1093/beheco/arq001
Campos CB, Esteves CF, Ferraz KMPMB, Crawshaw PG, Verdade LM (2007) Diet of free-ranging cats and dogs in a suburban and rural environment, south-eastern Brazil. J Zool 273:14–20. https://doi.org/10.1111/j.1469-7998.2007.00291.x
Chawla SK, Reece JF (2002) Timing of oestrus and reproductive behaviour in Indian street dogs. Vet Rec 150:450–451. https://doi.org/10.1136/vr.150.14.450
Cleaveland S, Kaare M, Knobel D, Laurenson MK (2006) Canine vaccination–Providing broader benefits for disease control. Vet Microbiol 117:43–50. https://doi.org/10.1016/j.vetmic.2006.04.009
Clutton-Brock TH, Parker GA (1992) Potential reproductive rates and the operation of sexual selection. Q Rev Biol 67:437–456. https://www.jstor.org/stable/2832015
Coppinger R, Coppinger L (2016) What is a dog? The University of Chicago Press, Chicago
Cutt H, Giles-Corti B, Knuiman M, Burke V (2007) Dog ownership, health and physical activity: a critical review of the literature. Health Place 13:261–272. https://doi.org/10.1016/j.healthplace.2006.01.003
Daniels TJ (1983) The social organization of free-ranging urban dogs. II. Estrous groups and the mating system. Appl Animal Ethol 10:365–373. https://doi.org/10.1016/0304-3762(83)90185-2
Daniels TJ, Bekoff M (1989) Population and social biology of free-ranging dogs, *Canis familiaris*. J Mammal 70:754–762. https://doi.org/10.2307/1381709
De Balogh KKIM, Wandeler AI, Meslin FX (1993) A dog ecology study in an urban and a semi-rural area of Zambia. Onderstepoort J Vet Res 60:437–443. http://hdl.handle.net/2263/41329
Doykin N, Popova E, Zlatanov V, Petrov P, Zlatanova D (2015) Preliminary data on the distribution of free-ranging dogs (*Canis familiaris* L.) in NP Vitosha, Bulgaria. Paper presented at the Youth Scientific Conference "Kliment's Days", Sofia, Bulgaria,
East ML, Burke T, Wilhelm K, Greig C, Hofer H (2003) Sexual conflicts in spotted hyenas: male and female mating tactics and their reproductive outcome with respect to age, social status and tenure. Proceedings of the Royal Society B-Biological Sciences 270:1247–1254. https://doi.org/10.1098/rspb.2003.2363
Emlen ST, Oring LW (1977) Ecology, sexual selection, and evolution of mating systems. Science 197:215–223. https://doi.org/10.1126/science.327542
Feng Z et al (2014) Dog ownership and physical activity in later life: a cross-sectional observational study. Prev Med 66:101–106. https://doi.org/10.1016/j.ypmed.2014.06.004
Font E (1987) Spacing and social organization: urban stray dogs revisited. Appl Anim Behav Sci 17:319–328. https://doi.org/10.1016/0168-1591(87)90155-9
Fox MW, Beck AM, Blackman E (1975) Behavior and ecology of a small group of urban dogs (*Canis familiaris*). Appl Animal Ethol 1:119–137. https://doi.org/10.1016/0304-3762(75)90082-6
Freedman AH et al (2016) Demographically-based evaluation of genomic regions under selection in domestic dogs. PLoS Genet 12:e1005851. https://doi.org/10.1371/journal.pgen.1005851
Ghosh B, Choudhuri DK, Pal B (1984) Some aspects of the sexual behaviour of stray dogs, *Canis familiaris*. Appl Anim Behav Sci 13:113–127. https://doi.org/10.1016/0168-1591(84)90057-1
Gray PB, Young SM (2011) Human–pet dynamics in cross-cultural perspective. Anthrozoös 24: 17–30. https://doi.org/10.2752/175303711X12923300467285
Hidalgo-Mihart MG, Cant-Salazar L, Lpez-Gonzlez CA, Fernandez EC, Gonzlez-Romero A (2004) Effect of a landfill on the home range and group size of coyotes (*Canis latrans*) in a tropical deciduous forest. J Zool 263:55–63. https://doi.org/10.1017/S0952836904004868
Hoffman CL, Ruiz-Lambides AV, Davila E, Maldonado E, Gerald MS, Maestripieri D (2008) Sex differences in survival costs of reproduction in a promiscuous primate. Behav Ecol Sociobiol 62:1711. https://doi.org/10.1007/s00265-008-0599-z

Home C, Bhatnagar YV, Vanak AT (2018) Canine conundrum: domestic dogs as an invasive species and their impacts on wildlife in India. Anim Conserv 21:275–282. https://doi.org/10.1111/acv.12389

Hughes J, Macdonald DW (2013) A review of the interactions between free-roaming domestic dogs and wildlife. Biol Conserv 157:341–351. https://doi.org/10.1016/j.biocon.2012.07.005

Hyeroba D, Friant S, Acon J, Okwee-Acai J, Goldberg TL (2017) Demography and health of "village dogs" in rural Western Uganda. Prev Vet Med 137:24–27. https://doi.org/10.1016/j.prevetmed.2016.12.009

Kitala P, McDermott J, Kyule M, Gathuma J, Perry B, Wandeler A (2001) Dog ecology and demography information to support the planning of rabies control in Machakos District, Kenya. Acta Trop 78:217–230. https://doi.org/10.1016/S0001-706X(01)00082-1

Knobel DL et al (2005) Re-evaluating the burden of rabies in Africa and Asia. B World Health Organ 83:360–368. https://doi.org/10.1590/S0042-96862005000500012

Koenig A (2002) Competition for resources and its behavioral consequences among female primates. Int J Primatol 23:759–783. https://doi.org/10.1023/A:1015524931226

Koster JM (2008) Hunting with dogs in nicaragua: an optimal foraging approach. Curr Anthropol 49:935–944. https://doi.org/10.1086/592021

Krauze-Gryz D, Gryz J (2014) Free-ranging domestic dogs (*Canis familiaris*) in Central Poland: density, penetration range and diet composition. Pol J Ecol 62:183–193. https://doi.org/10.3161/104.062.0101

Lakestani N, Donaldson ML, Verga M, Waran N (2011) Attitudes of children and adults to dogs in Italy, Spain, and the United Kingdom. J Vet Behav 6:121–129. https://doi.org/10.1016/j.jveb.2010.11.002

Linnell JDC et al (2002) The fear of wolves: a review of wolf attacks on humans, vol 731. NINA, Trondheim

Łomnicki A (2009) Scramble and contest competition, unequal resource allocation, and resource monopolization as determinants of population dynamics. Evol Ecol Res 11:371–380

Longpre KM, Koepfinger ME, Katz LS (2011) Female goats use courtship display as an honest indicator of male quality. Horm Behav 60:505–511. https://doi.org/10.1016/j.yhbeh.2011.07.019

Lukas D, Clutton-Brock TH (2013) The evolution of social monogamy in mammals. Science 341:526–530. https://doi.org/10.1126/science.1238677

Lupo KD (2011) A dog is for hunting. In: Albarella U, Trentacoste A (eds) Ethnozooarchaeology: the present and past of human-animal relationships. Oxbow Press, Oxford, pp 4–12. https://doi.org/10.2307/j.ctvh1dwvg.6

Lupo KD (2017) When and where do dogs improve hunting productivity? The empirical record and some implications for early Upper Paleolithic prey acquisition. J Anthropol Archaeol 47:139–151. https://doi.org/10.1016/j.jaa.2017.05.003

Macdonald DW, Carr GM (1995) Variation in dog society: between resource dispersion and social flux. In: Serpell J (ed) The domestic dog: its evolution, behaviour and interactions with people. Cambridge University Press, pp 199–216

Macdonald DW, Carr GM (2016) Variation in dog society: between resource dispersion and social flux. In: Serpell J (ed) The domestic dog. its evolution, behavior and interactions with people, vol PART IV–LIFE ON THE MARGINS. Cambridge University Press, pp 199–216. https://doi.org/10.1017/9781139161800.016

Macdonald DW, Johnson DDP (2015) Patchwork planet: the resource dispersion hypothesis, society, and the ecology of life. J Zool 295:75–107. https://doi.org/10.1111/jzo.12202

Majumder SS, Chatterjee A, Bhadra A (2014) A dog's day with humans–time activity budget of free-ranging dogs in India. Curr Sci 106:5. https://www.currentscience.ac.in/Volumes/106/06/0874.pdf

Manor R, Saltz D (2004) The impact of free-roaming dogs on gazelle kid/female ratio in a fragmented area. Biol Conserv 119:231–236. https://doi.org/10.1016/j.biocon.2003.11.005

Marino J, Sillero-Zubiri C, Johnson PJ, Macdonald DW (2012) Ecological bases of philopatry and cooperation in Ethiopian wolves. Behav Ecol Sociobiol 66:1005–1015. https://doi.org/10.1007/s00265-012-1348-x

Marshall-Pescini S, Cafazzo S, Virányi Z, Range F (2017) Integrating social ecology in explanations of wolf–dog behavioral differences. Curr Opin Behav Sci 16:80–86. https://doi.org/10.1016/j.cobeha.2017.05.002

McNicholas J, Gilbey A, Rennie A, Ahmedzai S, Dono J-A, Ormerod E (2005) Pet ownership and human health: a brief review of evidence and issues. BMJ 331:1252–1254. https://doi.org/10.1136/bmj.331.7527.1252

Miura A, Bradshaw JWS, Tanida H (2000) Attitudes towards dogs: a study of university students in Japan and the UK. Anthrozoös 13:80–88. https://doi.org/10.2752/089279300786999860

Morters MK, McKinley TJ, Restif O, Conlan AJK, Cleaveland S, Hampson K, Whay HR, Damriyasa IM, Wood JLN (2014) The demography of free-roaming dog populations and applications to disease and population control. J Appl Ecol 51(4):1096–1106

Nesbitt WH (1975) Ecology of a feral dog pack on a wildlife refuge. In: Fox MW (ed) The wild canids: their systematics, behavioral ecology, and evolution. Van Nostrand Reinhold, New York, pp 391–395

Newsome TM, Ballard G-A, Dickman CR, Fleming PJS, van de Ven R (2013) Home range, activity and sociality of a top predator, the dingo: a test of the Resource Dispersion Hypothesis. Ecography 36:914–925. https://doi.org/10.1111/j.1600-0587.2013.00056.x

Nicholson AJ (1954) An outline of the dynamics of animal populations. Aust J Zool 2:9–65. https://doi.org/10.1071/ZO9540009

Oppenheimer EC, Oppenheimer JR (1975) Certain behavioral features in the pariah dog (Canis familiaris) in West Bengal. Appl Animal Ethol 2:81–92. https://doi.org/10.1016/0304-3762(75)90067-X

Ortolani A, Vernooij H, Coppinger R (2009) Ethiopian village dogs: behavioural responses to a stranger's approach. Appl Anim Behav Sci 119:210–218. https://doi.org/10.1016/j.applanim.2009.03.011

Pal SK (1990) Some aspects of ecology and behaviour of urban free-ranging dogs (*Canis familiaris*). University of Burdwan

Pal SK (2001) Population ecology of free-ranging urban dogs in West Bengal, India. Acta Theriol 46:69–78. https://doi.org/10.1007/BF03192418

Pal SK (2003a) Reproductive behaviour of free-ranging rural dogs in West Bengal, India. Acta Theriol 48:271–281. https://doi.org/10.1007/BF03194167

Pal SK (2003b) Urine marking by free-ranging dogs (Canis familiaris) in relation to sex, season, place and posture. Appl Anim Behav Sci 80:45–59. https://doi.org/10.1016/S0168-1591(02)00178-8

Pal SK (2005) Parental care in free-ranging dogs, *Canis familiaris*. Appl Anim Behav Sci 90:31–47. https://doi.org/10.1016/j.applanim.2004.08.002

Pal SK (2008) Maturation and development of social behaviour during early ontogeny in free-ranging dog puppies in West Bengal, India. Appl Anim Behav Sci 111:95–107. https://doi.org/10.1016/j.applanim.2007.05.016

Pal SK (2011) Mating system of free-ranging dogs (*Canis familiaris*). Int J Zool 2011:314216. https://doi.org/10.1155/2011/314216

Pal SK (2015) Factors influencing intergroup agonistic behaviour in free-ranging domestic dogs (*Canis familiaris*). Acta Ethol 18:209–220. https://doi.org/10.1007/s10211-014-0208-2

Pal SK (2017) Free-ranging domestic dogs in India. LAP Lambert Academic Publishing

Pal SK, Ghosh B, Roy S (1998a) Agonistic behaviour of free-ranging dogs (*Canis familiaris*) in relation to season, sex and age. Appl Anim Behav Sci 59:331–348. https://doi.org/10.1016/S0168-1591(98)00108-7

Pal SK, Ghosh B, Roy S (1998b) Dispersal behaviour of free-ranging dogs (*Canis familiaris*) in relation to age, sex, season and dispersal distance. Appl Anim Behav Sci 61:123–132. https://doi.org/10.1016/S0168-1591(98)00185-3

Pal SK, Ghosh B, Roy S (1999) Inter- and intra-sexual behaviour of free-ranging dogs (*Canis familiaris*). Appl Anim Behav Sci 62:267–278. https://doi.org/10.1016/S0168-1591(98)00220-2

Paul M, Bhadra A (2017) Selfish pups: weaning conflict and milk theft in free-ranging dogs. PLoS One 12:e0170590. https://doi.org/10.1371/journal.pone.0170590

Paul M, Bhadra A (2018) The great Indian joint families of free-ranging dogs. PLoS One 13: e0197328. https://doi.org/10.1371/journal.pone.0197328

Paul M, Majumder SS, Bhadra A (2014) Grandmotherly care: a case study in Indian free-ranging dogs. J Ethol 32:75–82. https://doi.org/10.1007/s10164-014-0396-2

Paul M, Sau S, Nandi AK, Bhadra A (2017) Clever mothers balance time and effort in parental care: a study on free-ranging dogs. R Soc Open Sci 4:160583. https://doi.org/10.1098/rsos.160583

Paul M, Sen Majumder S, Sau S, Nandi AK, Bhadra A (2016) High early life mortality in free-ranging dogs is largely influenced by humans. Sci Rep 6:19641. https://doi.org/10.1038/srep19641

Pendleton AL, Shen F, Taravella AM, Emery S, Veeramah KR, Boyko AR, Kidd JM (2018) Comparison of village dog and wolf genomes highlights the role of the neural crest in dog domestication. BMC Biol 16:64. https://doi.org/10.1186/s12915-018-0535-2

Pilot M et al (2015) On the origin of mongrels: evolutionary history of free-breeding dogs in Eurasia. Proc R Soc B 282:20152189. https://doi.org/10.1098/rspb.2015.2189

Rao A, Range F, Kadletz K, Kotrschal K, Marshall-Pescini S (2018) Food preferences of similarly raised and kept captive dogs and wolves. PLoS One 13:e0203165. https://doi.org/10.1371/journal.pone.0203165

Reich D (2018) Who we are and how we got here. Ancient DNA and the new science of the human past. Oxford University Press, Oxford

Ritchie EG, Dickman CR, Letnic M, Vanak AT (2013) Dogs as predators and trophic regulators. In: Gompper ME (ed) Free-ranging dogs and wildlife conservation. Oxford University Press, pp 55–65

Rubin HD, Beck AM (1982) Ecological behavior of free-ranging urban pet dogs. Appl Animal Ethol 8:161–168. https://doi.org/10.1016/0304-3762(82)90141-9

Ruiz-Izaguirre E, van Woersem A, Eilers KHAM, van Wieren SE, Bosch G, van der Zijpp AJ de Boer IJM (2015) Roaming characteristics and feeding practices of village dogs scavenging sea-turtle nests. Anim Conserv 18:146–156. https://doi.org/10.1111/acv.12143

Ruusila V, Pesonen M (2004) Interspecific cooperation in human (*Homo sapiens*) hunting: the benefits of a barking dog (*Canis familiaris*). Ann Zool Fenn 41:545–549. http://www.jstor.org/stable/23735938

Scott JP, Fuller JL (1965) Genetics and the social behavior of the dog. The classic study. University of Chicago Press, Chicago, London

Sen Majumder S, Bhadra A (2015) When love is in the air: understanding why dogs tend to mate when it rains. PLoS One 10:e0143501. https://doi.org/10.1371/journal.pone.0143501

Serpell J (2016) The domestic dog: its evolution, behaviour and interactions with people, 2nd edn. Cambridge University Press

Shannon LM et al (2015) Genetic structure in village dogs reveals a Central Asian domestication origin. Proc Natl Acad Sci U S A 112:13639–13644. https://doi.org/10.1073/pnas.1516215112

Sillero-Zubiri C, Gottelli D (1995) Diet and feeding behavior of Ethiopian Wolves (*Canis simensis*). J Mammal 76:531–541. https://doi.org/10.2307/1382361

Sterck EHM, Watts DP, van Schaik CP (1997) The evolution of female social relationships in nonhuman primates. Behav Ecol Sociobiol 41:291–309. https://doi.org/10.1007/s002650050390

Tang X, Luo M, Zhang S, Fooks AR, Hu R, Tu C (2005) Pivotal role of dogs in rabies transmission, China. Emerg Infect Dis 11:1970–1972. https://doi.org/10.3201/eid1112.050271

Totton SC et al (2010) Assessing reproductive patterns and disorders in free-ranging dogs in Jodhpur, India to optimize a population control program. Theriogenology 74:1115–1120. https://doi.org/10.1016/j.theriogenology.2010.05.008

van Schaik CP (1989) The ecology of social relationships amongst female primates. In: Standen V, Foley RA (eds) Comparative socioecologie. The behavioural ecology of humans and other mammals. Blackwell Scientific Publications, Oxford, pp 195–218
Vanak AT, Gompper ME (2009) Dogs *Canis familiaris* as carnivores: their role and function in intraguild competition. Mammal Rev 39:265–283. https://doi.org/10.1111/j.1365-2907.2009.00148.x
Wandeler AI, Matter HC, Kappeler A, Budde A (1993) The ecology of dogs and canine rabies: a selective review. Revue Scientifique et Technique–Office international des épizooties 12:51–71. https://doi.org/10.20506/rst.12.1.663
Wang G-D et al (2014) Genetic convergence in the adaptation of dogs and humans to the high-altitude environment of the Tibetan Plateau. Genome Biol Evol 6:2122–2128. https://doi.org/10.1093/gbe/evu162
Ward A, Webster M (2016) Sociality: the behaviour of group-living animals. Springer. https://doi.org/10.1007/978-3-319-28585-6
WHO (World Health Organization) (2011) Rabies. Factsheet. 99. https://www.who.int/news-room/fact-sheets/detail/rabies
Young JK, Olson KA, Reading RP, Amgalanbaatar S, Berger J (2011) Is wildlife going to the dogs? Impacts of feral and free-roaming dogs on wildlife populations. BioSci 61:125–132. https://doi.org/10.1525/bio.2011.61.2.7

Social Dynamics Between Conspecifics 5

Sociality has always been a focus of interest for ethologists, in particular since even a cursory observation of different species can highlight the vast diversity in social organization and structures. In general, social groups can be classified along a continuum of gregariousness, from solitary individuals to unstable/fluctuating aggregations to stable and enduring social structures. Much of this is not 'species-specific' but depends on the ecological context. Sociality is particularly fitness-relevant in species depending on cooperative hunting and offspring rearing. Nevertheless, being social means balancing benefits and costs, for example, in competing with group members over coveted resources (food, denning, mating opportunities), yet at the same time maintaining positive, long-term valuable relationships. In fact, social dynamics between group members have thus mostly been studied with this challenge in mind.

As outlined in Chaps. 3 and 4, the process of domestication has altered the dogs' socio-ecological niche considerably, with a shift from wolves' high dependence on the pack for hunting and rearing of pups to a reliance on mostly itemized foods obtained solitarily and a system where pups are mostly mother reared and independently foraging already from a young age onwards. These observations would go in the direction of a decreased dependence on social cohesiveness in dogs, with potential consequences in terms of the social strategies employed within the group. However, the shift in the dogs' social ecology has also resulted in a potential for much higher density of dogs as compared to wolves. Living in small territories—if at all—may in turn have affected dog between-group social dynamics (if groups are formed).

In the present chapter, we focus on determining components of within-group dynamics including (1) dominance and aggression (2) affiliation, (3) play, and (4) 'conflict management'. For each topic, we briefly introduce the concepts, frameworks, and methods developed to investigate them, summarize what is known about these topics from studies of captive and free-living wolves and dogs, and present studies directly aimed at comparing these different aspects of wolves' and dogs' social behaviour. Finally, we summarize studies investigating how

F. Range, S. Marshall-Pescini, *Wolves and Dogs*, Fascinating Life Sciences,
https://doi.org/10.1007/978-3-030-98411-3_5

affiliative and dominance relationships may similarly (or differently) affect dogs' and wolves' behaviours in a number of different tests conducted at the Wolf Science Center where animals were confronted with novel, potentially 'dangerous' situations.

5.1 Dominance and Aggression

Dominance hierarchies emerge in virtually all group-living animals, including mammals (e.g. primates: Hausfater et al. 1982; carnivores: Holekamp and Smale 1991; ungulates: Clutton-Brock et al. 1986; Fairbanks 1994; and elephants: Archie et al. 2006; Moss and Poole 1983); they are characterized by asymmetric relationships between individuals, in which higher-ranking animals win in agonistic encounters and can consistently displace lower-ranking animals around resources. As discussed in Chap. 4, the steepness of hierarchies seems to scale with the distribution and thus defensibility of resources, which determines within-group competition (Isbell and Young 2002; Sterck et al. 1997; van Schaik 1989; Vehrencamp 1983; Wheeler et al. 2013; Wrangham 1980). If competition is mainly by contest, high-ranking animals have priority of access to critical resources such as food, mates, or high-quality habitat (Dunbar 1988; van Schaik 1983) and often experience fitness benefits such as enhanced reproductive success or higher offspring survival (e.g. red deer: Clutton-Brock et al. 1984; yellow baboons: Altmann and Alberts 2003; wild dogs, *Lycaon pictus*: Creel et al. 1997; spotted hyenas: Holekamp et al. 1996; and chimpanzees: Pusey et al. 1997). Moreover, high competition with neighbouring packs might partly inhibit competition within packs to allow for cooperative territorial defence. Finally, males and females rely on different key resources, which, depending on the ecological conditions, may lead to more or less conflicts between the sexes. Accordingly, differences and similarities observed between populations of the same, but also across species, might be largely driven by both past and present, ecological conditions.

At the physiological level, the proximate mechanisms involved are less clear; higher-ranking individuals show lower basal cortisol responses than lower ranked individuals when the group hierarchy is stable, but the opposite is true in situations of group instability and the links between physiological measures of stress and social status appear to be dependant also on other factors such as the breeding season and the overall frequencies of agonistic interactions in the group (Creel et al. 2013; Kaplan et al. 1982; Kotrschal et al. 1998; Sapolsky 1983, 2005; Vasconcellos et al. 2012).

Social dominance can be maintained through repeated agonistic interactions (agonistic dominance), but in stable social settings, it is rather maintained through species-specific ritualized signals (*formal dominance*), thereby avoiding the need for repeated conflicts (de Waal 1986). To be considered a formal indicator of status, a behaviour should (1) be unidirectional (i.e. be expressed more/only from one individual to the other and not vice versa); (2) show high coverage, i.e. a behaviour which is frequently observed among members of the group; (3) remain stable across different social contexts; (4) correlate, at least to some degree, with agonistic dominance relationships based on aggression; and (5) communicate relative status rather than overt aggression (i.e. it should occur in the absence of aggression) (de Waal 1986; de Waal and Luttrell 1985; Preuschoft 1999; Vervaecke et al. 2000b). Several findings suggest that rank positions can be rapidly identified through asymmetrical displays of formal dominance between individuals. For instance, a non-verbal cue as simple as gaze aversion has been shown to be used to indicate status roles among monkeys meeting for the first time (Deaner et al. 2005). Formal dominance is considered to occur mainly in societies characterized by high levels of both competition and cooperation/affiliation (de Waal 1986; East et al. 1993; Lu et al. 2008; Vervaecke et al. 2000a).

In operational terms, the dominance relationship between two individuals is calculated by analysing the extent to which individual A shows formal signals of dominance (and submission) to B and B to A. A is said to be dominant over B, if it frequently exhibits signals related to dominance and rarely or never signals of submission; conversely, B would show many displays of submission and no or rare signals of dominance. Importantly 'dominance' is not an individual trait but a reflection of an asymmetrical and dynamic relationship between two individuals that could vary over time and depend on the context.

At the group level, the dominance hierarchy can be characterized in terms of its *linearity, directional consistency index (DCI), and steepness*. The linearity of a hierarchy was initially proposed to give a measure of how consistently 'linear' the relationship between a group of individuals is, in other words, whether the animals can be arranged in such a way that if A is dominant over B, and B over C, then A is also dominant over C. The modified linearity index (h') ranges from 0 (indicating total lack of linearity) to 1 (perfect linearity), and its value is lowered by the presence of circular triads (relationships in which individual A dominates B, B dominates C, and C dominates A) (de Vries 1995). However, re-assessment of the linearity index has highlighted that it may rather represent a proxy of the quality and completeness of the data set. Indeed, this index (as also other measures below) is highly sensitive to missing data points, i.e. where only very few or no interactions between two individuals in a group were observed; where complete data sets are found, linearity is usually achieved (Klass and Cords 2011; Neumann et al. 2018).

The DCI gives the frequency with which a behaviour (e.g. dominant or submissive) occurred in its more frequent direction within each dyad relative to the total number of times the behaviour occurred (Vervaecke et al. 2000a). This index varies from 0 (complete bidirectionally) to 1 (complete unidirectionality) (van Hooff and Wensing 1987).

The steepness of a dominance hierarchy measures the degree to which individuals differ from each other in winning dominance encounters. If differences are large, the hierarchy is steep; if they are small, the hierarchy is shallow (de Vries et al. 2006). Linearity is based on the ordinal dyadic dominance relationships, whereas steepness requires a cardinal rank measure (Flack and de Waal 2004). The steepness of a pack's hierarchy is thought to be a more sensitive index than the pack's linearity, the reason being that it is calculated using each individual's David's score (nDS) (de Vries et al. 2006), i.e. the distance between each pack member considering each pack member's strength relative to all other members, rather than just its strength at the dyadic level. The nDS is also thought to better account for differences in pack size. Linearity and steepness are considered complementary measures which can more fully characterize the groups' dominance hierarchy (de Vries et al. 2006).

5.1.1 Formal Signals of Dominance in Wolves and Dogs

In both wolves and dogs, dominance relationships and resulting linear hierarchies have been observed (see section below for details). These dominance relationships seem to be mostly maintained through formal signals rather than aggressive interactions, and indeed such formal signals of dominance/submission have been described in both wolves and dogs.

In wolves such signals consist mostly of postural displays (viz. high posture, i.e. a rigid position of the body, the tail, and head held high as a signal of dominance, and low posture, i.e. crouching of the hindlegs, ears held back, head dipped, or lying on the back as signals of submission; see Fig. 5.1) (Schenkel 1967; Zimen 1982). Since these postures are exhibited consistently in a unidirectional manner from one partner to the other (and not vice versa), they have been suggested as indicators of formal dominance (van Hooff and Wensing 1987). In a study looking at more discrete body signals (such as tail and ear position and different expressions of the mouth) exhibited by wolves in social interactions, researchers found that the tail position seemed to be the most reliable indicator of social status, in that, taken alone, a tail held in a horizontal or above the horizontal position was a good predictor of higher dominance status, whereas a lowered tail consistently reflected submission (Fatjó et al. 2007; Harrington and Asa 2003). Although initial descriptions of hierarchical behaviours in wolves often included both aggressive behaviours (both threats and also attacks/biting) and the high postures described above (Jenks and Ginsburg 1987), researchers specifically investigating the use of these different signals have found that, in fact, aggressive behaviours are inadequate as indicators of social rank

Fig. 5.1 (**a**) Active submission and (**b**) passive submission in wolves (picture B: Walter Vorbeck)

Fig. 5.2 Tala is greeting Kaspar (photograph by Joachim Henckmann)

because they are more likely to occur in both directions within a dyad and may be elicited by different motivations (such as the more or less strong desire to obtain a resource—see below) (Cafazzo et al. 2016; van Hooff and Wensing 1987).

In a study we conducted together with Simona Cafazzo and Martina Lazzaroni, analysing the hierarchical structure of a family pack of 19 captive Arctic wolves, we found that although both high posture and low postures (submissive behaviours) revealed a clear linear hierarchy, the latter fulfilled the criteria of formal indicators better since they showed a higher linearity, a complete transitivity, and a rank order with no inconsistencies (Cafazzo et al. 2016). However, we need to interpret these results with caution for a number of reasons: (1) despite the captive settings, for a considerable number of dyads, we observed very few or no interactions, reducing our confidence in the results; (2) we saw a higher frequency of dominance than submissive behaviours (about a third more), resulting in a more complete data set for the one compared to the other; this further cautions us in relation to the conclusion in regard to the best indexes; and (3) we found some indications of potential sex differences, which are not fully accounted for in this analyses (due to the limitations of the data set). Greeting behaviours (i.e. mouth licking associated with tail-wagging—see Fig. 5.2), although displayed only by a subset of individuals, were consistently shown from subordinate to dominant individuals (except in the top-ranking breeding pair, where they were exhibited reciprocally by both partners), which would suggest they qualified as good indicators of formal dominance (Cafazzo et al. 2016).

In domestic dogs, signals of dominance and submission that have been used are largely the same as those described for wolves (Zimen 1970; Zimen 1971). However, often in the ethogram used by dog researchers, behaviours recorded are more discrete (e.g. muzzle-licking, tail-wagging, gaze aversion, crouch), whereas the wolf ethogram often relies on so-called behaviour complexes which describe the simultaneous occurrence of multiple discrete signals (e.g. active submission involves muzzle-licking in combination with a lowered posture). Nevertheless, similarly to studies on wolves, in both free-ranging and captive dogs, submissive behaviours have been shown to be better indicators of hierarchical rank than dominant or aggressive behaviours (Cafazzo et al. 2010; Van Der Borg et al. 2015). Furthermore, muzzle-licking or 'greeting' has repeatedly emerged as one of the best indicators of formal dominance in both free-ranging dog packs and groups of pet dogs (Cafazzo et al. 2010; Trisko and Smuts 2015; Van Der Borg et al. 2015), as well as tail-wagging (analysed only in pet dogs: Van Der Borg et al. 2015). As for wolves, also in dogs, aggressive behaviours are the least reliable indicators of dominance relationships, since they often occur bidirectionally (Cafazzo et al. 2010; Van Der Borg et al. 2015).

5.1.2 Dominance, Hierarchy, and Leadership in Captive and Free-Living Wolves and Dogs

Although dominance hierarchies have been described for a multitude of species and appear to solidly occur in both captive and wild settings (see Shizuka and McDonald David 2015 for an analytical review), more recently, there have been some controversies regarding the applicability of this concept to wolves (Mech 1999) and dogs (Bradshaw and Nott 1995; Bradshaw et al. 2009; O'Heare 2007; Schilder et al. 2014; Shepherd 2002; van Kerkhove 2004), albeit for different reasons.

Dominance relationships in wolf packs have invariably been described by researchers focusing on the social organization of this species (Jenks and Ginsburg 1987; Schenkel 1967; Schotté and Ginsberg 1987; van Hooff and Wensing 1987; Zimen 1982); where a more in-depth analysis has been made, they are normally considered to follow an 'age-graded model' (Packard 2003; Zimen 1982, pp. 54) presented either simply as separate linear hierarchies for each sex (Schenkel 1967; Zimen 1982) or as male dominance over females within each age class (Rabb et al. 1967; van Hooff and Wensing 1987; Zimen 1982). But as described in Chap. 3, the typical wild wolf social structure is based on a bonded male–female pair, with the tenure of the breeders usually lasting between 3 and 4 years (range of 1–10 years) (Mech et al. 1998). The offspring may forego dispersal for 2 to 3 years and remain with their native pack helping to raise later litters and resulting in a pack formed by siblings of various ages as well as the bonded pair (Mech 1999, 2000; Packard 2003). Thus, the 'age-graded' dominance hierarchy described in wolves within the typical 'nuclear' family pack could be alternatively viewed as just a reflection of the age, sex, and reproductive structure of the group (Mech 1999). Based on these facts, Mech (1999) suggested that one should refer to the breeding animals as the parents of the pack rather than alpha male and alpha female. Since any parent is dominant to its young offspring, alpha emphasizes the wrong information, while breeding pair denotes their role as pack progenitors (Mech 1999). However, given that packs can be composed of multiple siblings from several generations (up to 19–26 individuals (Bloch 2002; Smith et al. 2012; VonHoldt et al. 2008), the question still remains, if indeed the relationship between siblings is characterized by a linear hierarchical organization.

Furthermore, when watching a pack of wolves in a captive setting at least, it is quite striking to see how they continuously demonstrate their rank during social interaction. While the dominant wolves stand tall with their tail up or at least held horizontal, subordinate or submissive individuals might either show active submission by approaching another wolf excitedly, wagging the tail, lowering the ears, and licking the mouth of the other wolf or by displaying passive submission, where the submissive wolf rolls over on its side or back and the dominant wolf sniffs its groin or genitals. Active submission is thought to be derived from food begging, and the two behaviours cannot be distinguished (Mech 1999). The submissive postures are often displayed without provocation from the dominant animal and rather appear to promote friendly relations and/or reduce social distance. Thus, it helps if one does not think of dominance as the outcome of aggression but rather thinks of submissive behaviours including greeting being a display of respect and acknowledgement from certain individuals of the pack towards others—not unlike how we acknowledge the position and role of our boss or our parents without them needing to show aggression towards us.

One problem so far has been that although numerous studies have described a hierarchical organization in wild wolf packs (e.g. Baan et al. 2014; Bloch 2002; Mech 1999; Peterson et al. 2002; Sands and Creel 2004), a quantitative assessment and statistical validation of the existence of linear dominance relationships were not possible due to the limited possibility of collecting data on the interaction between individuals. In captivity, interactions between individuals, particularly the aggressive components, seem to be higher in comparison to wolf packs in the wild. However, studied packs mostly consisted of either 'complex families', i.e. composed of artificially assembled unrelated individuals (e.g. Packard 2003 for a review), or 'disrupted families', i.e. in which one or both of the original parents were missing (e.g. Cordoni and Palagi 2008; Packard 2003 for a review). In the two cases, where 'nuclear families' (with parents and multiple generations of offspring) and 'extended families' (consisting of parents plus one of more of their siblings and their direct offspring) were studied, a linear hierarchy was observed (Romero et al. 2014; van Hooff and Wensing 1987).

To explore this issue a bit further, we conducted a study on a captive family pack of 19 Arctic wolves. The pack was composed of two unrelated females (who in some, but not all, years both had pups), an unrelated male (the breeding male), and multiple generations of their offspring. Although the daughters of the two original females occasionally gave birth to pups over the years, these were often killed by the two highest ranking females or removed by the zookeepers. The year we observed the pack, the dominant female did not give birth, but two of her daughters did. However, the dominant female killed all the pups, when they were still in the den. The social structure of these wolves thus more closely reflects the wolf packs in the wild. In this study, we found that the hierarchical organization followed the age-graded model, with older siblings being more dominant than younger ones, but we did not find support for males being dominant over females in each age group (Cafazzo et al. 2016). However, one needs to keep in mind that this 'family' group was rather large in comparison to most packs in the wild and also that Arctic wolves live under different ecological conditions than other wolves, which may influence their interactions with each other as well.

To further investigate the relationship between siblings, together with Martina Lazzaroni and Martina Russignan, as well as enormous on-site support from Massimo Colombari, we initiated an observational study of wild wolves in the Italian Apennines. Over the course of 5 years, we were able to observe seven litters and collect sufficient data on social interactions between littermates in at least four litters to allow for an in-depth analysis of their rank relationships. All social interactions between littermates were observed at the rendezvous site from an estimated age of 5 to 8 months and mostly in the absence of the adult members of the pack. Preliminary results based on just two litters show that a clear hierarchical relationship can be detected between siblings, with males holding the higher rank positions.

Traditionally, the concept of dominance was also thought to be particularly relevant for dogs (Beach 1970; Pal et al. 1998). However, more recently, the validity of the concept to describe social relationships between dogs was called into question (Bradshaw and Nott 1995; Bradshaw et al. 2009; O'Heare 2007; Schilder et al. 2014; Shepherd 2002; van Kerkhove 2004; see Smuts 2014 for a review). In particular, Bradshaw et al. (2009) reported some asymmetrical agonistic relationships but no linear hierarchy in a group of 19 neutered male shelter dogs and concluded that dominance is not a useful construct for describing social relationships among domestic dogs. However, the fact that the animals were all male and neutered and that no tests required to detect the presence of a linear hierarchy were carried out questions the validity of the conclusion. Furthermore, authors added together different types of behaviours (dominant, submissive, and aggressive), which, as outlined above, may be problematic, since aggression has been repeatedly shown to be a bad indicator of dominance. In contrast to this study, a number of other studies applying the indexes of linearity and unidirectionality of behaviours have found clear linear dominance hierarchies in pet dog groups regularly meeting each other (Bonanni et al. 2017; Trisko and Smuts 2015; Van Der Borg et al. 2015), and this was the case also in a group of 24 neutered pet dogs (Trisko and Smuts 2015).

Reports of a hierarchical organization in free-ranging dogs have also revealed somewhat mixed results. Boitani and colleagues, based on studies of feral dogs in the Italian Apennines, describe their social organization as an aggregation of monogamous pairs with their sub-adult and juvenile offspring, with no stable hierarchical structure (Boitani and Ciucci 1995; Boitani et al. 1995, 2007, 2016). However, a number of other studies based on long-term observations of within-pack relationships have shown a clear hierarchical pattern in free-ranging dogs. For example, Pal et al. (1998) studied two packs of five and eight related individuals of free-ranging dogs in India; a re-analysis of this data to statistically determine the hierarchical structure of the two packs was presented in Bonanni and Cafazzo

(2014), showing that both groups showed a clear linear hierarchy. Cafazzo et al. (2016) studied a population of 27 free-ranging dogs, living on the outskirts of Rome, regularly provisioned by humans. Based on the knowledge of the relationships between individuals in the pack, it was thought to be composed mostly of related individuals (although genetic analyses were unfortunately missing). A clear age-graded dominance hierarchy in the pack emerged, with older individuals being dominant over younger ones. Furthermore, in the adults and sub-adults, males were dominant over females, although this sex-biased hierarchy was not as clear in juveniles. A social network analyses of the same data also revealed that aggressive interactions were more frequent in the central area of the hierarchy (i.e. mostly made up of sub-adults), where the positions were less certain (Silk et al. 2019). A further five packs of free-ranging dogs, in suburban areas of Rome, comprising 6, 11, 9, 15, and 3 individuals, respectively, were observed, and their within-pack agonistic and submissive interactions recorded (Bonanni et al. 2017). Based on the submissive signals shown by the individuals towards each other, a clear linear hierarchy could be established in four cases (the fifth pack did not have enough members to allow for statistical analyses), but there was substantial variability across packs, and a number of issues with this data set call for caution, with the main problem (as in other studies—see above the discussion of Cafazzo et al. 2016) being that there is a high percentage of unknown relationships (or dyads with only one or two interactions) which affect all measures of the hierarchy.

Another aspect closely linked to dominance is leadership. To maintain group cohesiveness and therefore benefit from sociality, animals need to coordinate their actions deciding what activities to perform and when to perform them (Conradt and Roper 2007). This process involves a reaching a 'consensus' of sorts. Leaders (or decision-makers) are those individuals that are more influential in determining the outcome of a decision (e.g. when to stop resting and start travelling, which direction to take while travelling, etc.). In wolves, Peterson et al. (2002) analysed leadership behaviour over 2 years in the Yellowstone National Park. The authors selected five different behaviours thought to indicate leadership: scent marking, being in the front of the pack while travelling, initiation of pack behaviour, and non-frontal leadership (e.g. leading from the back of the pack during travelling). Scent marking, including raised-leg urination, scratching, and double scent marking, was almost exclusively observed in the dominant breeding pair (155 of 158 observed events) (see also Mech 1999; Peterson et al. 2002). During the early winter, both breeding males and females did not differ in their scent marking activity, but during late winter, males scent marked significantly more often than females. Similarly, the dominant breeding females and males led the packs in 64% of the observed travelling bouts and did so more or less equally. The dominant breeding wolves were also more often responsible for initiating pack activities than non-breeders such as arousing the pack from rest, travelling after group rallies (greeting ceremonies), chasing prey, changing direction during travel, or defending the pack from trespassing wolves (70 vs. 25%). In 14 of 15 cases where non-frontal leadership (a wolf not in the lead broke ranks and led the pack in a new direction) was observed, a non-breeder was in the lead of the pack. In 12 of the 15 cases, breeding individuals changed the direction of the pack (Peterson et al. 2002). This observed pattern on leadership is easily explained when keeping the family structure of wolf packs in mind: in most cases, the breeding male and female are the oldest and most experienced animals in the pack and are actually the role models from which the youngsters learn.

A study involving two groups of free-ranging dogs of the Rome population was also carried out to investigate leadership in the context of group departure, i.e. when animals stopped resting and started a new activity (Bonanni et al. 2010a). The authors found that in both groups there were a limited number of habitual leaders, who succeeded in initiating a change of state from resting to moving in a specific direction. Similarly to wolf packs, the habitual leaders were mostly older, higher-ranking individuals who often received spontaneous displays of submission from the rest of the pack. This leadership of experienced individuals is widely spread across the animal kingdom including humans.

Taken together, but with the necessary cautionary notes relating to the small number of studies and the difficulty in obtaining complete and solid data sets for the hierarchical measures, it would appear that, where long-term studies including detailed observations of interactions are recorded, linear hierarchies do emerge, with adults clearly being dominant over sub-adults and juveniles and potentially with a male over female organization. The most experienced and thus usually also most dominant individuals seem to be leading the packs/groups. However, it should be noted that the social organization and thus also the dominance interactions of both wolves and dogs will be defined by the given ecological conditions (see above and Chaps. 3 and 4) and thus variability is expected.

5.1.3 Dominance and Aggression: Wolves and Dogs Compared

Pioneering studies of dogs' and wolves' conspecific social behaviour were conducted by Erik Zimen and Dorit Feddersen-Petersen at the University of Kiel, where packs of animals of both species were closely observed. Starting in the mid-1960s, observations of social interactions of packs of wolves, as well as packs of dogs of different breeds, and even mixed packs of wolves and dogs, were collected (Feddersen-Petersen 1991, 1994, 2004; Zimen 1971, 1982). A number of salient points emerged from these studies.

Zimen (1970) reports that in wolf puppies no rank order was apparent until they were between 6 and 10 months old when the young wolves became part of the adult hierarchy. In adults, two separate hierarchies emerged: one for males and one for females. Interestingly, in males, one higher-ranking individual emerged clearly, whereas the remaining males did not form a linear hierarchy. In females, the hierarchical position was clear also between the subordinate females. Once the hierarchy was established at the age of 1 year, there was still a continuous tension in the pack with the subordinates trying to climb the hierarchy and the dominants trying not to allow this. This tension might have potentially been caused by Zimen himself: being the 'super-alpha' in these packs, which might have destabilized the structure of the packs. However, these interactions, although expressed loudly, were mostly without serious aggression; in fact, interactions within the packs were 'restrained' with bite inhibition shown in most contexts. Moreover, if the higher-ranking individuals invaded the subordinate's 'personal space', the subordinate would often show signs of protest (and the closer the rank distance between the two, the more intense the subordinates' protest). Importantly, rank order was *not* the same in a social and feeding context in wolves.

In contrast, at 7 weeks, the poodles already showed a clear rank order based on the weight of the individuals, with sex not affecting the rank position at this age. At 8 to 12 weeks, young poodles were integrated in the adult hierarchy: the young animals were always subordinate to adults, and aggressive interactions were rare and mostly observed between the lowest ranking adults and the juveniles before they reached the age of 1 year. However, extreme aggression did occur on a number of occasions in the poodle packs. In one pack, the lowest ranking adult male killed two and severely injured one male puppy. This surviving pup, who then became the lowest ranking adult male, went on to kill two pups from a subsequent generation, at the age in which the juveniles start integrating into the adult hierarchy. Zimen observed that in the first year the tendency to submit in the subordinate dogs was not very pronounced, which he suggests may explain why the lowest ranking adult individual was observed seriously attacking and even killing juveniles. Zimen also carried out detailed analyses of the behaviours exhibited by the poodles and the wolves, concluding that compared to wolves, poodles showed a somewhat reduced variety and expressivity of facial expressions and body postures, suggesting this may further hinder dogs' abilities to regulate aggressive interactions compared to wolves (Zimen 1970).

However, in general, once the hierarchy was established, it was stable without much aggression even during the females' heat season, during which multiple males could mate with the females. Indeed, Zimen reports that, as adults, the characteristic feature of the social rank hierarchy in poodles was its stability, with fights for rank occurring only during development and stability in poodle packs being achieved much sooner than in wolf packs. No 'protest behaviours' were observed in the poodles when animals entered each other's personal space.

Based on his comparison of the poodles and wolves, Zimen concluded that natural selection was likely relaxed in poodles, particularly in some contexts, i.e. hunting, defence, behavioural repertoire for communication, choice of partners, social organization, and daily and yearly rhythms. In particular, some behaviours were altered in dogs such as food provisioning, alloparental care, and caring for juveniles in the pack. Aside from a relaxation of natural selective pressures in certain contexts, Zimen also highlights that some behaviours may be specific adaptations to life with humans including barking, subordination, decreased aggressiveness, decreased independence, easier socialization with other species, and lower flight distance, as well as a greater ease in activating them (when the human decides) and more flexible daily rhythms.

In general, Feddersen-Peteresen reports results in line with Zimen's. Developmentally, agonistic interactions (including submission, dominance, and aggression) appeared in dog packs earlier than in wolf packs (Feddersen-Petersen 2004). Indeed, in mixed dog (poodle)–wolf packs, by 4 months of age, male poodles were higher in rank than male wolves, with the former obtaining access to preferred resources (food and locations). Furthermore, the frequency of aggressive interactions in a pack of poodles during their first year of life was more similar to those of the solitary living jackals than wolves (Feddersen-Petersen 1991). However, poodles also appeared to have particularly high proportions of agonistic interactions in their first year of life, when compared to other breed groups, leading the authors to conclude that they were largely unable to live in a pack setting without human intervention to regulate their aggression (Feddersen-Petersen 1986; Zimen 1970).

A direct comparison of a pack of six wolves and six German shepherds (three females and three males in each pack, all related to each other), between the ages of 10 weeks and 6 months, revealed a number of interesting points. At 6–7 months, agonistic interactions formed about 10–15% of the total social interaction in dogs, whereas wolves did not show any agonistic behaviour at this age. As adults, wolves showed a number of behaviours, such as placing the snout in the other's mouth and gentle muzzle biting, that were completely absent in dogs. Agonistic behaviours in wolves tended to be limited to threats with no physical contact and never reached the highest 'escalation' level represented by biting. Although the rate of biting was also low in shepherds (1% of all agonistic interactions), they more often than wolves engaged in severe aggression involving body contact, which resulted in injuries (Feddersen-Petersen 2004).

Overall, striking differences between breed groups emerged, with poodles having the greatest difficulty of living in packs and malamutes, bull terrier, fila brasileiro, and German shepherds showing a greater capacity to do so. Nevertheless, wolves showed fewer dominant/aggressive behaviours within packs than any other dog breed (Feddersen-Petersen 2004). Furthermore, dogs showed a less differentiated communication system in regard to dominance than wolves, while the wolves' behaviours in this context were more fine-grained. In line with this, she found that dogs appeared to have greater difficulties in managing their conflicts, because they appeared to lack the communicative strategies necessary—or the willingness—to inhibit or resolve conflicts (Feddersen-Petersen 2007). She concludes that while wolves live in a social group, dogs live next to each other in a group (Feddersen-Petersen 1986).

Similar patterns were observed (although in a less systematic fashion) by Frank and Frank (1982) when comparing their hand-reared wolves and malamute pups. They observed that 'malamute pups exhibited unrestrained fighting from about 2 weeks of age' and continued to do so for longer than wolves. Indeed, this led authors to conclude that during the process of domestication, 'selection pressures against aggression have relaxed, and in most dog breeds, the wolf's highly predictable dominance ritual has disintegrated into an assortment of independent behavioral fragments' (pp. 519).

Differently from both the malamute–wolf and the poodle/shepherd–wolf comparison, Bekoff (1974) reported a comparison between beagle ($N = 4$), wolf ($N = 4$), and coyote ($n = 4$) pups on their agonistic and play behaviour. Relating to the

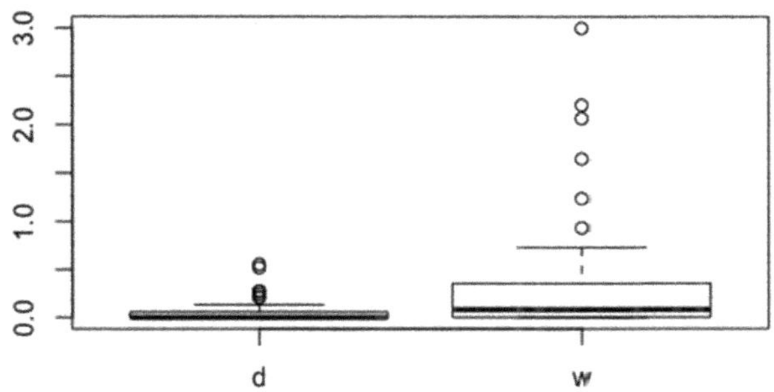

Fig. 5.3 Normalized frequency (rates per focal observation which lasted 10 minutes and controlling for pack size) of submissive behaviours exhibited by dogs and wolves

agonistic interactions, coyotes exhibited more intense agonistic interactions earlier, and by 4 weeks of age, they had established a clear hierarchy. In contrast, beagles showed rare agonistic interactions, and no hierarchical organization could be detected. Finally, wolf pups showed more frequent agonistic behaviours than the dogs, but they were limited to threats (not showing the intensity and severity of aggression shown by the coyotes).

Our own general experience at the Wolf Science Center has been that starting at the age of 6–9 months, it becomes increasingly difficult to keep the younger dogs in larger groups with peers and adults. As a consequence, most of our dog groups had to be split up into smaller groups of two to three individuals sometime after the first year. Moreover, during the heat of a female, the levels of dog aggression can become intense enough to put us on 'high alert'. In fact, in most dog packs, we need to separate our males when the females come into heat. Over the years, we have only considered this necessary in wolves once.

To better understand the differences, which may have emerged in dogs' and wolves' intraspecific social behaviour, we established a systematic data collection of the within-group social interactions between the animals at the Wolf Science Center starting in 2013. Focal observations, where one animal in the pack is observed for a set amount of time, recording all its interactions with other members of the pack, are combined with scan data (every 2 min), in which we record who is in proximity of our focal animal. We analysed the focal and scan observations over a 5-year period (2013–2017). This comprised a total of 21 dogs and 17 wolves, living in 12 dog and 11 wolf pack compositions, which over the years varied in size between three and six members.

In relation to agonistic social behaviours, we were mainly interested in seeing whether wolves and dogs differed in their rate of low- and high-intensity aggressions (i.e. with or without physical contact) and dominant, submissive, and greeting behaviours. We found that the relative frequency of behaviours for most categories (aggression, dominance, and greeting) did not differ between wolves and dogs. However, the frequency of submissive behaviours observed was higher in wolves than dogs (Fig. 5.3).

This result partially surprised us, since, based on a short-term data set (including data from 2013–2015 in both the breeding and non-breeding seasons), we had previously found that wolves engaged in conflicts significantly more often than dogs (0.80 aggressions per observation hour for wolves vs. 0.14 aggressions per observation hour for dogs). Interestingly, the proportion of high- vs. low-intensity aggression differed between the two species, with dogs showing significantly higher rates of contact (high-intensity) aggression than wolves (Cafazzo et al. 2018b). These results are in line with results from earlier studies, indicating that dogs' aggressive interactions were more likely to escalate into serious fighting compared to wolves (Feddersen-Petersen 2004). While we did not compare high- and low-intensity aggression in the 5-year data set, the similar aggression rates compared to the short-term study were puzzling.

One possibility is that results from the short-term data set were affected by the group dynamics playing out in our wolf and dog packs at the time. In fact, our so-called black wolf pack, which at the time consisted of five individuals, had recently seen a turnover in the rank position of the dominant female, with the young upstart Tala taking over from the 4-year older but not nearly as confident Shima. The turnover created tensions but still was relatively peaceful (with some minor injuries), and Shima conceded her position with a certain amount of savoir-faire. However, for quite some time, Tala would use any possible occasion to remind Shima who was in control. Moreover, we also had a period of instability in two of our dog packs due to severe incompatibilities between the animals, which resulted in the split of the packs and might have been responsible for the high-intensity aggression. Finally, the aggressive interactions in the dogs might be underestimated in the larger data set, because in the last 2 years (2015–2017) male dogs were separated when the females were in heat, and we refrained from feeding the dogs with clumped resources, thereby likely reducing their conflict rates.

Aside from the aggression rates, the results from our long-term data set surprised us also, because in line with the early studies by Zimen (1971, 1982) and Feddersen-Peterson (1991, 1994, 2004), our impression is that wolves are much more communicative with each other than dogs, engaging in more 'negotiations' via quarrelling, growling, pile erect, and greeting, which should not just be represented in the higher frequency of submissive behaviours. This discrepancy between our expectation and the observations might be due to the possibility that our perception of wolves being more communicative with one another is picking up on the intensity (or perhaps the greater variety) rather than the frequency of the behaviours they display; thus, if we would have measured durations rather than frequencies, the results might have been different. Moreover, compared to previous studies, the rate of submissive behaviours in dogs was also low, which may be due to differences in the ethograms adopted in our own and other studies. For example, Bonanni et al. (2017) report using 'avoiding eye contact', 'lowering head', 'flattening ears', and 'tail between hindlegs' as signals of submission, whereas all of these behaviours are included in more macroscopic descriptions in our ethogram (e.g. they are part of 'crouch' and 'active submission' etc.). It is possible that dogs exhibit these behaviours singularly, whereas wolves are more prone to combine them in more complex displays, potentially also making it easier for observers to see.

Still, based on the same 5-year data set, we then carried out a series of analyses at the pack level in order to explore potential difference in the hierarchical style of

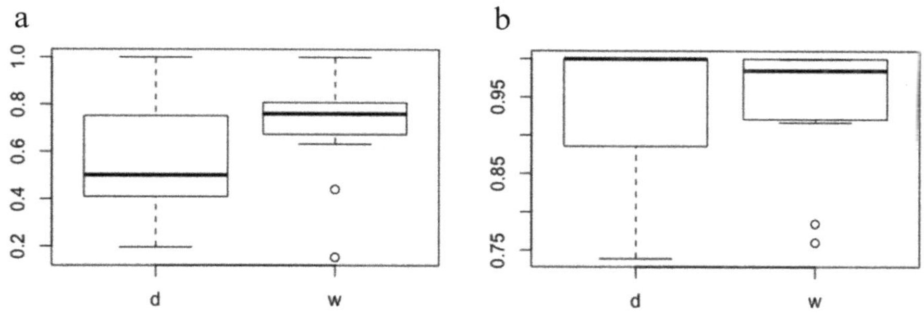

Fig. 5.4 DCI based on dominant (**a**) and submissive (**b**) behaviours, in wolf and dog packs at the Wolf Science Center. Note that the scale for submissive behaviours is very different than for dominant behaviours, since in general a better DCI was found for submissive than dominant behaviours. Nevertheless, the variability is higher for dog than wolf packs

wolves and dogs. In general, we found that both our wolf and dog packs formed linear hierarchies with good DCI, regardless of the behavioural category used (dominant, submissive, or greeting). However, dogs showed a much larger variability across packs compared to wolves (Fig. 5.4 shows the DCI based on dominant and submissive behaviours). Finally, no difference in steepness measures of our wolf and dog packs emerged when using dominant behaviours, but wolf packs showed a significantly steeper hierarchy than dogs when based on submissive behaviours. This latter difference is likely affected by the fact that we found wolves to carry out significantly more submissive behaviours than dogs; hence, it should be taken with a grain of salt (or three)!

Bonanni et al. (2017) carried out a comparison of the hierarchies based on submissive behaviours exhibited in packs of free-ranging dogs (N of packs = 7) and wild wolves (based on data by Bloch (2002), Bloch (2016), Mech (1999); N of pack = 12) showing that the DCI was higher in wolves compared to dogs, although no differences in steepness measures between the species emerged. However, as mentioned above, all these measures of group hierarchies are affected by (1) the percent of unknown relationships, (2) group size (typically with higher group size being more affected by unknown relationship), and (3) the presence of castrated individuals (which also increases unknown relationships, since typically such animals show fewer dominant displays). These difficulties make the data at present highly preliminary and comparisons between species in such diverse conditions difficult to interpret.

5.2 Dominance and Access to Resources

As mentioned above, rank position is thought to be important to allow priority of access to coveted, monopolizable resources (e.g. food and mates); thus, an individual's position in the hierarchy should accurately predict their access to food and mates.

In our study of the captive family pack of Arctic wolves, we observed their behaviour both in the presence and absence of food. Wolves were fed four to five times a week by a keeper driving into the enclosure and delivering a large quantity of chunks of meat, which was piled up on a low 2×2m table in the enclosure. During the delivery of the meat, the wolves were typically cowering at the opposite end of the enclosure, and only after the keeper left, the animals approached the table. Because the individual's level of fear of the situation may have had a strong influence on who approached the table first, rather than looking at the order of access, we collected the frequency of dominant, submissive, and aggressive behaviour both when food was available and when it was not. We found a strong correlation between the measured hierarchies based on dominant and submissive behaviours in the two contexts. Interestingly, however, whereas in the absence of food aggressive behaviours were observed to have a high unidirectionality (0.94), this was lower (0.8) as more inconsistencies (i.e. cases in which the subordinate is aggressive to the dominant) were observed in the feeding context (Cafazzo et al. 2016). The motivation for food thus increases the likelihood of aggressive behaviours from submissive towards more dominant individuals as discussed above. This has also been observed at the Wolf Science Center, where Aragorn always defended food against the higher-ranking Kaspar due to his high motivation for food.

Observations from wild wolves are in line with these results. Mech (1999) describes wolves as having an 'ownership zone', whereby regardless of rank the owner tries to defend its own food, and observed that wolves of any rank would try to steal food from others even of higher rank, although successful stealing may have been higher in higher-ranking individuals. He also describes higher-ranking animals as having only 'conditional' control over who has access to resources, in that with large carcasses the typical behaviour is that the whole pack is feeding side by side. But with smaller prey, the breeding pair may exert control over access (Mech 1999). Finally, Mech reports the often-observed behaviour of the male bringing food back for this breeding female when caring for pups.

In the only study, to our knowledge, that analysed the relationship between dominance and access to different resources in free-ranging dogs, Cafazzo et al. (2010) found that hierarchies based on submissive behaviours were highly correlated in the food and non-food context, and the higher the rank position, the more individuals 'displaced' or 'stole' food from others (in fact displacing only happened from higher- to lower-ranking individuals). Similarly, the hierarchies calculated based on submissive behaviours, when in the presence and absence of receptive females, were highly correlated. In fact, a study of the mating patterns in this pack showed that male copulation success and female reproductive success were positively influenced by a dominance rank (as well as age and leadership). Furthermore, both males and females showed a preference for higher-ranking mating partners (Cafazzo et al. 2014).

5.2.1 Dominance, Aggression, and Access to Resources: Wolves and Dogs Compared

Zimen (1970) reports that, in general, when food was placed in the wolf enclosure (in a big pile), all the wolves went to eat together. Aggression would only occur if an animal temporarily left the feeding area and then tried to come back. In this case, both dominant and submissive individuals would show threatening behaviours, and sometimes submissive individuals could keep dominant ones away. Access to the feeding area was not related to social rank but rather appeared to be affected by a number of factors including the hunger level of the animals, the food type, the number of animals present, and the relationship between them. Some animals did feed more than others but not necessarily the highest ranking individuals. Youngsters (until the age of 6 months) would often be the first at the feeding area, and they would be able to take food away from adults. In general, when aggressive interactions occurred in the feeding context, they rapidly ended once food was allocated, and no counter-aggression/retaliation occurred. Most aggressive interactions occurred between puppies/young wolves, and on occasions littermates would defend the feeding area together from other litters. Aggression between adults over food seemed to be connected to other problems in their relationship. If relatively small 'prey' (a hare or chicken carcass) were thrown into the enclosure, the wolves would take it and run off with it, and, in fact, if an animal succeeded in taking a piece, then no aggression was displayed by pack members regardless of status. We observed the same pattern in our wolf packs at the Wolf Science Center.

In poodles, higher food aggression was displayed even when enough food was available for the entire group. Access to the feeding area was affected by social rank, with the highest in rank feeding first and alone and the next in rank waiting (sitting) at a distance and approaching mostly when the dominant animal left the area. Until 12 weeks of age, poodle pups could approach and eat from the feeding area in the presence of the adult; however, when above 3 months, adults (including mothers) would displace pups from the food. Littermates displayed a level of tolerance with each other around food until 6–9 months of age, after which the rank position in the social hierarchy determined access to the resource. In mixed wolf–poodle packs, poodles had priority of access to the feeding area except for one wolf, the oldest female, who accessed the feeding area without showing any signs of aggression.

At the Wolf Science Centre, we conducted a series of experimental studies to investigate the potential differences between wolves and dogs in how rank regulates access to food resources in dyadic or group settings. In the dyadic test situation, a single bowl was placed in the enclosure, containing pieces of meat and dry food or a single large bone (edible together but easily monopolizable). This allowed simultaneous access by two pack members, whose relationship was known from prior, independent observations. In the first study, we found that while higher-ranking dogs were more likely to feed alone and did so for longer than lower-ranking dogs, in wolves no effect of rank emerged on the duration of feeding alone; see Fig. 5.5 (Range et al. 2015).

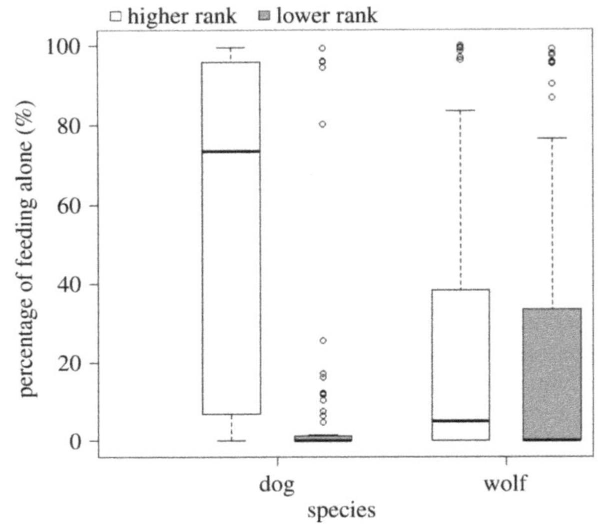

Fig. 5.5 A comparison of feeding alone based on rank between wolves and dogs in a dyadic test situation. Figure adapted from Range et al. (2015). https://doi.org/10.1098/rspb.2015.0220; CC BY 4.0; http://creativecommons.org/licenses/by/4.0/

Although agonistic behaviours were relatively infrequent in both wolf and dog dyads occurring in only 84 of the 260 test sessions (dogs, 36/134; wolves, 48/126), there was no effect of rank in wolves, but, in dogs, agonistic behaviours were exhibited more frequently by higher- than lower-ranking animals. When two animals were seen to both feed from the same resource, in dogs the higher-ranking individual was more likely to show threatening behaviours ('agonistic co-feeding') than the lower-ranking one, whereas in wolves no such rank effect emerged.

In a subsequent study (Dale et al. 2017b), using both the dyadic testing procedure described above and a group test, in which a carcass or large bone was provided to the animals, we analysed the effect of the affiliative relationship between the animals (see below for a more in-depth treatment of affiliative relationships) and their rank distance (Dale et al. 2017b) on their food-sharing behaviour. In the dyadic tests, we found that in both dogs and wolves, the duration or likelihood of peacefully co-feeding (hence eating from the same bowl without showing agonistic behaviours) scaled with the affiliative bond, i.e. the stronger the bond, the more sharing (Dale et al. 2017a). In the group context, when a carcass was placed in the enclosure for the entire pack, we found that resource sharing was affected mostly by rank and not affiliation in both species. The absence of an effect of affiliation in the pattern of food-sharing in the group test was likely due to the fact that since multiple individuals were present at the same time, the choice to share food with your preferred partner was made difficult by the presence of other non-preferred individuals. However, rank effects did emerge, for both wolves and dogs, and showed a similar pattern as the previous dyadic study.

Dominant as compared to subordinate wolves were more likely to arrive first at the carcass and defend it from others, but they were not more likely than subordinates to actually monopolize the resource. On the contrary, dominant dogs

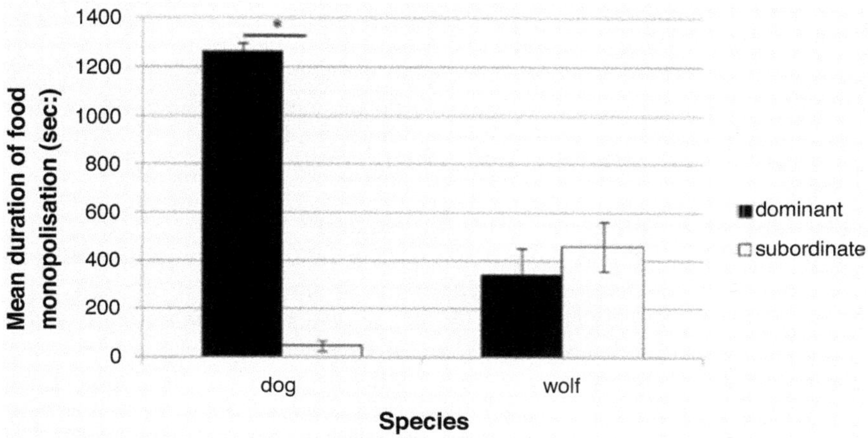

Fig. 5.6 Mean duration of food monopolization in dogs and wolves in the 'carcass test': dominant dogs spent significantly more time monopolizing the carcass than subordinate dogs. No such effect emerged in wolves. Note that the graph looks very similar to the previous one! Figure adapted from Dale et al. 2017b. https://link.springer.com/article/10.1007/s00265-017-2339-8/; CC BY 4.0; http://creativecommons.org/licenses/by/4.0/

were more likely than subordinate dogs to succeed in monopolizing and defending the resource, although they were not more likely to arrive first (maybe there was no need for them to hurry, since they could easily monopolize it once they arrived). Interestingly, proximity to the carcass was not affected by rank in wolves but was strongly related to rank in dogs (Fig. 5.6).

Taken together, it seems that in both wolves and dogs, the dominance relationships, in terms of the communicative interactions between animals, do not differ substantially, yet the way and the extent of how this affects the likelihood of gaining and maintaining access to resources differs between the two species. Indeed, the general strategy observed in the tests is that subordinate dogs would allow the most dominant individual to monopolize the resource and rather kept their distance from them while they ate. Subordinate wolves, on the other hand, showed no such reluctance in approaching the carcass when the dominant one was eating. As a consequence of this pattern, the dominant dog monopolized the carcass for much longer than any subordinate dog, but no such rank effect was found in wolves.

These experimental results of wolf behaviour are largely in line with descriptions by Mech (1999) on the pack feeding patterns, and results in our dogs appear in line with the higher likelihood of 'displacement and theft' of food by higher-ranking individuals than lower-ranking ones in free-ranging dogs (Cafazzo et al. 2010).

In a final study on food-sharing at the Wolf Science Center, we investigated whether the reproductive status of the females would affect both their own and the males' food-sharing behaviours. In dyadic tests, wolves and dogs showed fewer aggressive behaviours in the reproductive compared to non-reproductive period. Also reproductive females spent more time feeding from the bowl alone.

Furthermore, for both species, when the male was given a bone, females were more likely to show 'begging' (i.e. affiliative behaviours in proximity to the male and food) when reproductively active compared to non-active. Lastly, female dogs, but not wolves (maybe because it was not a problem anyhow—see results above), spent more time in proximity to the bone and had a shorter latency to gain access to it, when reproductively active (Dale et al. 2017a).

5.2.1.1 Summary Dominance and Aggression

In sum, the concept of dominance may still be a relevant feature to describe the social relationships of both wolves and dogs. In both species, submissive behaviours and greeting are the best indicators of formal dominance relationships, and in general a linear hierarchy can be detected in the packs. At present, we have some indication that wolf packs show a more consistent pattern of clear hierarchical relationships than dogs, but it also seems that subdominant animals are more willing to stand their ground to get access to resources, which is accepted by the higher-ranking wolves to some degree. But comparisons are difficult due to the many variables, which may affect group hierarchy measures. Nevertheless, if the higher consistency of hierarchical measures in wolf compared to dog packs holds up to closer scrutiny, then these results may suggest that whereas a cohesively structured group organization is the 'norm' for wolves (it could be essential for a species heavily relying on pack behaviour for hunting and pup-rearing), in dogs, the hierarchical organization is more strongly affected by the characteristics and the mindset of the animals in the pack and potentially, by as yet unidentified external, likely ecological factors. This highlights the need for more in-depth studies in particular with 'naturally occurring, free-ranging' packs to better understand the development and maintenance of such relationships and the functions they may have.

Rates of aggressive interactions in the pack (in the absence of resources) appear to be low in both wolf and dog packs. However, there are some inconsistencies in results at the moment, with some studies finding higher rates of high-level aggression in dogs and higher rates of low-level aggression in wolves, whereas other results suggest no difference between them (the 5-year data set at the Wolf Science Center presented here). Potential steps forward that are needed are a clearer differentiation between rates of aggression during the breeding and non-breeding periods and long-term, high-quality data sets in which the effect of destabilization in a single pack does not overly affect results.

In regard to access to resources, the effect of dominance appears to be a bit different in the two species. Whereas in dogs the most dominant individual monopolizes the resource (and the subordinates do not even approach the resource), in wolves, although the higher-ranking animal may be first to reach the food, they do not monopolize it (though they would be able to). This results in more food-sharing in wolves compared to dogs. Furthermore, the patterns of aggression are also different. Whereas in wolves (mild) aggressive interactions over food are as likely to occur from both dominant and subordinate individuals, in dogs, they are almost exclusively shown by higher-ranking animals. How and why these different patterns of conflict management may have evolved in wolves and dogs is at present largely an interpretative and speculative exercise. As we will outline in the last chapter of this book (Chap. 11), we think that these differences in wolves' and dogs' behaviours with conspecifics may have evolved as a result of their different ecological niches as well as human selection for and against specific traits.

5.3 Affiliative Relationships

Although Darwin (1871) already recognized that as in humans and other social species individuals are capable of forming preferential relationships with other individuals, the characterization and systematic investigation of such relationships has been problematic. One of the main problems in the field has been the need to define 'bonds' in both a meaningful and operationally adequate manner, and researchers often use the terms 'bonds', 'friendships', and 'affiliative' relationships interchangeable, though they may in fact be slightly different. Accordingly, in their seminal review, Shultz and Dunbar (2010) state the problem as follows: 'We know when we have such relationships, and we recognize them when we see them in others, but we do not have the vocabulary to describe them' (pp. 782). In human psychology, bondedness or 'closeness' is mostly characterized in both behavioural and emotional terms, so that the frequency and diversity of interactions as well as the 'emotional engagement' (as assessed by questionnaires) of the individual are considered (see also Chap. 6). However, in animal–animal interactions, the emotional layer is much harder to investigate.

At the behavioural level, affiliative relationships in studies of wild animals have mostly been quantified according to proximity-seeking/maintenance and the frequency/duration of socio-positive behaviours exchanged between individuals (e.g. grooming in primates) (Massen et al. 2010; Shultz and Dunbar 2010; Silk et al. 2013). However, a number of other parameters have been suggested to better characterize the potentially different facets of affiliative bonds. For example, Cords and Aureli (2000) noted that the quality of social relationships could be described by three components: value, compatibility, and security. The value of the bond is a measure of the different benefits that an individual gains from their social relationship: for example, in primates, these benefits might include grooming, tolerance around food, and support in conflicts against others, since they are thought to enhance the fitness of an individual. Compatibility would instead describe the shared history of social interactions exchanged within a dyad and the similarities in the temperament of the two partners towards each other. Security would rather be a measure of the consistency of the social relationship over time (Silk et al. 2006). Another aspect, which has been identified as potentially important, is the 'symmetry' in the affiliation, i.e. how skewed or conversely reciprocated the affiliative behaviours are within the dyad (Majolo et al. 2010; McFarland and Majolo 2011; Silk et al. 2013). In captivity, some studies have also focused on the potential physiological measures of bonding, finding, for example, that in many primate species the glucocorticoid response to a stressful event is less pronounced, when in the presence of an individual with whom the subject has a strong affiliative relationship ('social support' or 'social buffering'—see Kikusui et al. 2006).

Affiliative relationships have been shown to play an important role in regulating social life within a group, providing both immediate- and long-term benefits. For example, in the short term, grooming in primates has been shown to be associated with decreased cortisol and increased oxytocin levels (both related to reduced stress levels) (Crockford et al. 2013; Wittig et al. 2008). Furthermore, in a variety of species, partners with as strong affiliative relationship are more likely to support each other during conflicts (chimpanzees: Fraser and Aureli 2008; ravens: Fraser and Bugnyar 2010), thereby reducing the likelihood of injury and increasing the access to resources (hyenas: Holekamp et al. 2007). Not surprisingly, the ability to form affiliative relationships with others has not just short-term benefits but also fitness consequences (Ostner and Schülke 2018): in primates long-term, strong social relationships between individuals have been shown to increase longevity (Silk et al. 2010) and predict future cooperation (Berghänel et al. 2011) and reproductive success (Schülke et al. 2010; Silk et al. 2003).

In dogs and wolves, the characterization of conspecific relationships in terms of affiliation has not received much attention. In his description of wolf behaviour, Zimen mentions several 'friendly' behaviours observed between wolves, in particular greeting and grooming (Ziemen 1981, pp. 64, 76), as well as muzzle-to-muzzle contact (pp. 176). In more recent studies relating to other topics, researchers have assessed the closeness of the affiliative relationship between individuals by looking at the frequency of socio-positive behaviours exchanged. Interestingly, the kind of behaviours used varied somewhat across studies. In a captive pack of wolves, Romero et al. (2014) used licking, sniffing, playing, and head rubbing and found that animals exchanging more affiliative relationships also showed an increased likelihood of contagious yawning in comparison to less affiliative pairs. In another study on play, Cordoni (2009) used the frequency of body contact interactions (i.e. contact sitting, touching, social licking) as an index of a good affiliative relationship. In two studies looking at conflict management in the same population, different measures of closeness were used. In the first study, the above definition by the same author with the addition of frequency of social play (Palagi and Cordoni 2009) was applied. In the second study (Cordoni and Palagi 2008), they defined the social bond by a single behaviour, i.e. the frequency of body contact. Interestingly, with the first two measures of affiliation, the authors found no effect of relationship quality on time spent playing or likelihood of reconciliation, while the third measure revealed a positive relationship between closeness and third-party affiliative interaction (consolation) after a conflict was found (see Section 'Conflict Management'), begging the question whether playing and licking might represent good measures of affiliative relationships.

In dogs, Trisko et al. (2016) investigated the social relationships between pet dogs in a 'doggy care centre' ($N = 24$) and included the 'affiliative dimension' as well as dominance. The affiliative relationships were measured by looking at the frequency a dyad (1) spent play fighting and (2) showing gentle affiliation, which included the following: nose nudge, coat lick, genital lick, head rub, body rub, and nibble.

Interestingly, only approximately 50% of the studied dyads showed any affiliative behaviour towards each other. In a minority of dyads, both these behaviour categories were exhibited; however, in most cases, dyads exhibited either one or the other of these behaviour types. Furthermore, in 64% of dyads, gentle affiliation was skewed, in that it was exhibited more often by one partner than the other, and in 80% of such cases it was shown 'up the hierarchy', i.e. from the subordinate towards the dominant.

Few studies in free-ranging dogs have attempted to measure the affiliative dimension. In a free-ranging pack on the outskirts of Rome, Bonanni et al. (2010a) measured the relationship between individuals in terms of the frequency with which they stayed in close proximity (within 1 m of each other). Researchers found that 'pack leaders' (animals which successfully initiated a change of state in the pack, e.g. from resting to foraging) were more likely to have higher proximity scores with many individuals than non-leaders. In a second study on the same population, Bonanni et al. (2010a) were interested in what would affect the patterns of participation in intergroup conflicts. In this case, affiliative partners were considered those with whom an individual frequently displayed the following activities: tail-wagging, playing, play invitation, allogrooming, and resting in physical contact.

What appears to emerge quite clearly from these studies in both wolves and dogs is that differently from studies in other species (in particular primates), where measures of affiliation have been investigated and refined to even include different nuances of the relationship (e.g. value, compatibility, and security; Cords and Aureli 2000), in wolves and dogs, there is not even a consistent use of the same behaviours that are considered 'affiliative'. In general, close proximity and resting in body contact often appear as measures of affiliation, as do rubbing against (head or body), grooming (licking), and other forms of gentle physical contact (nose nudge, muzzle-to-muzzle). 'Greeting', i.e. muzzle-licking in combination with tail-wagging and both its components separately, has been shown to have good unidirectionality, i.e. exhibited mostly from the subordinate to the dominant (Cafazzo et al. 2010; Van Der Borg et al. 2015) and can hence be considered a clear indicator of status in both, wolves (Cafazzo et al. 2016) and dogs (Cafazzo et al. 2010). However, they may be exhibited more in some dyads compared to others, suggesting it may indicate another dimension of the relationship aside from dominance.

Finally, play has also been considered a potential measure of affiliative relationship; however, since it has also been suggested to have a multitude of other functions, we will treat it separately in a section below.

5.3.1 Affiliation: Wolves and Dogs Compared

Zimen (1970) reports that when sleeping, but also when moving around, adult wolves kept a distance of at least 50 cm from one another, whereas poodles slept close together and did not maintain distances when moving around. Very little more emerges from his observations in relation to the affiliative relationships between the animals and differences between wolf and dog packs.

At the Wolf Science Center, based on previous ethograms of dogs and wolves, we outlined a number of behaviours considered to express an affiliative interaction (see Appendix for Ethogram). These behaviours were collected during focal samples in order to have a clear rate of affiliative behaviours exchanged over specific time periods of observation. Based on this, we calculated an 'affiliation score' for each dyad, which we then used as a measure of the socio-positive relationship between individuals in most studies (but see last section of this chapter, where different 'friendship' components emerged in one study).

Proximity measures posed a 'conceptual' challenge for us. From our observations of dogs and wolves, we noted that one behaviour exhibited by both wolves and dogs in an agonistic context, especially when a change in rank between individuals was taking place, was the 'close following and blocking' of another individual's movements. Thus, we reasoned that proximity, as a measure of affiliative interactions, may be unduly affected by proximity events of the kind described above, which appeared to have an agonistic (and not at all affiliative) connotation. We nevertheless decided to record the proximity measures in the classic manner (i.e. during focal sampling, we would note every 2 min which individuals were in close proximity (one body length) to our focal animal) and rely on the subsequent statistical analyses to evaluate whether in fact the resulting 'association index' (i.e. the proportion of scans in which two individuals were observed close to one another) would correlate with the 'affiliation score' across dyads. Based on the 5-year data set outlined above, we found that the affiliative and association scores were highly correlated in both wolves and dogs: animals that frequently interacted in an affiliative manner were also more often observed in close proximity to one another. Likely, the less often observed 'following and controlling' behaviours did not impact the more affiliative-driven proximity-seeking and maintenance over the long period of data collection. Nevertheless, we consider the affiliative score a more accurate measure of relationship quality, and in most studies we have used this measure to investigate how affiliation may regulate the animals' behaviours in other tasks (see final section of this chapter and Chap. 8).

Based on our 5-year observational data set, we found no differences between species in the frequency of affiliative interactions exchanged, nor the time spent in close proximity to other pack members. In both wolves and dogs, affiliative behaviours were exhibited more frequently among the younger animals and more often by females than males. We further analysed how symmetrical the dyads were, i.e. to what extent the affiliative behaviours were exchanged reciprocally. The symmetry index goes from 0 (completely symmetrical exchange) to 1 (completely asymmetrical), and in both wolves and dogs, we found some individuals with complete symmetrical and complete asymmetrical affiliative relationships, and overall there was no difference between wolves and dogs (mean for wolves = 0.38, mean for dogs = 0.31). In most dyads, the affiliative exchanges were quite skewed, since on average one partner showed twice as many affiliative interactions than the other partner. We will need to dig a bit deeper to better understand what variables are affecting the pattern of affiliative behaviours exchanged.

5.3.1.1 Summary Affiliation

Based on the present data, and in line with other studies, it appears that affiliative relationships are important for both wolves and dogs, and they show similar age- and sex-related patterns in terms of the distribution of the behaviours (see Table 5.1*). For both species, measures of proximity and affiliative rates correlate, suggesting both can be used as indexes of bondedness, although, arguably, exchange of affiliative interactions provides a more accurate measure. A number of open questions still remain in particular relating to how patterns of affiliation relate to rank.*

5.4 Play

Social play includes competitive and cooperative elements and mostly action patterns used also in other contexts (e.g. agonistic, predatory, and sexual contexts), albeit often exhibited in an exaggerated form. Species-specific play signals have been thought to have evolved to communicate the 'playful context' of a specific

Table 5.1 A comparison of the normalized frequencies of behaviours shown by wolves and dogs when controlling for pack size revealed only one significant difference: wolves were observed carrying out submissive behaviours more frequently than dogs (based on the 5-year WSC data set). Additional results from published papers are included in the table to give the whole picture emerging so far: [1]Results relating to the rates of aggression from Cafazzo et al. (2018b). [2]Results in relation to access to resource refer to Range et al. (2015)

Behavioural category	Description	Species comparison
Aggressive	Low: threat, jaw spar, snap, lunge	D = W, D < W[1]
	High: attack, knock down, pin, fight, chase	D = W, D > W[1]
	Access to food resources:	
	High rank higher aggression	D[2]
	Rank no effect on aggression	W[2]
Greeting	To interact in a friendly and relaxed manner holding the ears back, showing much tail-wagging and licking of the other's mouth/muzzle. The subject however does not show crouching/lowered hindquarters nor is the tail tucked between the legs	D = W
Dominant	Stand tall, stand over, approach dominant, paw-on, head-on, ride up, muzzle bite	D = W
	Access to (food) resources dependent on dominance	D > W[2]
Submissive	Active submission, passive submission, crouch, avoid, withdraw, approach submissive	W > D
Affiliative	Lie friendly, stand/approach friendly, body rubbing, BC, grooming, social sniff	D = W
Play		D = W
Proximity (one body length from any partner)		D = W

interaction, for example, the 'play face' in bonobos (Palagi 2006, 2008) and in rhesus macaques (Yanagi and Berman 2014), 'toss head' in pigs (Newberry et al. 1988), 'neck and head rotations' in gazelles (Gomendio 1988), and 'play bows' in dogs (Bekoff 1995). Based on Bekoff's seminal paper on play in canids (Bekoff 1974), play bows have been considered a 'metacommunication' signal, allowing animals to better understand the non-agonistic and playful intent of exhibited behaviours.

In a set of studies on different canids, Byosiere and colleagues carried out a more in-depth analysis of this behaviour to better understand its function. In adult pet dogs, play bows were shown to be carried out mostly when there was a break in a play bout, and they were then followed by highly active play, such as a chase. The function of the play bow thus appears to be the re-initiation of play when a pause has occurred (Byosiere et al. 2016). In our dog and wolf puppy cohorts, similarly to pet dogs, bows were performed only when the partner animal was in visual contact with the bower, suggesting the pups used the bow as a communicative visual signal, and appreciated the necessity of the other individual's visual attention and barked when this was not the case! However, whereas both wolf and dog pups performed bows when there was a pause in their game, only in dogs did the bows successfully re-initiate play. Wolf pups were no more likely to adopt active roles after the bow than before. At present, it is not clear whether this species difference may be affected by other factors such as the different male vs. female composition of the puppy cohorts and/or the proportion of time spent in different locations of the enclosure (Byosiere et al. 2016). However, in a subsequent study, Byosiere and colleagues (2018) carried out the same study on a similarly aged cohort of ten dingo pups. Interestingly, results for dingoes were somewhere in between those found for wolf and dog pups. In general, wolf pups' play bows had the shortest duration, followed by dingo pups, dog pups, and adult dogs. The frequency of play bows followed the same pattern, with the fewest being exhibited by wolf pups, followed by dingo pups and then dogs.

Since play is a ubiquitous behaviour in social mammals, it has been suggested that it may be a critical component of ontogeny. Many functional hypotheses have been formulated for its presence including training motor skills and physical agility (Byers and Walker 1995; Smith 1982), developing cognitive skills (Dolhinow 1999; Fagen 1981; Loizos 1967; Martin and Caro 1985; Poirier et al. 1978), improving fighting abilities and/or achieving social status or assessing mate quality (Pellis and Iwaniuk 1999a; Pellis and Iwaniuk 2000; Thompson 1998), training for the unexpected (Špinka et al. 2001), and limiting aggression and increasing tolerance around food (Palagi 2004, 2006; Palagi 2007). These hypotheses are not mutually exclusive, and despite extensive research, the actual benefits of animal social play and its evolutionary significance remain unclear (Caro 1988; Pellis and Iwaniuk 1999b; Graham and Burghardt 2010; Palagi 2018).

In wolves and dogs, two main hypotheses have been suggested. One hypothesis views social play as a cooperative interaction between two individuals. Based on this idea, for play to continue, neither of the playmates should be consistently at a disadvantage, and individuals should rather adapt to each other's behaviours during play (Bekoff 2001). In other words, individuals should alternate roles within play, for example, by following the '50:50' rule, whereby both playing individuals show 'winning' behaviours (e.g. play-biting or chasing) equally often. It has been suggested that behaviours such as 'role-reversing' and so-called self-handicapping can be used to maintain equality and thus continue social play (Bekoff and Allen 2009).

In support of this hypothesis, Bekoff observed that in coyote puppies, individuals are less likely to engage in play with others who do not play in a 'symmetrical' way, suggesting that individuals which do not play 'fairly' may be excluded from play groups (Bekoff 2004). However, a number of studies on dogs did not support this idea. Bauer and Smuts (2007) found that adult dogs regularly meeting up at a dog park (and thus being familiar with each other) did not adhere to the '50:50' rule but rather showed varying degrees of asymmetry in their play. For example, older and more dominant dogs engaged in higher levels of offensive (= winning) behaviours and lower levels of self-handicapping behaviours and thus did not give up their advantage to facilitate play with their disadvantaged partners, but might have 'showed off'. Similarly, a study with dog puppies found no adherence to the '50: 50' rule and rather found that the asymmetrical play increased with age (Ward et al. 2008).

Zimen (1970) also observed differences in the roles wolves and dogs adopted during play. In wolves, any individual could play any role (the winner or the looser), but the highest ranking individuals were those that more often adopted the loosing role. The animals that initiated play were also those who more often adopted the loosing role. Lower-ranking animals often adopted the winning role, but if they were playfully attacked during the play session, the interaction often escalated into aggression. Mid-ranking animals adopted the winning and losing roles equally often. In contrast, in dogs there appeared to be no link between the rank position and the roles adopted during play.

To further elucidate the cooperative nature of play, we investigated bouts of dyadic play occurring between 3- and 5-month-old wolf puppies (two cohorts, six pups each) and between mixed dyads of juvenile (6 months of age) and adult wolves (four packs: six juvenile + three adults; two juvenile + three adults; 2 + 2 and 2 + 2) carrying out an in-depth analysis of the postures and behaviours displayed by each individual during play interactions (Essler et al. 2016). We found no evidence for the 50:50 rule at any time during our study period; however, dyads consisting of two puppies had significantly more equal play than dyads consisting of one puppy and one adult. These results suggest that wolf puppies are most likely to play on equal terms with similarly aged play partners. However, regardless of age pairings, roles in play largely mirrored those outside of the play context in terms of the dominance relationship and thus supported another hypothesis that has been proposed for the function of play.

Namely, that play (especially in adults) is rather used as a way to obtain information about the competitive ability of potential opponents in a 'safe' context or potentially to appease a tense situation (domestic dogs: Scott and Fuller 1965; Bekoff 1972; Pal 2010; wolves: McLeod and Fentress 1997). Similar arguments for the 'safe' context have been presented also for other species (Palagi et al. 2004a, 2006; Palagi and Paoli 2007; Pellis and Iwaniuk 1999a; Tanner et al. 2007). Somewhat in line with this second hypothesis, in a pack of adult captive wolves, Cordoni (2009) found that play did not correlate with the two measures of bond strength (i.e. frequency of body contact and coalitionary support); however, animals closer in rank played more often leading the author to support the idea that the wolves were using play as a means of testing each other's competitive skills. The study also found that rates of play were not affected by the sex of the animal, but they were higher in the absence of competitive stimuli (i.e. food and receptive mates).

Zimen also mentions a number of observations that support Cordoni's results (Zimen 1970): adult wolves only played with partners of similar rank (the lowest ranking females did not play), and males that were usually less aggressive outside of play showed more frequent aggression during play. In young adults and older animals, harmless aggressive fights within the play context occurred often. However, if the aggression between two individuals outside the play context increased, they would generally stop engaging in play bouts, suggesting that play was not used to reduce tension within relationships. In poodles, most play occurred between the higher-ranking animals, and they generally showed substantial aggressive tendencies in play, both as youngsters and as adults.

At the Wolf Science Center, in the same dyads described above (Essler et al. 2016), we analysed the extent to which behaviours within the play context mirrored the social relationships outside it (Cafazzo et al. 2018a). Dyadic relationships between individuals in terms of affiliation and dominance rank, as well as the rate of aggressive interactions exhibited, were established from observations outside the play context. Differently from most other studies in wolves, in our second study, we analysed two different types of play categories: (1) competitive play, characterized by offensive behaviours (such as play bites, chase and charge, sitting, or lying on top of the partner) and self-handicapping behaviours (muzzle-licking and voluntary downs), and (2) relaxed play, characterized by both partners sitting or lying on the ground pawing each other and rubbing against each other often accompanied by a 'play face' (relaxed open mouth). The latter category is characterized by an absence of a 'winning' position by one of the partners. We hypothesized that, in wolves, social play may help individuals assess both affiliative and dominance relationships and that different types of play (competitive and relaxed) would mirror different aspects of the animals' social relationships. Indeed, in mixed aged packs, we found that dyads spending more time in relaxed play engaged in more affiliative and fewer aggressive interactions outside the play context. Moreover, the frequency and direction of competitive behaviours displayed during competitive play were positively correlated with the rank relationships outside of play. Interestingly, dyads that played more (regardless of type of play) exchanged fewer aggressive behaviours. In the puppy packs, the pattern of results was not as clear. In fact, although the dominance relationships exhibited in the play context matched those shown outside of play, play did not have a moderating effect on aggression, but the more time each dyad spent in competitive play, the more aggressive interactions were exchanged outside of play. Perhaps even more telling, we found that individuals with a less clear dominance relationship (i.e. greater symmetrical exchange of dominant and submissive behaviours) spent more time playing in a competitive way. Based on these results, we support the idea that especially competitive play in wolves may be used to establish and maintain dominance relationships as suggested by the second hypothesis outlined above, whereas relaxed play might be a good indicator of the affiliative relationships at least in older animals.

One other aspect that is important in regard to canines and their play is the idea that domestication is responsible for the retention of juvenile behaviours into adulthood. This process has been referred to as neoteny or pedomorphism (Clutton-Brock 1981; Coppinger and Smith 1983; Darwin 1859; Fox 1968; Morey 1994; Price 1999). Based on this idea, it has been suggested that domestic animals may retain play behaviours, typical of the juvenile period, into adulthood, to a greater extent than their wild counterparts. This would predict that puppies would be more prepared to play than adults and dogs more than wolves. In his comparative analyses of play behaviour between wolf, beagle, and coyote pups, Bekoff reports the highest frequency of play in beagles, followed by wolf pups and lastly coyotes (Bekoff 1974). In contrast, Zimen (1970) reports that, overall, adult poodles played less often than wolves. In wolves, the frequency of running games increases with age, and overdevelopment play goes from being more explorative to being more social and organized. In contrast, object play in wolves remains largely the same unless it is integrated in a social context, in which case it becomes a 'tug-of-war'--type game. In general, object play is less frequent in wolves than dogs. In wolves, older adults play with the same intensity as younger animals but less often and with shorter bouts. In both species, an increase in play was observed during the breeding season. Feddersen-Petersen (2007) reports a similar pattern of play interactions in her wolves and German shepherds. In their early months of life, an increase in frequency of play was evident in both. However, in wolves this increase continued so that between 6 and 12 months of age, play represented 60–70% of all social interactions. In contrast, in dogs the frequency of play dropped significantly, when females first started coming into heat (around 6–7 months), decreasing from 55% at 7 months to 28% at 12 months.

To further explore the difference in play between wolves and dogs, we analysed the potential effect of species, sex, and age on the frequency of play in our packs based on the 5-year observational data set describe above. This included juveniles from the age of 6 months when they entered the packs to adults of 8 years of age. We found that the frequency of play interactions at the individual level did not differ between wolves and dogs (Fig. 5.7) nor between males and females. However, in both species, an inverse correlation was found between age and play (the younger the individual, the more frequently they played).

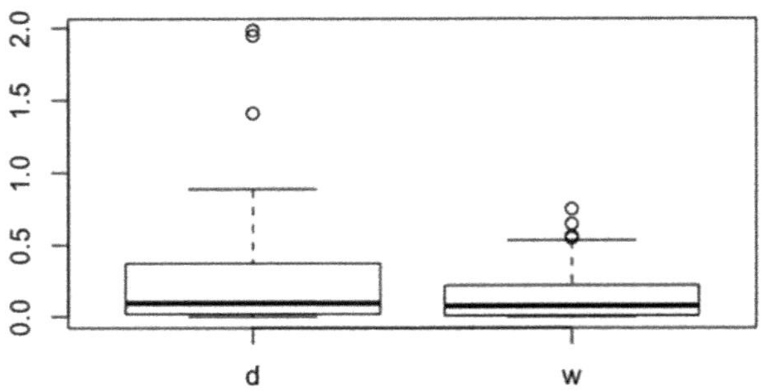

Fig. 5.7 Normalized frequency of play (mean bouts of play during 10-min focal observations) in dog and wolf packs at the Wolf Science Center, over a 5-year period, from their first integration in adult packs (at 6 months of age) to adulthood. Frequency of play did not differ between dogs and wolves

5.4.1 Summary Play

Although very few studies have directly compared wolves and dogs play, and hence conclusions may be premature, it appears that, in both species, behaviours shown in the play contexts in terms of competitive vs. *more gentle play reflect the social relationships the animals exhibit outside of play. In contrast to reports by Zimen, we found that winning positions during dyadic play are exhibited more by the dominant individual in the dyad, and in general there seems little evidence of 'fair play' in terms of adherence to the 50:50 rule. The function of the play bow has also been shown to be largely similar in the two species, being not so much a clarification of intent but rather a spur to re-initiate active movement. This may be why bows are also observed in the hunting context, when wolves try to elicit movement from their ungulate prey or when border collies try to herd sheep. In both wolves and dogs, careful analyses of behaviours show that individuals appear to have some awareness of the need for the partner's visual attention, when exhibiting such signals (particularly interesting in the context of more experimental studies looking at the animals' understanding of gazing behaviour; see Chap. 7).*

5.5 Reconciliation

Established social dominance relationships may help regulate conflicts over resources and as such could be considered as one of the tools social animals can use to manage potential conflicts. However, hierarchical status may also be the cause of conflicts, since lower-ranking animals may strive for higher status allowing greater access to limited resources (e.g. feeding and mating opportunities). Thus, some conflicts are bound to occur within social groups, with consequences ranging from injury, to loss of status, to disruption of social relationships and potential effects on group cohesion.

When conflicts occur, one potential strategy open to opponents is to attempt to repair the damaged relationship through 'reconciliation'. The term 'reconciliation' first used by de Waal and van Roosmalen (1979) refers to affiliative contacts between former opponents, which occur soon after a conflict. Although the study of reconciliation initially focused on primates, it has since been found in many species from goats to hyenas to ravens (reviewed in Aureli and de Waal 2000). Furthermore, in a number of species, reconciliation has been shown to occur more often between animals that are closely bonded suggesting that one of the functions of reconciliation may be to repair 'valued' bonds (valuable relationship hypothesis; de Waal and Aureli 1997; Kappeler and van Schaik 1992). For example, in a number of primate species, dyads characterized by more frequent affiliation, e.g. as indexed by grooming and proximity measures, are known to exhibit higher conciliatory tendencies than less affiliative dyads (reviewed in Romero and Aureli 2017), and in chimpanzees, Wittig and Boesch (2003) found that dyads that supported one another and shared food reconciled more often than those that did not. However, the extent to which species use reconciliation is thought to be affected by their dependence on cooperation (Aureli et al. 2002; de Waal 1989; Thierry 2000; Weaver 2003).

Species highly dependent on cooperation are expected to engage in frequent reconciliation following conflicts (e.g. Arnold and Barton 2001; de Waal and Luttrell 1989; Sterck et al. 1997) in order to re-establish peaceful relationships and maintain pack cohesiveness, since this may be essential for their cooperative success (and hence fitness) (Aureli et al. 2002; Kutsukake and Clutton-Brock 2008). In contrast, in species with low reliance on cooperation, reconciliation is expected to be less frequent (e.g. Demaria and Thierry 2001; Kutsukake and Castles 2001; Lu et al. 2008).

Operationally, the standard method to investigate reconciliation is to compare the likelihood of an affiliative interaction occurring between two individuals after a conflict with the likelihood of an affiliative behaviour occurring between the same two individuals in a 'control' period, where no prior conflict has occurred. This is commonly known as the PC-MC (post-conflict, matched control) method. Species that show systematic affiliative behaviours soon after a conflict (usually within 2 min) would thus be said to show 'reconciliation' (or in more neutral terms 'post-conflict affiliative interactions').

A number of studies have shown reconciliation in both captive (Cordoni and Palagi 2008; Lazzaroni et al. 2017) and wild wolves (Baan et al. 2014). Overall the mean conciliatory tendency (i.e. the willingness to reconcile after conflicts) in wolves ranged from 44% to 53% (Baan et al. 2014; Cordoni and Palagi 2008; Lazzaroni et al. 2017), which is relatively high compared with other species (e.g. hyenas, 16.6%: Wahaj et al. 2001; chimpanzees, 16.3%: Webb et al. 2014; bonobo, 35.6%: Palagi et al. 2004b; ravens, 0.16%: Fraser and Bugnyar 2011; common marmosets, 31%: Westlund et al. 2000; tamarins, from 24.7% to 48.3%: Peñate et al. 2009; and meerkats, 0%: Kutsukake and Clutton-Brock 2008).

Post-conflict affiliation in two of the three studies (Baan et al. 2014; Lazzaroni et al. 2017) was more likely to be initiated by the victim (which in 92% of all conflict interactions was the subordinate in Lazzaroni et al. 2017). In the third study, reconciliation was initiated equally often by both the aggressor and the victim (Cordoni and Palagi 2008), and no effect of rank was found on the likelihood of reconciliation. In contrast, the closeness of the affiliative relationship (as measured by the frequency of affiliative interaction exchanged) was not found to affect the likelihood of reconciliation in neither of the two studies considering this measure (Cordoni and Palagi 2008; Lazzaroni et al. 2017). However, the frequency of coalitionary support (i.e. how often two individuals supported each other against a third) positively predicted post-conflict affiliation in wolves (Cordoni and Palagi 2008).

Taken together, the few studies conducted on reconciliation in wolves suggest that indeed this conflict management strategy is frequently observed. However, similarly to results from other species heavily reliant on cooperative activities such as group hunting (e.g. hyenas: Wahaj et al. 2001) or cooperative breeding (callitrichids: Schaffner and Caine 2000, lemurs: Norscia and Palagi 2011; Palagi et al. 2005; and meerkats: Kutsukake and Clutton-Brock 2008, but see opposite results in pair-bonded ravens: Fraser and Bugnyar 2011), measures of affiliation between individuals do not seem to predict reconciliation. It is possible that in cooperative species, such as wolves, the disruption to group cohesion, regardless of relationship quality, functions as an incentive to pursue reconciliation (Aureli 1997). However, future studies considering different measures of relationship quality (e.g. cooperative support or relationship security) may help to better disentangle these issues.

In contrast, only one study has been conducted on reconciliation patterns in dogs. The dogs were living in small groups in a kennelled environment. Their mean conciliatory tendency ranged between 16.6% and 29.7% (Cools et al. 2008) and was affected by how familiar opponents were with one another (the greater the familiarity, the more likelihood of reconciliation).

To investigate potential wolf–dog differences in conflict management in more standardized conditions, we conducted a study in which we compared the conflict management behaviour of four wolf and four dog packs at the Wolf Science Center (Cafazzo et al. 2018b) and found remarkable differences: wolf dyads were more likely to exhibit an affiliative interaction after a conflict than in the control period, and they showed a mean reconciliatory tendency of 22% (with a range from 25% to 63.16%). In contrast, dog dyads in our population did not show more affiliative behaviours after conflicts than in the control periods, suggesting that they did not use affiliative interactions to repair the relationship after the potential disruption. Although because of the much smaller number of conflicts observed in dogs ($N = 55$) compared to wolves ($N = 177$), these results should be taken with caution. Perhaps the more interesting result was not so much the absence of reconciliation in dogs but the use of a different post-conflict strategy. In fact, a careful analysis of the animals' behaviour after conflicts showed that, in dogs, former opponents were more likely to spend time at a greater distance from one another, and this was not the case in wolves. Although more studies are needed on this topic, it would seem from this data that, after a conflict, dogs avoid each other for some time, whereas wolf opponents reconcile.

The likelihood of reconciliation in wolves was linked to a number of factors. Supporting the idea that reconciliation functions to repair relationships by reducing the probability of renewed aggression ('relationships repair hypothesis', e.g. Japanese macaques: Kutsukake and Castles 2001; olive baboons: Castles and Whiten 1998; long-tailed macaques: Aureli and van Schaik 1991; and wolves: Lazzaroni et al. 2017), we found that in wolf dyads rank distance scaled with tendency to reconcile, and the intensity of the reconciliation was higher between wolves where affiliative behaviours were more asymmetrical. Furthermore, reconciliation was more likely to occur and was more intense (the affiliative interaction lasted longer) after high (compared to low)-intensity aggression suggesting that relationships between opponents may be increasingly disturbed with increasing conflict intensity and that the occurrence of reconciliation and its intensity may help to restore the disrupted relationships.

5.5.1 Summary Conflict Management

Taken together, results suggest that wolves and dogs are adopting two different post-conflict strategies: whereas dogs appear to adopt an avoidance strategy, using spatial distance potentially to reduce the risk of re-aggression, wolves seek their partner out to re-establish the relationship, and they do so more if the relationship is highly skewed (high rank distance and unbalanced affiliative exchanges). This is not to say that dogs may not also use reconciliation as a conflict management strategy. In the pack of free-ranging dogs studied by Cafazzo and colleagues, the reconciliatory tendency was actually quite high (Cafazzo—personal communication), although from our own preliminary data in another pack of dogs in Morocco, very few conflicts were observed and, hence, almost no reconciliations (Lazzaroni—personal communication). Kinship, pack stability, captivity vs. *free-living conditions, and age structure of the pack are just a few of the variables, which may affect these.*

Although more studies are needed, it is likely that the larger variability observed in reconciliation behaviour in dog packs may be linked to the larger variability observed in the dogs' social structures as well as genetic variability between dog populations. Indeed, free-ranging dogs show a much more fluid social structure, from solitary to pack-living, leading authors to consider them a 'facultatively social' species (Majumder et al. 2014*) potentially adapting to the fast-changing anthropogenic resources at their disposal. In contrast, while wolves also show considerable variability in their social structure, they appear to show a higher dependence on their pack mates for both pup-raising and food acquisition, potentially requiring more ingrained social mechanisms to help both avoid and resolve conflicts, to ensure ongoing group cohesiveness.*

5.6 Do the Affiliative and Dominance Dimensions Affect Wolves' and Dogs' Behaviour in Different Contexts?

A number of studies reviewed already above have shown that in both wolves and dogs, dominance status and affiliative bonds may modulate leadership, group coordination, and hunting (wolves: MacNulty et al. 2012; Mech and Boitani 2003; Peterson et al. 2002; dogs: Bonanni et al. 2010a), marking behaviour and territorial defence (wolves: Peterson et al. 2002; dogs: Bonanni et al. 2010b), as well as play (wolves: Cafazzo et al. 2018a). In a next chapter (Chap. 8), we will also see how it affects cooperation, prosociality, and access to resources (Dale et al. 2017b, 2020; Marshall-Pescini et al. 2017, 2018; Range et al. 2015). In the final section of this chapter, we describe how dominance and affiliation may also affect wolves' and dogs' behaviours in various tests where animals were presented with 'strange', novel, or somewhat distressful situations (separation from their pack mates) (see Table 5.2 for a summary).

For example, in a study at the Wolf Science Center, it emerged that when a higher-ranking individual left the group, the frequency of howling in the remaining individuals was significantly higher than when a lower-ranking individual left. However, an even stronger effect emerged in relation to the affiliative bond between leaver and howler, with significantly more howls being produced the closer the bond between them (Mazzini et al. 2013). Interestingly, salivary cortisol measures did not show a corresponding increase, which may suggest that rather than being just a result of separation distress, howling may be used by pack members to re-establish contact with members of their pack that are considered of particular importance to them.

As part of testing wolves' and dogs' neophobic and explorative tendencies (see Chap. 9), dogs and wolves at the WSC were presented with novel objects in an outdoor enclosure in a dyadic and a group setting and also one solitary condition. As always, independent data (outside the testing context) was collected to establish the dominance relationship between individuals. What emerged is that in the solitary condition, for both wolves and dogs, the higher-ranking individuals were faster in approaching within 1 m of the object, but no effect emerged in relation to contact latency and manipulation time. Higher-ranking dogs, but not wolves, also spent more time investigating the object from a distance (sniffing and looking at the object—but not manipulating it). In dyadic tests, for wolves, the smaller the rank distance, the more time the dyad spent investigating (sniffing and looking at) the novel object, but no such effect was found in dogs (Moretti et al. 2015). No other effect of rank distance was found on latency to approach and contact nor on time spent manipulating the object. Both wolves and dogs approached within 1 m of the object faster when tested with an unrelated partner than with a sibling; however, relatedness did not influence contact latency. In addition, we found that in both wolves and dogs, sibling pairs investigated the object for longer than non-siblings, but they did not manipulate it for longer. No effect of rank emerged in the group condition.

In another study, we compared how the affiliative and dominant dimensions of the animal's relationships would affect their behaviour in an experimental setting

Table 5.2 The effect of the affiliative and dominance dimensions of relationships on the animals' behaviours in different contexts

Test	Species	Affiliation	Rank distance/ Dominance	References
Reconciliation	Wolf	High asymmetry (more insecure) > intense reconciliation	Larger RD > reconciliation	Cafazzo et al. (2018b)
Food-sharing (dyadic test)	Wolf	High > food-sharing	Dom = sub duration feeding alone	Range et al. (2015)
	Dog	High > food-sharing	Dom > sub duration feeding alone	
Food-sharing (group test)	Wolf	No effect	Dominant arrives first Dom = sub duration feeding alone	Dale et al. (2017b)
	Dog	No effect	Dominant does not arrive first Dom > sub duration feeding alone	
	Dog	Not tested	Larger > higher aversion	
Howling after separation	Wolf	High > howling	High > howling	Mazzini et al. (2013)
Neophobia/ exploration (alone condition)	Wolf	Not tested	Higher rank > faster approach	Moretti et al. (2015)
	Dog	Not tested	Higher rank > faster approach Higher rank > object investigation	
Neophobia/ exploration (dyadic condition)	Wolf	Not tested	Smaller RD > object investigation	
	Dog	Not tested	No effect of rank	
Social support	Wolf	High > synchronization (exploration, novel object), BC (social threat), < stress (separation), > stress (social threat), < cortisol reactivity, > escaping (separation, social threat)	Higher RD > escaping (separation, social threat), < cortisol reactivity, < proximity (reunion), < BC initiation (social threat)	Cimarelli et al. (2021)
	Dog	High > synchronization (exploration, novel object), BC (social threat), < stress (separation), > stress (social threat), < cortisol reactivity, > escaping (separation, social threat)	Higher RD > escaping (separation, social threat), < cortisol reactivity, < proximity (exploration), > proximity (reunion, novel object), > BC initiation (social threat)	

Fig. 5.8 Wolf and dog dyads were tested in a series of 'episodes' including a separation and reunion phase and the presentation of somewhat scary options, such as a flag-type object moving on tree in the 'novel object' test and a person, wearing a carnival costume, approaching in a threatening manner. Figure adapted from Cimarelli et al. (2021). https://www.sciencedirect.com/science/article/pii/S0003347221001755; CC BY 4.0; http://creativecommons.org/licenses/by/4.0/

designed to introduce a potentially scary and uncomfortable situation. The test included 14 dogs tested in 15 dyads and 11 wolves tested in 14 dyads, and it comprised different phases mostly lasting approximately 3 min: an 'exploration' phase, in which animals were introduced to an unfamiliar enclosure; a 'separation' phase, in which animals were placed in separate compartments with no visual contact; a 'reunion' phase in which they were allowed back together in the same enclosure; a 'novel object' phase, in which a strange novel object appeared in the enclosure; and a 'social threat' phase (lasting 30 sec), in which a humanoid (a person dressed in a scary costume) approached the animals (located behind a fence—for security reason) while flapping their arms around and jumping in strange ways (Fig. 5.8).

The first interesting result was that when running a 'principal component analyses' (PCA), i.e. a statistical method which allows us to evaluate which behaviours group together in terms of the frequency of their occurrence, we found that what we categorized as 'affiliative behaviours' actually appears to fall into two separate groupings, the first comprising the frequency and duration that animals spent in body contact (BC) and the second including the other affiliative behaviours we recorded, i.e. grooming and the categories lie, approach, and stand friendly (i.e. close while tail-wagging) (we dubbed this friendly contact, FC).

Similar to results found in our other studies, the effect of affiliation was the same in wolves and dogs.

The FC measure of affiliation correlated with a reduction in cortisol levels but also with a higher frequency of stress-related behaviours. This component could therefore be linked to a stress management strategy with animals seeking each other out and coordinating their behaviour to deal with the external events. In fact, in both wolves and dogs, dyads that scored high on this friendship component synchronized their movements more and showed lower cortisol increases than dyads with a lower score. The BC measure of affiliation was instead mostly correlated with other measures of proximity-seeking between individuals and appeared to be less related to stress management.

Overall, for both wolves and dogs, the stronger the affiliative bond, the more animals (1) synchronized with each other in the exploration and novel object phase, (2) initiated BC with each other in the social threat phase, (3) attempted to escape when separated or when approached by the scary humanoid, and, finally, (4) the less their cortisol increased!

The effect patterns relating to rank distance however were not always the same for wolves and dogs. For both species, the greater the rank distance between them, the less the cortisol increased after the test, and the more the animals attempted to escape both during the separation and the social threat phase. However, rank distance differently influenced proximity-seeking behaviours in the two species: while in wolves the smaller the rank distance, the more time they spent in proximity in the reunion phase and the more likely they initiated BC in the social threat phase, the opposite was true in dogs.

Thus, it seems that whereas affiliation regulates how both wolves and dogs confront stress situation in similar ways, the dominance relationships may be perceived differently by wolves and dogs: dogs were more likely to seek for social support in the presence of a partner distant in rank; wolves searched more for support if the companion being tested with was closer in rank. In a later chapter (Chap. 8), we will again find differences in how the dominance relationships regulate wolves' and dogs' behaviour in a cooperative task, suggesting that this aspect really needs more of our attention in the future.

5.7 Summary

The two dimensions of dyadic relationships so far used in dogs and wolves, i.e. affiliation and dominance, affect the animals´ behaviour in multiple ways, from sharing resources, to showing different patterns of behaviours when confronted with novel situations/objects, to the way they play with each other. In a later chapter, we will also see their influence on multiple aspects of cooperation. Taken together, although there are still questions in relation to how best to measure the affiliative relationship between partners and in which contexts dominance relationships affect wolves' and dogs' behaviours, it emerges that both these aspects are important and need to be taken into account when trying to explain their behaviour.

References

Altmann J, Alberts SC (2003) Variability in reproductive success viewed from a life-history perspective in baboons. Am J Hum Biol 15:401–409. https://doi.org/10.1002/ajhb.10157

Archie EA, Morrison TA, Foley CAH, Moss CJ, Alberts SC (2006) Dominance rank relationships among wild female African elephants, *Loxodonta africana*. Anim Behav 71:117–127. https://doi.org/10.1016/j.anbehav.2005.03.023

Arnold K, Barton RA (2001) Postconflict behavior of spectacled leaf monkeys (*Trachypithecus obscurus*). I Reconciliation. Int J Primatol 22:243–266. https://doi.org/10.1023/A:1005623631652

Aureli F (1997) Post-conflict anxiety in nonhuman primates: the mediating role of emotion in conflict resolution. Aggress Behav 23:315–328. https://doi.org/10.1002/(SICI)1098-2337(1997)23:5<315::AID-AB2>3.0.CO;2-H

Aureli F, Cords M, Van Schaik CP (2002) Conflict resolution following aggression in gregarious animals: a predictive framework. Anim Behav 64:325–343. https://doi.org/10.1006/anbe.2002.3071

Aureli F, de Waal FBM (2000) Natural conflict resolution. University of California Press, Berkely

Aureli F, van Schaik CP (1991) Post-conflict behaviour in long-tailed macaques (*Macaca fascicularis*). I The social events. Ethology 89:89–100. https://doi.org/10.1111/j.1439-0310.1991.tb00296.x

Baan C, Bergmüller R, Smith DW, Molnar B (2014) Conflict management in free-ranging wolves, *Canis lupus*. Anim Behav 90:327–334. https://doi.org/10.1016/j.anbehav.2014.01.033

Bauer EB, Smuts BB (2007) Cooperation and competition during dyadic play in domestic dogs, *Canis familiaris*. Anim Behav 73:489–499. https://doi.org/10.1016/j.anbehav.2006.09.006

Beach FA (1970) Coital behaviour in dogs. VIII. Social affinity, dominance and sexual preference in the bitch. Behaviour 36:131–147. https://doi.org/10.1163/156853970X00088

Bekoff M (1972) The development of social interaction, play, and metacommunication in mammals: an ethological perspective. Q Rev Biol 47:412–434. https://doi.org/10.1086/407400

Bekoff M (1974) Social play in coyotes, wolves and dogs. BioSci 24:225–230. https://doi.org/10.2307/1296803

Bekoff M (1995) Play signals as punctuation: the structure of social play in canids. Behaviour 132: 419–429. https://doi.org/10.1163/156853995x00649

Bekoff M (2001) Social play behaviour. Cooperation, fairness, trust, and the evolution of morality. J Conscious Stud 8:81–90. https://www.ingentaconnect.com/contentone/imp/jcs/2001/00000008/00000002/1075

Bekoff M (2004) Wild justice, cooperation and fair play: minding manners, being nice, and feeling good. In: Sussman R, Chapman A (eds) The origins and nature of sociality. Aldine, Chicago, pp 53–79

Bekoff M, Allen C (2009) Intentional communication and social play: how and why animals negotiate and agree to play. Cambridge University Press, Cambridge. https://doi.org/10.1017/CBO9780511608575.006

Berghänel A, Ostner J, Schröder U, Schülke O (2011) Social bonds predict future cooperation in male barbary macaques, *Macaca sylvanus*. Anim Behav 81:1109–1116. https://doi.org/10.1016/j.anbehav.2011.02.009

Bloch G (2002) Alpha-concept, dominance & leadership in wolf families; Report to Parks Canada. Wolf! Magazine 20

Bloch G (2016) The pipestone wolves. The rise and fall of a wolf family. Rocky Mountain Books Ltd

Boitani L, Ciucci P (1995) Comparative social ecology of feral dogs and wolves. Ethol Ecol Evol 7: 49–72. https://doi.org/10.1080/08927014.1995.9522969

Boitani L, Ciucci P, Ortolani A (2007) Behaviour and social ecology of free-ranging dogs. In: Jensen P (ed) The behavioural biology of dogs. Cromwell Press, Trowbridge, pp 147–165

Boitani L, Francisci F, Ciucci P (1995) Population biology and ecology of feral dogs in Central Italy. In: Serpell J (ed) The domestic dog: its evolution, behaviour and interactions with people. Cambridge University Press, Cambridge, pp 218–245

Boitani L, Francisci F, Ciucci P, Andreoli G (2016) The ecology and behavior of feral dogs: a case study from Central Italy. In: Serpell J (ed) The domestic dog: its evolution, behaviour and interactions with people, 2nd edn. Cambridge University Press, Cambridge, pp 342–368. https://doi.org/10.1017/9781139161800.017

Bonanni R, Cafazzo S (2014) The social organisation of a population of free-ranging dogs in a suburban area of Rome: a reassessment of the effects of domestication on dogs' behaviour. In: Kaminski J, Marshall-Pescini S (eds) The Social Dog. Behaviour and Cognition. Elsevier, pp 65–104. https://doi.org/10.1016/B978-0-12-407818-5.00003-6

Bonanni R, Cafazzo S, Abis A, Barillari E, Valsecchi P, Natoli E (2017) Age-graded dominance hierarchies and social tolerance in packs of free-ranging dogs. Behav Ecol 28:1004–1020. https://doi.org/10.1093/beheco/arx059

Bonanni R, Cafazzo S, Valsecchi P, Natoli E (2010a) Effect of affiliative and agonistic relationships on leadership behaviour in free-ranging dogs. Anim Behav 79:981–991. https://doi.org/10.1016/j.anbehav.2010.02.021

Bonanni R, Valsecchi P, Natoli E (2010b) Pattern of individual participation and cheating in conflicts between groups of free-ranging dogs. Anim Behav 79:957–968. https://doi.org/10.1016/j.anbehav.2010.01.016

Bradshaw JW, Nott HMR (1995) Social and communication behaviour of companion dogs. In: Serpell J (ed) In the domestic dog. Cambridge University Press, Cambridge, GB, pp 115–133

Bradshaw JWS, Blackwell EJ, Casey RA (2009) Dominance in domestic dogs–useful construct or bad habit? J Vet Behav 4:135–144. https://doi.org/10.1016/j.jveb.2008.08.004

Byers JA, Walker C (1995) Refining the motor training hypothesis for the evolution of play. Am Nat 146:25–40. https://doi.org/10.1086/285785

Byosiere S-E, Espinosa J, Smith BP (2018) The function of play bows in *Canis lupus* and its variants: a comparison of dingo (*Canis lupus dingo*), dog (*Canis lupus familiaris*) and wolf puppies (*Canis lupus*). Behaviour 155:369–388. https://doi.org/10.1163/1568539X-00003495

Byosiere SE, Espinosa J, Marshall-Pescini S, Smuts B, Range F (2016) Investigating the function of play bows in dog and wolf puppies (*Canis lupus familiaris, Canis lupus occidentalis*). PLoS One 11:e0168570. https://doi.org/10.1371/journal.pone.0168570

Cafazzo S, Bonanni R, Valsecchi P, Natoli E (2014) Social variables affecting mate preferences, copulation and reproductive outcome in a pack of free-ranging dogs. PLoS One 9:e98594. https://doi.org/10.1371/journal.pone.0098594

Cafazzo S, Lazzaroni M, Marshall-Pescini S (2016) Dominance relationships in a family pack of captive arctic wolves (*Canis lupus arctos*): the influence of competition for food, age and sex. PeerJ 4:e2707. https://doi.org/10.7717/peerj.2707

Cafazzo S, Marshall-Pescini S, Essler JL, Virányi Z, Kotrschal K, Range F (2018a) In wolves, play behaviour reflects the partners' affiliative and dominance relationship. Anim Behav 141:137–150. https://doi.org/10.1016/j.anbehav.2018.04.017

Cafazzo S, Marshall-Pescini S, Lazzaroni M, Virányi Z, Range F (2018b) The effect of domestication on post-conflict management: wolves reconcile while dogs avoid each other. R Soc Open Sci 5:171553. https://doi.org/10.1098/rsos.171553

Cafazzo S, Valsecchi P, Bonanni R, Natoli E (2010) Dominance in relation to age, sex and competitive contexts in a group of free-ranging domestic dogs. Behav Ecol 21:443–455. https://doi.org/10.1093/beheco/arq001

Caro TM (1988) Adaptive significance of play: are we getting closer? Trends Ecol Evol 3:50–54. https://doi.org/10.1016/0169-5347(88)90048-1

Castles DL, Whiten A (1998) Post-conflict behaviour of wild olive baboons. I. Reconciliation, Redirection and Consolation. Ethology 104:126–147. https://doi.org/10.1111/j.1439-0310.1998.tb00057.x

Cimarelli G, Marshall-Pescini S, Range F, Berghänel A, Virányi Z (2021) Relationship quality affects social stress buffering in dogs and wolves. Anim Behav 178:127–140. https://doi.org/10.1016/j.anbehav.2021.06.008

Clutton-Brock J (1981) Domesticated animals from early times. W. Heinemann, London

Clutton-Brock TH, Albon D, Guiness FE (1986) Great expectations: dominance, breeding success and offspring sex ratios in red deer. Anim Behav 34:460–471. https://doi.org/10.1016/S0003-3472(86)80115-4

Clutton-Brock TH, Albon SD, Guinness F (1984) Maternal dominance, breeding success and birth sex ratios in red deer. Nature 308:358–360. https://doi.org/10.1038/308358a0
Conradt L, Roper TJ (2007) Democracy in animals: the evolution of shared group decisions. Proc R Soc B 274:2317–2326. https://doi.org/10.1098/rspb.2007.0186
Cools AKA, van Hout AJM, Nelissen MHJ (2008) Canine reconciliation and third-party-initiated Postconflict affiliation: do peacemaking social mechanisms in dogs rival those of higher primates? Ethology 113:53–63. https://doi.org/10.1111/j.1439-0310.2007.01443.x
Coppinger RP, Smith CK (1983) The domestication of evolution. Environ Conserv 10:283–292. https://doi.org/10.1017/S0376892900012984
Cordoni G (2009) Social play in captive wolves (*Canis lupus*): not only an immature affair. Behaviour 146:1363–1385. https://doi.org/10.1163/156853909x427722
Cordoni G, Palagi E (2008) Reconciliation in wolves (*Canis lupus*): new evidence for a comparative perspective. Ethology 114:298–308. https://doi.org/10.1111/j.1439-0310.2008.01474.x
Cords M, Aureli F (2000) Reconciliation and relationship qualities. In: Aureli F, de Waal FBM (eds) Natural conflict resolution. University of California Press, Berkeley, pp 177–198
Creel S, Creel NM, Mills MGL, Montfort SL (1997) Rank and reproduction in cooperatively breeding African wild dogs: behavioral and endocrine correlates. Behav Ecol 8:298–306. https://doi.org/10.1093/beheco/8.3.298
Creel S, Dantzer B, Goymann W, Rubenstein DR (2013) The ecology of stress: effects of the social environment. Funct Ecol 27:66–80. https://doi.org/10.1111/j.1365-2435.2012.02029.x
Crockford C, Wittig RM, Langergraber K, Ziegler TE, Zuberbühler K, Deschner T (2013) Urinary oxytocin and social bonding in related and unrelated wild chimpanzees. Proc R Soc B 280: 20122765. https://doi.org/10.1098/rspb.2012.2765
Dale R, Marshall-Pescini S, Range F (2017a) Do females use their sexual status to gain resource access? Investigating food-for-sex in wolves and dogs. Curr Zool 63:323–330. https://doi.org/10.1093/cz/zow111
Dale R, Marshall-Pescini S, Range F (2020) What matters for cooperation? The importance of social relationship over cognition. Sci Rep 10:11778. https://doi.org/10.1038/s41598-020-68734-4
Dale R, Range F, Stott L, Kotrschal K, Marshall-Pescini S (2017b) The influence of social relationship on food tolerance in wolves and dogs. Behav Ecol Sociobiol 71:107–131. https://doi.org/10.1007/s00265-017-2339-8
Darwin C (1859) On the origin of species by means of natural selection, or the preservation of favoured races in the struggle for life. John Murray, London
Darwin C (1871) The descent of man, and selection in relation to sex. John Murray, London
de Vries H (1995) An improved test of linearity in dominance hierarchies containing unknown or tied relationships. Anim Behav 50:1375–1389. https://doi.org/10.1016/0003-3472(95)80053-0
de Vries H, Stevens JMG, Vervaecke H (2006) Measuring and testing the steepness of dominance hierarchies. Anim Behav 71:585–592. https://doi.org/10.1016/j.anbehav.2005.05.015
de Waal FBM (1986) The integration of dominance and social bonding in primates. Q Rev Biol 61: 459–479. https://doi.org/10.1086/415144
de Waal FBM (1989) Dominance 'style' and primate social organization. In: Standen V, Foley RA (eds) Comparative Socioecology: the Behavioural ecology of humans and other mammals. Blackwell Scientific Publications, Oxford, pp 243–263
de Waal FBM, Aureli F (1997) Conflict resolution and distress alleviation in monkeys and apes. In: Carter CS, Kirkpatric B, Lenderhendler I (eds) The integrative neurobiology of affiliation, vol 807. Annals of the New York Academy of Sciences, New York, pp 317–328
de Waal FBM, Luttrell LM (1985) The formal hierarchy of rhesus macaques: an investigation of the bared-teeth display. Am J Primatol 9:73–85. https://doi.org/10.1002/ajp.1350090202
de Waal FBM, Luttrell LM (1989) Toward a comparative Socioecology of the genus *Macaca*: different dominance styles in rhesus and Stumptail monkeys. Am J Primatol 19:83–110. https://doi.org/10.1002/ajp.1350190203

de Waal FBM, van Roosmalen A (1979) Reconciliation and consolation among chimpanzees. Behav Ecol Sociobiol 5:55–66. https://doi.org/10.1007/BF00302695

Deaner RO, Khera AV, Platt ML (2005) Monkeys pay per view: adaptive valuation of social images by rhesus macaques. Curr Biol 15:543–548. https://doi.org/10.1016/j.cub.2005.01.044

Demaria C, Thierry B (2001) A comparative study of reconciliation in rhesus and tonkean macaques. Behaviour 138:397–410. https://doi.org/10.1163/15685390152032514

Dolhinow P (1999) Play: a critical process in the developmental system. In: Dolhinow P, Fuentes A (eds) The nonhuman primates. Mayfield Publishing Co., Mountain View, pp 231–236

Dunbar RIM (1988) Primate social systems. Croom Helm studies in behavioural adaptation, Croom Helm, London, Sydney

East M, Hofer H, Wickler W (1993) The erect penis is a flag of submission in a female-dominated society: greetings in Serengeti spotted hyenas. Behav Ecol Sociobiol 33:355–370. https://doi.org/10.1007/BF00170251

Essler JL, Cafazzo S, Marshall-Pescini S, Virányi Z, Kotrschal K, Range F (2016) Play behavior in wolves: using the '50:50' rule to test for egalitarian play styles. PLoS One 11:e0154150. https://doi.org/10.1371/journal.pone.0154150

Fagen R (1981) Animal play behavior. Oxford University Press, New York

Fairbanks WS (1994) Dominance, age and aggression among female pronghorn, *Antilocapra americana* (family: Antilocapridae). Ethology 97:278–293. https://doi.org/10.1111/j.1439-0310.1994.tb01047.x

Fatjó J, Feddersen-Petersen D, Ruiz de la Torre JL, Amat M, Mets M, Braus B, Manteca X (2007) Ambivalent signals during agonistic interactions in a captive wolf pack. Appl Anim Behav Sci 105:274–283. https://doi.org/10.1016/j.applanim.2006.11.009

Feddersen-Petersen D (1986) Observations on social play in some species of canidae. Zool Anz 217:130–144

Feddersen-Petersen D (1991) The ontogeny of social play and agonistic behaviour in selected canid species. Bonn Zool Beitr 42:97–114. https://www.zobodat.at/pdf/Bonner-Zoologische-Beitraege_42_0097-0114.pdf

Feddersen-Petersen DU (1994) Fortpflanzungsverhalten beim Hund. Enke

Feddersen-Petersen DU (2004) Hundepsychologie. Sozialverhalten und Wesen, Emotionen und Individualität, vol 4, Kosmos, Stuttgart

Feddersen-Petersen DU (2007) Social behaviour of dogs and related canids. In: Jensen P (ed) The behavioural biology of dogs. Cromwell Press, Trowbridge, pp 105–119

Flack JC, de Waal FBM (2004) Dominance style, social power, and conflict management: a conceptual framework. In: Thierry B, Singh M, Kaumanns W (eds) Macaque societies. A model for the study of social organization. Cambridge studies in biological and evolutionary anthropology. Cambridge University Press, New York, pp 157–182

Fox MW (1968) The influence of domestication upon behaviour of animals. In: Fox MW (ed) Abnormal behavior in animals. W.B. Saunders, Philadelphia, pp. 64–76

Frank H, Frank MG (1982) On the effects of domestication on canine social development and behavior. Appl Animal Ethol 8:507–525. https://doi.org/10.1016/0304-3762(82)90215-2

Fraser ON, Aureli F (2008) Reconciliation, consolation and Postconflict behavioral specificity in chimpanzees. Am J Primatol 70:1–10. https://doi.org/10.1002/ajp.20608

Fraser ON, Bugnyar T (2010) The quality of social relationships in ravens. Anim Behav 79:927–933. https://doi.org/10.1016/j.anbehav.2010.01.008

Fraser ON, Bugnyar T (2011) Ravens reconcile after aggressive conflicts with valuable partners. PLoS One 6:e18118. https://doi.org/10.1371/journal.pone.0018118

Gomendio M (1988) The development of different types of play in gazelles: implications for the nature and functions of play. Anim Behav 36:825–836. https://doi.org/10.1016/S0003-3472(88)80165-9

Graham KL, Burghardt GM (2010) Current perspectives on the biological study of play: signs of Progress. Q Rev Biol 85:393–418. https://doi.org/10.1086/656903

Harrington FH, Asa CS (2003) Wolf communication: behavior, ecology, and conservation. In: Mech LD, Boitani L (eds) Wolves: behavior, ecology and conservation. University of Chicago Press, Chicago & London, pp 96–99

Hausfater G, Altmann J, Altmann S (1982) Long-term consistency of dominance relations among female baboons (*Papio cynocephalus*). Science 217:752–755. https://doi.org/10.1126/science.217.4561.752

Holekamp KE, Sakai ST, Lundrigan BL (2007) Social intelligence in the spotted hyena (*Crocuta crocuta*). Phil Trans R Soc B 362:523–538. https://doi.org/10.1098/rstb.2006.1993

Holekamp KE, Smale L (1991) Dominance acquisition during mammalian social development: the "inheritance" of maternal rank. Am Zool 31:306–317. https://doi.org/10.1093/icb/31.2.306

Holekamp KE, Smale L, Szykman M (1996) Rank and reproduction in the female spotted hyaena. J Reprod Fert 108:229–237. https://doi.org/10.1530/jrf.0.1080229

Isbell LA, Young TP (2002) Ecological models of female social relationships in primates: similarities, disparities, and some directions for future clarity. Behaviour 139:177–202. https://doi.org/10.1163/156853902760102645

Jenks SM, Ginsburg BE (1987) Socio-sexual dynamics in a captive wolf pack. In: Frank H (ed) Man and wolf. Dr. W Junk Publishers, Dordrecht, NL, pp 375–400

Kaplan JR, Manuck SB, Clarkson TB, Lusso FM, Taub DM (1982) Social status, environment, and atherosclerosis in Cynomolgus monkeys. Arteriosclerosis 2:359–368. https://doi.org/10.1161/01.atv.2.5.359

Kappeler PM, van Schaik CP (1992) Methodological and evolutionary aspects of reconciliation among primates. Ethology 92:51–69. https://doi.org/10.1111/j.1439-0310.1992.tb00948.x

Kikusui T, Winslow JT, Mori Y (2006) Social buffering: relief from stress and anxiety. Phil Trans R Soc Lond B 361:2215–2228. https://doi.org/10.1098/rstb.2006.1941

Klass K, Cords M (2011) Effect of unknown relationships on linearity, steepness and rank ordering of dominance hierarchies: simulation studies based on data from wild monkeys. Behav Proc 88: 168–176. https://doi.org/10.1016/j.beproc.2011.09.003

Kotrschal K, Hirschenhauser K, Möstl E (1998) The relationship between social stress and dominance is seasonal in Greylag geese. Anim Behav 55:171–176. https://doi.org/10.1006/anbe.1997.0597

Kutsukake N, Castles DL (2001) Reconciliation and variation in post-conflict stress in Japanese macaques (*Macaca fuscata fuscata*): testing the integrated hypothesis. Anim Cogn 4:259–268. https://doi.org/10.1007/s10071-001-0119-2

Kutsukake N, Clutton-Brock TH (2008) Do meerkats engage in conflict management following aggression? Reconciliation, submission and avoidance. Anim Behav 75:1441–1453. https://doi.org/10.1016/j.anbehav.2007.09.018

Lazzaroni M, Marshall-Pescini S, Cafazzo S (2017) Post-conflict opponent affiliation reduces victim re-aggression in a family group of captive arctic wolves (*Canis lupus arctos*). PLoS One 12:e0187450. https://doi.org/10.1371/journal.pone.0187450

Loizos C (1967) Play behaviour in higher primates: a review. In: Morris D (ed) Primate ethology. Anchor Books, Chicago, pp 226–282

Lu A, Koenig A, Borries C (2008) Formal submission, tolerance and socioecological models: a test with female Hanuman langurs. Anim Behav 76:415–428. https://doi.org/10.1016/j.anbehav.2008.04.006

MacNulty DR, Smith DW, Mech LD, Vucetich JA, Packer C (2012) Nonlinear effects of group size on the success of wolves hunting elk. Behav Ecol 23:75–82. https://doi.org/10.1093/beheco/arr159

Majolo B, Ventura R, Schino G (2010) Asymmetry and dimensions of relationship quality in the Japanese macaque (*Macaca fuscata yakui*). Int J Primatol 31:736–750. https://doi.org/10.1007/s10764-010-9424-4

Majumder SS et al (2014) To be or not to be social: foraging associations of free-ranging dogs in an urban ecosystem. Acta Ethol 17:1–8. https://doi.org/10.1007/s10211-013-0158-0

Marshall-Pescini S, Basin C, Range F (2018) A task-experienced partner does not help dogs be as successful as wolves in a cooperative string-pulling task. Sci Rep 8:16049. https://doi.org/10.1038/s41598-018-33771-7

Marshall-Pescini S, Schwarz JFL, Kostelnik I, Virányi Z, Range F (2017) Importance of a species' socioecology: wolves outperform dogs in a conspecific cooperation task. Proc Natl Acad Sci U S A 114:11793–11798. https://doi.org/10.1073/pnas.1709027114

Martin P, Caro TM (1985) On the functions of play and its role in behavioral development. Advances in the Study of Behavior 15:59–103. https://doi.org/10.1016/S0065-3454(08)60487-8

Massen JJM, Sterck EHM, de Vos H (2010) Close social associations in animals and humans: functions and mechanisms of friendship. Behaviour 147:1379–1412. https://doi.org/10.1163/000579510x528224

Mazzini F, Townsend SW, Virányi Z, Range F (2013) Wolf howling is mediated by relationship quality rather than underlying emotional stress. Curr Biol 23:1677–1680. https://doi.org/10.1016/j.cub.2013.06.066

McFarland R, Majolo B (2011) Exploring the components, asymmetry and distribution of relationship quality in wild barbary macaques (*Macaca sylvanus*). PLoS One 6:e28826. https://doi.org/10.1371/journal.pone.0028826

McLeod PJ, Fentress JC (1997) Developmental changes in the sequential behavior of interacting timber wolf pups. Behav Proc 39:127–136. https://doi.org/10.1016/S0376-6357(96)00056-3

Mech LD (1999) Alpha status, dominance, and division of labor in wolf packs. Can J Zool 77: 1196–1203. https://doi.org/10.1139/cjz-77-8-1196

Mech LD (2000) Leadership in wolf, *Canis lupus*, packs. Can Field-Nat 114:259–263. https://www.biodiversitylibrary.org/page/34236757

Mech LD, Adams LG, Meier TJ, Burch JW, Dale BW (1998) The wolves of Denali. University of Minnesota Press, Minneapolis

Mech LD, Boitani L (2003) Wolf social ecology. In: Mech LD, Boitani L (eds) Wolves: behavior, ecology, and conservation. The University of Chcago Press, Chicago, London, pp 1–35

Moretti L, Hentrup M, Kotrschal K, Range F (2015) The influence of relationships on neophobia and exploration in wolves and dogs. Anim Behav 107:159–173. https://doi.org/10.1016/j.anbehav.2015.06.008

Morey DF (1994) The early evolution of the domestic dog. Am Sci 82:336–347 http://www.jstor.org/stable/29775234

Moss CJ, Poole JH (1983) Relationships and social structure in African elephants. In: Hinde RA (ed) Primate social relationships. Sinauer, Sunderland, MA, pp 315–325

Neumann C, McDonald DB, Shizuka D (2018) Dominance ranks, dominance ratings and linear hierarchies: a critique. Anim Behav 144:e1–e16. https://doi.org/10.1016/j.anbehav.2018.07.012

Newberry RC, Wood-Gush DGM, Hall JW (1988) Playful behaviour of piglets. Behav Proc 17: 205–216. https://doi.org/10.1016/0376-6357(88)90004-6

Norscia I, Palagi E (2011) Do wild brown lemurs reconcile? Not always. J Ethol 29:181–185. https://doi.org/10.1007/s10164-010-0228-y

O'Heare J (2007) Social dominance: useful construct or quagmire? J Appl Comp Anim Behav 1:56–83

Ostner J, Schülke O (2018) Linking sociality to Fitness in primates: a Call for mechanisms. In: Naguib M, Barrett L, Healy SD, Podos J, Simmons LW, Zuk M (eds) Advances in the study of behavior, vol 50. Elsevier Academic Press, pp 127–175. https://doi.org/10.1016/bs.asb.2017.12.001

Packard JM (2003) Wolf behavior: reproductive, social and intelligent. In: Mech LD, Boitani L (eds) Wolves: behaviour, ecology and conservation. University of Chicago Press, Chicago, pp 35–65

Pal SK (2010) Play behaviour during early ontogeny in free-ranging dogs (*Canis familiaris*). Appl Anim Behav Sci 126:140–153. https://doi.org/10.1016/j.applanim.2010.06.005

Pal SK, Ghosh B, Roy S (1998) Agonistic behaviour of free-ranging dogs (*Canis familiaris*) in relation to season, sex and age. Appl Anim Behav Sci 59:331–348. https://doi.org/10.1016/S0168-1591(98)00108-7

Palagi E (2004) Adaptive role of social play behaviour and use of play signals in *pan troglodytes* and *pan paniscus*: a comparative study. PhD Dissertation, University of Pisa

Palagi E (2006) Social play in bonobos (*pan paniscus*) and chimpanzees (*pan troglodytes*): implications for natural social systems and Interindividual relationships. Am J Phys Anthropol 129:418–426. https://doi.org/10.1002/ajpa.20289

Palagi E (2007) Play at work: revisiting data focusing on chimpanzees (*pan troglodytes*). J Anthropol Sci 85:63–81

Palagi E (2008) Sharing the motivation to play: the use of signals in adult bonobos. Anim Behav 75: 887–896. https://doi.org/10.1016/j.anbehav.2007.07.016

Palagi E (2018) Not *just for fun*! Social play as a springboard for adult social competence in human and non-human primates. Behav Ecol Sociobiol 72:90. https://doi.org/10.1007/s00265-018-2506-6

Palagi E, Cordoni G (2009) Postconflict third-party affiliation in *Canis lupus*: do wolves share similarities with the great apes? Anim Behav 78:979–986. https://doi.org/10.1016/j.anbehav.2009.07.017

Palagi E, Cordoni G, Borgognini Tarli SM (2004a) Immediate and delayed benefits of play behaviour: new evidence from chimpanzees (*pan troglodytes*). Ethology 110:949–962. https://doi.org/10.1111/j.1439-0310.2004.01035.x

Palagi E, Paoli T (2007) Play in adult bonobos (*pan paniscus*): modality and potential meaning. Am J Phys Anthropol 134:219–225. https://doi.org/10.1002/ajpa.20657

Palagi E, Paoli T, Borgognini Tarli S (2004b) Reconciliation and consolation in captive bonobos (*pan paniscus*). Am J Primatol 62:15–30. https://doi.org/10.1002/ajp.20000

Palagi E, Paoli T, Borgognini Tarli S (2006) Short-term benefits of play behavior and conflict prevention in *pan paniscus*. Int J Primatol 27:1257–1270. https://doi.org/10.1007/s10764-006-9071-y

Palagi E, Paoli T, Tarli SB (2005) Aggression and reconciliation in two captive groups of *Lemur catta*. Int J Primatol 26:279–294. https://doi.org/10.1007/s10764-005-2925-x

Pellis SM, Iwaniuk AN (1999a) The problem of adult play fighting: a comparative analysis of Playand courtship in primates. Ethology 105:783–806. https://doi.org/10.1046/j.1439-0310.1999.00457.x

Pellis SM, Iwaniuk AN (1999b) The roles of phylogeny and sociality in the evolution of social play in muroid rodents. Anim Behav 58:361–373. https://doi.org/10.1006/anbe.1999.1141

Pellis SM, Iwaniuk AN (2000) Adult–adult play in primates: comparative analyses of its origin, distribution and evolution. Ethology 106:1083–1104. https://doi.org/10.1046/j.1439-0310.2000.00627.x

Peñate L, Peláez F, Sánchez S (2009) Reconciliation in captive cotton-top tamarins (*Saguinus oedipus*), a cooperative breeding primate. Am J Primatol 71:895–900. https://doi.org/10.1002/ajp.20721

Peterson RO, Jacobs AK, Drummer TD, Mech LD, Smith DW (2002) Leadership behavior in relation to dominance and reproductive status in gray wolves, *Canis lupus*. Can J Zool 80:1405–1412. https://doi.org/10.1139/z02-124

Poirier FE, Bellisari A, Haines L (1978) Functions of primate play behavior. In: Smith EO (ed) Social play in primates. Academic Press, New York, pp 143–168

Preuschoft S (1999) Are primates behaviorists? Formal dominance, cognition, and free-floating rationales. J Comp Psychol 113:91–95. https://doi.org/10.1037/0735-7036.113.1.91

Price EO (1999) Behavioral development in animals undergoing domestication. Appl Anim Behav Sci 65:245–271. https://doi.org/10.1016/S0168-1591(99)00087-8

Pusey AE, Williams J, Goodall J (1997) The influence of dominance rank on the reproductive success of female chimpanzees. Science 277:828–830. https://doi.org/10.1126/science.277.5327.828

Rabb GB, Woolpy JH, Ginsburg BE (1967) Social relationships in a group of captive wolves. Am Zool 7:305–312. https://doi.org/10.1093/icb/7.2.305

Range F, Ritter C, Virányi Z (2015) Testing the myth: tolerant dogs and aggressive wolves. Proc R Soc B 282:20150220. https://doi.org/10.1098/rspb.2015.0220

Romero T, Aureli F (2017) Conflict resolution. In: Call J (ed) APA handbook of comparative psychology. Basic concepts, methods, neural substrate, and behavior, vol 1. American Psychological Association, Washington, DC, pp 1–9

Romero T, Ito M, Saito A, Hasegawa T (2014) Social modulation of contagious yawning in wolves. PLoS One 9:e105963. https://doi.org/10.1371/journal.pone.0105963

Sands J, Creel S (2004) Social dominance, aggression and faecal glucocorticoid levels in a wild population of wolves, *Canis lupus*. Anim Behav 67:387–396. https://doi.org/10.1016/j.anbehav.2003.03.019

Sapolsky RM (1983) Endocrine aspects of social instability in the olive baboon (*Papio anubis*). Am J Primatol 5:365–379. https://doi.org/10.1002/ajp.1350050406

Sapolsky RM (2005) The influence of social hierarchy on primate health. Science 308:648–652. https://doi.org/10.1126/science.1106477

Schaffner CM, Caine NG (2000) The peacefulness of cooperatively breeding primates. In: Aureli F, de Waal FBM (eds) Natural conflict resolution. University of California Press, Berkeley, CA, pp 155–175

Schenkel R (1967) Submission: its features and function in the wolf and dog. Am Zool 7:319–329. https://doi.org/10.1093/icb/7.2.319

Schilder MBH, Vinke CM, van der Borg JAM (2014) Dominance in domestic dogs revisited: useful habit and useful construct? J Vet Behav 9:184–191. https://doi.org/10.1016/j.jveb.2014.04.005

Schotté CS, Ginsberg G (1987) Development of social organization and mating in a captive wolf pack. In: Frank H (ed) Man and Wolf. advances, issues, and problems in captive wolf research. Dr. W. Junk Publishers, NL, Dordrecht, pp 349–374

Schülke O, Bhagavatula J, Vigilant L, Ostner J (2010) Social bonds enhance reproductive success in male macaques. Curr Biol 20:2207–2210. https://doi.org/10.1016/j.cub.2010.10.058

Scott JP, Fuller JL (1965) Genetics and the social behavior of the dog. The classic study. University of Chicago Press, Chicago, London

Shepherd K (2002) Development of behaviour, social behaviour and communication in dogs. In: Horwitz DF, Mills DS, Heath S (eds) BSAVA Manual of Canine and Feline Behavioural Medicine British Small Animal Veterinary Association (BSAVA), Gloucester, pp. 8–20

Shizuka D, McDonald David B (2015) The network motif architecture of dominance hierarchies. J R Soc Interface 12:20150080. https://doi.org/10.1098/rsif.2015.0080

Shultz S, Dunbar R (2010) Bondedness and sociality. Behaviour 147:775–803. https://doi.org/10.1163/000579510X501151

Silk J, Cheney D, Seyfarth R (2013) A practical guide to the study of social relationships. Evol Anthropol 22:213–225. https://doi.org/10.1002/evan.21367

Silk JB, Alberts SC, Altmann J (2003) Social bonds of female baboons enhance infant survival. Science 302:1231–1234. https://doi.org/10.1126/science.1088580

Silk JB, Alberts SC, Altmann J (2006) Social relationships among adult female baboons (*Papio cynocephalus*) II. Variation in the quality and stability of social bonds. Behav Ecol Sociobiol 61: 197–204. https://doi.org/10.1007/s00265-006-0250-9

Silk JB et al (2010) Strong and consistent social bonds enhance the longevity of female baboons. Curr Biol 20:1359–1361. https://doi.org/10.1016/j.cub.2010.05.067

Silk MJ, Cant MA, Cafazzo S, Natoli E, McDonald RA (2019) Elevated aggression is associated with uncertainty in a network of dog dominance interactions. Proc R Soc B 286:20190536. https://doi.org/10.1098/rspb.2019.0536

Smith DW et al. (2012) Yellowstone Wolf Project: annual report, 2011. National Park Service, Yellowstone Center for Resources, YCR-2012-01. Yellowstone National Park, Wyoming, USA. https://www.nps.gov/yell/learn/nature/upload/Wolf_AR_2011.pdf

Smith PK (1982) Does play matter? Functional and evolutionary aspects of animal and human play. Behav Brain Sci 5:139–155. https://doi.org/10.1017/S0140525X0001092X
Smuts B (2014) Social behaviour among companion dogs with an emphasis on play. In: Kaminski J, Marshall-Pescini S (eds) The social dog. Behaviour and cognition. Academic Press, San Diego, pp 105–130. https://doi.org/10.1016/B978-0-12-407818-5.00004-8
Špinka M, Newberry RC, Bekoff M (2001) Mammalian play: training for the unexpected. Q Rev Biol 76:141–168. https://doi.org/10.1086/393866
Sterck EHM, Watts DP, van Schaik CP (1997) The evolution of female social relationships in nonhuman primates. Behav Ecol Sociobiol 41:291–309. https://doi.org/10.1007/s002650050390
Tanner JB, Smale L, Holekamp KE (2007) Ontogenetic variation in the play behavior of spotted hyenas (*Crocuta crocuta*). J Dev Proc 2:5–30
Thierry B (2000) Covariation of conflict management patterns across macaque species. In: Aureli F, de Waal FBM (eds) Natural conflict resolution. University of California Press, Berkeley, pp 106–128
Thompson KV (1998) Self-assessment in juvenile play. In: Bekoff M, Byers JA (eds) Animal play: evolutionary, comparative, and ecological perspectives. Cambridge University Press, Cambridge, pp 183–204
Trisko RK, Sandel AA, Smuts B (2016) Affiliation, dominance and friendship among companion dogs. Behaviour 153:693–725. https://doi.org/10.1163/1568539X-00003352
Trisko RK, Smuts BB (2015) Dominance relationships in a group of domestic dogs (*Canis lupus familiaris*). Behaviour 152:677–704. https://doi.org/10.1163/1568539X-00003249
Van Der Borg JAM, Schilder MBH, Vinke CM, De Vries H, Petit O (2015) Dominance in domestic dogs: a quantitative analysis of its Behavioural measures. PLoS One 10:e133978. https://doi.org/10.1371/journal.pone.0133978
van Hooff JARAM, Wensing JAB (1987) Dominance and its behavioral measures in a captive wolf pack. In: Frank H (ed) Man and wolf. Advances, issues, and problems in captive wolf research. Dr. W. Junk Publishers, Dordrecht, The Netherlands, pp 219–252
van Kerkhove W (2004) A fresh look at the wolf-pack theory of companion-animal dog social behavior. J Appl Anim Welf Sci 7:279–285. https://doi.org/10.1207/s15327604jaws0704_7
van Schaik CP (1983) Why are diurnal primates living in groups? Behaviour 87:120–144. https://doi.org/10.1163/156853983X00147
van Schaik CP (1989) The ecology of social relationships amongst female primates. In: Standen V, Foley RA (eds) Comparative socioecology: the behavioural ecology of humans and other mammals. Blackwell Scientific Publications, Oxford, pp 195–218
Vasconcellos AS, Kotrschal K, Ades C (2012) Social stress in wolves: biology, behavior and conservation. In: Maia AP, Crussi HF (eds) Wolves. biology, behavior and conservation. Animal science, issues and professions. Nova Science Publishers, Hauppauge, NY, pp 157–175
Vehrencamp SL (1983) A model for the evolution of despotic versus egalitarian societies. Anim Behav 31:667–682. https://doi.org/10.1016/S0003-3472(83)80222-X
Vervaecke H, de Vries H, van Elsacker L (2000a) Dominance and its behavioral measures in a captive Group of Bonobos (*Pan paniscus*). Int J Primatol 21:47–68. https://doi.org/10.1023/A:1005471512788
Vervaecke H, De Vries H, Van Elsacker L (2000b) The pivotal role of rank in grooming and support behavior in a captive group of bonobos (*Pan paniscus*). Behaviour 137:1463–1485. https://doi.org/10.1163/156853900502673
VonHoldt BM, Stahler DR, Smith DW, Earl DA, Pollinger JP, Wayne RK (2008) The genealogy and genetic viability of reintroduced Yellowstone grey wolves. Mol Ecol 17:252–274. https://doi.org/10.1111/j.1365-294X.2007.03468.x
Wahaj S, Guse KR, Holekamp KE (2001) Reconciliation in the spotted hyena (*Crocuta crocuta*). Ethology 107:1057–1074. https://doi.org/10.1046/j.1439-0310.2001.00717.x

Ward C, Bauer EB, Smuts BB (2008) Partner preferences and asymmetries in social play among domestic dog, *Canis lupus familiaris,* littermates. Anim Behav 76:1187–1199. https://doi.org/10.1016/j.anbehav.2008.06.004

Weaver A (2003) Conflict and reconciliation in captive bottlenose dolphins, *Tursiops truncatus*. Mar Mamm Sci 19:836–846. https://doi.org/10.1111/j.1748-7692.2003.tb01134.x

Webb CE, Franks B, Romero T, Higgins ET, de Waal FBM (2014) Individual differences in chimpanzee reconciliation relate to social switching behaviour. Anim Behav 90:57–63. https://doi.org/10.1016/j.anbehav.2014.01.014

Westlund K, Ljungberg T, Borefelt U, Abrahamsson C (2000) Post-conflict affiliation in common marmosets (*Callithrix jacchus jacchus*). Am J Primatol 52:31–46. https://doi.org/10.1002/1098-2345(200009)52:1<31::AID-AJP3>3.0.CO;2-Z

Wheeler BC, Scarry CJ, Koenig A (2013) Rates of agonism among female primates: a cross-taxon perspective. Behav Ecol 24:1369–1380. https://doi.org/10.1093/beheco/art076

Wittig RM, Boesch C (2003) The choice of post-conflict interactions in wild chimpanzees (*Pan troglodytes*). Behaviour 140:1527–1559. https://doi.org/10.1163/156853903771980701

Wittig RM, Crockford C, Lehmann J, Whitten PL, Seyfarth RM, Cheney DL (2008) Focused grooming networks and stress alleviation in wild female baboons. Horm Behav 54:170–177. https://doi.org/10.1016/j.yhbeh.2008.02.009

Wrangham RW (1980) An ecological model of female-bonded primate groups. Behaviour 75:262–300. https://doi.org/10.1163/156853980X00447

Yanagi A, Berman CM (2014) Body signals during social play in free-ranging rhesus macaques (*Macaca mulatta*): a systematic analysis. Am J Primatol 76:168–179. https://doi.org/10.1002/ajp.22219

Ziemen E (1981) The wolf: his place in the natural world. Souvenir Press, London

Zimen E (1970) Vergleichende Verhaltensbeobachtungen an Wölfen und Königspudeln. Universität Kiel, Dissertation

Zimen E (1971) Wölfe und Königspudel. Vergleichende Verhaltensbeobachtungen. Ethologische Studien, Piper, München

Zimen E (1982) A wolf pack sociogram. In: Harrington FH, Paquet PC (eds) Wolves of the world: perspectives of behaviour, ecology and conservation. Noyes Publications, Park Ridge, NY, pp 282–322

Social Behaviour with Humans 6

Perhaps because of the dog's special place in Western society as a family member, in the last 20 years, a conspicuous part of dog research has aimed at investigating dogs' understanding of human communication (see Chap. 7) and the dog-human social bond with the aim of characterizing this interspecific relationship. Rather than coming from an 'ethological' perspective, where, as outlined in the previous chapter, dominance and affiliation are the two basic dimensions used to define dyadic relationships between individuals, the dog-human bond has largely been investigated by adopting paradigms and terminologies deriving from the infant-mother literature in developmental psychology—probably because of the close relationships of dogs and humans.

In the current chapter, we outline the main concepts and paradigms adopted to investigate the dog-human bond, and then turn to the main question addressed with the wolf-dog comparison, i.e. has domestication qualitatively affected the bond that dogs form with humans (or turning the question around: can wolves and dogs establish similar bonds with humans given similar conditions?). In order to elucidate potential differences in the emotional reaction of the animals towards humans, aside from a comparison of the behaviour, some studies have also included physiological and hormonal measures; thus, in the final section, we review these as well.

6.1 Attachment and Sociability

A number of terms are used in studies investigating the dog-human bond, and a brief review of these may be helpful in clarifying later results.

An 'attachment' bond in the psychology literature refers to an 'affectional tie' displayed by one individual towards another specific individual (Ainsworth and Bell 1970). It is typified by the parent-infant bond (Collins and Feeney 2000). The original concept of attachment was derived from developmental psychology, whereby Bowlby (1982, 1988) argued that attachment behaviours in young infants are regulated by an innate behavioural system that functions to enhance survival by

F. Range, S. Marshall-Pescini, *Wolves and Dogs*, Fascinating Life Sciences,
https://doi.org/10.1007/978-3-030-98411-3_6

maintaining a child's proximity to their caretaker. Thus, the attachment system is thought to be activated when an infant is frightened, or in unfamiliar surroundings. Four specific behavioural criteria emerged as crucial in discriminating a true attachment bond from other affectional bonds, i.e.:

(a) Staying near to and resisting separation from the attachment figure (proximity maintenance)
(b) Feeling distress upon involuntary separation from the attachment figure (separation distress)
(c) Using the attachment figure as a base for exploring the environment free of anxiety (secure base)
(d) Seeking out the attachment figure for contact and assurance in times of emotional distress (safe haven) (Bowlby 2012; Bretherton 1985)

Although the concept of an attachment bond was initially adopted to characterize the infant-caregiver bond, Bowlby (1979/1994) believed that attachment is an important component of human experience 'from the cradle to the grave' (p. 129) suggesting the concept could also be used in relation to forms of bonding between adults. Indeed, in subsequent years the 'attachment' framework was adopted also to investigate romantic relationships (Fraley and Shaver 2000; Hazan and Shaver 1987) with the same brain-physiological mechanisms and the same mental tools (social representation) being used for partner-partner relationships.

A classic test, dubbed the Strange Situation Test (SST), was designed by Mary Ainsworth for human infants up to 19 months and their caregivers to investigate the quality of the attachment bond. The test consists of an alternating sequence of nine brief (typically 2 min) episodes, in which the subject is in a novel environment (an unfamiliar room), first with their caregiver, then joined by a stranger, then alone with the stranger, then completely alone in the room, and then again with either stranger or caregiver—with the goal to progressively increase the stress load of the infant to trigger attachment behaviours.

In more recent years, the same experimental paradigm has been adopted to investigate the bond between non-human animals (chimpanzees, dogs, cats, etc.) and their human caregivers although, importantly, the coding system to analyse the behaviour differs (Bard 2007, 1991; Topál et al. 1998; Vitale et al. 2019). Studies using this paradigm with pet dogs have largely shown the predicted results, with dogs showing (1) distress behaviours at separation from the human caregiver but not the stranger, (2) increased explorative and play behaviours when in the presence of their human caregiver, and (3) increased affiliative behaviours towards the caregiver compared to the stranger upon reunion (Palmer and Custance 2008; Prato-Previde et al. 2003; Topál et al. 1998; see Payne et al. 2015).

Although, as we will see in the current chapter, the attachment framework has mostly been used to investigate animals' bond with humans, an interesting perspective would be to also apply this concept to the characterization of conspecific bonds. Indeed, at present it is unclear to what extent the attachment framework can be applicable to species outside the primate family. It is interesting to note, for example, that in most canid species, young infants are left for long periods of time in their den and/or in 'rendezvous sites' while parents forage and hunt. This pattern of infant-parent separation is not found in any primate species where infants are either with one or other of their parents at all times. This is not to say that mother-infant bonds are not of fundamental importance also to canids, but rather to highlight that the species' ecological context may alter the behavioural manifestations of this bond.

Similarly, given wolves' monogamous mating system, it is perhaps not surprising that some of our female-male pairs, which have been kept together for several years, show behaviours similar to the attachment criteria when the partner is taken out of the enclosure for longer time periods (either for tests, walks, or medical care). Although we have not systematically tested this yet, the long-term male-female dog pairs seem to care a lot less if they are separated. Moreover, we also find that same-sex wolf dyads are less than happy when separated, suggesting that in wolves, the dependency on pack members may be particularly strong. However, whether these bonds can be considered 'attachment' (i.e. whether satisfying the attachment criteria) will require a more directed investigation.

Aside from studies using the attachment construct, a considerable body of literature has investigated dogs' 'sociability' towards humans. The term 'sociability' is used mostly to refer to the extent to which a dog seeks and maintains contact with humans. Typically, tests will involve an animal given the choice to interact with a person (which may vary in terms of the familiarity with the subject) or between two persons varying in familiarity, and the dog's behaviour will be monitored in terms of contact-seeking, affiliative/aggressive interactions and avoidance towards the person. Tests that have been conducted with pet dogs include investigations of breed differences both as puppies and adults (Barnard et al. 2017), but also tests with shelter dogs as a means to potentially evaluate their adoptability (Valsecchi et al. 2011) and with free-ranging dogs to investigate their spontaneous reaction to unfamiliar humans (Ortolani et al. 2009). The term 'sociability' has also been used to describe a dog's personality trait in questionnaire-based studies (Gosling et al. 2003; Hsu and Serpell 2003; Kubinyi et al. 2009; Mirkó et al. 2012). Hence, a dog deemed by the evaluator (shelter worker, owner, researcher) to be of high sociability would typically receive a high score in relation to their openness to interact with new people in a friendly manner and their curiosity, but low scores in relation to their 'aloofness' and how 'reserved' they are towards people.

Although 'sociability tests' do not specifically fall within the 'attachment framework', they cover overlapping grounds and can be used to compare how animals interact with unfamiliar vs. 'bonded' individuals, and as such they can inform us about how the animals attend and react to humans with whom they are bonded in comparison with those they are not.

6.2 Wolves and Dogs Compared

As mentioned earlier, in a number of early studies, wolves were raised with extensive human interaction. In relation to sociality towards humans, Fentress (Fentress 1967) reports the behaviour of a hand-raised wolf over a 3-year period. He highlights that both as a puppy and even more so when reaching adulthood, the wolf showed cautious interactions with strangers while being more enthusiastic in his greeting of people he knew well. Nevertheless, this behaviour was described as seeking the presence, rather than the contact, of people.

More recently, a few studies have more systematically compared attachment to humans in wolf and dog puppies using the SST. Sociability towards humans has also been tested in wolves and dogs (both pups and adults). In most studies, wolves and dogs are confronted both with humans they have a close bond with (typically their hand-raisers or caregivers) and other less familiar or completely unfamiliar individuals, and their reactions to each of these are analysed.

6.2.1 Puppy Tests

6.2.1.1 Attachment

Sixteen-week-old hand-reared wolves ($N = 13$), hand-reared dogs ($N = 11$), and conventionally reared dogs (i.e. mother nursed in human homes; $N = 11$) were tested by the Budapest group in a canine version of the SST, and some noticeable differences emerged. After separation, dog pups (hand-reared and pet) showed more following, greeting, and play behaviour towards the caregiver than the stranger (Topál et al. 2005), whereas wolves did not show a statistically significant preference. However, wolf pups spent more time in bodily contact with the human partner than hand-reared dogs (regardless of identity), and both hand-reared dogs and wolves spent more time in contact with the unfamiliar individual than the caregiver (and this was not the case for pet dogs). The researchers concluded from these results that, whereas dogs (both hand-reared and pets) showed a clear preferential bond to the owner (greeting and following them more than the stranger), this was not the case for wolves, suggesting this difference may be a result of the domestication process. However, some reported results do not fit the idea of dogs being more 'human oriented' than wolves, since indeed it was the wolf pups that spent more time in physical contact with the humans than the dogs, and both spent more time in contact with the stranger than the caregiver. Hence, it seems that differences are rather more specifically related to greeting and following. Furthermore, by the time of testing, at least some of the wolves were no longer with their human caregiver but had been returned to the farm where they were born, seeing their caregivers no more than once or twice per week (see introduction).

Hall et al. (2015) tested a different population of ten hand-raised wolf pups from two litters (Wolf Park, Indiana). They were tested using the SST at 3, 5, and 7 weeks of age, with the caregiver remaining constant throughout this time period and in daily contact with the puppies outside of testing. Overall, the pups were shown to exhibit a pattern of behaviours consistent with an attachment bond to their caregivers (Hall et al. 2015); for example, they showed more intense greeting behaviour towards the caregiver than towards the stranger upon reunion, as well as an increase in exploration of the novel person and environment when the caregiver was present (known as the 'safe haven' effect). One difference between the two studies (apart from the fact that some of the wolves were no longer in daily contact with their caregivers in Budapest) is that the puppies in Wolf Park were a lot younger when tested. While at the age of 4 months, the wolves are actually already very mobile, and under natural conditions spend much time alone close to the rendezvous site, at 5–7 weeks they are still a lot more dependent on their mothers, and this may be reflected in the behavioural responses seen in these tests at different ages.

6.2.1.2 Sociability

The Budapest wolf ($N = 13$) and dog ($N = 11$) pups were also tested at 3, 4, and 5 weeks of age in a sociability test. They were simultaneously presented with their hand-raiser sitting quietly on the floor vs. another stimulus. More specifically, the second stimulus could be (a) a full milk bottle, (b) an unknown human also quietly sitting on the floor, (c) an unknown adult dog lying down, or (d) a known, sleeping conspecific puppy. Both wolf and dog pups showed a preference for the hand-raiser vs. the unknown human, but this preference was evident at 4 weeks, but not 5 weeks, in dogs and vice versa in wolf pups (Gácsi et al. 2005). Both at 4 and 5 weeks, dog pups preferred staying in proximity to the caregiver rather than close to the conspecific (pup and adult), but no such preference emerged in wolves. Interestingly, from this study it emerges that both dogs and wolves recognized their caregiver and showed a preference for them, but such preferences appeared to also be modulated by how explorative the animals were (see Chap. 9). Wolves at 4 weeks, when first confronted with the choice between a novel person and the caregiver, distributed their time equally between the two. In the second such test a week later, potentially because it was no longer a novel situation, the wolves showed a clear preference for the caregiver over the new person. Accordingly, the greater interest in novelty of wolves compared to dogs (Chap. 10) could drive these differences rather than the sociability of the animals. This explanation could very well also be valid for the differences observed in the attachment tests summarized above.

Aside from the different preferences for staying close to the two persons, a number of other wolf-dog behavioural differences emerged during the test. Wolf pups showed more aggressive behaviours (growls and/or attempted bites occurring in 26% of trials by nine individuals in contrasts to no dog showing any such behaviours) and avoidance behaviours towards the caregiver. The authors suggest that both the aggressiveness and the avoidance were likely in response to the wolf pup's dislike at being touched/handled by the experimenter. Dog pups also carried out significantly more vocalizations and tail wagging than wolf pups. The latter is particularly interesting since tail wagging both in wolves and dogs is associated with greeting rituals towards conspecifics, which are largely shown from the submissive towards the dominant individual (Cafazzo et al. 2010, 2016). Taking this into consideration, it would suggest that dog pups are more submissive towards their human caregivers than wolf pups.

Response to humans, in this case an unfamiliar human, was also partially tested in a different context in our wolf and dog pups, in a study aimed at investigating their response to novel objects. At 6 and 8 weeks, pups ($N = 17$ wolves, $N = 18$ dogs) were placed in an unfamiliar room with a toy car or toy dog that could be activated so that it moved (see Chap. 10 for more details of the animals' response to the objects). The tests included the presence of an unfamiliar female person standing quietly to one side of the room and activating the toys once the pup was placed in the room. Moreover, in some cases a camera person was also present, but hiding behind a barrier. We found no effect of species on the likelihood of contacting the experimenter: at 6 and 8 weeks, respectively, on average 35% and 40% of dogs contacted the experimenter, while 58% and 59% of wolves did so. In regard to the time spent in proximity of people (experimenter and camera person summed), a different pattern of results emerged for the two species depending on the age of testing. At 6 weeks, wolves had a tendency to spend more time close to the people than dogs, but no such difference was evident at 8 weeks. The lack of a difference at 8 weeks is explained by dogs increasing the time they spent in proximity to the person from 6 to 8 weeks, while the wolves showed no such increase (Marshall-Pescini et al. 2017).

Although wolf puppies raised in the Budapest project did not appear to show differential behaviours between their hand-raiser and strangers in the SST when tested at 4 months (Topál et al. 2005), they did show a preference for their caregiver when tested in the choice task at a younger age (Gácsi et al. 2005), suggesting that some bonding process had occurred, and the animals did differentiate between the people present based on their familiarity/role. Researchers hence retested the same animals at the age of 6 months in a 'Greeting Situation Test', i.e. a test in which the animals could meet and interact with a person as they pleased.

At 6 months, wolf puppies ($N = 8$), who at this point were living with their same-aged cohort mates and adult animals at an animal park, were tested together in groups of four individuals. The test consisted of a person entering the enclosure and, in a first phase, standing quietly for 5 seconds at the entrance and in a second phase actively calling the pups and squatting down to pet them if they approached (45 seconds). Four 'types' of people visited the pups on repeated occasions: (a) a foster parent (of one of the pups); (b) a complete stranger, i.e. someone they had never seen before; (c) an 'almost' stranger person, i.e. a person they had met once before in the same context; and (d) a close acquaintance, i.e. a person they had known since 4 weeks of age but seeing them only occasionally in social contexts (i.e. the foster parents of the other pups).

Wolves jumped up more, stayed significantly longer in proximity and physical contact, and tail wagged more towards their foster parents than towards any of the other visitor types. They also jumped up more and stayed longer in proximity and tail wagged more with close acquaintances than strangers. They also physically contacted close acquaintances more than people they had met only once before. Interestingly, no significant difference was found between close acquaintances and total strangers in respect to physical contact, suggesting that again interest in novelty might have influenced the animal's behaviour. Thus, overall as 6-month-old youngsters, the wolves were clearly able to differentiate between visitors with different familiarity levels, and as displayed by their more intense greeting behaviours, they favoured their hand-raisers over all other visitor types (Ujfalussy et al. 2017).

6.2.1.3 Tractability

The hand-raised wolf ($N = 16$) and dog ($N = 11$) pups in Budapest were also compared in their reaction to common situations for a pet such as retrieving an object, responding to being called, sitting on request, allowing themselves to be brushed, and walking while wearing a muzzle. The tests were carried out between the ages of 3 and 24 weeks. Wolf puppies were less likely than dogs to fetch a ball and, at the age of 4 and 6 months, to come when called or walk with a muzzle (Ujfalussy et al. 2020). Similarly, sitting on command was highly dependent on the age at test. The only clear difference was that wolves were more likely to show aggression when the person tried to take the ball away from them than dogs. In general, wolves' willingness to comply with the human caregiver depended more on age and context than dogs suggesting that the dogs may be more 'tractable' than wolf pups (Ujfalussy et al. 2020).

6.2.2 Summary Studies with Puppies

Taken together (see also Table 6.1*), results from studies looking at potential differences in social behaviour towards caregivers and strangers in wolves and dogs seem to suggest that both have a clear capacity to distinguish between the two figures already at an early age. However, across a number of studies (with different animals), wolf pups also appear to show greater curiosity/explorative tendencies towards strangers, which may have, on occasions, masked the animals' ability to distinguish between the bonded and non-bonded individual.*

Comparing wolf and dog pups in their 'tractability' to restraining situations and responses to command also shows a varied picture, with clear differences showing dogs' greater tractability in some context but not in others, and variations depending on age.

Taken together the results highlight the need to carry out multiple tests, using different paradigms to address the same question, thereby helping us to tease apart, or at least have a more inspired guess, about the motivation underlying the animal's behaviours.

6.2.3 Adult Tests

The Budapest wolves were submitted to a Greeting Test Situation ($N = 9$) at 12 and 24 months of age. In this case the test was conducted with each animal separately, and results showed no significant differences between foster parents and close

Table 6.1 Summary of studies conducted with wolf and dog puppies on aspects of sociality towards humans

Test	Behaviour	Age	Species	Result	References
Attachment: Strange Situation Test	Exploration Greet/follow/play	16 weeks	D&W	W > D D: Caregiver>stranger W: Caregiver = stranger	(Topál et al. 2005)
	Greet Proximity/contact	3, 5, 7 weeks	W	Caregiver>stranger Caregiver>stranger	(Hall et al. 2015)
Sociability	Proximity caregiver vs. stranger	3, 4, 5 weeks	D&W	D=W Caregiver>stranger (4 wk. dog; 5 wk. wolves)	(Gácsi et al. 2005)
	Proximity caregiver vs. conspecific	3, 4, 5 weeks	D&W	D: Caregiver>conspecific W: Caregiver = conspecific	(Fentress 1967)
	Proximity & greeting to caregiver, very familiar, almost stranger, stranger	6 months	W	Caregiver> others Very familiar = stranger	(Ujfalussy et al. 2017)
Novel object test	Proximity stranger	6 weeks 8 weeks	W&D	W > D (trend) W=D	(Marshall-Pescini et al. 2017)
Tractability	Retrieving Aggression and carrying away an object	6,9 weeks	W&D	W < D W > D	(Ujfalussy et al. 2020)
	Speed of responding to calling	3, 4, 5, 6, 7, 8, 12, 16, 24 weeks	W&D	W < D at 16 and 24 weeks	
	Speed of sitting on request	7, 9, 12, 16, 24 weeks	W&D	W < D at 7 and 16 weeks	
	Biting during brushing	12 weeks 16 weeks	W&D	W > D D=W	
	Acceptance of muzzle	16, 24 weeks	W&D	W < D	

acquaintances in any of the behaviours related to approach and contact seeking (proximity, contact, tail wagging, jumping up). However, significantly more such behaviours occurred towards foster parents than the two stranger types. Again, adult wolves behaved differently with strangers when meeting them a first or second time, in that they approached, contacted, and jumped up more during the first visit compared to the second (Ujfalussy et al. 2017), suggesting again that curiosity towards novelty might strongly influence such tests in wolves (but probably not in dogs).

A preliminary study (conducted by Büttner 2016 for her Bachelor thesis) was also conducted at the Wolf Science Center, comparing wolves' and dogs' greeting behaviour towards differently bonded people (hand-raiser vs. a newer trainer at the Wolf Science Center). To keep the situation as 'normal' as possible for our animals and to avoid any potential elements of stress affecting the greeting behaviours, rather than testing each individual separately from its pack mates, we conducted the tests with the whole pack. Hence, a trainer would approach the packs' home enclosure and kneel down close to fence calling the animals to her. Using a five-point Likert scale, the intensity of the greeting behaviour was assessed from a video by five independent observers, who did not know the relationship type of the people with the animals. In general dogs' greeting behaviours towards humans (regardless of relationship type) were ranked as more intense than those exhibited by wolves. The dogs' individual participation in the greeting ritual was also more consistent (all dogs in the pack were generally rated as showing high-intensity greeting behaviours), whereas ratings of wolves' greeting were much more variable, with some wolves' greeting being ranked as low in intensity and others rated as intensely as the dogs (Büttner 2016). Although results are preliminary, since more in-depth analysis of more variables would be more revealing than the more general evaluation used here, these results are in line with our everyday experience of the animals. Whereas greeting behaviour in wolves tends to occur frequently when the animals' hand-raisers arrive at their home enclosure (sometime including also extensive howling while the hand-raiser is still quite far away), dogs will tend to show such behaviours also with less well-known individuals (e.g. PhD students that observe the animals but have had no part in raising them).

A number of other aspects relating to attachment, specifically separation stress, contact seeking, and secure base towards the handler, were investigated in adult hand-raised wolves in Budapest ($N = 11$) and compared to normally raised pet dogs ($N = 9$) (Lenkei et al. 2020). An outdoor separation test similar to the SST was designed, where the presence of a stranger in an unfamiliar location was included to elicit a moderate stress response that might activate the attachment behaviour. The basic paradigm involved the dog/wolf being on leash in an unfamiliar location, with two people present: the handler/owner and a stranger. The handler/owner and stranger took it in turns to hold the wolf on the leash, while the other person walked away and out of sight for 3 min (separation phase) and then returned and greeted and petted the animals (reunion phase). Both wolves and dogs pulled at the leash in the direction of the leaving person more (proximity-seeking) and showed more signals of stress such as panting, whining, agitated moving, and lips licking (separation distress) when it was the handler/owner leaving than the stranger. Both also showed more exploration of the surrounding novel environment when the handler/owner was holding them on the leash than the stranger (secure base effect).

A different paradigm, aimed at investigating the 'safe haven' effect of the caregiver, was used with both adult wolves and dogs in the Budapest project. In this paradigm, animals were confronted with a potentially difficult/stressful situation while in the company of their owner/caregiver (the attachment figure) or a less well-known individual. The main underlying idea is that animals will show fewer stress-like behaviours and higher social support seeking when they are in the presence of their attachment figure compared to another person.

At the age of 2 years, the 13 hand-reared wolves were tested while in the presence of their caregiver as a stranger approached them while exhibiting either signals of friendliness (approaching while gently looking into the animal's eyes and talking in a gentle tone) or of mild threat (approaching haltingly with upper body thrust forward, in silence and holding the animal's gaze). In this study, (unfortunately) the wolves' response was compared to that of 13 pet dogs tested in the presence of their owner. Wolves and dogs both responded in an affiliative manner to the friendly approach (dogs greeted mostly by tail wagging and wolves by licking the experimenter's hand or face). The response to the threatening approach was much more varied. Most pet dogs looked at their owner (suggesting they may have been seeking social support from them), and their responses to the stranger varied from friendly, to avoidant, to aggressive. Wolves appeared more homogenous in their response in that they mostly displayed either passive behaviours or active avoidance, stepping back and interrupting eye contact with the stranger. In contrast to dogs, no wolf showed signs of aggression towards the stranger (Gácsi et al. 2013). The authors conclude that the more varied responses by the dogs, together with their looking behaviour towards the owner, support 'the role of domestication in shaping flexible and socially competent behaviours in interactions with humans' (p. 113). However, the conclusions are somewhat surprising, since the wolves also showed different responses towards the friendly and threatening strangers. In response to the friendly approach, six wolves responded in a friendly manner, five were passive, and two were avoidant, whereas in response to the threatening approach, all but one (whose response was friendly) behaved in an avoidant manner. Hence, it would appear that all the wolves adapted their behaviour in response to the different approaches. Furthermore, wolves actually averted their gaze faster than dogs to the threatening approach (within 2 seconds) suggesting they were able to 'read' the human's behaviour rapidly and respond to it (see Chap. 7 for more about gazing behaviour). Thus, it would seem that if anything, it is wolves that showed the more socially competent behaviour of avoiding the threatening human, in contrast to some of the dogs being friendly and some being aggressive. However, as said above, of course it would be interesting to know how dogs with the same experience as these wolves would behave, since there is certainly an effect of the owner presence as demonstrated by the pet dogs' looking behaviour towards the owner.

The same wolves and dogs were also tested in a 'playful' context. While in the presence of their caregiver, the experimenter (a familiar person the animals had met a few times and played with) initiated a 'guarding' game with the animals. The experimenter had a neutral object (an empty bag) which neither contained food nor was encountered by the animal in the past. Following an introductory phase in which the animals were invited to interact with the object after the experimenter showed playful interest in it, four wolves and three dogs were excluded since they were completely disinterested in the object/situation. Another wolf was excluded for health reasons. The remaining animals were tested in the second phase, which involved the experimenter approaching in a pseudo-threatening way (see above), but at the same time stretching out her hand, and jumping back and laughing if the animal made any move to guard the object. In this phase, a further three wolves and two dogs lost interest. In the final phase, the experimenter changed her demeanour and now approached the animal in a calm way talking to it gently and stroking it, and then tried to remove the bag. Of the nine dogs that could be involved in the game, six showed some type of guarding when challenged for the bag in phase two (mostly barking towards the experimenter). In the final phase, all dogs allowed the experimenter to remove the bag within the first 3 seconds. Of the five wolves that could be tested since they showed some interest, two responded to the approach with aggression (one growled and one snapped and attacked), and three responded with a friendly/submissive manner (tail wagging, face licking) in the second phase. Two of the latter animals also displayed play-like behaviours such as pawing and rolling over while pawing. None of the wolves allowed the experimenter to take the bag from them in the final phase. Similarly, to the previous test, whereas the dogs often looked at their caregiver, wolves did not exhibit this type of behaviour during the test (Gácsi et al. 2013).

As discussed above, it is important to recall that the dogs were all *pet* dogs that have probably had a much more extensive experience of meeting strangers and especially with playing guarding games. They have most likely been reinforced, or discouraged, to behave in specific ways in such contexts and have in most cases been actively trained to not guard any objects. Thus, how much the comparison of the wolves' and dogs' behaviour in this case can tell us about the effects of domestication remains unclear.

The Wolf Park (Indiana) wolves were also tested in a sociability study, but again the comparison was with a pet dog population (Bentosela et al. 2016). Six adult wolves and six adult pet dogs were tested with both a familiar and unfamiliar individual. An additional five individuals from each population were tested only with the familiar person (due to security reason, not all the wolves could be tested with unfamiliar people). The tests were carried out in a familiar outdoor enclosure for wolves, and in a familiar indoor ($N = 9$) or outdoor ($N = 2$) environment for pet dogs. The test consisted of one person remaining seated for an initial 2-min 'passive phase', in which the person looked down, and only briefly interacted with the animal if the latter sought contact. A subsequent 2-min 'active phase' consisted of the person standing and actively calling the animal, petting it if it approached, and calling them again if they moved away. Results showed that regardless of familiarity, pet dogs were faster at approaching the person than wolves. Both wolves and dogs spent more time with the familiar than the unfamiliar individual, but overall dogs spent significantly more time with the person than wolves. Considering all 11 wolves and pet dogs tested with the familiar person, the same results emerged, with pet dogs approaching sooner and spending longer with the person than wolves (Bentosela et al. 2016).

The same sociability test together with a solvable task test (Udell 2015), in which a small box containing food could be opened by pulling a rope while the experimenter was standing nearby (see Chap. 7 for details), was presented to the Wolf Park wolves and a different cohort of pet dogs. In this second study, the researchers aimed to investigate both the behavioural and genetic differences related to social behaviour (vonHoldt et al. 2017). Regarding the sociability test, researchers compared wolves and dogs on three measures:

(1) An 'attentional bias to social stimuli' measured as the ratio of time spent looking at the experimenter during the solvable task and the sum of the time spent looking at the experimenter plus the time spent looking at the puzzle box
(2) 'Hypersociability', i.e. the willingness to engage with humans (regardless of familiarity) measured here as the sum of the time spent in proximity to the experimenter in each phase of the sociability test
(3) 'Social interest in strangers' measured as the sum of the time spent in proximity to the experimenter in the two unfamiliar phases of the sociability test

Overall, dogs showed a higher 'attentional bias', spending more time looking at the person than wolves, and showed higher 'hypersociability', although there was no difference between wolves and dogs in their 'social interest towards strangers'.

Authors related these results to observed structural variants (SVs) linked to the Williams-Beuren syndrome in humans. The Williams-Beuren syndrome is characterized by delayed development, cognitive impairment, behavioural abnormalities, and hypersociability. In fact, in a prior study involving a genome-wide association scan of more than 48,000 single-nucleotide polymorphism (SNP) genotypes from 701 dogs of 85 breeds and 92 grey wolves, one of the most striking difference between wolves and dogs was indeed observed in the WBSCR17 gene (the principle gene associated with Williams-Beuren syndrome in humans) (vonHoldt et al. 2010). Indeed, a candidate region association test including the behavioural data from both wolves ($N = 8$) and dogs ($N = 16$) described above showed that structural variants on genes associated with WBS (GTF2I and GTF2IRD1) were significantly associated with human-directed social behaviour (vonHoldt et al. 2017). Based on this gene-behaviour association and the large-scale comparison of the structural variants associated with this gene in multiple wolf and dog populations, authors conclude that selection for hypersociability during the course of domestication resulted in the differential distribution of the structural variants in the canine WBS locus observed. In other words, during domestication there was a selection for increased sociability in dogs, which is evident in their genetic make-up.

It is important to note a number of issues relating to the behaviour part of the study. First, as will be discussed in a later chapter (Chap. 7: Communication), the measure of looking behaviour towards the person, when a problem-solving task is simultaneously available, may be problematic, since there is no way to disentangle to what extent the looking behaviour is a by-product of reduced persistence in trying to get at the food vs. a spontaneous social interest in the person. Indeed, in studies where the looking behaviour and persistence in problem solving are linked, wolves take longer than dogs to look at humans, but in studies where they are required to communicate with a human to 'show' where food is located, their behaviour does not differ from that of dogs (Heberlein et al. 2016; summarized in Chap. 7).

Furthermore, the 'hypersociability' measure includes both the behaviours towards the known individual and the stranger, and this measure is the one that is statistically different between wolves and dogs, with dogs spending more time in proximity with the people than wolves. However, there is no difference between wolves in the 'social interest in stranger' measure. This would suggest that the difference in 'hypersociability' is driven by dogs spending more time with the familiar person than wolves. In the case of dogs, the familiar person is embodied by the owner, whereas for wolves this is the caregiver. This is of course problematic, since although the wolves were hand-raised, there is no specification whether the familiar person was indeed the hand-raiser. But even if the familiar person was indeed a hand-raiser, the assumption that the caregiver in an animal facility can establish the same kind of bond with the wolves as an owner can with it`s dog is a big leap given that the animals live with conspecifics and are visited by humans only for a fraction of that time and in specific situations.

Also, the fact that the wolves are more interested in strangers than the dogs, as has been shown in a number of different studies reviewed above, would further suggest that maybe the differences in experiences between wolves and dogs in this study did influence their behaviours in the direction predicted by the 'hypersociability' hypothesis.

At the Wolf Science Center, we conducted a conceptually similar study with our adult animals to investigate potential difference in their interest/willingness to interact with people (as well as potential differences in their physiological responses in terms of oxytocin and cortisol—see section below) (Gwendolyn Wirobski et al. 2021). We set up a test situation in which each of our animals ($N = 10$ wolves, $N = 11$ dogs) was tested with a hand-raiser (bonded person) and a person they were familiar with, i.e. long-term students at the Wolf Science Center, who had been present during testing sessions of the animals or had observed them, but who did not have a close affiliative bond with them.

The interaction test was carried out with the animal being alone in their home enclosure, and the person approaching and calling them from the other side of a fence. The use of the fence in between the human and the animal allowed us to test all our animals in exactly the same way with both the familiar and bonded person. The test consisted of two phases; in the first phase, the person approached the fence, called the animal and (if the animal approached) spent up to 5 min gently petting, stroking, and/or talking to the dog/wolf. This phase was followed by a brief 'training' session, in which the person asked the animal to perform well-known behaviours on cue (e.g. 'sit' in response to an index finger held high in front of the animal; 'touch' a nose-touch to the outstretched closed fist) and then rewarded them

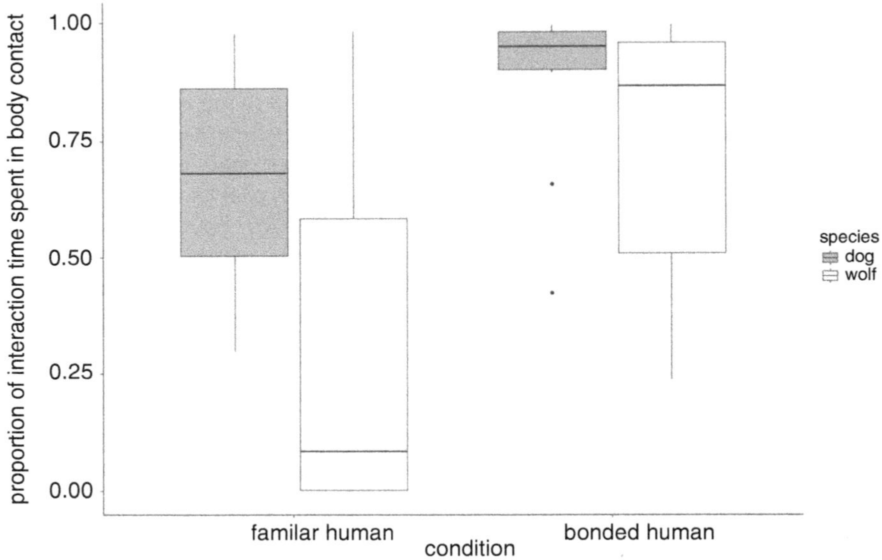

Fig. 6.1 Proportion of time spent cuddling with the bonded and just familiar persons in Wolf Science Center wolves and dogs. Figure adapted from Gwendolyn Wirobski et al. (2021). https://doi.org/10.1038/s41598-021-93922-1; CC BY 4.0; http://creativecommons.org/licenses/by/4.0/

with sausage (exactly 15 pieces of sausage were given to the experimenter to reward the animal).

All dogs approached, and then also interacted, with both categories of humans. All wolves approached and interacted with the bonded person, whereas nine of the ten wolves approached the familiar human, and of these, only seven interacted with her. Regardless of the person's identity, dogs spent more time in proximity and more time in body contact with the person than wolves (dogs with the bonded person spent 87.7% of the trial time in contact with them, and 68.6% of time in contact with the familiar person; wolves spent 73% of the trial time in contact with the bonded person and 31.7% with the familiar person). However, both wolves and dogs spent more time in proximity and body contact with the bonded compared to the familiar individual (Fig. 6.1). Interestingly, wolves showed a much larger variability in the time they spent close to, and interacting with, the human (especially with the familiar person) than dogs, with some animals spending their entire time being petted no matter by whom, whereas others hardly approached.

Surprisingly, dogs showed more frequent so-called self-directed behaviours (formerly called 'conflict behaviours', ambivalent behaviours) which are thought to be related to stress (lips licking, yawning, body shaking) than wolves in both social conditions requiring interaction with the human partner, whereas no such difference emerged in a food-only condition (where no interaction with the human occurred). When looking at the social conditions more closely, in dogs we found a positive correlation between the frequency of self-directed behaviours and the duration of body contact with the familiar person (Fig. 6.2a). Wolves showed fewer of these behaviours overall and a positive association (but a lot weaker) between the time spent cuddling the bonded person and the frequency of self-directed behaviours (Fig. 6.2b). We carried out exactly the same test also with pet dogs ($N = 10$ for comparability with the above), and as for pack dogs, there was a positive association

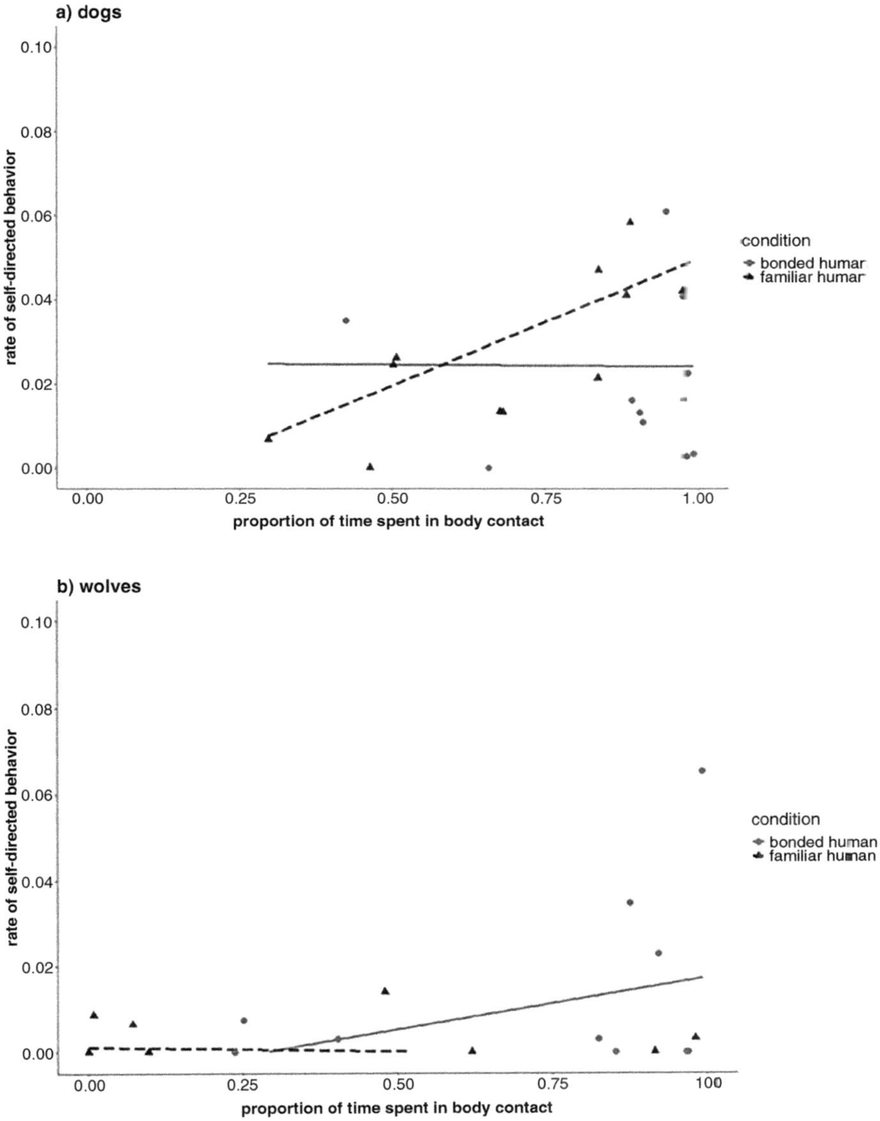

Fig. 6.2 Association between the rate of self-directed behaviours and the time spent cuddling either the bonded or familiar person in Wolf Science Center dogs and wolves. Figure adapted from Gwendolyn Wirobski et al. (2021). https://doi.org/10.1038/s41598-021-93922-1; CC BY 4.0; http://creativecommons.org/licenses/by/4.0/

between the duration of body contact and the frequency of stress-related behaviours emitted, although here the association was not statistically significant, so it does not appear to be so strong.

The importance of the social bond in wolf-dog-human interaction emerged also in another study conducted at the Wolf Science Center, in which the behaviour of wolves and dogs was compared when on leash walks. At the Wolf Science Center, we train our animals to walk on the leash starting at the age of 6–8 weeks. While we did not compare their performance at that age, we ran leash-walking tests when the animals (11 wolves and 9 dogs) were around 10 months old. The same test was re-run with six wolves and six dogs when they were adults (older than 2 years). The animals were led by different human partners, who had had variable contact with them over the previous months, but were all highly familiar to the animals. The human partners walked the animals on a forest road using a 10-metre-long leash. At certain points during the walk, the animals were asked to follow commands such as sit and lay down. A number of behaviours were coded such as mutual attention, tension on the leash, who initiated the change of direction (leadership), and conflicts relating to these, as well as how often animals needed to be called and how fast they of responded to commands. Interestingly, we found that as youngsters, wolves were faster at responding to some commands (e.g. the down command) than dogs, although as adults the speed of responding between the two species was similar (Auer et al. 2011). There was also no difference between wolves and dogs on their cooperativeness on the leash, in terms of how often they pulled. However, compared to dogs, wolves more often disagreed with the trainer's change of direction (i.e. wanted to go in the opposite direction) and were more likely to make their own changes of direction without waiting for the trainers to do so. This was the case in juvenile wolves more than in the adults.

Another way that has been used to evaluate the potential differences between dogs' and wolves' sociability towards humans is to assess the value that such a contact may have for individual wolves or dogs.

In one study, researchers assessed whether brief social contact with a known individual would be as successful as a titbit of food in reinforcing a novel behaviour for wolves and dogs (Feuerbacher and Wynne 2012). If animals allocate a higher 'value' to human social contact than wolves, then they should be more sensitive to this kind of reinforcer and hence learn faster with it than wolves. Nine (and then six for a second study) shelter dogs, six owned pet dogs, and nine hand-raised wolves (Wolf Park, Indiana) were tested comparing their response rates in 'nose-touching' the experimenter's (a familiar person) hand, when reinforced with a piece of sausage, a 4-second social contact, or nothing (extinction). For all the populations, the latency and the rate of performance of the target behaviour was much higher with the food than the social reward, and although not statistically compared, the pattern of results appeared pretty much identical in wolves and dogs (Feuerbacher and Wynne 2012).

In a comparison of our animals at the Wolf Science Center, free-ranging dogs from Morocco and pet dogs, we addressed a similar question (Lazzaroni et al. 2020). All tests were conducted in an outside area such as the testing enclosure at the Wolf Science Center, the beach in Morocco, dog zones in Vienna, and a 'doggy day-care centre' in Italy. The test involved two phases: in the pre-test phase, animals were exposed to two people in succession. One person invited the animal for a social/cuddle session, squatting down and gently petting the animal for up to 30 seconds (contact provider); the other fed the animal five pieces of dry food for the same duration of time (food provider). In the test phase, animals were then presented with both individuals simultaneously, and they could choose which of the two persons to approach.

Because at the Wolf Science Center the test could only be carried out with our trainers acting the role of the contact or food provider, each animal was tested twice with the two people alternating roles across the two tests. This allowed us to exclude the possibility that the animals' preference was linked to identity rather than role of the person. Pet dogs and free-ranging dogs received the same test, but in this case, the two people were complete strangers and therefore the test was run only once. We found that in the first phase, whereas all animals at the Wolf Science Center approached the food provider, there were differences in their behaviour towards the contact provider. While all 13 dogs approached the contact provider, 1 wolf in the first test and 1 in the second test (out of 16 wolves) avoided contact. Furthermore, whereas dogs spent almost the entire 30 seconds (88% of the time) in contact with the contact provider, wolves spent only about half the time with them (52.2%) although there was a considerably greater variability in the latter (Fig. 6.3).

In the first test, when animals could choose whether to approach the food or the contact provider, we found that all but one Wolf Science Center dog approached at least one of the people present, but six wolves approached neither person (instead they spent their time walking around the enclosure exploring). In the second test, two

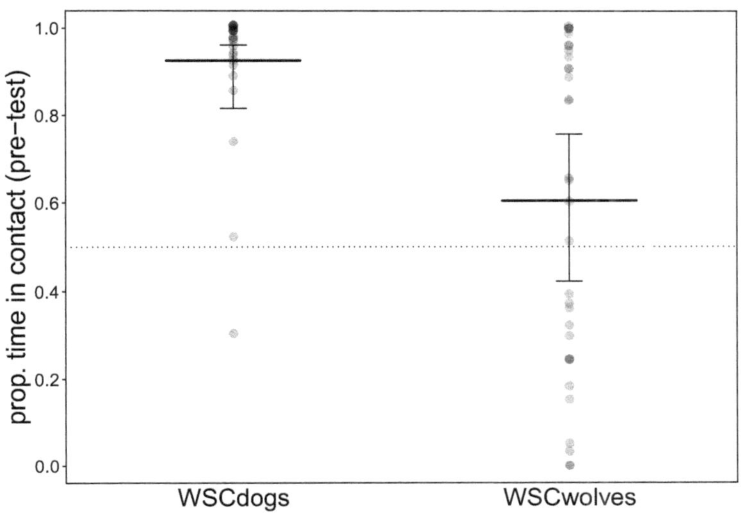

Fig. 6.3 Time spent in contact with the contact provider for wolves and dogs. Figure adapted from Lazzaroni et al. (2020). https://doi.org/10.3389/fpsyg.2020.00785; CC BY 4.0; http://creativecommons.org/licenses/by/4.0/

dogs and nine wolves did not make a choice. However, neither wolves nor dogs showed a clear preference for the food vs. contact provider (Test 1, seven dogs chose the food and five chose the contact provider; six wolves chose the food and four the contact provider; Test 2, three dogs chose the food and seven the contact provider; five wolves chose the food and two the contact provider).

Interestingly, comparing free-ranging and pet dogs tested in the dog areas, we found that in the pre-test phase, 31 of 53 pet dogs and 39 of 45 free-ranging dogs accepted the cuddle session. In terms of the time spent in contact with the cuddler, pet dogs spent less time in contact than free-ranging dogs (pet dogs, mean = 9.6 s, dev.st = 11.83; free-ranging dogs, mean = 24.43 s, dev.st = 11.32). However, no differences were found between the two groups in the test phase. We reasoned that one possibility why pet dogs spent less time being cuddled by the person was because the dog areas may have provided a very strong incentive to spend time exploring. In effect when we take our dogs to these areas to let them off leash and allow them to 'do their own thing', these areas may also be quite new for the dogs providing even more motivation to explore. So we decided to test a second group of pet dogs at a doggy day-care centre which they visited on a very regular and consistent basis. This would then allow us to assess whether the familiarity of the environment had affected our results. Indeed, we found that free-ranging dogs and pet dogs tested at the dog care centre did not differ in the time they spent being cuddled by the person in the pre-test, and both spent more time with the cuddler than pet dogs tested in the dog areas.

Overall, studies with adults show that both wolves and dogs can clearly distinguish between individuals with whom they have a long history (caregivers, hand-raisers) and are more likely to seek and accept physical contact with them rather than with complete strangers, and/or people who are just familiar (i.e. not closely bonded) (Table 6.2). The pattern of wolves showing more curiosity towards strangers than dogs that emerged in puppy studies no longer seems to hold when the animals are adults. In general, dogs seek out, and are more acceptant of, physical contact (cuddles), from both bonded and 'just familiar' people than wolves; however, it is interesting to note that the variability in the wolves is much higher, with some animals seemingly more looking forward to, and enjoying cuddle sessions than others. The higher rates of self-directed behaviours in dogs warrant a more in-depth investigation, since at the moment it is not clear what these ambivalent behaviours really convey. The fact that they are probably related to mild stress does not reveal their ultimate causation. Are they so-called 'appeasement' or 'submissive' behaviours? Are they used by dogs specifically with humans? And are they an indication of relationship uncertainty? Or are they shown equally with conspecifics? We hope to answer these questions in future studies.

Only two studies have looked at the potential differences in wolves' and dogs' willingness to respond to learned 'commands', employing the same animals, but in somewhat different settings. In an environment in which the animals could explore the room and move around freely, dogs were slightly more responsive to human commands than wolves; however, on a leash walk outdoor, the opposite pattern emerged.

Table 6.2 Summary of studies conducted comparing adult wolves and dogs on aspects of sociality towards humans

Test	Behaviour	Species	Result	References
Attachment	Proximity seeking	D (pet) &W	W=D handler/owner > stranger	(Lenkei et al. 2020)
	Separation distress		W=D handler/owner > stranger	
	exploration (safe haven) when with handler/ owner vs. stranger		W=D handler/owner > stranger	
Sociability	Proximity & greeting: to caregiver, very familar, almost familiar, almost stranger, stranger Greeting to caregiver, very familiar, almost stranger, stranger	W	Caregiver = very familiar Caregiver > both stranger types	(Ujfalussy et al. 2017)
	Greeting intensity to	W D	Caregiver>familiar Caregiver = familiar	(Büttner 2016)
	Caregiver		D > W	
	Familiar		D > W	
	Friendly stranger approach	W D (pet)	Licking Tail wagging	(Márta Gácsi et al.2013)
	Threatening stranger approach	W D (pet)	Avoidant Friendly, avoidant, aggressive	
	Proximity & contact to	D (pet) &W	D=W familiar>stranger	(Bentosela et al. 2016)
	Familiar		D (pet) > W	
	Stranger		D (pet) > W	
	Proximity & contact to	D&W	D=W bonded>familiar	(Wirobski et al. 2021)
	Hand-raiser		D > W	
	Familiar		D > W	
	Self-directed behaviours (social conditions)		D = positive correlation between contact duration with familiar and frequency of cues	
	Proximity, contact, choice: Cuddle provider vs. food provider	D&W	No preference in choice for food vs. cuddle provider D > W time being cuddled D=W in choice behaviour	(Lazzaroni et al. 2020)
	Social contact vs. food as reinforcer	D (shelter & pet) & W	Food > social contact	(Feuerbacher and Wynne 2012)
Response to training situation	Leash walking and response to commands	D&W	W > D but dependent on familiarity with trainer	(Auer et al. 2011)
	Response to cues, proximity, gaze at face	D&W	D > W	(Vasconcellos et al. 2016)
	Stress-related cues		D=W decrease post-training	

6.3 Physiological Correlates of the Human-Animal Social Bond

As has been suggested by different hypotheses, it is thought that domesticated animals in comparison with their wild ancestors are mainly characterized by a decrease in their fear and aggressive response to humans and increased docility (Trut et al. 2009), as well as an increased tolerance/willingness to closely interact with humans (hypersociability; vonHoldt et al. 2017; see Chap. 11 for further discussion of these hypotheses). These behavioural changes parallel modifications of the domesticated animals' physiological response in particular involving their autonomic nervous system (ANS), glucocorticoid (GC), and oxytocin (OT) systems (Büttner 2016; Herbeck et al. 2016; Kaiser et al. 2015; Künzl and Sachser 1999; Zipser et al. 2014).

The autonomic nervous system (ANS) regulates the heart and other visceral functions (Shields Jr 1993) and modulates the expression of emotions in the social contexts (Porges 2003). A measure of the ANS can be estimated using cardiac activity (Kreibig 2010; Porges 2001, 2003, 2009; Rajendra Acharya et al. 2006) and more specifically by considering heart rate (HR) (i.e. the number of heartbeats per time unit) and heart rate variability (HRV) (i.e. the variation in time between consecutive heartbeats). Whereas both physical activity and mental states modulate HR (Maros et al. 2008; Visser et al. 2002), HRV appears to be less dependent on physical activity, and it has been shown to generally decrease with an increase in states of stress and increase during states of relaxation (Luque-Casado et al. 2013).

The glucocorticoid system is the main metabolic system regulating an individual's physiological response to external and internal stressors and helps the body to maintain homeostasis in changing environments (Sapolsky 2002). As such, glucocorticoids (GCs) are responsive to psychosocial stimuli, and GC levels may increase or decrease depending on the context and the individual's perception of a specific situation (Raulo and Dantzer 2018), and this may also be the case in relation to social bonding. For example, GC concentrations decreased following a cooperation task in macaque dyads that had a strong socio-positive bond, but not in dyads with a less positive relationship (Stocker et al. 2020).

In recent years, oxytocin (OT) has been identified as a potentially crucial hormone involved in social bonding and cognition (see Feldman 2012; Crockford et al. 2014, for a review). In humans, OT has been shown to be a key mediator of emotional and social behaviours such as attachment (Insel and Young 2001) and social cognition (Graustella and MacLeod 2012; see Horta et al. 2020 for a recent review). A similar involvement of OT in mother-infant as well as sexual-partner bonding has been shown in a number of non-human species (Anacker and Beery 2013; Broad et al. 2006; Coria-Avila et al. 2014). More recently, OT has also been shown to be involved in affiliative relationships (friendships). For example, in wild chimpanzees oxytocin increased after grooming bouts, but this effect was mediated by the quality of the affiliative bond between partners, i.e. the closer the affiliative bond, the higher the OT increase (Crockford et al. 2013).

Heart rate (and its variability), cortisol, and oxytocin have all been associated with dog owner interactions. In one study, when petted by their owners during a veterinary examination, dogs' heart rate increased less than when dogs had to undergo the vet procedure alone (Csoltova et al. 2017). Another study showed that pet dogs' heart rate variability generally increased when being cuddled by their owners (Katayama et al. 2016). As regards GC concentrations, results are a bit more mixed. In one study, GC actually increased in dogs when they were being petted by their owners (Handlin et al. 2011); however, in subsequent studies, results suggested that the style of the interaction and how the dogs were petted significantly mediated the GC response. Overall, a calm and gentle type of physical contact carried out by a familiar individual (Rehn et al. 2014) or a truly playful interaction in a tug of war game with a handler, with no 'disciplining' involved (Horváth et al. 2008), has been shown to decrease dogs' GC concentrations. In contrast, a more stimulating/ activating touch by the owner (Petersson et al. 2017) or play, which however involves some form of conflict with their handler, may increase dogs' GC (Horváth et al. 2008). Further, strengthening the potential stress-buffering role of a known human, in one study with kennel dogs, researchers found that when exposed to a novel environment either alone or with a conspecific, dogs showed an increase in GC concentrations; however, no such increase occurred when they were introduced to the novel location in the presence of their caretaker (Tuber et al. 1996).

Finally, a number of studies have found that oxytocin levels in dogs and their human partners increased after an exchange of friendly interactions (Handlin et al. 2011, 2012; MacLean et al. 2017; Miller et al. 2009; Odendaal and Meintjes 2003). Moreover, OT administration in dogs increased affiliative and proximity-seeking behaviours towards the owner (Romero et al. 2014; but see also Thielke et al. 2017), as well as dogs' ability to follow human-given cues (Macchitella et al. 2016; Oliva et al. 2015).

Taken together, these results have led a number of authors to suggest that changes in the cortisol and oxytocin system may have played a crucial role in the dog domestication process (Herbeck and Gulevich 2019; Herbeck et al. 2016; Kikusui et al. 2019; Nagasawa et al. 2015). However, so far only a few studies have compared wolves' and dogs' heart rate, oxytocin, and cortisol response when engaged in human interactions.

6.3.1 Wolves and Dogs Compared

In a recent Wolf Science Centter study, wolves were more relaxed than dogs showing a lower HR and higher HRV during periods of rest and inactive wakefulness (Kortekaas and Kotrschal 2019). The authors suggested that this could be a specific adaptation dogs have evolved for living in a human environment, since we usually want our dogs to be ready and active at any given time, but to relax in between. However, in this pilot study, an only marginally familiar person, who had no relationship with the animals, was present during the recording of the animals' behaviour, which might have differently affected the wolves and the dogs.

Therefore, in a more in-depth investigation of this topic, we conducted a study comparing the HR and HRV of wolves ($N = 6$) and dogs ($N = 7$), while they either rested or were alert (but still lying down) either alone, in proximity of a bonded human (animal trainer), or close to a conspecific pack member (Jean-Joseph et al. 2020). In contrast to Kortekaas and Kotrschal (2019), when resting, dogs were generally more relaxed than wolves (i.e. showed lower HR and higher HRV), but the cardiac parameters between dogs and wolves when awake and inactive were similar. However, dogs and wolves did seem to respond somewhat differently when in the presence of the bonded trainer, a conspecific, and being alone. Compared to when resting alone, dogs showed a lower HR and higher HRV both in the presence of the bonded human and the conspecific, in other words, both the human and their pack mate acted as social support for dogs, increasing their state of relaxation. Wolves relaxed predominantly when the conspecific was with them (compared to when they were alone) and not so much when the bonded human was present. These results suggest that while wolves can form close relationships with humans, the bond they have with them does not equal the one they built up with conspecifics providing less security (i.e. haven of safety), whereas dogs might seem to see humans on a similar level as conspecifics even if they do not have as much contact with them as with their peers.

Interestingly, a different pattern emerged when wolves and dogs were alert in regard to the presence of a trainer. During this condition, the dogs had distinctly higher HRs and lower HRVs than when alone or with pack members. Wolves were similarly aroused as dogs (similar HR); however, their HRV was lower. Since HRV has been linked to cognitive processes (Luque-Casado et al. 2013; Maros et al. 2008), we speculated that the presence of a trainer led the anticipation of interesting things to come, e.g. going for a walk, a test, or training interaction, whereas the dogs were in general just excited that the trainer was around. This interpretation suggests that our wolves may attribute a different meaning or valence (HRV is frequently used to assess affective state; Kreibig 2010) to the trainers than the dogs. However, we still need to be careful about the interpretation of the results since, first, the sample size was rather small, second, there was a huge inter-individual variability and, last, the wolves were older than the dogs. Further studies will hopefully shed some light on the observed differences. However, it is exciting to realize that we can get an insight about how the animals perceive their environment by looking at their heart rates and heart rate variability!

In a first study addressing the potential differences between wolves' and dogs' cortisol responses to interactions with a human partner, we measured the cortisol levels (in saliva) of nine wolves and nine dogs at the Wolf Science Center before and after a short training session, in which animals were asked to respond to familiar cues (sit, lie, etc.) given by a closely bonded partner (Vasconcellos et al. 2016). Each animal was tested in 15 5-min training sessions, involving 5 trainers (all people, who had been equally involved in hand-raising the animal from puppyhood). From a behavioural perspective, there were no differences in the frequency of stress-related behaviours exhibited by wolves and dogs during the training session. However, dogs spent longer in close proximity (1 m, 99% vs. 89% of trial time) and oriented with their face towards the trainer (98% vs. 83% of trial) than wolves, whereas wolves spent longer exploring the room (7.2% vs. 0.8% of trial time). While these differences between wolf and dog performance were statistically significant, the effect size is relatively minor. This is supported by the fact that wolves seem to pay more attention to details and take faster decisions (see Chap. 10).

Dogs also responded to a greater number of cues and did so faster than wolves (79% vs. 65% of cues). This difference is not surprising especially considering that also in other contexts wolves appear to be less forthcoming when asked to follow commands (see Chaps. 7 and 8). As regards the cortisol levels, both wolves and dogs showed a significant decrease after the training session, suggesting that individuals from both species viewed this as a positive social interaction, although in general dogs' salivary cortisol levels were much higher than those of wolves (Oliva et al. 2015).

Aside from cortisol, the other hormone which, as mentioned above, has attracted quite some interest in the dog domestication literature is oxytocin.

In a high-impact study published in Science in 2015, Nagasawa and colleagues (2015) investigated which characteristics of the dog/wolf-human interaction (e.g. physical contact, gazing, etc.) may elicit oxytocin release (as measured in urine). They found that oxytocin levels increased in a subset (8/28) of the dogs tested (i.e. those in which dogs showed a longer gazing towards their owner), and this pattern of results was not found in the wolves and their caretakers. Based on these results, the authors conclude that a positive oxytocin-mediated loop evolved in dogs during the domestication process (Maros et al. 2008). However, a number of problematic issues make results from this study difficult to evaluate (see also Kekecs et al. 2016), among which:

(1) The comparison between the 28 pet dogs tested with their owners and 5 wolves and their caretakers (wolves were kept in outdoor enclosures).
(2) The testing of animals in a novel environment, which may have stressed out the wolves more than the dogs given the wolves' reduced exposure to such situations (oxytocin is a cortisol antagonist).
(3) A suspiciously high pretreatment oxytocin level in wolves (three times higher than that found in dogs), which could have been caused by the greeting of the caretaker before the experiment and may have made a further increase due to the experimental procedure unlikely.
(4) Only about 8 of the 28 dogs showed an effect, but only 5 wolves were tested.

To investigate these matters further, we conducted a 4-year project, specifically on the potential differences in wolves' and dogs' hormonal responses when interacting with humans. This was perhaps one of the most challenging projects we have carried out so far, requiring extreme levels of persistence both in the lab, where validation of hormonal methods was required to properly conduct our studies, and perhaps even more so in the 'field', where the first challenge we faced was to reliably obtain urine from our animals to allow for oxytocin extraction (see Box 6.1 for a description of the trials and tribulations of Gwen Wirobski, our P(ee)hD student on the project!) (Fig.6.4).

Once accomplished our 'mission impossible', largely thanks to Gwen and the trainer team's humongous efforts, we were able to conduct a number of studies comparing wolves and dogs.

Fig. 6.4 (**a**) Marianne Heberlein and (**b**) Kathi Kriegler busy 'catching' pee from a wolf and dog: speed, precision, and perseverance in enormous quantities proved essential requirements for success! Picture A was taken by Sarah Marshall-Pescini; Picture B adapted from Wirobski et al. 2020, CC BY 4.0; https://creativecommons.org/licenses/by/4.0/

In a first study, we found that baseline (unstimulated) oxytocin levels were in fact different in wolves and dogs but... in the opposite direction to that observed by Nagasawa et al. (2015). Overall dogs showed higher unstimulated oxytocin levels than wolves, but it is male dogs in particular that show height levels compared to the other groups (Wirobski et al. 2020). We also found that cortisol levels were significantly higher in dogs than wolves, which is also contrary to most hypotheses suggesting a dampening of the cortisol response in domestic animals compared to their wild counterparts. Similar results in regard to cortisol responses were already obtained in previous studies at the Wolf Science Center (Vasconcellos et al. 2016).

These first results on oxytocin and cortisol are not at all in line with previous thinking on the topic, and this has made us question what may be the underlying cause of these hormonal differences. At this moment we still don't have a firm answer. We suspect that the different mating systems of wolves and dogs may be at least partly responsible for these differences. In the study above, we focused our data collection in the non-breeding season; however, since male dogs are sexually active all year-round (and ready to mate with females with un-synchronous oestrus), this may be reflected in a different hormonal profile than wolves, where both males and females are sexually active only during a specific season from January to March. Thus, it could be that the there is an interplay between different hormones (in particular testosterone, oxytocin, and cortisol) and the sexual status of the animals, which are manifested differently in wolves and dogs. We are currently collecting data on our animals' more complete hormonal profile across a 1-year period, thereby including the different stages of their breeding season, to evaluate this 'working hypothesis'.

In a second study (including $N = 10$ wolves and $N = 11$ dogs), we investigated the effect of human-animal interactions on oxytocin and cortisol levels, in relation to the quality of the bond with the human. This was very much the 'heart' of the project!

As described above, the test involved a cuddle session between each animal and either a bonded (hand-raiser) or a familiar (a student) person, followed by a brief training session. Two further control conditions were carried out to better tease apart the relationship between behaviour and potential physiological effects, i.e. a 'food-only' condition, in which the animals were given the same quantity of food by a third person as in the social tests, but this time with no social interaction, and a 'baseline' condition in which after an hour of observation of the focal animal in the pack (to check that no excessive locomotion or social interaction with conspecifics occurred), the urine was collected for OT and cortisol analyses.

Again, we had some surprising results from this study!

The oxytocin response of wolves and dogs to the human contact not only did not significantly differ, but it also did not show an increase following the positive interaction with humans. In fact, the food-only condition, i.e. when no social interaction was offered but just food, was where the highest OT levels were found in both wolves and dogs! Also, the cortisol response was no different in wolves and dogs across conditions. However, the time spent cuddling with the familiar person (but not with the bonded hand-raiser) was associated with an increase in cortisol, for both wolves and dogs, suggesting that for both species equally, having direct contact with someone they do not know so well can cause a mild stress/arousal (Fig. 6.5). We talk here of just mild stress because it is important to note that cortisol levels were no different than those in the baseline (where no particular event occurred prior to sample collection) nor the condition in which they got food without social interaction. So, the experience per se did not seem to be particular stressful; however, the part involving being in direct body contact with the unfamiliar person both for

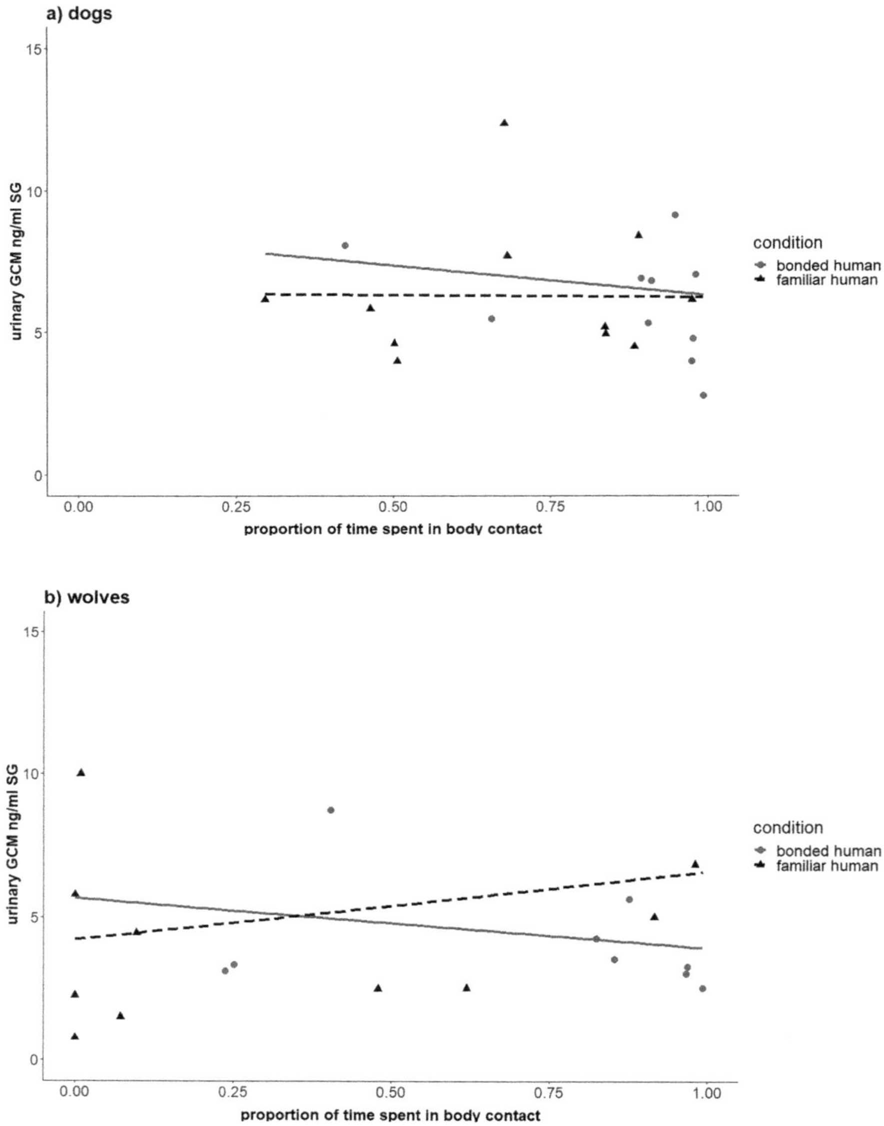

Fig. 6.5 Dogs' (**a**) and wolves' (**b**) cortisol response in relation to the time spent cuddling with the bonded and the familiar person (familiar person triangles and dotted line, bonded person circles and full line). Figure adapted from Gwendolyn Wirobski et al. (2021) https://www.nature.com/articles/s41598-021-93922-1/; CC BY 4.0; http://creativecommons.org/licenses/by/4.0/

wolves and dogs did make them somewhat uncomfortable. So again, our dogs and wolves seem to respond physiologically in similar ways to the differently bonded humans!

These results really challenge those of the previous Science study.... Could it be then that it is 'being a pet dog' that makes the difference? In fact, it was pet dogs with

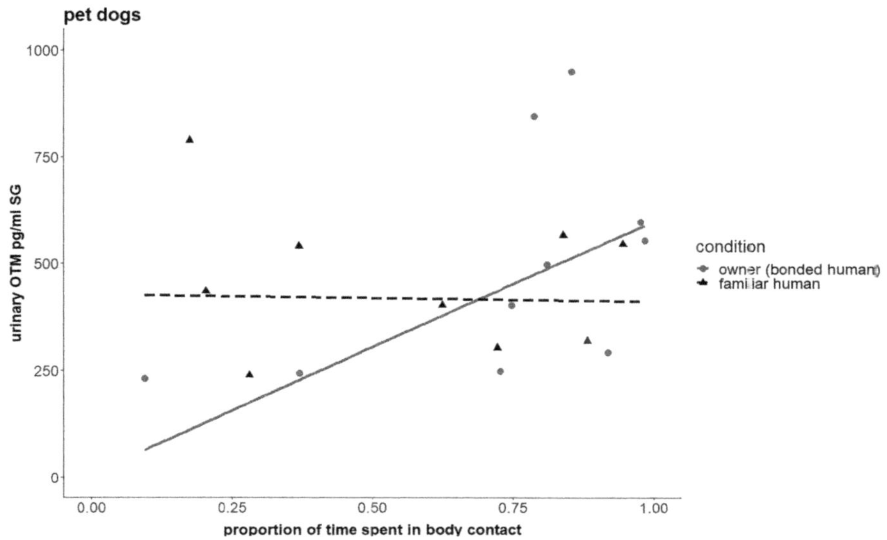

Fig. 6.6 In pet dogs higher OT levels were associated with longer cuddle sessions with the owner (but not with the familiar person). Figure adapted from Gwendolyn Wirobski et al. (2021) https://www.nature.com/articles/s41598-021-93922-1/figures/2; CC BY 4.0; http://creativecommons.org/licenses/by/4.0/

their owners that were tested and compared to the wolves in the Science study. Maybe the bond between the owner and dog is different from the relationship our trainers have with the dogs and wolves at the Wolf Science Center. To test this further, we did a follow-up study, in which we conducted the same conditions described above, with pet dogs and their owners. And yes! Here we did find that the more time spent cuddling with the owner, the more pet dogs' oxytocin increased (Fig. 6.6).

6.3.2 Summary Bonding and Physiology

Relatively few studies have been conducted investigating the physiological correlates of bonding, and having faced the many challenges that such studies generate, we are not surprised! But the results were worth the effort. Firstly, what emerges is that for both wolves and dogs, when raised in a similar way, the bond between the trainers and the animals may still differ at a very subtle level with dogs maybe seeing the bonded trainers more as a reliable social/emotional supporter than wolves. The fact that dogs nevertheless choose to interact longer with human partners than wolves may suggest that we have selected dogs to accept and tolerate such situations, whereas wolves literally 'walk away'. Rates of self-directed behaviours in dogs were not associated with cortisol levels. Such results question whether these behaviours are in fact purely an expression of stress. At the moment we really don't know much about them; however, these results may point at a communicative function, potentially used by dogs in their human-directed communication.

Secondly, the differential physiological response to bonded or just 'familiar' people further supports the behavioural studies showing that both wolves and dogs can clearly differentiate between people that are closely bonded to them (have raised them) and those who are just 'familiar'. However, for neither species, when raised in a similar manner (and hence with the possibility of establishing similar bonds with the people around them), do we see a strong association between social interaction and oxytocin levels (regardless of relationship type). Pet dogs, however, show a different response pattern. Duration of social contact with the owners (but not the familiar person) was found to be associated with high OT levels, whereas interacting with a 'just familiar' person was not. The latter results suggest that pet dogs that have extensive contact to many humans from an early age respond differently to such interaction compared to our pack animals. Results also suggest the type of bond our pack animals establish with the hand-raisers is somewhat different to the one pet dogs establish with their owners and further supports the need to take comparison between populations of animals that are not raised in the same way with extreme caution (Table 6.3).

Table 6.3 Summary of results on the physiological measures related to bonding

Test	Physiological measure	Species	Result	References
Resting and alert (but inactive) if Alone (A), with conspecific pack mate (C), with human (H)	Heart rate (HR)	D&W	D < W (A, H); D=W (C) (rest) W > D (A); D=W (C, H) (alert)	(Jean-Joseph et al. 2020)
		D	C=H < A (rest); A = C < H (alert)	
		W	C < A = H (rest); C < A = H (alert)	
	Heart rate variability (HRV)	D&W	D=W (A); D > W (C, H) D > W (rest); W=D (alert)	
Response to cues	Cortisol, stress-related cues	D&W	D=W decrease post-training	(Vasconcellos et al. 2016)
Baseline	Oxytocin Cortisol	D&W	D > W D > W (non-breeding season)	(Wirobski et al. 2020)
Sociability	Oxytocin	D (WSC & pet) & W	D (WSC) = W food condition > social interaction D (pet): Duration owner interaction associated with higher OT	(Wirobski et al. 2021)
	Cortisol	D (WSC & pet) & W	D (WSC) = W duration contact with 'just familiar' person associated with increased cortisol	

6.4 Overall Summary

Taken together results show that initial differences between wolves and dogs in their interactions with humans, when revisited, did not often hold up to closer inspection. Moreover, since the first comparisons almost 15 years ago, research on different dog populations has highlighted the strong effect of experience and raising on the human-related behaviours under investigation. Accordingly, when comparisons are carried out on similarly raised, intensely socialized wolves and dogs, differences are either smaller than expected or disappear altogether. Both wolves and dogs have the capacity to establish close preferential bonds with specific people and this is evident in tests both with puppies and adults. Nevertheless, some differences do emerge. Both in studies at the Wolf Science Center and in other populations, dogs appear to seek out, and stay in physical contact with humans (whether bonded, familiar, or strangers) more than wolves, even if in some cases this is associated with higher behavioural and physiological signals of stress. Pilot results also suggest that dogs show more greeting, tail wagging, and submissive behaviours towards humans than wolves. Such results (together with results from cooperative interactions between humans and wolves/dogs—see Chap. 8) point to the possibility that the hierarchical elements of the relationship with humans may differ in wolves and dogs, and this needs further, more direct investigation.

Box 6.1 Life of a P(ee)hD student at the Wolf Science Center (by Gwendolyn Wirobski)

I had just returned to Austria from the UK, where I was working as a zoo-based researcher studying the social behaviour of a baboon troop, when I saw the announcement of an open PhD position at the Wolf Science Center. I knew I wanted to pursue a career in animal behaviour science, but, having trained as a veterinarian, my first job after graduation was in a small animal clinic. When I got the call that I was accepted and could join the research team to study the wolves and dogs at the Wolf Science Center, I was over the moon! I vividly remember the first meeting with the wolves and dogs because the two species behaved strikingly different. To me, then still naïve to the hypotheses about their domestication process, it seemed as if the dogs were longing for any human contact, happily interacting even with complete strangers, while the wolves acted much more reserved and selected whom they approached and whom they stayed away from. My PhD project was centred on the idea that a particular hormonal system, namely, the oxytocin system, had changed over the course of domestication which helped dogs to become exceptionally sociable (even 'hypersocial') and capable of forming long-lasting social bonds with their human partners. To investigate this, we first needed to figure out a way to analyse wolves' and dogs' oxytocin levels. Being a peptide hormone produced in the brain, oxytocin is not easy to get hold of, especially if working non-invasively. However, researchers had already shown that oxytocin could be measured from urine samples, and thus we started looking for ways to collect pee (or as we like to say at the Wolf Science Center, the 'liquid gold') samples from our animals. This

(continued)

Box 6.1 (continued)

was easy enough with the dogs but the wolves needed some convincing.... During my first few months at the Wolf Science Center, I came up with several different stationary 'pee collection devices' (pictures) which I placed in different spots in the enclosures and the surrounding park. Most of them did not work. Then, I tried to entice the animals to mark their territory by using dog or wolf urine, or urine collected from females in heat, but they still did not care to pee for science (or at least, most were not very reliable in their marking behaviour). Luckily, the animal trainers had started habituating the animals to a 'pee collection stick' (an expandable metal stick with a plastic cup attached to it—picture) during their leash walks through the park. Finally we were able to collect the liquid gold from most of our animals and relate those samples back to observations of their social behaviour and experimental conditions. The next challenge arose from the difficult task of choosing and validating an assay system that would allow us to reliably and precisely measure urinary oxytocin concentrations. With the help of our collaborators at the hormone lab of the Max Planck Institute in Leipzig, Germany, this also eventually succeeded. Almost 4 years later, we are approaching the end of the project. By now I have analysed social behaviour and corresponding oxytocin concentrations of more than a thousand urine samples of wolves, dogs, and humans. Currently, we are working out the patterns that emerge from our data, and we are excited to learn what they might tell us about the underlying hormonal mechanisms of dog domestication.

Photos 1 and 2 taken by Gwendolyn Wirobski. Photo 3 adapted from Wirobski et al. 2021 (https://www.nature.com/articles/s41598-021-92356-z; CC BY 4.0 http://creativecommons.org/licenses/by/4.0/.)

References

Ainsworth MDS, Bell SM (1970) Attachment, exploration, and separation: illustrated by the behavior of one-year-olds in a strange situation. Child Dev 41(1):49–67. https://doi.org/10.2307/1127388

Anacker AMJ, Beery AK (2013) Life in groups: the roles of oxytocin in mammalian sociality. Front Behav Neurosci 7:185. https://doi.org/10.3389/fnbeh.2013.00185

Auer M, Wedl M, Range F, Virányi Z, Belényi B, Kotrschal K (2011) Leash walking as a model for cooperation between humans and wolves: the effects of personality and intensity of contact. J Vet Behav 6:100–101. https://doi.org/10.1016/j.jveb.2010.08.002

Bard K (2007) Studies of attachment in nursery-reared chimpanzees: from peers to responsive caregivers. Paper presented at the American Journal of Primatology

Bard KA (1991) Distribution of attachment classifications in nursery chimpanzees. Am J Phys Anthropol 24(2):88

Barnard S, Marshall-Pescini S, Pelosi A, Passalacqua C, Prato-Previde E, Valsecchi P (2017) Breed, sex, and litter effects in 2-month old puppies' behaviour in a standardised open-field test. Sci Rep 7(1):1802. https://doi.org/10.1038/s41598-017-01992-x

Bentosela M, Wynne CDL, D'Orazio M, Elgier A, Udell MAR (2016) Sociability and gazing toward humans in dogs and wolves: simple behaviors with Broad implications. J Exp Anal Behav 105(1):68–75. https://doi.org/10.1002/jeab.191

Bowlby J (1979/1994) The making & breaking of affectional bonds. Routledge, New York

Bowlby J (1982) Attatchment and loss (2nd Vol. 1: Attachment). New York: Basic Books

Bowlby J (1988) A secure base, 2nd edn. Basic Books, New York

Bowlby J (2012) A secure base. Routledge

Bretherton I (1985) Attachment theory: retrospect and Prospect. Monogr Soc Res Child Dev 50(1/2):3–35. https://doi.org/10.2307/3333824

Broad KD, Curley JP, Keverne EB (2006) Mother-infant bonding and the evolution of mammalian social relationships. Philos Trans R Soc Lond B 361(1476):2199–2214. https://doi.org/10.1098/rstb.2006.1940

Büttner, C. S. (2016). Relationship specific greeting excitement in wolves and dogs. FREIE UNIVERSITAET BERLIN

Cafazzo S, Lazzaroni M, Marshall-Pescini S (2016) Dominance relationships in a family pack of captive arctic wolves (*Canis lupus arctos*): the influence of competition for food, age and sex. PeerJ 4:e2707. https://doi.org/10.7717/peerj.2707

Cafazzo S, Valsecchi P, Bonanni R, Natoli E (2010) Dominance in relation to age, sex and competitive contexts in a group of free-ranging domestic dogs. Behav Ecol 21(3):443–455

Collins N, Feeney B (2000) A safe haven: an attachment theory perspective on support seeking and caregiving in intimate relationships. J Pers Soc Psychol 78:1053–1073. https://doi.org/10.1037/0022-3514.78.6.1053

Coria-Avila GA, Manzo J, Garcia LI, Carrillo P, Miquel M, Pfaus JG (2014) Neurobiology of social attachments. Neurosci Biobehav Rev 43:173–182. https://doi.org/10.1016/j.neubiorev.2014.04.004

Crockford C, Deschner T, Ziegler TE, Wittig RM (2014) Endogenous peripheral oxytocin measures can give insight into the dynamics of social relationships: a review. Front Behav Neurosci 8:68. https://doi.org/10.3389/fnbeh.2014.00068

Crockford C, Wittig RM, Langergraber K, Ziegler TE, Zuberbühler K, Deschner T (2013) Urinary oxytocin and social bonding in related and unrelated wild chimpanzees. Proc R Soc B 280: 20122765. https://doi.org/10.1098/rspb.2012.2765

Csoltova E, Martineau M, Boissy A, Gilbert C (2017) Behavioral and physiological reactions in dogs to a veterinary examination: owner-dog interactions improve canine Well-being. Physiol Behav 177:270–281

Feldman R (2012) Oxytocin and social affiliation in humans. Horm Behav 61(3):380–391. https://doi.org/10.1016/j.yhbeh.2012.01.008

Fentress JC (1967) Observations on behavioral development of a hand reared male timber wolf. Am Zool 6(3):333–351. https://doi.org/10.1093/icb/7.2.339
Feuerbacher EN, Wynne CDL (2012) Relative efficacy of human social interaction and food as reinforcers for domestic dogs and hand-reared wolves. J Exp Anal Behav 98(1):105–129. https://doi.org/10.1901/jeab.2012.98-105
Fraley R, Shaver P (2000) Adult romantic attachment: theoretical developments, emerging controversies, and unanswered questions. Rev Gen Psychol 4:132–154. https://doi.org/10.1037/1089-2680.4.2.132
Gácsi M, Győri B, Miklósi Á, Virányi Z, Kubinyi E, Topál J, Csányi V (2005) Species-specific differences and similarities in the behavior of hand-raised dog and wolf pups in social situations with humans. Dev Psychobiol 47(2):111–122. https://doi.org/10.1002/dev.20082
Gácsi M, Vas J, Topál J, Miklósi Á (2013) Wolves do not join the dance: sophisticated aggression control by adjusting to human social signals in dogs. Appl Anim Behav Sci 145(3–4):109–122. https://doi.org/10.1016/j.applanim.2013.02.007
Gosling SD, John OP, Kwan VSY (2003) A Dog's got personality: a cross-species comparative approach to personality judgements in dogs and humans. J Pers Soc Psychol 85:1161–1169. https://doi.org/10.1037/0022-3514.85.6.1161
Graustella AJ, MacLeod C (2012) A critical review of the influence of oxytocin nasal spray on social cognition in humans: evidence and future directions. Horm Behav 61(3):410–418
Hall NJ, Lord K, Arnold AMK, Wynne CDL, Udell MAR (2015) Assessment of attachment behaviour to human caregivers in wolf pups (*Canis lupus lupus*). Behav Process 110:15–21. https://doi.org/10.1016/j.beproc.2014.11.005
Handlin L, Hydbring-Sandberg E, Nilsson A, Ejdebäck M, Jansson A, Uvnäs-Moberg K (2011) Short-term interaction between dogs and their owners: effects on oxytocin, cortisol, insulin and heart rate—an exploratory study. Anthrozoös 24:301–315. https://doi.org/10.2752/175303711X13045914865385
Handlin L, Nilsson A, Ejdebäck M, Hydbring-Sandberg E, Uvnäs-Moberg K (2012) Associations between the psychological characteristics of the human–dog relationship and oxytocin and cortisol levels. Anthrozoös 25:215–228. https://doi.org/10.2752/175303712X13316289505468
Hazan C, Shaver P (1987) Romantic love conceptualized as an attachment process. J Pers Soc Psychol 52:511–524. https://doi.org/10.1037//0022-3514.52.3.511
Heberlein MTE, Turner DC, Range F, Virányi Z (2016) A comparison between wolves, *Canis lupus*, and dogs, *Canis familiaris*, in showing behaviour towards humans. Anim Behav 122:59–66. https://doi.org/10.1016/j.anbehav.2016.09.023
Herbeck YE, Gulevich RG (2019) Neuropeptides as facilitators of domestication. Cell Tissue Res 375(1):295–307. https://doi.org/10.1007/s00441-018-2939-2
Herbeck YE, Gulevich RG, Shepeleva DV, Grinevich VV (2016) Oxytocin: coevolution of human and domesticated animals. Russian J Genetics: Appl Res 7(3):235–242. https://doi.org/10.18699/VJ16.145
Horta M, Pehlivanoglu D, Ebner NC (2020) The role of intranasal oxytocin on social cognition: an integrative human lifespan approach. Curr Behav Neurosci Rep:1–18
Horváth Z, Dóka A, Miklósi Á (2008) Affiliative and disciplinary behavior of human handlers during play with their dog affects cortisol concentrations in opposite directions. Horm Behav 54(1):107–114
Hsu Y, Serpell JA (2003) Development and validation of a questionnaire for measuring behavior and temperament traits in pet dogs. J Am Vet Med Assoc 223(9):1293–1300. https://doi.org/10.2460/javma.2003.223.1293
Insel TR, Young LJ (2001) The neurobiology of attachment. Nat Rev Neurosci 2(2):129–136
Jean-Joseph H, Kortekaas K, Range F, Kotrschal K (2020) Context-specific arousal during resting in wolves and dogs: effects of domestication? Front Psychol 11
Kaiser S, Hennessy MB, Sachser N (2015) Domestication affects the structure, development and stability of biobehavioural profiles. Front Zool 12(1):S19. https://doi.org/10.1186/1742-9994-12-S1-S19

Katayama M, Kubo T, Mogi K, Ikeda K, Nagasawa M, Kikusui T (2016) Heart rate variability predicts the emotional state in dogs. Behav Process 128:108–112

Kekecs Z, Szollosi A, Palfi B, Szaszi B, Kovacs KJ, Dienes Z, Aczel B (2016) Commentary: oxytocin-gaze positive loop and the coevolution of human–dog bonds. Front Neurosci 10(155). https://doi.org/10.3389/fnins.2016.00155

Kikusui T, Nagasawa M, Nomoto K, Kuse-Arata S, Mogi K (2019) Endocrine regulations in human–dog coexistence through domestication. Trends in Endocrinology & Metabolism 30(11):793–806

Kortekaas K, Kotrschal K (2019) Does socio-ecology drive differences in alertness between wolves and dogs when resting? Behav Process 166:103877

Kreibig SD (2010) Autonomic nervous system activity in emotion: a review. Biol Psychol 84(3): 394–421. https://doi.org/10.1016/j.biopsycho.2010.03.010

Kubinyi E, Turcsán B, Miklósi Á (2009) Dog and owner demographic characteristics and dog personality trait associations. Behav Process 81(3):392–401. https://doi.org/10.1016/j.beproc.2009.04.004

Künzl C, Sachser N (1999) The behavioral endocrinology of domestication: a comparison between the domestic Guinea pig (*Cavia apereaf. porcellus*) and its wild ancestor, the cavy (*Cavia aperea*). Horm Behav 35(1):28–37

Lazzaroni M, Range F, Backes J, Portele K, Scheck K, Marshall-Pescini S (2020) The effect of domestication and experience on the social interaction of dogs and wolves with a human companion. Front Psychol 11(785). https://doi.org/10.3389/fpsyg.2020.00785

Lenkei R, Újváry D, Bakos V, Faragó T (2020) Adult, intensively socialized wolves show features of attachment behaviour to their handler. Sci Rep 10(1):1–8

Luque-Casado A, Zabala M, Morales E, Mateo-March M, Sanabria D (2013) Cognitive performance and heart rate variability: the influence of fitness level. PLoS One 8(2):e56935. https://doi.org/10.1371/journal.pone.0056935

Macchitella L, Stegagno T, Giaconella R, Polizzi Di Sorrentino E, Schino G, Addessi E (2016) Oxytocin improves the ability of dogs to follow informative pointing: a neuroemotional hypothesis. Rendiconti Lincei 1-11. https://doi.org/10.1007/s12210-016-0579-6

MacLean EL, Gesquiere LR, Gee N, Levy K, Martin WL, Carter CS (2017) Effects of affiliative human–animal interaction on dog salivary and plasma oxytocin and vasopressin. Front Psychol 8:1606. https://doi.org/10.3389/fpsyg.2017.01606

Maros K, Dóka A, Miklósi Á (2008) Behavioural correlation of heart rate changes in family dogs. Appl Anim Behav Sci 109(2–4):329–341. https://doi.org/10.1016/j.applanim.2007.03.005

Marshall-Pescini S, Virányi Z, Kubinyi E, Range F (2017) Motivational factors underlying problem solving: comparing wolf and dog Puppies' explorative and neophobic behaviors at 5, 6, and 8 weeks of age. Front Psychol 8(180). https://doi.org/10.3389/fpsyg.2017.00180

Miller SC, Kennedy C, DeVoe D, Hickey M, Nelson T, Kogan L (2009) An examination of changes in oxytocin levels in men and Women Before and After Interaction with a Bonded Dog. *Anthrozoos 22*(1):31–42. https://doi.org/10.2752/175303708X390455

Mirkó E, Kubinyi E, Gácsi M, Miklósi Á (2012) Preliminary analysis of an adjective-based dog personality questionnaire developed to measure some aspects of personality in the domestic dog (*Canis familiaris*). Appl Anim Behav Sci 138(1–2):88–98. https://doi.org/10.1016/j.applanim.2012.02.016

Nagasawa M, Mitsui S, En S, Ohtani N, Ohta M, Sakuma Y et al (2015) Oxytocin-gaze positive loop and the coevolution of human-dog bonds. Science 348(6232):333–336. https://doi.org/10.1126/science.1261022

Odendaal JSJ, Meintjes RA (2003) Neurophysiological correlates of affiliative behaviour between humans and dogs. Vet J 165(3):296–301. https://doi.org/10.1016/S1090-0233(02)00237-X

Oliva JL, Rault JL, Appleton B, Lill A (2015) Oxytocin enhances the appropriate use of human social cues by the domestic dog (*Canis familiaris*) in an object choice task. Anim Cogn 18:767–775. https://doi.org/10.1007/s10071-015-0843-7

Ortolani A, Vernooij H, Coppinger R (2009) Ethiopian village dogs: behavioural responses to a stranger's approach. Appl Anim Behav Sci 119(3–4):210–218. https://doi.org/10.1016/j.applanim.2009.03.011
Palmer R, Custance D (2008) A counterbalanced version of Ainsworth's strange situation procedure reveals secure-base effects in dog-human relationships. Appl Anim Behav Sci 109:306–319. https://doi.org/10.1016/j.applanim.2007.04.002
Payne E, Bennett PC, McGreevy PD (2015) Current perspectives on attachment and bonding in the dog–human dyad. Psychol Res Behav Manag 8:71–79. https://doi.org/10.2147/PRBM.S74972
Petersson M, Uvnäs-Moberg K, Nilsson A, Gustafson L-L, Hydbring-Sandberg E, Handlin L (2017) Oxytocin and cortisol levels in dog owners and their dogs are associated with behavioral patterns: an exploratory study. Front Psychol 8:1796
Porges SW (2001) The polyvagal theory: phylogenetic substrates of a social nervous system. Int J Psychophysiol 42(2):123–146. https://doi.org/10.1016/S0167-8760(01)00162-3
Porges SW (2003) The polyvagal theory: phylogenetic contributions to social behavior. Physiol Behav 79(3):503–513. https://doi.org/10.1016/S0031-9384(03)00156-2
Porges SW (2009) The polyvagal theory: new insights into adaptive reactions of the autonomic nervous system. Cleve Clin J Med 76(Suppl 2):S86–S90. https://doi.org/10.3949/ccjm.76.s2.17
Prato-Previde E, Custance DM, Spiezio C, Sabatini F (2003) Is the dog-human relationship an attachment bond? An observational study using Ainsworth's strange situation. Behaviour 140(2):225–254. https://doi.org/10.1163/156853903321671514
Rajendra Acharya U, Paul Joseph K, Kannathal N, Lim CM, Suri JS (2006) Heart rate variability: a review. Med Biol Eng Comput 44(12):1031–1051. https://doi.org/10.1007/s11517-006-0119-0
Raulo A, Dantzer B (2018) Associations between glucocorticoids and sociality across a continuum of vertebrate social behavior. Ecol Evol 8(15):7697–7716. https://doi.org/10.1002/ece3.4059
Rehn T, Handlin L, Uvnäs-Moberg K, Keeling LJ (2014) Dogs' endocrine and behavioural responses at Reunion are affected by how the human initiates contact. Physiol Behav 124:45–53. https://doi.org/10.1016/j.physbeh.2013.10.009
Romero T, Nagasawa M, Mogi K, Hasegawa T, Kikusui T (2014) Oxytocin promotes social bonding in dogs. Proc Natl Acad Sci U S A 111:9085–9090. https://doi.org/10.1073/pnas.1322868111
Sapolsky, R. M. (2002). Endocrinology of the stress-response
Shields RW Jr (1993) Functional anatomy of the autonomic nervous system. J Clin Neurophysiol 10(1):2–13
Stocker M, Loretto M-C, Sterck EH, Bugnyar T, Massen JJ (2020) Cooperation with closely bonded individuals reduces cortisol levels in long-tailed macaques. R Soc Open Sci 7(5):191056
Thielke LE, Rosenlicht G, Saturn SR, Udell MAR (2017) Nasally-administered oxytocin has limited effects on owner-directed attachment behavior in pet dogs (*Canis lupus familiaris*). Front Psychol 8:1699. https://doi.org/10.3389/fpsyg.2017.01699
Topál J, Gácsi M, Miklósi Á, Virányi Z, Kubinyi E, Csányi V (2005) Attachment to humans: a comparative study on hand-reared wolves and differently socialized dog puppies. Anim Behav 70:1367–1375. https://doi.org/10.1016/j.anbehav.2005.03.025
Topál J, Miklósi Á, Csányi V, Dóka A (1998) Attachment behavior in dogs (*Canis familiaris*): a new application of Ainsworth's (1969) strange situation test. J Comp Psychol 112(3):219–229. https://doi.org/10.1037/0735-7036.112.3.219
Trut L, Oskina I, Kharlamova A (2009) Animal evolution during domestication: the domesticated fox as a model. BioEssays 31:349–360. https://doi.org/10.1002/bies.200800070
Tuber DS, Hennessy MB, Sanders S, Miller JA (1996) Behavioral and glucocorticoid responses of adult domestic dogs (*Canis familiaris*) to companionship and social separation. J Comp Psychol 110(1):103
Udell MAR (2015) When dogs look back: inhibition of independent problem-solving behaviour in domestic dogs (*Canis lupus familiaris*) compared with wolves (*Canis lupus*). Biol Lett 11(9): 20150489. https://doi.org/10.1098/rsbl.2015.0489

Ujfalussy DJ, Kurys A, Kubinyi E, Gácsi M, Virányi Z (2017) Differences in greeting behaviour towards humans with varying levels of familiarity in hand-reared wolves (*Canis lupus*). R Soc Open Sci 4:160956. https://doi.org/10.1098/rsos.160956

Ujfalussy DJ, Virányi Z, Gácsi M, Faragó T, Pogány Á, Bereczky BM, Kubinyi E (2020) Comparing the tractability of young hand-raised wolves (*Canis lupus*) and dogs (*Canis familiaris*). Sci Rep 10(1):1–14

Valsecchi P, Barnard S, Stefanini C, Normando S (2011) Temperament test for re-homed dogs validated through direct behavioral observation in shelter and home environment. J Vet Behav 6(3):161–177. https://doi.org/10.1016/j.jveb.2011.01.002

Vasconcellos A, Virányi Z, Range F, Ades C, Scheidegger JK, Möstl E, Kotrschal K (2015) Training reduces stress in human-socialised wolves to the same degree as in dogs. PLoS One 11(9):e0162389. https://doi.org/10.1371/journal.pone.0162389

Visser EK, van Reenen CG, van der Werf JTN, Schilder MBH, Knaap JH, Barneveld A, Blokhuis HJ (2002) Heart rate and heart rate variability during a novel object test and a handling test in young horses. Physiol Behav 76(2):289–296. https://doi.org/10.1016/S0031-9384(02)00698-4

Vitale KR, Behnke AC, Udell MAR (2019) Attachment bonds between domestic cats and humans. Curr Biol 29(18):R864–R865. https://doi.org/10.1016/j.cub.2019.08.036

vonHoldt BM, Pollinger JP, Lohmueller KE, Han E, Parker HG, Quignon P et al (2010) Genome-wide SNP and haplotype analyses reveal a rich history underlying dog domestication. Nature 464(7290):898–902. https://doi.org/10.1038/nature08837

vonHoldt BM, Shuldiner E, Koch IJ, Kartzinel RY, Hogan A, Brubaker L et al (2017) Structural variants in genes associated with human Williams-Beuren syndrome underlie stereotypical hypersociability in domestic dogs. Sci Adv 3(7):e1700398. https://doi.org/10.1126/sciadv.1700398

Wirobski G, Range F, Schaebs F, Palme R, Deschner T, Marshall-Pescini S (2020) Endocrine changes related to dog domestication: comparing urinary cortisol and oxytocin in hand-raised, pack-living dogs and wolves. Horm Behav 128:104901

Wirobski G, Range F, Schaebs FS, Palme R, Deschner T, Marshall-Pescini S (2021) Life experience rather than domestication accounts for dogs' increased oxytocin release during social contact with humans. Sci Rep 11(1):1–12. https://doi.org/10.1038/s41598-021-93922-1

Zipser B, Schleking A, Kaiser S, Sachser N (2014) Effects of domestication on biobehavioural profiles: a comparison of domestic Guinea pigs and wild cavies from early to late adolescence. Front Zool 11(1):1–14

Social Cognition Part I: Communication with Humans

7

Every day, animals face a variety of challenges including coping with conspecifics, avoiding predators, and foraging. These challenges are thought to act as selection pressures and drive the evolution of strategies to deal with them. Over the last decades, several hypotheses have been proposed as to which specific challenges might have driven the evolution of sophisticated—but energetically costly—cognitive abilities (the brain is really expensive! see, e.g. Kotrschal et al. 2013, 2019).

One of the more popular hypotheses is the 'social brain hypothesis', which suggests that the brain evolved to cope with social complexity (Dunbar 1998; Dunbar and Shultz 2007). The hypothesis is based on the correlations of brain size with group size and the realization of researchers studying non-human primates that the social world of these species can be rather complex, arguably even more complex than their physical world. According to Tomasello and Call (1997), this is mainly due to three aspects that make the social world cognitively very demanding.

First, interacting with other group members can be fairly complex, since they are often unpredictable. Being able to recognize group members individually, as many animals can do, allows animals to predict how others might behave in previously encountered situations. Moreover, based on continuous interactions between individuals, relationships with certain characteristics might emerge, e.g. that individual A is always the loser in aggressive interactions with individual B or that A and B are always in close proximity. Accordingly, in non-human animals, relationships have been characterized in terms of affiliation, dominance, and relatedness (see Chap. 5). Interestingly, individuals in at least some primate species recognize not just the relationships they themselves have with others but also the relationships that various other group members form with each other. This 'knowledge', whether implicit or explicit, allows animals to predict the behaviours of others quite well (Bergman et al. 2003) and should be taken into account when cooperating or competing with others.

F. Range, S. Marshall-Pescini, *Wolves and Dogs*, Fascinating Life Sciences,
https://doi.org/10.1007/978-3-030-98411-3_7

Second, in order to interact and manipulate the behaviour of others, individuals may employ various social and communicative strategies. For example, in between group encounters, animals may lag behind to avoid the risk of getting injured. Moreover, the better they understand the communicative signals of others, the better they can predict those individuals' behaviours and can use the information provided by them to achieve their own goals (Byrne 1997).

Finally, learning socially from others about how to interact with the physical and social environment can be cognitively demanding depending on the underlying mechanisms (see Chap. 8). Moreover, deciding which novel skills to acquire and from whom likely poses additional cognitive demands on the individuals.

Some of the aspects are also relevant for wolves despite the fact that they live in relatively small family packs. For example, being able to predict the behaviour of and clearly communicative with the other pack member when cooperating in hunting, raising the young and between group competition is likely advantageous. Moreover, learning how to hunt and do so cooperatively may require some high-level social learning skills.

In free-ranging dogs, cognitive challenges might be more similar to those faced by non-human primates, since they seem to live in very flexible social structures (including coherent packs, fission fusion, and solitary) (see Chap. 4). However, further research has to reveal whether they have a similar knowledge about their group members as non-human primates. In addition, due to dogs' close cohabitation with humans, it has been suggested that dogs may have evolved some special social cognitive skills to thrive in the human environment that may even be functionally equivalent to our own social skills (Hare et al. 2002; Miklósi et al. 2003; Topál et al. 2009a, b). Alternatively, or in addition, life-long experiences with humans may facilitate the ontogeny of special cognitive skills (Udell et al. 2010). This development can result from a variety of mechanisms ranging from conditioning (Bentosela et al. 2008) to more complex modifications of cognitive and emotional processes, such as the ontogenetic process of 'enculturation' that has been proposed to result in enhanced cognitive abilities in non-human primates raised by humans as compared to individuals raised by their own kind (Hare and Tomasello 2005b).

The comparison of social cognitive skills of wolves and dogs, which have had similar experiences throughout their lifetime, can shed light on some of these hypotheses. Moreover, comparisons of different populations of domesticated dogs—pet dogs, shelter dogs, dogs living in packs without much human interference, and free-ranging dogs—can help to disentangle how much life-long experiences can drive the social strategies employed by dogs.

In the previous two chapters, we already explored what *kind* of relationships wolves and dogs form with their conspecific and human partners. In this and the next chapter, we will explore what they understand about their social world and how they may use this knowledge to make appropriate decisions in a given situation.

In the current chapter, we look in detail at the communicative abilities of wolves and dogs, especially in their interactions with humans. We will first summarize the studies comparing wolves' and dogs' abilities in comprehending human and, to a lesser extent, conspecific communicative gestures (gazing and pointing patterns), their sensitivity towards human ostensive cues, and how well they attend to the attentional states of humans. In the second part of this chapter, we focus on the few studies investigating how wolves and dogs use gazing to communicate with humans, especially those aspects of communication that are related to potential wolf-dog differences. For a more general, recent review of dog communicative abilities with humans, please see Siniscalchi (2018).

In Chap. 8, we will then review what we know about wolves' and dogs' acquisition of novel skills by interacting with conspecifics (social learning) and their competence in performing joint cooperative actions both with conspecific and human partners.

7.1 Understanding (Human) Communicative Cues

Humans convey information not just about themselves but also regarding ongoing external events or objects towards others by means of behaviour or language. This capacity of directing the attention of an observer to a distal object or entity is termed referential signalling. While humans often use language to achieve this goal, human infants cannot rely on words to exchange information. Instead they use gazing and pointing gestures in connection with acoustic communication from very early on, to gather and convey information. For example, human infants may pay attention to the mother's gaze direction to figure out to what object she is referring to when she talks (and actually learn the words for these objects by doing so). Or, since they do not yet have the motoric skills to freely explore their environment, they will often try to communicate using utterances, gazing, and pointing cues to indicate what they find interesting or what kind of object they would like to obtain. These gazing and pointing gestures of human infants are interpreted as communicative referential behaviours (Senju and Csibra 2008). They are considered a developmental precursor of linguistic communication and have been the focus of intensive research programmes in developmental psychology (Brooks and Meltzoff 2005; Colonnesi et al. 2010).

Interestingly, also non-human animals have been shown to communicate information regarding ongoing external events or objects, and these forms of non-verbal referential communication might constitute the evolutionary origins of language. Most research in this field has focused on vocal communication in non-human primates, with studies, for example, showing that many species have distinct vocalizations referring to specific predators (Seyfarth et al. 1980; Townsend and Manser 2013). However, another line of research has focused on visual communicative cues such as gazing and whether animals understand the referentiality of gazing and/or pointing gestures of humans, an aspect that is especially relevant in daily interactions between dog-owner pairs.

7.1.1 Gaze Following

Monitoring others' head and eye orientation (gaze) (Brooks and Meltzoff 2002) and coordinating with others to look in the same direction (gaze following) or at a specific target (joint attention) are considered to be a key step towards an understanding of mental states like attention and intentions (Baron-Cohen 1995; Tomasello et al. 2005). Three paradigms have been used to investigate gaze following behaviour in non-human animals. While at the behavioural level, the different forms of gaze following measured in these paradigms look very similar, the underlying mechanisms are thought to be different.

Following others' gaze into distant space (i.e. turning one's head in the same direction someone is looking towards; Fig. 7.1) may simply be a reflex-like response that occurs in various animal species including ravens, (Bugnyar et al. 2004), apes (Bräuer et al. 2005), monkeys (Amici et al. 2009), domestic goats (Kaminski et al. 2005), northern bold ibises, (Loretto et al. 2009), and red-footed tortoises (Wilkinson et al. 2010). The presence of gaze following into distant space in a wide range of species and its general early emergence in the developing organism supports the notion that it is driven by a partially reflexive mechanism (Shepherd 2010). The benefit of a reflex-like response would be that one can spot interesting things in the environment that you would have otherwise missed (see Fiset 2010 for a review). However, the quality of one's social relationships, as well as the facial expressions of the producer, seems to also influence the gaze following response of the observer indicating that it is not a purely reflexive response but to some degree flexible and context-specific (Goossens et al. 2008; Micheletta and Waller 2012). Surprisingly, pet dogs have been shown to perform rather badly in this paradigm (Agnetta et al. 2000). Only recently, we found in two studies that dogs can actually follow gaze cues if they are preceded either by a surprised expression on the face of the experimenter (Wallis et al. 2015) or when the experimenter calls the animals attention before presenting the cue (Duranton et al. 2017; Téglás et al. 2012).

The reason for the previous negative results may be due to dogs having been trained to pay attention to humans (either during life-long experiences with them or during formal training) leading to a fixation on the human face, which then interferes with the dog's response to the cue given by the experimenter (Wallis et al. 2015). Also, the communicative context might be important for dogs to not ignore the cue given by the experimenter (Duranton et al. 2017; Téglás et al. 2012). After all, dogs probably see many head turns during daily life that are not relevant for them, and they learn, through experience, not to pay attention to them. Thus, creating a communicative context or by increasing the relevance of the cue (i.e. looking at a door instead of into distant space) that increases the saliency of the situation and thus, the cue, might be important for dogs to follow the gaze of others (Fig. 7.1).

Following gaze around a barrier deals with the question whether subjects successfully follow another's gaze geometrically behind a visual barrier (see Fig. 7.2). In this task, the subject needs to understand that if another individual's eyes are directed towards a location behind a visual barrier, it must alter its own

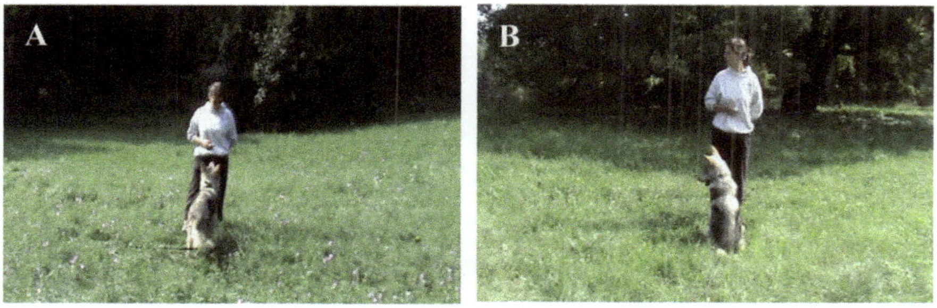

Fig. 7.1 Gaze following into distance space in a wolf puppy: The experimenter first tried to get the animal to look into her face (**a**) and then looks to the side (**b**) to see if the animal is following her gaze, which Geronimo nicely does here. Figure adapted from Range and Virányi (2011) (https://doi.org/10.1371/journal.pone.0016888; CC BY 4.0; http://creativecommons.org/licenses/by/4.0/)

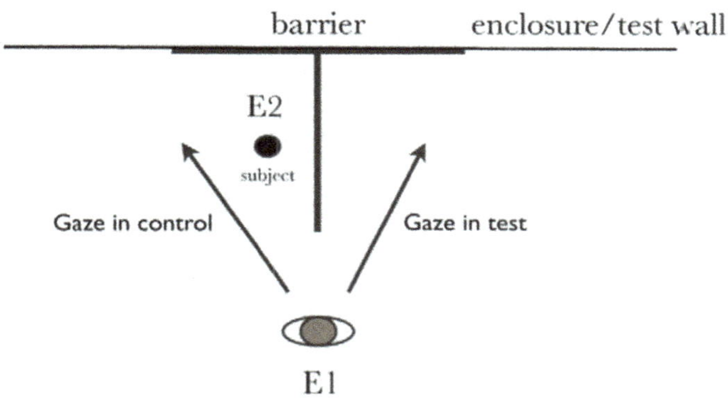

Fig. 7.2 Set-up of an experiment investigating the ability to follow gaze around a barrier (Range and Virányi 2011) (https://doi.org/10.1371/journal.pone.0016888; CC BY 4.0; http://creativecommons.org/licenses/by/4.0/)

position in order to see the object of its interest. This repositioning requires them to assess the difference in visual perception between the cue-giver and themselves. This may either be achieved by mentally representing the looker's visual perspective (Povinelli and Eddy 1996) or by learning how visual barriers impair perceptions (Tomasello et al. 1999). It has been hypothesized that this ability to track another's gaze around a barrier might be less widespread in the animal kingdom than following the gaze into distant space and limited to species with high levels of cooperative and competitive interactions (Schloegl et al. 2007). In accordance with this idea, only few species have been shown to master this task including great apes (Bräuer et al. 2005), some monkey species (Amici et al. 2009), two corvid species (ravens (Bugnyar et al. 2004), rooks (Schloegl et al. 2008), and pet dogs (Met et al. 2014). Two studies reported negative results for geometrical gaze following in the Northern bald ibis, (Loretto et al. 2009), and the common marmoset, (Burkart and Heschl 2007).

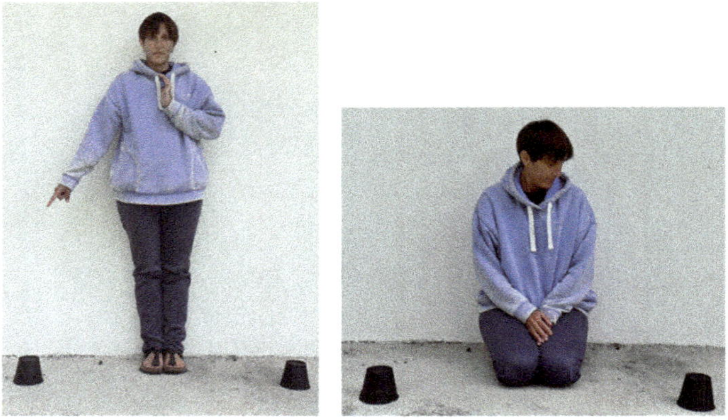

Fig. 7.3 Gazing (**a**) or pointing (**b**) in object-choice tasks (pictures taken by Friederike Range)

Finally, tasks which allow researchers to investigate if subjects *follow others' gaze to locate hidden food* in object-choice tasks (Fig. 7.3) seem to be quite difficult for animals to solve. Interestingly, in contrast to the mostly negative results of dogs solving the gaze following into distant space paradigm, there is a widespread belief that dogs readily follow human gaze in object-choice tasks (Hare and Tomasello 2005b; Kaminski and Nitzschner 2013).

However, most studies claiming that dogs can follow the gaze cues of the human experimenter to the baited container had exposed subjects first to an intensive training with easier social cues such as pointing, bowing, and touching (Agnetta et al. 2000; Miklósi et al. 1998; Soproni et al. 2001). When tested without prior training, dogs failed to follow the cue to the correct container and instead either chose randomly or even avoided the container the experimenter looked at (Duranton et al. 2017; Kaminski et al. 2012; Oliva et al. 2015; Soproni et al. 2001). To date, two studies have found that dogs spontaneously followed human gaze cues in an object-choice task, but in these studies the gaze cue was given repeatedly, and the experimenter attracted the subject's attention by calling its name (Duranton et al. 2017; Téglás et al. 2012). In both studies, dogs followed the gaze of the experimenter looking themselves at the correct container. However, the choice behaviour was recorded only in the more recent study, and here, interestingly, the animals chose at random if the gazing was accompanied by communicative cues (calling, repeated looking), but below chance (i.e. preferentially choosing the object that was not gazed at) when presented without communicative cues (Duranton et al. 2017).

Together these results suggest that dogs do not spontaneously comprehend human gaze as a cooperative communicative signal intended to share information, but rather they may see it as a cue that the experimenter is interested in the food contained in the object. The dogs' avoidance of the object (and choice of the empty alternative) might be explained as them trying to avoid a potential conflict with the human. The communicative cues—calling the dogs' name in a friendly tone—may change the dogs' perception of the gaze somewhat, by emphasizing the positive, encouraging nature of the interaction, and hence giving the dogs 'permission' to take the food. However, since the dogs are at chance level even with communicative cues, this communication does not seem to be very effective.

7.1.2 Following Pointing

If the human experimenter does not just look at the baited container, but points towards it with their outstretched arm and finger, the situation is very different. In fact, dogs actually perform rather well in this case, outperforming most other species that have been tested (Kaminski and Nitzschner 2013; Krause et al. 2018). However, while human infants understand such pointing gestures as a means to share information (e.g. declarative pointing) (Tomasello 2010; Tomasello et al. 2005), several more recent studies suggest that dogs understand it rather as following the human directives/commands (e.g. imperative pointing) (see Kaminski and Nitzschner 2013 for a review). However, the latter interpretation would be in line with the results found in regard to gazing in the two-choice task.

A number of researchers have hypothesized that dogs' willingness to engage in eye contact with humans and their abilities to follow human communicative cues may be a result of the domestication process (Hare et al. 2002; Hare and Tomasello 2005a, b; Miklósi et al. 2003). Accordingly, many studies have compared wolves' and dogs' performance in communicative tasks at different ages, with humans as the demonstrators.

7.1.3 Do Dogs and Wolves Differ in Regard to Understanding Communicative Cues?

7.1.3.1 Establishing Eye Contact

In order to communicate with each other, getting into eye contact is usually the first step. Therefore, most studies using communicative gestures first try to establish eye contact with the subject. Learning to establish eye contact with a human is rather easy for dogs, and it has been argued that the willingness of dogs to look at the human face differentiates dogs from wolves, allowing for complex dog-human communication to be established (Miklósi et al. 2003; Chap. 11).

Gácsi et al. (2005) reinforced three dog and four wolf puppies for establishing eye contact with a familiar experimenter at the ages of 5 and 9 weeks. A plate of food was placed on a table beside the experimenter out of the subjects' reach. The animal could move freely around, and whenever it made eye contact with the experimenter, she enforced the correct response with a clicker and immediately threw a piece of food to the animal. At 5 weeks of age, most of the wolves—but none of the dogs—used the test to take a nap so that data could not be analysed. But at the age of 9 weeks, while wolves and dogs did not differ in the first minute in terms of the number of times they established eye contact with the experimenter, dogs looked more often into the experimenter's eyes during the fourth minute. Similarly, in a study investigating pointing in a two-choice task, wolves were more reluctant to establish eye contact with the experimenter at 8 weeks, 4 months, and as adults compared to dogs (Gácsi et al. 2009; see also Virányi et al. 2008). Similar results also emerged in the training context with adult animals tested at the Wolf Science Center, with wolves spending less time looking at their handler than dogs (Vasconcellos et al. 2016—see Chap. 6).

7.1.3.2 Gaze Following into Distant Space

At the WSC we tested wolves ($N = 9$) and dogs ($N = 10$) in both the 'gazing into distance space' and the 'gazing around a barrier' tasks at different ages to look at potential differences between species in the developmental trajectory of these abilities. Moreover, in relation to the first paradigm, we also carried out extensive observations among conspecifics in their enclosures to investigate if the experimental results mirror the gaze following responses between conspecifics in their day-to-day life, i.e. if they follow the gaze of their pack members or not.

Before starting the experiments, we trained all our animals to establish eye contact with an experimenter using clicker training. In other words, we asked the puppies to sit in front of us, and if they looked up into our face, we clicked and then rewarded them. After this initial step, we started to reward them only if they actually got in eye contact with us and not just for looking up. And finally, we added a command 'look' so that we could ask for the behaviour. After this initial training, we started the experiment.

The looking into distant space paradigm consisted of one look-aside trial and one control trial. At the start of each trial (test and control), the human experimenter (E) established eye contact with the dogs and wolves by giving the trained command 'look', rewarding the animal when it established eye contact. After 2–4 such warm-up trials, the experimenter asked once again for establishing eye contact. As soon as the subject looked into the eyes of the E, E gave a gaze cue. In the look-aside trials, E looked to a point at a 90 ° angle (Fig. 7.1). In control trials, E looked to a point at the ground just next to the animal instead of looking to the side. Importantly, in the prior looking training, the animals were not rewarded for looking away. We found that already at the age of 14 weeks, wolves followed the gaze of the human into distant space (Range and Virányi 2011), while the dogs did not reliably follow the human gaze direction even at 23 weeks of age (Werhahn et al. 2016).

At the end of the study, we also did a habituation experiment, where we repeated the gaze cue ten times in a row. Interestingly, the wolves did not habituate to the gaze cue, e.g. there was no difference in their latency to follow the human gaze in the first and the last five trials. This latter test was not conducted with the dogs.

In an observational study, we videorecorded our animals (11 wolves, 13 dogs) in their living enclosures. We recorded a *gaze cue* if an animal (= the demonstrator) turned its head with a continuous movement to gaze into a new direction that neither the model nor the potential observer had looked into within the previous 10s. Then, we recorded whether another animal, which could see this gaze cue, turned its head in the same direction as the demonstrator within 2 s and did not look in another direction first. As control conditions we used similar situations with the difference that the potential observer actually could not see the demonstration. So any gaze behaviour by this 'observer' would be independent of the demonstrator's gaze cue. Finally, we also recorded synchronous gazing reactions, when multiple animals *simultaneously gazed* into the same direction, independently reacting to the same external stimulus, providing a measure of the general vigilance towards external stimuli in the two species. Geraldine, Gabriela, and Andrea observed the various packs for over 29 h and observed 25 *simultaneous gazing* events and 35 *gaze cues* in the dogs and 19 *simultaneous gazing* events and 18 *gaze cues* in the wolves. In the dogs, the 35 gaze cues resulted in 11 *gaze followings*; in the wolves, the 18 cues resulted in 10 *gaze followings*. These results showed that the dogs and wolves did not significantly differ in their general responsiveness towards external stimuli and there was also no difference in how often they followed their conspecifics' gaze into distant space (Werhahn et al. 2016).

7.1.3.3 Following Gaze Around a Barrier

In the barrier task, we used a wooden board (90° angle to the stone wall of the testing room or puppy enclosure) or a door as a visual barrier for the test subjects. The experiment started when the animals were 16 weeks old. They were tested three times in each condition over the first 6 months of their lives. Each animal was tested with a human and a trained pet dog as demonstrators in a test and a control trial. The pet dogs used as demonstrators had been present during the hand-raising, and therefore had a close social relationship with the wolves and dogs. In each trial, the demonstrator (human or dog) was positioned in line with the barrier, while an animal trainer held the subject by the collar and stood on one side of the barrier so that the subject's view to the other side of the barrier was blocked (Fig. 7.2). In the test trials, the demonstrator then gave a gaze cue to the other side of the barrier, while in the control trials, they gazed at the subject's side of the barrier, without looking directly at the animal. To 'encourage' the dog demonstrators to actually look to the other side of the barrier, unobserved by the subject, we put some food there in both test and control trials. Before the test trials, we showed the food explicitly to the pet dog demonstrator. In the control trials, we also hid food on the other side of the barrier to control for potential odour cues, but now the food was hidden in the absence of the demonstrator so that they would not gaze to the hidden food. No food was present in the human trials. As soon as the cue was given, the animal trainer let go of the collar. Gaze following was coded if the subject moved around the barrier to look to the other side. The wolves started to walk around the barrier and look on the other side more often in the test trials than in the control trials only at the age of 6 months during the dog demonstration and tended to do the same in the human demonstration (Range and Virányi 2011). In contrast to the study conducted with adult pet dogs (Met et al. 2014), we found no indication that the dogs followed the gaze of the dog or human demonstrator around the barrier at any age (unpublished data). The different results between the two dog studies may be due to age differences in the tested subjects (puppies versus adult animals) or procedural differences. In the study with the adult pet dogs, experimenters carried out a warm-up phase, in which dogs were either given the experience that food could be found in the experimental room or experienced social contact with the experimenter. Dogs did better in the food than the social context, suggesting that raising the animal's expectation that there was something of interest to them affected their gaze following tendencies. This was not done in the Wolf Science Center study. Finally, experience might matter, as has been mentioned repeatedly in this book, i.e. the pet dogs' everyday experience of similar contexts with their human companions may strongly affect such behaviours in the experimental context.

Considering the results from the two gaze following paradigms (into distant space and around a barrier), we were somewhat surprised that the wolves performed so much better than the dogs. There could be a number of reasons for this. Potentially the wolves' developmental trajectory for this kind of tasks is faster, whereas dogs only show such abilities when they are a bit more mature (tests ended when animals were only about 6 months of age). Another aspect to consider is wolves' and dogs' ability to pay attention to details. In social learning tasks, for example (see Chap. 8) we have found that wolves at 6 months appear to be much better at focusing on the relevant parts of a demonstration, which may also be relevant in gaze following tasks. A final possibility is that the 'look' command we gave the animals in the gaze following task into distant space overrode all else in the dogs. In fact, when given the command, both wolves and dogs responded promptly, but then the dogs just continued looking at the person, seemingly unaware (or choosing to ignore) that the human had turned their head in another direction. Potentially, dogs were simply continuing to do what they were asked to do.... 'Look'! This is in line with the results found by Lisa Wallis and colleagues (Wallis et al. 2015), showing that training the command to establish eye contact actually reduces gaze following in pet dogs.

7.1.3.4 Gazing and Pointing in the Two-Choice Task

Due to human infants doing so well with pointing gestures and the idea that dogs follow pointing because they perceive it as information sharing and cooperative, the first studies comparing wolves and dogs in their communicative abilities with humans focused on the two-object-choice task using the pointing gesture (Fig. 7.3). In this paradigm two opaque containers, one of them containing food, are placed on the ground at a distance of ca. 1.5 metres, and an experimenter is standing slightly behind them, at an equal distance to both. The experimenter, after calling the dog's attention, points at the baited container to indicate where the food is hidden. Then, the subject is released to make a choice and only receives the reward if it follows the cue. Researchers have used different pointing gestures, increasing in difficulty depending on the distance to the containers and where the arms are positioned when the animal is released. During proximal pointing, the finger of the experimenter is rather close (5–10 cm), whereas in distal pointing the distance is at least 50 cm. In the case of stationary pointing, the arm is still pointing at the container when the animal is released, while in the momentary pointing, the hands are placed back to the starting position (chest or side of the body) *before* the subject is released and allowed to make a choice. Hence, the animal cannot use the outstretched arm as a local cue to find the food. This most difficult version (momentary distal pointing) is thought to inform us whether the subjects can comprehend referential communication (Anderson et al. 1996; Povinelli et al. 1997).

We will now summarize first the various pointing studies, since these were the first ones conducted, and then look at the wolves' and dogs' performance in the gaze following two-object-choice tasks.

Proximal (Momentary) Pointing

During proximal pointing, the index finger of the kneeling human is approximately 5–10 cm away from the container. This task has the advantage that it can easily be tested with puppies, since they have difficulty establishing eye contact with a standing human experimenter.

Brian Hare et al. (2002) compared the success of seven pet dogs and seven non-socialized sanctuary-kept wolves using different cues including proximal pointing (unclear whether it was momentary or stationary) both with and without a head turn (gaze cue) performed simultaneously. At the group level, wolves were only above chance in the proximal pointing cue when the experimenter also looked at the pointed container, while the dogs were significantly above chance in all conditions. Also, at the individual level, while none of the wolves were above chance in the 'gaze + pointing' cue, five dogs used that cue successfully above chance.

Moreover, two 1-year old, non-socialized arctic wolves (one male, one female) were tested in the Leipzig Zoo using a pointing cue (unclear if proximal/distal and stationary/momentary) accompanied by looking at the baited container, just a gaze cue involving a clear and distinct head movement and a control with no cue. Each animal received 50 trials in each condition. The animals had to position themselves in the right location behind the fence to receive a reward for a correct choice. Neither wolf used either of the two communicative cues, but the female did learn to pay attention to the look + pointing cue as she was successful on seven of the last ten trials (Agnetta et al. 2000).

However, as explained in the introduction, the raising history of the wolves and dogs in these studies was quite different, disadvantaging the wolves. To remedy this shortcoming, the Budapest group tested their similarly raised wolves and dogs using this paradigm. Two of 4 wolves hand-raised in 2001, which were tested in a total of 40 trials at the age of 4 months (2 $\times$10 trials, 2 weeks apart), performed above chance in this task, but no comparison was made with similarly raised dogs (Miklósi et al. 2003). In a study comparing 8-week-old hand-raised dog ($N = 8$) and wolf ($N = 13$) puppies, the wolf pups had a higher latency for eye contact with the pointing experimenter, struggling more often and even biting the handler who was holding them in the starting position (Gácsi et al. 2009). As a result of this struggling, seven wolf pups had to be excluded from this study. The other six, however, performed at the same level as the dogs. Moreover, 2 of 7 wolves hand-raised in 2002 that were tested repeatedly with distal pointing cues at 4 months of age and then with different pointing cues again between 7 and 11 months of age performed above chance in the 20 momentary proximal pointing cues (Virányi et al. 2008). So overall, it seems that if wolves are hand-raised and are willing to cooperate with the handler on this task, they can solve it at a similar level as dogs. But as explained above, this task is relatively easy since the distance between the finger and the bowl is relatively close and so the animals can use the pointing finger as a location cue.

Distal Pointing

During the distal pointing task, the experimenter is standing (rather than kneeling) between the containers, and the distance between the tip of the pointing finger and the bowl in this version is thus normally more than 50 cm. Several studies have been carried out by the Budapest group, comparing their hand-raised wolves and different populations of dogs at different ages. Here we will summarize the results according to the age of the animals that were tested.

Nine 4-month-old wolf pups hand-raised in 2001 (four animals) and 2002 (five animals) were compared to eight hand-raised dogs and nine normally socialized pet dogs in Budapest. All animals were tested using the momentary distal pointing gesture at the same age in the same unfamiliar room. While the pet dogs and hand-raised dogs performed above chance and at similar levels, the wolf pups were at chance and took significantly longer to establish eye contact with the experimenter (Virányi et al. 2008).

Four naïve wolf pups hand-raised in 2002 were tested from 5 to 7 months of age using the momentary distal pointing (total number of trials: 220) partly intermixed with proximal pointing and touching. While all wolves were at chance level in the first session, one wolf reached 80% correct performance after the first 100 trials. No information is provided for the other three wolves in the paper. However, based on the graph in the paper, they all did improve over sessions and did better in the last 20 trials than in the first 20 trials (Miklósi et al. 2003).

Furthermore, Viranyi and colleagues tested seven hand-raised wolves from 7 to 11 months in a total of five different conditions (see also proximal pointing) (Virányi et al. 2008). Five of these wolves had already participated in the distal pointing study at 4 months of age. The other two had participated in the distal pointing study by Miklósi et al. (2003). Each pointing cue was tested 20 times in total, and at the group level, the animals were above chance level with each cue including momentary distal pointing and static distal pointing. A single wolf performed above chance level when presented with all three cues. Moreover, 4 wolves received intense exposure to the various pointing cues (momentary distal pointing, 220 trails; stationary distal pointing, 60 trials). All wolves performed above chance in the momentary distal pointing, and two did so in the stationary distal pointing task at the individual level, showing that they learned to use the cues over the sessions. The extensively trained wolves performed similarly well in their last session as naïve pet dogs tested with the same cues. However, both wolves and dogs showed a high individual variability in their ability to use the provided cues (Virányi et al. 2008).

Finally, testing eight *adult* pet dogs and eight hand-raised, pointing naïve, wolves of 3–5 years old, Gacsi and colleagues (Gácsi et al. 2009) could show that the extensive training was not necessary for the wolves to follow the distal pointing cues, since wolves and dogs did not differ in their performance, suggesting that the ability to follow the human cueing is rather a developmental effect. This latter evidence gained support from a study with eight adult wolves socialized and tested outdoors at Wolf Park in the USA (Udell et al. 2008). These wolves were as successful as eight pet dogs tested indoors with a familiar person (and even more so at the individual level), and they were more successful than pet dogs tested outside and shelter dogs. However, the study has been criticized since, in contrast to the definition for momentary pointing, the pointing gesture was still visible when the subjects were released by the handler and thus could use the location of the hand/finger to make their choice. Moreover, they applied a special training technique involving a secondary reinforcer (clicker), which likely had an effect on the performance (Pongrácz et al. 2013). Overall, the better performance of adult wolves in the pointing task compared to younger animals has been thought to be due to a higher willingness of the older animals to cooperate with humans, reducing their struggling with the hand-raiser (Gácsi et al. 2009) and likely also due to some learning.

Based on the failure of young wolves in the momentary distal pointing task, researchers concluded that dogs indeed are special in comprehending this referential communication, likely as a result of the process of domestication which may have enhanced (1) their humanlike socio-cognitive skills (Hare et al. 2002), (2) their inhibition (Gácsi et al. 2009), (3) their cooperativeness due to a selection against fear and aggression (Hare and Tomasello 2005a), or (4) their attentiveness towards the face (Miklósi et al. 2003). However, several alternative mechanisms have been proposed since then, suggesting that following the pointing gesture might rely on (1) a local enhancement effect (the pointing hand is closer to the baited than to the non-baited container), (2) an association of the hand with food, or (3) perceiving the pointing gesture as a command to go to the indicated direction (Kaminski 2008). Given these possible alternative mechanisms, it is unlikely that the pointing studies alone can inform us about comprehension of referential communication.

In contrast to pointing, following others' gaze to identify a third object in the environment is not subject to these alternative explanations, and is used, at least in humans, as a referential communicative gesture. Moreover, while pointing with the hand is an arbitrary, human-specific gesture that animals never use with their conspecifics, gazing might actually be the more relevant cue for the animals to use to understand their partners' behaviours. In line with this, we tested wolf and dog puppies at the WSC with both a human and a conspecific using their gaze to indicate the food location in the two-choice task (Fig. 7.4).

Gazing

We tested 11 3- to 4-month-old wolves and 10 dogs at the WSC in 3 different conditions.

1) Conspecific gaze: Again, we used the pet dogs that had been present during the hand-raising as demonstrators. During the test, the pet dog was sitting in the 'experimenter's' position between the two containers and indicated the baited one by looking at it. The pet dogs had been trained to look at the container indicated by the owner with a hand signal, but not approach it. The demonstrator's owner was standing behind the handler who was holding the subject, so that the subject

Fig. 7.4 Additional condition used at the Wolf Science Center: a dog provides the gaze cue (picture taken by Friederike Range)

did not see the signal. After the pet dog had looked at the container, it remained seated while the subject approached one of the two containers. If the subject chose the baited one, it was allowed to eat the food; if not, the handler moved forward and confiscated the baited one.

2) Human gaze: A familiar human experimenter knelt between the two containers and, once eye contact was established, looked at the baited container with a head turn. The animal's name was used to call their attention in case no eye contact was forthcoming.
3) Momentary distal pointing: A familiar human experimenter stood between the containers and, once eye contact was established, pointed to the baited container in the manner described above.

Interestingly, at the age of 3–4 months, the wolves were indeed able to use the dog's gaze to find the food above chance level but neither of the other cues. In contrast, the dogs followed the momentary distal pointing of the human experimenter above chance level but none of the other cues. At the individual level, two wolves and eight dogs followed the human pointing cue. None of the animals, not even a dog, followed the human gaze cue at the individual level. Interestingly, one of our male wolves, Nanuk, significantly avoided the container the human experimenter gazed at, demonstrating that he took the cue into account but not as an invitation to go and get the food, but rather has an indication that the food should be avoided.

Thus, our results are in line with other studies showing that dogs do not reliably use the human or conspecific gaze cue to choose a baited container (human demonstrator: Kaminski et al. 2012) or dog demonstrator (video projection; Bálint et al. 2015), but may even avoid it if tested without communicative cues (Duranton et al. 2017) or, in case of conspecifics, if the demonstrator gazed for a long time at the baited container (Bálint et al. 2015). One possible explanation that might account for these results is that the dogs try to avoid conflicts with the demonstrator—conspecific or human—who is indicating their interest in the container by looking at it. This interpretation is in line with our studies on conflict management and competition in the dogs, where the dogs often seem to adopt an avoidance strategy (Chap. 5). It also would explain the very submissive behaviour that we observed in our dogs at the Wolf Science Center when they were tested with a conspecific gaze cue: they often seemed to want to go to the baited container, but then turned around and went to the empty one instead and displayed submissive behaviours towards the demonstrator at the same time.

Wolves, on the other hand, do not seem to have this problem and take what they want either because they are more confident or because they are not afraid to stand their ground when approaching a resource that is not necessarily theirs, as we also see in our conspecific food sharing experiments (see Chap. 5).

In contrast to gazing, pointing might be perceived as a command, which the dogs easily comply with whereas the wolves think twice about it. This latter idea is again in line with our studies showing that dogs are more likely than wolves to wait and follow the human lead in a cooperative task (see Chap. 8) and hence may have been selected for more deferential and compliant behaviours (Range et al. 2019; Chap. 11). The fact that wolves get better with age at following the pointing gesture could be partly due to their general training experience with humans. In fact, although no specific training on pointing is imparted, all our animals, especially in the first years, undergo positive reinforcement training in various situations in which they learn that the human is a cooperative partner and usually provides some incentives (aka sausages!) if one complies with their commands. This type of training regime occurs at the Wolf Science Center, but is also a component of the wolf routine in other facilities where adult wolves have been tested. This then would explain why wolves at the WSC, but also in other facilities, are more willing to follow the pointing once they are a bit more mature.

We have started to run additional studies to test these ideas with pet dogs. For example, we tested whether it helps the dogs if the gazing is accompanied with clear ostensive communication by the human experimenter, e.g. making eye contact with the dog, calling its name, and so forth. The idea is that this communication is perceived by the dogs as a clear sign that the human actually allows the dog to take the food since she actively draws the dogs' attention to the container (Duranton et al. 2017). While the ostensive communication did increase the likelihood that the dogs looked at the indicated container, dogs still did not choose the correct container above chance level. However, at least they did not actively avoid the container, as they did if no ostensive cues were given (Fig. 7.5).

While of course further studies are needed to see if the interpretations of the wolves' and dogs' behaviours in the pointing and gazing tasks are correct, our results

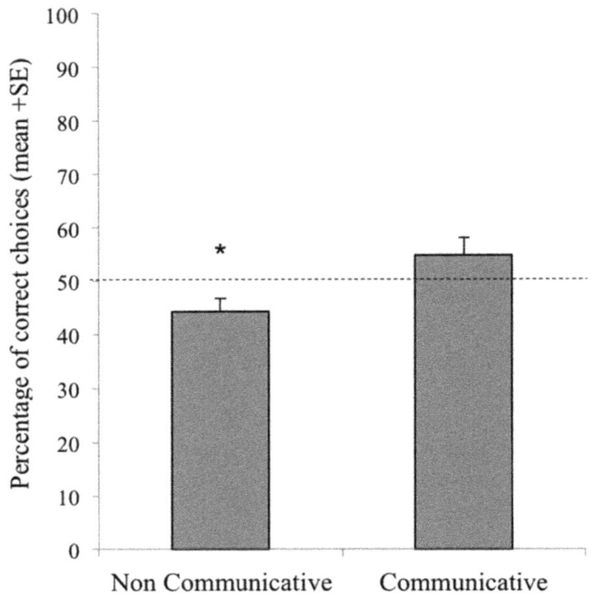

Fig. 7.5 Pet dog's choice in a pointing task with communicative and without communicative cues. Figure adapted from Duranton et al. 2017. https://royalsocietypublishing.org/cms/asset/b90bb768-a9d5-4dd4-a33e-2ccbc9de02d3/rsos170349f04.jpg; CC BY 4.0; http://creativecommons.org/licenses/by/4.0/

challenge the idea that dogs are better at understanding referential communication than wolves.

Finally, in a task based on a study by Bräuer and colleagues (Bräuer et al. 2006), the wolves and dogs at the Wolf Science Center, as well as pet dogs, were tested on gazing, pointing, and behavioural cues at an adult age again using a two-choice task (Lampe et al. 2017). However, the set-up differed a bit from previous studies in that the experimenter was sitting behind a table that was positioned behind the fence and the animals were allowed to move around freely. In the *gaze* condition, the experimenter established eye contact with the animal, nodding once and then looking at the baited container for 2 seconds. In the *pointing* condition, the experimenter was seated behind the empty container, again establishing eye contact and nodding before pointing four times (each time for 3 seconds) at the baited container with her index finger on her ipsilateral hand. The experimenter kept looking at the animal while conducting the pointing gestures. In the *reach* condition, the experimenter briefly looked at the empty container and then desperately stretched her arms trying to grab the baited container that was out of reach and then fell off the chair in the direction of the empty cup. In the *open* condition, the experimenter first picked up the empty container and sniffed it for 10 seconds, putting it back with a disappointed face, and then took the baited container, again sniffing it and then trying to open it for 10 seconds. In all conditions the experimenter finished the demonstration and looked down, not to give an involuntary cue, before the animal was allowed to make a choice by touching a target fixed on each side of the table or by touching the end of the table directly.

While none of the wolves, pack, or pet dogs chose the baited container significantly above chance in the behaviour cues conditions (reach and open), all performed above chance in the communicative cues (gaze and pointing). No difference emerged between the different groups indicating that at an adult age, given the intensive socialization at the Wolf Science Center and previous experience with pointing and gazing studies, the wolves were as proficient at using the communicative cues as dogs.

7.1.4 Sensitivity to Ostensive Cues

As mentioned above, in a number of studies, dogs were shown to follow the human gaze more easily if this was preceded by getting the dog's attention by displaying so-called ostensive cues. The term 'ostensive cues' derives from developmental psychology and more specifically from the theory of 'natural pedagogy' proposed by Gergely and Csibra (Csibra and Gergely 2009). Its central tenant is that there are specific cues, namely, direct gaze accompanied by infant-directed speech, that, when used, alert young infants to the fact that something is about to happen which is intentionally relevant to them (a 'teaching context'), which is likely to involve some form of referential communication (Csibra 2010). A prominent example of this putative process is that gaze following is demonstrated by 6-month-old infants only if gaze is preceded by ostensive cues such as direct gaze and infant-directed greeting, and not when these cues are replaced with a non-ostensive and non-social animation (Senju and Csibra 2008). Further support also comes from a classic paradigm in the developmental psychology literature, first designed by Piaget, the A-not-B task. In this task an infant (typically of around 9 months) has two screens or boxes in front of him/her. The experimenter repeatedly hides a toy under box 'A' within the babies' reach. Infants at this age have no problem reaching under the box to retrieve the toy. After a few repetitions, in which the infant is successful, the experimenter hides the toy under box 'B', just next to it. Until the age of 12 months, infants will make 'perseverative search' errors, i.e. they will continue to search under box 'A' despite box 'B' being as easily reachable. It is thought that until this age, the babies' object permanence abilities are not fully developed. This test is one of the most reliable and replicated phenomenon in infant psychology. However, Gergley and Csibra noted that this test is carried out inevitably in an 'ostensive' environment, in that the researcher hiding the toy interacts with the babies by looking at them and using baby talk. They reasoned that, possibly, infants were committing the error because they were drawing some 'rules' from the experimenter's ostensive behaviours and the repetition of the same task, for example, that 'the toy lives under A' (and it is just an accident that it then ends up in B).

To test this, the researchers carried out the same experiment, but with three different conditions. In the ostensive condition, the experimenter, as usual, used gazing and infant speech when interacting with the babies. In the non-communicative condition, the researcher performed the experiment exactly as usual, but without any of the ostensive-referential cues. While all their actions were completely visible, the experimenters took care never to make eye contact and speak to the infant during the test. In the social condition, the experimenter was removed from the scene, in that the movement and hiding of the object were the same, but the experimenter conducted the entire experiment from behind a curtain. As usual, in each condition, the toy was first hidden four times under A, and then three times under B. Results strongly showed that once the communicative cues were removed, infants were much more likely to successfully reach for the toy under the 'B' box (Topál et al. 2008). Humans are not the only species, who commit the A-not-B error. Apes (Mathieu and Bergeron 1981; Poti 1989), monkeys (de Blois et al. 1998; Kis et al. 2012; Neiworth et al. 2003), birds (Pepperberg et al. 1997; Pollok et al. 2000; Zucca et al. 2007), and dogs (Watson et al. 2001; but see Gagnon and Doré 1992, 1994) also show evidence of similar errors in this task.

Intriguingly, Téglás et al. (2012), using the experimental paradigm developed for human infants (Senju and Csibra 2008), found that dogs followed an experimenter's gaze to an object more often when preceded by ostensive eye contact and directed speech, and in a canine version of the A-not-B task, dogs would commit far fewer perseverative errors if no communicative cues (name-calling and direct eye gaze) were directed at the dog prior to the object hiding episodes.

This has led some researchers to hypothesize that this sensitivity to ostensive communication may have been selected for in dogs during domestication (Topál et al. 2009a, b). To test this hypothesis, the Budapest group carried out the A-not-B task with pet dogs and socialized wolves. Two conditions were presented to the same animals on different days: a social-communicative condition in which the experimenter called the animal's name and repeated the word 'watch' both when starting the demonstration and when hiding the food behind the 'A' barrier (but not the 'B' barrier) and a non-social condition in which the food was hidden repeatedly behind the A and then finally the B barrier, through a set of invisible strings, so that no human was actually present in the test scenario.

Differently from wolves that largely did not commit the perseverative search error in either condition, dogs showed significantly more errors in the social-communicative than the non-social condition, leading authors to conclude that the selective processes in the course of domestication of dogs have led to their sensitivity to human ostensive and referential signals, 'an evolutionary novel skill in the Canis genus providing a typical case for convergent social evolution (as the consequence of domestication) between man and dog' (pag. 1271, Topál et al. 2009a, b).

The extent to which ostensive cues can really be considered anything more than a powerful attention-getting signal has been questioned extensively both in studies relating to infants (Szufnarowska et al. 2014) and dogs (Fiset 2010), as have been the details of the procedures in these tasks (Marshall-Pescini et al. 2010). However, in the context of the wolf-dog comparison, three aspects are important to note. First, whereas pet dogs may have had a long history of associating gazing and their name with specific positive outcomes, this is not the case for wolves that were socialized as puppies, but then went on to live in enclosures with a much more limited amount of experience of human communication. Hence, it is not so unexpected that dogs may bias their searching in a place that has been extensively 'enhanced' by the person repeatedly calling attention to it with cues that have been repeatedly rewarded during the course of their lives. Wolves' sensitivity to such cues may be considerably less, simply due to a much more limited history of reinforcement. The second aspect worth considering is motivation. Considering wolves' greater persistence in tasks involving food (see both below and Chap. 9) as compared to dogs, one possibility is that wolves in general have a higher food motivation than dogs, and this may indeed be affecting their searching pattern in this task. Indeed, in a later study just with pet dogs, dogs with a very high motivation for toys (used in the test) were far less likely to commit errors, providing a direct link between motivation and error patterns (Sümegi et al. 2014).

The third aspect is that calling the name of a dog can likely be considered a 'command' for the animal to come over to the indicated place. Indeed, in later studies, dogs' higher error rate in communicative contexts in this task was suggested to be a good example of their 'readiness-to-please' (Sümegi et al. 2014). As suggested also in other context, what emerges is that wolves are somewhat less likely to comply with, or accommodate, a person's wishes (see Chap. 11). So instead of evolving a greater sensitivity to human communicative skills during domestication, dogs may differ fundamentally from wolves in their willingness to give up their own ideas and follow someone else's (see Chap. 8 on cooperation with humans, for a similar pattern, and Chap. 6 on the leash walking with humans).

7.1.5 Summary

Taken together results from studies on gaze following of conspecifics and humans suggest that wolves appear to be able to follow both humans' and conspecifics' gaze in various situation, but dogs fail to do so at least in some situations (see Table 7.1*). At present it remains unclear why the dogs perform so badly in some of these tasks. One possibility is that the initial 'attention calling' by the human overrides the later cues, which would explain why they follow the gaze of conspecifics into distant space when in their packs, but not when in an experimental condition. Another possibility is that the developmental trajectory is different for wolves and dogs. Additional studies, especially with adult animals, are needed, given that adult pet dogs do follow the gaze of humans into distance space if ostensive cues are given or certain facial expressions are exhibited before the demonstration (*Duranton et al. 2017*;* Wallis et al. 2015*) and that they can also follow the gaze geometrically around barriers in certain contexts (*Catala et al. 2017*;* Met et al. 2014*).*

Dogs outperform wolves at an early age with pointing cues (see Table 7.1*), suggesting that they are better at using human communicative cues. Wolves are sometimes able to use the information conveyed by a human gazing, albeit not necessarily as a cooperative cue, and furthermore can use gazing cues of conspecific partners. Dogs in contrast have difficulties using gaze cues both of conspecifics and human partners. These results suggest that the pointing cues might be perceived as imperative and dogs may just be better in following the suggestions made by humans from an early age on, while wolves can learn to follow them with age probably based*

Table 7.1 Summary of the two-choice pointing and gazing studies

Behaviour	Task	Age	Species: success	References
Gaze following into distant space	Humans	14 w	W: Yes D: No	(Range and Virányi 2011; Werhahn et al. 2016)
	Conspecifics	Adults	W = D	(Werhahn et al. 2016)
Gaze following around a barrier	Humans and conspecifics	16 w	W: Yes D: No	(Range and Virányi 2011; Werhahn et al. 2016)
Establishing eye contact	Training for eye contact	9 w	W < D	(Gácsi et al. 2005)
	Pointing study	8 w	W < D	(Gácsi et al. 2009)
	Training context	Adults	W < D	(Vasconcellos et al. 2016)
Pointing	Proximal pointing (+ look)	Adults (pet dogs, non-socialized wolves)	W = D	(Hare et al. 2002)
	Proximal pointing	2 m + 4 m + 11 m (lots of experience)	W = D	(Miklósi et al. 2003) (Gácsi et al. 2009) (Virányi et al. 2008)
	Momentary distal pointing	4 m	W < D	(Virányi et al. 2008) (Gácsi et al. 2009) WSC data (unpublished)
		5–7 m	W = 50%	(Miklosi et al. 2003)
		7–11 m (lots of experience)	W > 50% also at individual level	(Virányi et al. 2008)
		Adults	W = D	(Gácsi et al. 2009) (Udell et al. 2008)
Gazing	Conspecific	4 m	W > D	Not published
	Human	4 m	W = D (50%)	Not published

w = weeks, m = months, y = years

on positive experience (learning). In contrast, the gazing cues might be perceived more as indicating 'this is mine'. Due to wolves being more likely to question the authority of others, they might actually follow the cues and see if they cannot get access to the food, while the dogs rather try to avoid conflicts and thus refrain from approaching the indicated bowl. Accordingly, both species do not seem to see the pointing or gazing gesture as a cooperative, informative interaction as has been suggested previously, but their different responses might hinge on their social conflict strategies and their temperament. However, also in this case, further studies teasing apart the potential factors which may facilitate following the gazing gesture to a baited container, or even inhibit the choice, would help our understanding of how wolves and dogs perceive such tasks.

7.2 Communicating with Humans

While humans use the information provided by others through gazing or pointing, we also actively point in many social situations to share information or our interest with others. Not surprisingly, gazing and pointing are among the first communicative gestures that human infants use to share needs, interests, and intentions (Butterworth and Simion 2013). Also, non-human animals often need to direct a conspecific's attention to certain aspects of the social and physical environment. While non-human primates have fingers and actually use the pointing gesture to communicate (for a review see Krause et al. 2018), this is not possible for many other species due to their morphology. However, all species may use gaze cues to indicate specific aspects of their environment. For example, if a monkey wants to recruit support from a higher-ranking animal against a potential opponent, they rapidly alternate the gaze between the two while threatening the opponent. Similarly, dogs often use gazing to communicate with us. Indeed, most of us realize when they want to go out, want something to eat, or want a cuddle—dogs can be very specific about these needs and also very, very persistent if we ignore them. This cute way to communicate what they want or do not like has led people to regularly over-interpret their behaviours, giving them often deeper meaning than is warranted (e.g. see the study on the 'guilty gaze' as a good example of this; Horowitz 2009b). Interestingly, domestication has selected for facial anatomy with a muscle responsible for intensely raising the inner eyebrow being present in dogs but absent in wolves (Kaminski et al. 2019). This movement of the eyebrow looks similar to the human expression when sad.

7.2.1 Sensitivity to the Attentional States of Others

In order to effectively communicate visually, one first needs to be sensitive to what others are attending to (see Woodruff and Premack 1979), since if, for example, the owner is engaged in a conversation with another person, he/she might not realize that the dog needs to go out for a walk if the dog is only gazing towards the door. Thus, communication via gazing makes little sense in that situation. The ability to recognize others' attentional states is thought to reduce the cost of the communicative signal by providing information about when one should start to (visually) communicate with others.

In regard to dogs understanding of other dogs' attentional state, one 'naturalistic study' was conducted by analysing the likelihood of dogs displaying a play bow towards a conspecific partner, depending on the latter's orientation (Horowitz 2009a). Dogs invariably displayed play bows when the partner was visually attentive towards them, but accompanied the bow with a bark, or preceded it with a nudge, if the partner was positioned in such a way so as to not see them. This study suggests that dogs are sensitive to the visual access that other dogs have of them. In a study investigating the function of play bows in young wolves and dogs (<1 year of age) at the Wolf Science Center, we found a similar pattern of results as found in pet dogs (Byosiere et al. 2016) (Chap. 5).

The most used paradigm to investigate whether animals take the *human's* attention into account when communicating is the begging paradigm. The rationale of this task is that animals should preferentially beg from an attentive person in contrast to an inattentive partner, which would reflect their ability to recognize that the attentive partner is in the position to perceive the beggar's visual signals and thus provide food, whereas the inattentive partner, whose eyes are invisible, will miss the relevant signals and therefore will not be able to fulfil the beggar's request (Gácsi et al. 2004; Povinelli and Eddy 1996). In order to receive a share of food, the beggar needs to initiate communication by either using visual signals towards the partner that is looking at the subject or by using acoustic or tactile attention-getting signals in case the partner is inattentive (Theall and Povinelli 1999). When examining whether the subject recognizes who can see her and who cannot, it is thus important that the partner does not provide additional cues that may indicate their attentiveness, since this is likely to render the visibility of the eyes as an unreliable signal of attention.

In their study, Udell and colleagues (2011) asked 8 hand-raised wolves at Wolf Park, pet dogs (1 group tested outside ($N = 18$) and 1 inside ($N = 17$)), and 17 shelter dogs (tested indoors) to choose between an attentive human and a human whose visual attention was blocked ('blind') in one of 4 ways: the human had her back turned, she held a book over her face, she had a bucket over her head, or she held a camera over her eyes (Fig. 7.6). The results showed that the outdoor tested pet dogs and the wolves performed rather well in the back-turned condition with a median of 9/10 trials correct, whereas shelter dogs performed worse. However, all groups performed above chance in this condition. Contrarily, in the camera and bucket

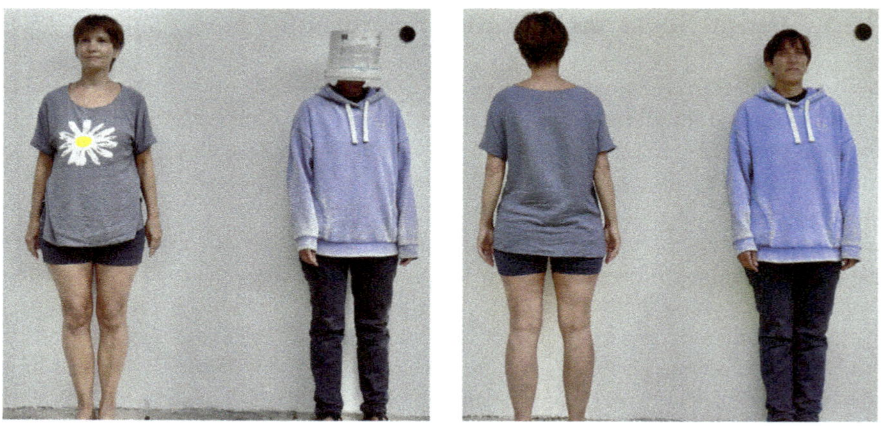

Fig. 7.6 Begging paradigm to study perspective taking. Pictures taken by the camera

conditions, all groups performed at chance level. In the book condition, the two pet dog groups performed above chance but not the other groups. In a subsequent study, the authors tested whether animals would be faster and more reliable when learning to beg from the attentive vs. inattentive (bucket wearing) person. They tested 12 naïve pet dogs and retested 8 wolves with half of the animals being reinforced for begging from a 'seeing' experimenter and the other half of the animals being tested with the 'bucket on the head' person. The results revealed that the animals learned faster when the attentive rather than the inattentive person was reinforcing them (3/6 dogs and 1/4 wolves in the attentive (seer) group and 1/6 and 0/4 wolves in the inattentive group performed above chance in the first 10 trials).

While the authors have claimed that this study presents the first evidence that also non-domesticated animals like wolves are sensitive to human attentional states under certain conditions, the methodology has been criticized. Indeed we have argued (Z. Virányi and Range 2011) that, since both partners expressed similar attentiveness towards the subjects by calling their name, Udell and colleagues' conclusion does not refer to the sensitivity of canines to others' attentiveness, but instead it might reflect that the animals better obey a familiar command in a familiar context (e.g. in the familiar situations where the human is actually facing them when addressing them) rather than the unfamiliar one (e.g. when the human is turning their back towards them, which does not happen during normal training/interaction situations).

7.2.2 Showing Behaviour

In an experiment by Miklósi and colleagues (2000), pet dogs could observe the hiding of a piece of food or a favourite toy in a place that they could not access themselves. After the hiding, the dogs engaged in showing behaviour defined as communicative behaviours consisting of a directional component referring to an external target and an attention getting component that attracts the receiver's attention to the sender towards the owner (e.g. looking and/or barking at the person and then looking at the place in which the object was hidden). The showing behaviour was exhibited more often towards the object of desire than empty locations and only in the presence and not in the absence of the owner. Based on these results, it was argued that dogs engage in functional referential communication, as has been shown in many other animal species (e.g. alarm calling behaviour of monkeys). However, the dogs' looking behaviour in this situation could easily be explained also by an association that the dogs have formed during life-long experiences between food/toys and their owners, e.g. we often give food or toys that are out of reach to our dogs, especially if they are looking at it. A similar experiment was carried out by Virányiet al. (2006) with 1-year-old wolves in Budapest and pet dogs, but never published (see Kubinyi et al. 2007). In that study, both wolves and pet dogs communicated the food location to a person, but pet dogs did so more than the wolves. However, the same simple explanation might account for these results, and moreover, the comparison between wolves and dogs is problematic due to the different life experiences.

To test whether wolves and dogs indeed engage in showing behaviour, Marianne Heberlein improved the showing paradigm to exclude the simpler association explanation by testing the animals with two familiar partners playing either a cooperative or competitive role during the experiment (Heberlein et al. 2016). If the animals indeed want to communicate to their partner to get the food, they should do so only with the cooperative partner, since the competitive partner ate the food herself. The two partners were similarly familiar with the animals and had worked with them in a cooperative manner in the past; thus, previous associations alone were unlikely to be driving the animals' response.

The experiment started with several training sessions where the animals learned that if a third person offered food to the cooperative partner, she would give it to the animal, while if the food was given to the competitive person, she would herself enjoy the fantastic sausage in full view of the subject. At the risk of anthropomorphizing, some of the animals had a very confused look on their face, sometimes even approaching the competitive partner as if they could not believe that the person, that they knew so well, ate the food herself instead of handing it over. Anyhow, over several sessions they learned that the competitive partner would not share, while the cooperative one was nice. After this pretraining, the experiment started. Three out of reach containers were mounted in the testing room, and the subject could observe in which of the containers a human experimenter hid a food reward. The experimenter then left the room, and either the cooperative or competitive partner, who did not know where the food was hidden, came in and sat down. After 1 minute of observing the animal, based on its behaviour, the human went to the location she thought the food was hidden at. If the food was in that location, the cooperative partner gave the food to the subject, whereas the competitive partner ate the food. If the subject had indicated an empty box, the person showed the empty box after checking it. The subjects were tested twice with each partner.

The dogs and wolves behaved very similarity in this experiment, both being more likely to show the correct food location to the cooperative than the competitive partner from the very first trial on. However, whereas the wolves spent also less time with the competitive person than with the cooperative one, dogs were not so resentful and spent the same amount of time with both experimenters. Interestingly, similar results were also found in a study with pet dogs with the difference that the pet dogs even actively showed an empty location to the competitive partner more than to the cooperative one (Heberlein et al. 2017), highlighting that the experience of living with a human partner can further increase the sophistication of the animals' communicative behaviours.

7.2.3 'Looking Back' Behaviour in the Impossible Task

Another paradigm that has been used to highlight differences between wolves and dogs in their human-directed communication is the 'impossible task' (Miklósi et al. 2003). In this paradigm, the subject is faced with a problem-solving task in the context of a food reward. While in the first trials the animal can independently solve the apparatus to obtain the reward (possible trials), after a few successful attempts, the apparatus is blocked (impossible trial), so that the animal's previously successful strategy becomes useless.

In a seminal study by the Budapest group, seven hand-raised wolf and nine pet dog puppies at the age of 4 months participated in a rope-pulling task and a bin-opening task. In both tasks, the animals were first presented with six possible trials and then an impossible one. The wolves and pet dogs did not differ in their latencies to get access to the reward in the possible trials, indicating that they had the same motivation and abilities to solve the problem. However, in the impossible trials, their behaviours differed markedly. While the wolves continued to try and solve the problem during the 2-minute period, the dogs soon looked at their owner. They did so faster and longer than the wolves, which, instead, continued to engage with the problem. This difference in behaviour has been interpreted by the authors as dogs being more willing to look at the human's face, a precondition for complex animal-human communication.

Ever since, the impossible task paradigm has been extensively used to study the looking-back behaviour in many additional contexts: training (D'Aniello et al. 2015; Marshall-Pescini et al. 2009, 2016); breed differences (Brodd 2014; Konno et al. 2016; Marshall-Pescini et al. 2016); genetic bases and heritability (Hori et al. 2013; Persson et al. 2015); ageing (Brodd 2014; Passalacqua et al. 2011); effect of experience (D'Aniello and Scandurra 2016; Marshall-Pescini et al. 2017); reputation (Piotti et al. 2017); and comprehension of the other's attentional stance (Marshall-Pescini et al. 2013). In quite a number of these studies (Brodd 2014; D'Aniello and Scandurra 2016; Hori et al. 2013; Konno et al. 2016; S. Marshall-Pescini et al. 2016; Passalacqua et al. 2011; Persson et al. 2015; Piotti et al. 2017), the authors hypothesized that the looking back was not only determined by 'human's attractiveness' to dogs, but that dogs might use this looking behaviour as an alternative problem-solving strategy (i.e. asking the human for help). Some studies have also shown an effect of experience with humans, which seems to strongly affect some aspects of the looking-back behaviour. For instance, highly trained dogs (agility, retrieving, search and rescue) showed more independent problem-solving abilities than untrained pet dogs, who conversely looked towards the owner longer (Marshall-Pescini et al. 2009) in such tasks. Moreover, adult dogs looked longer at the owner and/or experimenter than juvenile dogs (Brodd 2014; Passalacqua et al. 2011), and pet dogs looked back longer than dogs living in kennels (which thus had far less experience with human interactions; D'Aniello and Scandurra 2016). The results of these studies using the impossible task paradigm are in line with the original wolf-dog comparison suggesting the selection of dogs to look at the human, which would facilitate the development of complex socio-communicative skills given the right experience. This could include requesting help in specific situations and would be in line with the hypothesis that both domestication and an individual's experience play a role in the emergence of a dog's socio-cognitive skills (Miklósi and Topál 2013; Reid 2009).

However, while dogs' behaviour in such tasks has traditionally been interpreted as a clever, socially advanced, response to this problem, it can also be seen as the dogs just simply giving up faster than the wolves. Several recent studies by two independent groups have provided evidence for this latter (and more parsimonious) explanation.

Monique Udell (2015) tested ten pet dogs, ten shelter dogs, and ten hand-raised wolves from Wolf Park (all adults) in different conditions with a *solvable* puzzle box. First, all animals were tested when alone or with a neutral human caregiver in a counterbalanced order. Independent of the condition, eight of the ten wolves solved the problem within the allotted time period of 2 min, whereas only one pet dog solved the problem and only in the human-present condition. Importantly, all animals were initially interested in the box and approached it quickly, suggesting that the basic motivation to engage with the box was the same for wolves and dogs. Moreover, in the human-present condition, while the wolves basically never looked at the human, but were rather engaged with the box, both shelter and pet dogs gazed at the human on average 10 and 16% of the testing time, respectively. The dogs were then also tested with an encouraging human. Four shelter dogs and one pet dog solved the problem when the human was encouraging them, and they spent more time gazing at the box and engaging with the box than in the neutral human condition.

Similar results were obtained when reusing a subset of the wolf data, adding an additional 3 wolves and comparing their behaviour in the human-present condition with data collected on 2 different pet dog populations (1 tested indoors with a familiar experimenter ($N = 20$) and 1 tested outdoors with an unfamiliar experimenter ($N = 20$)) and 20 free-ranging dogs in the human-present situation (Brubaker et al. 2017). Interestingly, the free-ranging dogs gazed even longer at the human experimenter than the pet dogs that were tested outside with an unfamiliar experimenter, while they were at a similar level as those tested inside with a familiar person. Wolves were more interested in the box and thus did not gaze at the human experimenter. Brubaker et al. (2017) and Udell (2015) gave a twofold interpretation of the difference in behaviour between the populations: (1) it's a result of greater persistence in the wolves compared to the various dog populations, allowing them to solve the problem, and (2), possibly, more social inhibition of pet and shelter dogs compared to wolves, influencing their willingness to engage in problem-solving behaviour unless encouraged by humans. This later aspect is suggested to imply generalized dependence on, or deference to, human action.

While these results suggest that, indeed, wolves are more persistent than dogs in problem-solving tasks (see Chap. 9) but dogs might engage more in looking at the human, the results have to be considered carefully. Firstly, because, although not intended by the authors, the problem was actually solvable and some animals, especially the wolves, solved the task during the 2 min, this begs the question why they should ever look at the human if they can perfectly manage by themselves. Secondly, the pet dogs and shelter dogs, as well as the free-ranging dogs, had vastly different experiences with humans compared to the wolves. For example, while when socializing wolves, one usually avoids getting into conflicts, all dog populations likely have ample experience of being inhibited by humans—even the free-ranging dogs. This experience, as also realized by Udell (2015), probably influenced their propensity to look at the human experimenter.

Another study investigated the performance of 15 wolves and 14 pack dogs at the Wolf Science Center in the classical version of the impossible task, comparing their behaviour also to 11 pet dogs and 19 free-ranging dogs in India (Sarah Marshall-Pescini et al. 2017). Unlike previous studies, we measured the degree of persistence in trying to solve the task in the impossible trial and added it as a potential explanatory factor in our statistical analyses. Not surprisingly, wolves were more persistent than all dog groups (see Chap. 9). However, interestingly, it turned out that the less persistent animals—whether pet dogs, street dogs, pack dogs, or wolves—looked back sooner and for longer at the human experimenter (Fig. 7.7). In other words, the effect of population on the looking-back behaviour, as has been found previously, disappeared once the persistence in solving the task was considered. These results suggest that since wolves are just a lot more motivated and persistent when trying to solve problems, not giving up as quickly as most dogs, they also do not look back at humans or, at least, do so a lot later. Maybe, when giving the animals a lot more time, they would all look back towards the human at one point. Actually, in this later study, the trial duration was increased to 3 instead of 2 min, and 11 of the 15 wolves did look back at the human (all dogs looked back). Another very interesting aspect is that the experience of dogs with humans did not really influence the looking-back behaviour of dogs neither in the Brubaker et al. (2017) nor in our own study. This is another indicator that the 'looking-back' behaviour does not represent a 'social problem-solving strategy' (asking for help), since then, we would have expected that animals with more experience with humans (e.g. pet dogs) would

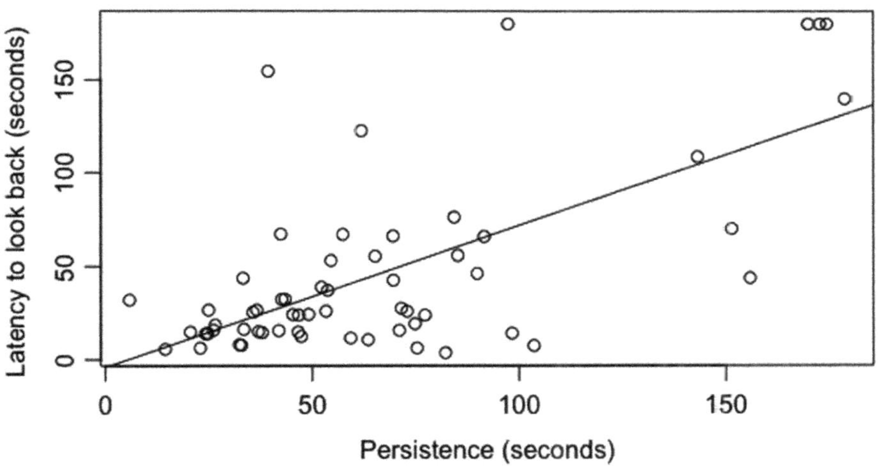

Fig. 7.7 Linear positive relationship between persistence and the latency to look back at the person. The more animals spent interacting with the apparatus, the longer they took to look back at the person. Figure adapted from Marshall-Pescini et al. 2017. https://www.nature.com/articles/srep46636/; CC BY 4.0; http://creativecommons.org/licenses/by/4.0/

use this behaviour more often and more intensely than street dogs or pack dogs that seldom have the help of humans to solve their problems (Fig. 7.7).

These findings encouraged us to run another study to tease apart the looking-back behaviour of the dogs (Lazzaroni et al. 2020). Since, although the previous study suggests that the differences in looking-back behaviour (at least between adult wolves and dogs) are likely due to persistence, there are still two alternative explanations—first, the dogs may give up earlier *in order* to look at the human, and second, due to the lower persistence, the dogs stop manipulating—and then, since there is nothing else to do, they look at the most salient object/person in their vicinity, which usually is the experimenter in such set-ups.

Accordingly, Martina Lazzaroni designed a modified version of the impossible task, in which the subjects simultaneously faced three possible and one impossible trial. This set-up allowed us to bait the apparatus for the entire experiment beforehand in the absence of the subjects, so that they never observed a person interacting with the apparatus and handling the rewards (thus avoiding the direct food-person association in the test). We then tested adult pet dogs in Vienna and free-ranging dogs in Morocco, in four different conditions: in the social condition, the experimenter was present during the experiment standing behind the apparatus. In the asocial condition, the subject was alone (the experimenter hid). In the 'dummy' human condition, a cardboard live-size version of Han Solo (of Star Wars fame) was present (sufficiently realistic to make colleagues startle when they first walk into our office where he has now retired). Finally, in the 'cardboard' condition, a rectangular piece of cardboard the size of Han Solo was present next to the apparatus (this latter condition was not run with the free-ranging dogs). The rationale for these conditions was that if indeed looking back functions as a problem-solving strategy, i.e. the dogs purposefully know that humans can help them, then they should give up trying themselves earlier when the actual human is present than when they are alone. If this is not the case, e.g. they persist for a certain amount of time independent of the condition (potentially indicating a personality trait; see Chap. 10) and then give up and just look at the most salient object in their vicinity, they should look at the experimenter, Han Solo, and the large cardboard rectangle equally soon after giving up manipulation. Furthermore, pet dogs with much more extensive experience of humans helping them obtain desired objects and food should look back much sooner, be more discriminative between conditions and potentially be more persistent in their looking behaviours than street dogs (Fig. 7.8).

We found that the subjects of both groups were equally persistent in interacting with the impossible task in all conditions, and there was no difference between groups in how soon they looked back at the human in the social condition. Moreover, the condition had no effect on how quickly the dogs looked back, i.e. they took the same amount of time to look back at the human, at Han Solo, and at the cardboard rectangle after attempting the impossible task (Fig. 7.9). In contrast, the frequency and duration of looking in pet dogs decreased from the human to Han Solo to the object, whereas the free-ranging dogs looked at the human and Han Solo equally. This main difference observed between pet dogs and free-ranging dogs could be because the street dogs may have been a bit more worried about the Star

Fig. 7.8 (**a**) Impossible/possible apparatus designed by Martina Lazzaroni. Three of the bottles can be moved to get access to the food, whereas the last bottle was fixed to the wooden board. (**b**) Experimental condition with Han Solo. Figure adapted from Lazzaroni et al. 2020. https://link.springer.com/article/10.1007/s10071-020-01345-8/; CC BY 4.0; http://creativecommons.org/licenses/by/4.0/

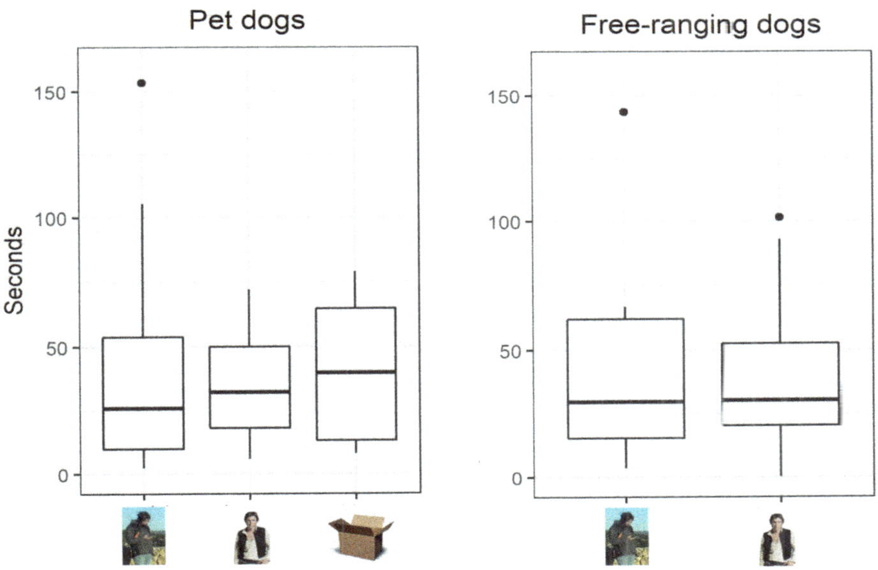

Fig. 7.9 Latency to look back after interacting with the impossible bowl for pet dogs ($N = 20$) and free-ranging dogs ($N = 26$) in the social, Han Solo, and, for the pet dogs, card box condition. Figure adapted from Lazzaroni et al. 2020 (https://link.springer.com/article/10.1007/s10071-020-01345-8/; CC BY 4.0; http://creativecommons.org/licenses/by/4.0/)

Wars hero landing in their quiet fishing village. Interestingly all dog (pet and street dogs) initially treated Han as a social entity (approaching and tail wagging towards it), but pet dogs may have realized more quickly that this was not a 'real human'. In the social condition, we also found that pet dogs looked at the human for longer than street dogs; this could be because pet dogs may be generally more attracted to humans, and/or have a stronger association of humans with food than free-ranging dogs.

7.2.4 Summary Communication with Humans

Socialized wolves generally see humans as a collaborative partner worth communicating with. They engage in communicating where food is hidden, paying also attention to the role of the human, i.e. *whether she is competitive or cooperative. Moreover, wolves as well as dogs realize to some extent also if the human is attentive or not. Partly, these results—also with the dogs—seem to be dependent on their experience,* e.g. *shelter dogs and wolves do not really seem to know what it means if a human is reading, whereas pet dogs probably have experienced this quite a lot during their day-to-day lives with humans. Finally, the impossible task does not seem to measure what was originally intended. The latest results show that the dogs are less persistent than wolves when trying to solve a challenging task, thus giving up earlier and then turning towards the most salient object in their environment. This interpretation was supported by a study showing that persistence and the looking-back behaviour is intrinsically linked—no matter of wolves or dogs. The less persistent animals look back earlier than the more persistent animals. Thus, overall, differences between wolves and dogs seem to be minor in how they communicate with humans and rather lay in other temperament traits,* e.g. *persistence (Table 7.2).*

7.3 Summary Communication

Overall, based on current knowledge, the differences in communicative abilities with humans between wolves and dogs are probably rather small and, if existing at all, quantitative rather than qualitative in nature. Rather it seems that other traits, such as conflict avoidance, proneness to 'obey', and persistence, may be responsible for some of the differences observed in the various tests. Indeed, although we design experiments with the idea that we are testing one specific aspect of the animal's behaviour, it is evident from their results that a number of factors may affect how the animals perceive the testing situation and disentangling these factors is difficult.

Results in the two-choice task could be driven by the fact that neither wolves nor dogs seem to perceive these situations as cooperative, but rather as competitive. Since the wolves are more confident (and also more tolerant towards lower-ranking animals; see Chap. 5), they follow the gaze cue regardless of the potential conflict inherent in doing so, whereas dogs hesitate and only dare to follow if humans clearly tell them that it is fine to go by using more explicit (ostensive) communication or a more explicit gesture which may be viewed instead as a command, i.e. *the point. Wolves, on the contrary, either are not as prone as dogs to follow 'commands' or at least need some training/experience/reinforcement to do so (hence their willingness*

Table 7.2 Summary of studies to investigate communication with humans

Task	Groups	Success	References
Sensitivity towards human's attentional states (begging paradigm)	Pet dogs (PD), shelter dogs (SD), wolves (W)	Back turned: W = PD > SD Camera + bucket: all at chance Book: PD > W = SD	(Udell et al. 2011)
Getting help (showing paradigm)	Pet dogs (PD), 1-year-old wolves (W)	D > W	(Virányi et al. 2006)
⇨ Differentiating between cooperative and competitive partners	Pet dogs (PD), WSC pack dogs (WD), wolves (W)	WD = W < PD	(Heberlein et al. 2016)
Looking at human face (possible task)	Pet dogs (PD), shelter dogs (SD), wolves (W)	Looking duration: W < PD & SD Wolves solved the task	(Udell 2015)
	Pet dogs tested outdoors (PDO) and indoors (PDI), free-ranging dogs (FRD), wolves (W)	Looking duration: W < PDO < FRD = PDI Wolves solved the task	(Brubaker et al. 2017)
Looking at human face (impossible task)	Pet dogs (PD), wolves (W)	Latency: PD < W Looking duration: W < PD	(Miklósi et al. 2003)
	WSC pack dogs (WD), free-ranging dogs (FRD), wolves (W) controlling for persistence	Looking behaviour dependent on individual persistence: W = WD = FRD	(Marshall-Pescini et al. 2017)

to follow the pointing gesture increasing with age/experience, but is not there from the start). Accordingly, other factors such as the social dynamics between the humans and the animals, the potential for conflict, as well as their individual confidence levels may be playing a role in a seemingly 'simple' test of communicative gazing/pointing.

Similarly, the studies on the looking-back paradigm may be more telling about the animals' persistence than their willingness or interest in communicating with us. This seems to be particularly evident when considering results from the showing experiment, which at present is perhaps the cleanest test available for 'gazing behaviour'. Here wolves and dogs again did not differ and clearly showed that given the right setting and motivation, they can communicate referentially with a human partner, even taking the cooperativeness of the partner into account.

However, from these studies it is also clear that the level of experience with humans can heavily affect the animals' communicative behaviour. Dogs living in the human social environment and being extensively exposed to our human communication may achieve more sophisticated skills than reviewed here (see Siniscalchi 2018*). How far dogs can go with extensive socialization and experience is a very interesting avenue of research—but one that will be difficult to follow with wolves,* *since some of their defining characteristics (their stubbornness/persistence in trying to get things they want, a cooperativeness which is not based on submission and the fact that they are not as prone as dogs to avoid conflicts) make them pretty impossible pets.*

References

Agnetta B, Hare B, Tomasello M (2000) Cues to food location that domestic dogs (*Canis familiaris*) of different ages do and do not use. Anim Cogn 3(2):107–112. https://doi.org/10.1007/s100710000070

Amici F, Aureli F, Visalberghi E, Call J (2009) Spider monkeys (*Ateles geoffroyi*) and capuchin monkeys (*Cebus apella*) follow gaze around barriers: evidence for perspective taking? J Comp Psychol 123(4):368–374. https://doi.org/10.1037/a0017079

Anderson JR, Montant M, Schmitt D (1996) Rhesus monkeys fail to use gaze direction as an experimenter-given cue in an object choice task. Behav Process 37:47–55. https://doi.org/10.1016/0376-6357(95)00074-7

Bálint A, Faragó T, Meike Z, Lenkei R, Miklósi Á, Pongrácz P (2015) "Do not choose as I do!" – dogs avoid the food that is indicated by another dog's gaze in a two-object choice task. Appl Anim Behav Sci 170:44–53. https://doi.org/10.1016/j.applanim.2015.06.005

Baron-Cohen S (1995) Mindblindness: an essay on autism and theory of mind. MIT Press, Cambridge, MA

Bentosela M, Barrera G, Jakovcevic A, Elgier AM, Mustaca AE (2008) Effect of reinforcement, reinforcer omission and extinction on a communicative response in domestic dogs (*Canis familiaris*). Behav Process 78(3):464–469. https://doi.org/10.1016/j.beproc.2008.03.004

Bergman TJ, Beehner JC, Cheney DL, Seyfarth RM (2003) Hierarchical classification by rank and kinship in baboons. Science 302(5648):1234–1236

Bräuer J, Call J, Tomasello M (2005) All great ape species follow gaze to distant locations and around barriers. J Comp Psychol 119(2):145–154. https://doi.org/10.1037/0735-7036.119.2.145

Bräuer J, Kaminski J, Riedel J, Call J, Tomasello M (2006) Making inferences about the location of hidden food: social dog, causal ape. J Comp Psychol 120(1):38–47. https://doi.org/10.1037/0735-7036.120.1.38

Brodd L (2014) The help-seeking behaviour of dogs (*Canis familiaris*), BSc Examensarbeit. Linköpings universitet, Linköping

Brooks R, Meltzoff AN (2002) The importance of eyes: how infants interpret adult looking behavior. Dev Psychol 38:958–966. https://doi.org/10.1037//0012-1649.38.6.958

Brooks R, Meltzoff AN (2005) The development of gaze following and its relation to language. Dev Sci 8:535–543. https://doi.org/10.1111/j.1467-7687.2005.00445.x

Brubaker L, Dasgupta S, Bhattacharjee D, Bhadra A, Udell MAR (2017) Differences in problem-solving between canid populations: do domestication and lifetime experience affect persistence? Anim Cogn 20(4):717–723. https://doi.org/10.1007/s10071-017-1093-7

Bugnyar T, Stöwe M, Heinrich B (2004) Ravens, *Corvus corax*, follow gaze direction of humans around obstacles. Proc R Soc Lond B 271(1546):1331–1336. https://doi.org/10.1098/rspb.2004.2738

Burkart JM, Heschl A (2007) Understanding visual access in common marmosets, *Callithrix jacchus*: perspective taking or behaviour reading? Anim Behav 73:457–469. https://doi.org/10.1016/j.anbehav.2006.05.019

Butterworth G, Simion F (2013) What is special about pointing in babies. The development of sensory, motor and cognitive capacities in early infancy: from perception to cognition, 171–190

Byosiere SE, Espinosa J, Marshall-Pescini S, Smuts B, Range F (2016) Investigating the function of play bows in dog and wolf puppies (*Canis lupus familiaris, Canis lupus occidentalis*). PLoS One 11(12):e0168570. https://doi.org/10.1371/journal.pone.0168570

Byrne RW (1997) Machiavellian intelligence. Evol Anthropol 172-180. https://doi.org/10.1002/(SICI)1520-6505(1996)5:5<172::AID-EVAN6>3.0.CO;2-H

Catala A, Mang B, Wallis L, Huber L (2017) Dogs demonstrate perspective taking based on geometrical gaze following in a Guesser–Knower task. Anim Cogn 20(4):581–589. https://doi.org/10.1007/s10071-017-1082-x

Colonnesi C, Stams GJJM, Koster I, Noom MJ (2010) The relation between pointing and language development: a meta-analysis. Dev Rev 30(4):352–366. https://doi.org/10.1016/j.dr.2010.10.001

Csibra G (2010) Recognizing communicative intentions in infancy. Mind Lang 25(2):141–168. https://doi.org/10.1111/j.1468-0017.2009.01384.x

Csibra G, Gergely G (2009) Natural pedagogy. Trends Cogn Sci 13(4):148–153. https://doi.org/10.1016/j.tics.2009.01.005

D'Aniello B, Scandurra A (2016) Ontogenetic effects on gazing behaviour: a case study of kennel dogs (Labrador retrievers) in the impossible task paradigm. Anim Cogn 19(3):565–570. https://doi.org/10.1007/s10071-016-0958-5

D'Aniello B, Scandurra A, Prato-Previde E, Valsecchi P (2015) Gazing toward humans: a study on water rescue dogs using the impossible task paradigm. Behav Process 110:68–73. https://doi.org/10.1016/j.beproc.2014.09.022

de Blois ST, Novak MA, Bond M (1998) Object permanence in orangutans (*Pongo pygmaeus*) and squirrel monkeys (*Saimiri sciureus*). J Comp Psychol 112(2):137–152. https://doi.org/10.1037/0735-7036.112.2.137

Dunbar RIM (1998) The social brain hypothesis. Evol Anthropol 6:178–190. https://doi.org/10.1002/(SICI)1520-6505(1998)6:5<178::AID-EVAN5>3.0.CO;2-8

Dunbar RIM, Shultz S (2007) Evolution in the social brain. Science 317(5843):1344–1347. https://doi.org/10.1126/science.1145463

Duranton C, Range F, Virányi Z (2017) Do pet dogs (*Canis familiaris*) follow ostensive and non-ostensive human gaze to distant space and to objects? R Soc Open Sci 4(7):170349. https://doi.org/10.1098/rsos.170349

Fiset S (2010) Comment on "differential sensitivity to human communication in dogs, wolves, and human infants". Science 329(5988):142–142. https://doi.org/10.1126/science.1184045

Gácsi M, Győri B, Miklósi Á, Virányi Z, Kubinyi E, Topál J, Csányi V (2005) Species-specific differences and similarities in the behavior of hand-raised dog and wolf pups in social situations with humans. Dev Psychobiol 47(2):111–122. https://doi.org/10.1002/dev.20082

Gácsi M, Győri B, Virányi Z, Kubinyi E, Range F, Belényi B, Miklósi A (2009) Explaining dog wolf differences in utilizing human pointing gestures: selection for synergistic shifts in the development of some social skills. PLoS One 4(8):e6584. https://doi.org/10.1371/journal.pone.0006584

Gácsi M, Miklósi A, Varga O, Topál J, Csányi V (2004) Are readers of our face readers of our minds? Dogs (*Canis familiaris*) show situation-dependent recognition of human's attention. Anim Cogn 7(3):144–153. https://doi.org/10.1007/s10071-003-0205-8

Gagnon S, Doré FY (1992) Search behavior in various breeds of adult dogs (*Canis familiaris*): object permanence and olfactory cues. J Comp Psychol 106(1):58–68. https://doi.org/10.1037/0735-7036.106.1.58

Gagnon S, Doré FY (1994) Cross-sectional study of object permanence in domestic puppies (*Canis familiaris*). J Comp Psychol 108(3):220–232. https://doi.org/10.1037/0735-7036.108.3.220

Goossens BMA, Dekleva M, Reader SM, Sterck EHM, Bolhuis JJ (2008) Gaze following in monkeys is modulated by observed facial expressions. Anim Behav 75:1673–1681. https://doi.org/10.1016/j.anbehav.2007.10.020

Hare B, Brown M, Williamson C, Tomasello M (2002) The domestication of social cognition in dogs. Science 298:1634–1636. https://doi.org/10.1126/science.1072702

Hare B, Tomasello M (2005a) The emotional reactivity hypothesis and cognitive evolution. Trends Cogn Sci 9(10):464–465. https://doi.org/10.1016/j.tics.2005.08.010
Hare B, Tomasello M (2005b) Human-like social skills in dogs? Trends Cogn Sci 9(9):439–444. https://doi.org/10.1016/j.tics.2005.07.003
Heberlein MTE, Manser MB, Turner DC (2017) Deceptive-like behaviour in dogs (*Canis familiaris*). Anim Cogn 20(3):511–520. https://doi.org/10.1007/s10071-017-1078-6
Heberlein MTE, Turner DC, Range F, Virányi Z (2016) A comparison between wolves, *Canis lupus*, and dogs, *Canis familiaris*, in showing behaviour towards humans. Anim Behav 122:59–66. https://doi.org/10.1016/j.anbehav.2016.09.023
Hori Y, Kishi H, Inoue-Murayama M, Fujita K (2013) Dopamine receptor D4 gene (*DRD4*) is associated with gazing toward humans in domestic dogs (*Canis familiaris*). Open J Anim Sci 03: 54–58. https://doi.org/10.4236/ojas.2013.31008
Horowitz A (2009a) Attention to attention in domestic dog (*Canis familiaris*) dyadic play. Anim Cogn 12(1):107–118. https://doi.org/10.1007/s10071-008-0175-y
Horowitz A (2009b) Disambiguating the "guilty look": salient prompts to a familiar dog behaviour. Behav Process 81(3):447–452. https://doi.org/10.1016/j.beproc.2009.03.014
Kaminski J (2008) Dogs (*Canis familiaris*) are adapted to receive human communication. In: Berthoz A, Christen Y (eds) Neurobiology of "Umwelt": how living beings perceive the world, research and perspectives in neurosciences. Springer, Heidelberg, pp 103–107
Kaminski J, Nitzschner M (2013) Do dogs get the point? A review of dog–human communication ability. Learn Motiv 44(4):294–302. https://doi.org/10.1016/j.lmot.2013.05.001
Kaminski J, Riedel J, Call J, Tomasello M (2005) Domestic goats, *Capra hircus*, follow gaze direction and use social cues in an object choice task. Anim Behav 69:11–18. https://doi.org/10.1016/j.anbehav.2004.05.008
Kaminski J, Schulz L, Tomasello M (2012) How dogs know when communication is intended for them. Dev Sci 15(2):222–232. https://doi.org/10.1111/j.1467-7687.2011.01120.x
Kaminski J, Waller BM, Diogo R, Hartstone-Rose A, Burrows AM (2019) Evolution of facial muscle anatomy in dogs. Proc Natl Acad Sci 116(29):14677–14681
Kis A, Gácsi M, Range F, Virányi Z (2012) Object permanence in adult common marmosets (*Callithrix jacchus*): not everything is an "A-not-B" error that seems to be one. Anim Cogn 15(1):97–105. https://doi.org/10.1007/s10071-011-0436-z
Konno A, Romero T, Inoue-Murayama M, Saito A, Hasegawa T (2016) Dog breed differences in visual communication with humans. PLoS One 11(10):e0164760. https://doi.org/10.1371/journal.pone.0164760
Kotrschal A, Corral-Lopez A, Kolm N (2019) Large brains, short life: selection on brain size impacts intrinsic lifespan. Biol Lett 15(5):20190137
Kotrschal A, Rogell B, Bundsen A, Svensson B, Zajitschek S, Brännström I et al (2013) Artificial selection on relative brain size in the guppy reveals costs and benefits of evolving a larger brain. Curr Biol 23(2):168–171
Krause MA, Udell MAR, Leavens DA, Skopos L (2018) Animal pointing: changing trends and findings from 30 years of research. J Comp Psychol 132(3):326–345. https://doi.org/10.1037/com0000125
Kubinyi E, Virányi Z, Miklósi A (2007) Comparative social cognition: from wolf and dog to humans. Comparative Cognition & Behavior Reviews 2:26–46. https://doi.org/10.3819/ccbr.2008.20002
Lampe M, Bräuer J, Kaminski J, Virányi Z (2017) The effects of domestication and ontogeny on cognition in dogs and wolves. Sci Rep 7:11690. https://doi.org/10.1038/s41598-017-12055-6
Lazzaroni M, Marshall-Pescini S, Manzenreiter H, Gosch S, Přibilová L, Darc L et al (2020) Why do dogs look back at the human in an impossible task? Looking back behaviour may be over-interpreted. Anim Cogn:1–15
Loretto MC, Schloegl C, Bugnyar T (2009) Northern bald ibises follow others' gaze into distant space but not behind barriers. Biol Lett 6(1):14–17. https://doi.org/10.1098/rsbl.2009.0510

Marshall-Pescini S, Colombo E, Passalacqua C, Merola I, Prato-Previde E (2013) Gaze alternation in dogs and toddlers in an unsolvable task: evidence of an audience effect. Anim Cogn 16(6): 933–943. https://doi.org/10.1007/s10071-013-0627-x

Marshall-Pescini S, Frazzi C, Valsecchi P (2016) The effect of training and breed group on problem-solving behaviours in dogs. Anim Cogn 19(3):571–579. https://doi.org/10.1007/s10071-016-0960-y

Marshall-Pescini S, Passalacqua C, Barnard S, Valsecchi P, Prato-Previde E (2009) Agility and search and rescue training differently affects pet dogs' behaviour in socio-cognitive tasks. Behav Process 81(3):416–422. https://doi.org/10.1016/j.beproc.2009.03.015

Marshall-Pescini S, Passalacqua C, Valsecchi P, Prato-Previde E (2010) Comment on "differential sensitivity to human communication in dogs, Wolves, and Human Infants". Science 329(5988). https://doi.org/10.1126/science.1187748

Marshall-Pescini S, Rao A, Virányi Z, Range F (2017) The role of domestication and experience in 'looking back' towards humans in an unsolvable task. Sci Rep 7:46636. https://doi.org/10.1038/srep46636

Mathieu M, Bergeron G (1981) Piagetian assessment on cognitive development in chimpanzee (*Pan troglodytes*). Paper presented at the primate behavior and sociobiology, Berlin, Heidelberg

Met A, Miklósi Á, Lakatos G (2014) Gaze-following behind barriers in domestic dogs. Anim Cogn 17(6):1401–1405. https://doi.org/10.1007/s10071-014-0754-z

Micheletta J, Waller BM (2012) Friendship affects gaze following in a tolerant species of macaque, *Macaca nigra*. Anim Behav 83(2):459–467. https://doi.org/10.1016/j.anbehav.2011.11.018

Miklósi Á, Kubinyi E, Topál J, Gácsi M, Virányi Z, Csányi V (2003) A simple reason for a big difference: wolves do not look Back at humans, but dogs do. Curr Biol 13(9):763–766. https://doi.org/10.1016/s0960-9822(03)00263-x

Miklósi Á, Polgárdi R, Topál J, Csányi V (1998) Use of experimenter-given cues in dogs. Anim Cogn 1(2):113–121. https://doi.org/10.1007/s100710050016

Miklósi Á, Polgárdi R, Topál J, Csányi V (2000) Intentional behaviour in dog-human communication: an experimental analysis of "showing" behaviour in the dog. Anim Cogn 3(3):159–166. https://doi.org/10.1007/s100710000072

Miklósi Á, Topál J (2013) What does it take to become 'best friends'? Evolutionary changes in canine social competence. Trends Cogn Sci 17:287–294. https://doi.org/10.1016/j.tics.2013.04.005

Neiworth JJ, Steinmark E, Basile BM, Wonders R, Steely F, DeHart C (2003) A test of object permanence in a new-world monkey species, cotton top tamarins (*Saguinus oedipus*). Anim Cogn 6:27–37. https://doi.org/10.1007/s10071-003-0162-2

Oliva JL, Rault JL, Appleton B, Lill A (2015) Oxytocin enhances the appropriate use of human social cues by the domestic dog (*Canis familiaris*) in an object choice task. Anim Cogn 18:767–775. https://doi.org/10.1007/s10071-015-0843-7

Passalacqua C, Marshall-pescini S, Barnard S, Lakatos G (2011) Human-directed gazing behaviour in puppies and adult dogs, *Canis lupus familiaris*. Anim Behav 82:1043–1050. https://doi.org/10.1016/j.anbehav.2011.07.039

Pepperberg IM, Willner MR, Gravitz LB (1997) Development of Piagetian object permanence in a Grey parrot (*Psittacus erithacus*). J Comp Psychol 111(1):63–75. https://doi.org/10.1037/0735-7036.111.1.63

Persson ME, Roth LSV, Johnsson M, Wright D, Jensen P (2015) Human-directed social behaviour in dogs shows significant heritability. Genes Brain Behav 14(4):337–344 https://doi.org/10.1111/gbb.12194

Piotti P, Spooner RM, Jim H-L, Kaminski J (2017) Who to ask for help? Do dogs form an opinion on humans based on skilfulness? Appl Anim Behav Sci 195:93–102. https://doi.org/10.1016/j.applanim.2017.05.024

Pollok B, Prior H, Gunturkun O (2000) Development of object permanence in food-storing magpies (*Pica Pica*). J Comp Psychol 114(2):148–157. https://doi.org/10.1037/0735-7036.114.2.148

Pongrácz P, Gácsi M, Hegedüs D, Péter A, Miklósi Á (2013) Test sensitivity is important for detecting variability in pointing comprehension in canines. Anim Cogn 16(5):721–735. https://doi.org/10.1007/s10071-013-0607-1

Poti P (1989) Early sensimotor development in macaques (*Macaca fuscata*, *Macaca fuscicularis*). In: Antinucci F (ed) Cognitive structure and development in nonhuman primates. Erlbaum, Hillsdale, NJ, pp 97–112
Povinelli DJ, Eddy TJ (1996) What young chimpanzees know about seeing. Monogr Soc Res Child Dev 61(3):1–152. https://doi.org/10.2307/1166159
Povinelli DJ, Reaux JE, Bierschwale DT, Allain AD, Simon BB (1997) Exploitation of pointing as a referential gesture in young children, but not adolescent chimpanzees. Cogn Dev 12(4): 327–365. https://doi.org/10.1016/S0885-2014(97)90017-4
Range F, Marshall-Pescini S, Kratz C, Virányi Z (2019) Wolves lead and dogs follow, but they both cooperate with humans. Sci Rep 9(1):3796. https://doi.org/10.1038/s41598-019-40468-y
Range F, Virányi Z (2011) Development of gaze following abilities in wolves (*Canis Lupus*). PLoS One 6(2):e16888. https://doi.org/10.1371/journal.pone.0016888
Reid PJ (2009) Adapting to the human world: dogs' responsiveness to our social cues. Behav Process 80(3):325–333. https://doi.org/10.1016/j.beproc.2008.11.002
Schloegl C, Kotrschal K, Bugnyar T (2007) Gaze following in common ravens, *Corvus corax*: ontogeny and habituation. Anim Behav 74:769–778. https://doi.org/10.1016/j.anbehav.2006.08.017
Schloegl C, Schmidt J, Scheid C, Kotrschal K, Bugnyar T (2008) Gaze following in non-human animals: the corvid example. In: Weber EA, Krause LH (eds) Animal behaviour. New research. Nova Science Publishers, New York, pp 73–92
Senju A, Csibra G (2008) Gaze following in human infants depends on communicative signals. Curr Biol 18:668–671. https://doi.org/10.1016/j.cub.2008.03.059
Seyfarth R, Cheney D, Marler P (1980) Vervet monkey alarm calls: semantic communication in a free-ranging primate. Anim Behav 28:1070–1094
Shepherd SV (2010) Following gaze: gaze-following behavior as a window into social cognition. Front Integr Neurosci 4:5. https://doi.org/10.3389/fnint.2010.00005
Siniscalchi M (2018) Communication in Dogs. Anim Behav Monograph 8(8):e131. https://doi.org/10.3390/ani8080131
Soproni K, Miklósi Á, Topál J, Csányi V (2001) Comprehension of human communicative signs in pet dogs (*Canis familiaris*). J Comp Psychol 115(2):122–126. https://doi.org/10.1037//0735-7036.115.2.122
Sümegi Z, Kis A, Miklósi Á, Topál J (2014) Why do adult dogs (*Canis familiaris*) commit the A-not-B search error? J Comp Psychol 128(1):21–30. https://doi.org/10.1037/a0033084
Szufnarowska J, Rohlfing KJ, Fawcett C, Gredebäck G (2014) Is ostension any more than attention? Sci Rep 4(1):5304. https://doi.org/10.1038/srep05304
Téglás E, Gergely A, Kupán K, Miklósi Á, Topál J (2012) Dogs' gaze following is tuned to human communicative signals. Curr Biol 22(3):209–212. https://doi.org/10.1016/j.cub.2011.12.018
Theall LA, Povinelli DJ (1999) Do chimpanzees tailor their gestural signals to fit the attentional states of others? Anim Cogn 2:207–214. https://doi.org/10.1007/s100710050041
Tomasello M (2010) Origins of human communication. MIT Press, Cambridge, MA; London
Tomasello M, Call J (1997) Primate cognition. Oxford University Press
Tomasello M, Carpenter M, Call J, Behne T, Moll H (2005) Understanding and sharing intentions: the origins of cultural cognition. Behav Brain Sci 28(05). https://doi.org/10.1017/S0140525X05000129
Tomasello M, Hare B, Agnetta B (1999) Chimpanzees, *pan troglodytes*, follow gaze direction geometrically. Anim Behav 58:769–777. https://doi.org/10.1006/anbe.1999.1192
Topál J, Gergely G, Erdőhegyi Á, Csibra G, Miklósi Á (2009a) Differential sensitivity to human communication in dogs, wolves, and human infants. Science 325(5945):1269–1272. https://doi.org/10.1126/science.1176960
Topál J, Gergely G, Miklósi Á, Csibra G (2008) Infants' perseverative search errors are induced by pragmatic misinterpretation. Science 321:1831–1834. https://doi.org/10.1126/science.1161437

Topál, J., Miklósi, Á., Gácsi, M., Dóka, A., Pongrácz, P., Kubinyi, E., . . . Csányi, V. (2009b). The dog as a model for understanding human social behavior. In Advances in the study of behavior (Vol. 39, pp. 71–116): Academic Press

Townsend SW, Manser MB (2013) Functionally referential communication in mammals: the past, present and the future. Ethology 119(1):1–11. https://doi.org/10.1111/eth 12015

Udell MAR (2015) When dogs look back: inhibition of independent problem-solving behaviour in domestic dogs (*Canis lupus familiaris*) compared with wolves (*Canis lupus*). Biol Lett 11(9): 20150489. https://doi.org/10.1098/rsbl.2015.0489

Udell MAR, Dorey NR, Wynne CDL (2008) Wolves outperform dogs in following human social cues. Anim Behav 76(6):1767–1773. https://doi.org/10.1016/j.anbehav.2008.07.028

Udell MAR, Dorey NR, Wynne CDL (2010) What did domestication do to dogs? A new account of dogs' sensitivity to human actions. Biol Rev 85(2):327–345. https://doi.org/10.1111/j.1469-185X.2009.00104.x

Udell MAR, Dorey NR, Wynne CDL (2011) Can your dog read your mind? Understanding the causes of canine perspective taking. Learn Behav 39(4):289–302. https://doi.org/10.3758/s13420-011-0034-6

Vasconcellos A, Virányi Z, Range F, Ades C, Scheidegger JK, Möstl E, Kotrschal K (2016) Training reduces stress in human-socialised wolves to the same degree as in dogs. PLoS One 11(9):e0162389. https://doi.org/10.1371/journal.pone.0162389

Virányi Z, Gácsi M, Kubinyi E, Topál J, Belényi B, Ujfalussy D (2006) Comparative analysis of 'showing' behavior of human-reared dogs and wolves. Unpublished manuscript

Virányi Z, Gácsi M, Kubinyi E, Topál J, Belényi B, Ujfalussy D, Miklósi Á (2008) Comprehension of human pointing gestures in young human-reared wolves (*Canis lupus*) and dogs (*Canis familiaris*). Anim Cogn 11(3):373–387. https://doi.org/10.1007/s10071-007-0127-y

Virányi Z, Range F (2011) Evaluating the logic of perspective-taking experiments. Learn Behav 39(4):306–309. https://doi.org/10.3758/s13420-011-0040-8

Wallis LJ, Range F, Müller CA, Serisier S, Huber L, Virányi Z (2015) Training for eye contact modulates gaze following in dogs. Anim Behav 106:27–35. https://doi org/10.1016/j.anbehav.2015.04.020

Watson JS, Gergely G, Csanyi V, Topal J, Gacsi M, Sarkozi Z (2001) Distinguishing logic from Association in the Solution of an invisible displacement task by children (*Homo sapiens*) and dogs (*Canis familiaris*): using negation of disjunction. J Comp Psychol 115(3):219–226. https://doi.org/10.1037/0735-7036.115.3.219

Werhahn G, Virányi Z, Barrera G, Sommese A, Range F (2016) Wolves (*Canis lupus*) and dogs (*Canis familiaris*) differ in following human gaze into distant space but respond similar to their Packmates' gaze. J Comp Psychol 130(3):288–298. https://doi.org/10.1037/com0000036

Wilkinson A, Mandl I, Bugnyar T, Huber L (2010) Gaze following in the red-footed tortoise (*Geochelone carbonaria*). Anim Cogn 13(5):765–769. https://doi.org/10.1007/s10071-010-0320-2

Woodruff G, Premack D (1979) Intentional communication in the chimpanzee: the development of deception. Cognition 7:333–362. https://doi.org/10.1016/0010-0277(79)90021-0

Zucca P, Milos N, Vallortigara G (2007) Piagetian object permanence and its development in Eurasian jays (*Garrulus glandarius*). Anim Cogn 10(2):243–258. https://doi.org/10.1007/s10071-006-0063-2

Social Cognition Part II: Social Learning and Cooperation

8

Social learning and cooperation are at the adaptive core of complex social systems. Social learning allows individuals to profit from the experience of others and to establish group-specific traditions. Cooperation allows individuals to reach goals that are not easily achievable alone. Not surprisingly, much research has focused on trying to understand the underlying mechanisms of social learning and cooperation as well as the factors and conditions that may affect the kind of expression of these abilities. In this chapter, we will thus first review what is known about differences in social learning between wolves and dogs and then focus on cooperative interactions.

8.1 Social Learning

In social species, one of the most relevant sources of information for individuals is the behaviour and interactions of group mates. For example, at the age of about 6 months, wolves start to accompany their pack mates in hunting expeditions without engaging in the hunt themselves; it has been proposed that they learn at least some aspects of how to hunt large game from observing their parents and older siblings during these activities. When our first wolves were about 6–7 months old, we, indeed, observed that they started to 'copy' (at least it looked like it) how we opened the doors to the human parts of the enclosures, and we had to take additional measures to secure the doors; otherwise, they would have gone for a stroll through the kitchen and, possibly, the game park.

There are many situations in which individuals can learn from social partners, for example, where to find food and about potential threats, how to behave with other group members, whether or not to avoid humans, or how to open doors (Fig. 8.1). In general, it is assumed that learning from others is less costly than if each individual has to gather the necessary information independently, because it reduces the energy and risks that needs to be invested to collect the information by oneself. Thus, social learning allows individuals to profit from the experience of others and, also, to establish group-specific traditions.

F. Range, S. Marshall-Pescini, *Wolves and Dogs*, Fascinating Life Sciences,
https://doi.org/10.1007/978-3-030-98411-3_8

Fig. 8.1 Cherokee had observed us open the gate to the testing house in our still tentative enclosures when we started the Wolf Science Center in Ernstbrunn several times and then had a go. We were not at all happy with his success

Not surprisingly, many studies in the past decades have shown that animals do benefit from learning by observing others. Whereas ethologists and ecologists have been interested in understanding the contribution of social learning to adaptive variations of behaviour within and between groups in terms of tradition forming (reviewed in Laland 2004), psychologists have mainly been interested in experimentally identifying the mental processes responsible for such learning (reviewed in Heyes 1994). At one end of the continuum, local and stimulus enhancement occurs when the observer's attention is drawn to where the model is, or to what the model is acting on, thus increasing the probability of approaching or contacting the place or object compared to if no model would have been present (e.g. Spence 1996) and, thus, focuses on individual learning. This simple mechanism offers a plausible explanation as to how animals might learn where to find food and what to do with potential food items (Galef and Giraldeau 2001; Renevey et al. 2013). Local or stimulus enhancement is mainly perceptual or attentional in nature (Thorpe 1956) and does not require learning about new behaviours (imitation), object properties (affordance learning), or the goals of the demonstrator (emulation) (Whiten and Ham 1992; Zentall 2004). Imitation refers to instances when an observer learns to 'do an act by seeing it done' and copies the behaviours shown by the demonstrator (Thorndike 1898; Thorpe 1956). Imitation has received much attention in the studies of animal behaviours since it is thought to be an important mechanism in the development of human culture.

To date, we have run three experiments to compare the social learning abilities of dogs and wolves at the Wolf Science Center. In the first study, a local enhancement study, we tested whether wolves and dogs may learn from a demonstrator where to find food by paying attention to a place or location where a conspecific is showing a species-specific behaviour (Giraldeau 1997). In the second study, we expanded this approach and had a human demonstrator hide several chickens, so that the animals had to remember several locations, and in the third study, we investigated whether the animals would imitate the action of a demonstrator dog.

8.1.1 Local Enhancement

Eleven wolves and 14 dogs were tested in 5 different conditions on meadows close to their home enclosures that they were familiar with from previous walks. Being held on a leash by a hand-raiser at a defined starting location, they could observe a familiar pet dog that was involved in their hand-raising to either hide a food reward (1-day-old chick) in one of three possible locations (Fig. 8.2) or to 'fake hide' a food reward (Range and Virányi 2013). In the second condition, the demonstrator dog received a chick from its handler, but directly returned it to her, and went to the chosen location without a chick in the mouth. This condition was carried out to investigate if the subjects actually paid attention to the details of the demonstration and realized that actually no food was present in the latter, and thus looking for it would not make any sense. The same two conditions were carried out also with a familiar human demonstrator. Finally, in the control condition, no demonstration was carried out, but the food reward was hidden before the subject was led to the meadow; this would control for the possibility that animals could find the food reward based on olfaction alone. The sequence of conditions was randomized between individuals. Importantly, before all experimental trials and before the subjects were taken to the meadow, the demonstrator (pet dog or human) walked twice along the direct line from the starting position to each of the three potential hiding locations to control for possible odour effects in relation to the trail they left. Since we were also interested in the development of the subjects' abilities, we tested every animal in all five experimental conditions at the age of 4, 5, and 7 months.

We found that both dogs and wolves were more likely to find the chick after observing a demonstrator hiding the chick than if no demonstration had been provided, but that actually dogs outperformed wolves in the demonstration and the control conditions (Fig. 8.3). This suggests that while both clearly profited from the demonstration, dogs either were more motivated to find the chick or additionally used their sense of smell to find the chick. Since we found no difference in how fast wolves and dogs found the chicks when successful, the latter explanation of dogs relying more on smell than the wolves seems more likely. Interestingly, as the animals got older, they relied more on the visual demonstration than on olfactory cues since the likelihood of finding the chick increased with age in the demonstration condition, but decreased in the smell control.

While the dogs did not search for the chick as often after the 'fake hiding' condition as after the real dog demonstration, the wolves did not differentiate between the two different demonstrations. This indicates a difference between wolves and dogs in the attention they paid to the demonstrator dog (experimental vs. fake hiding condition).

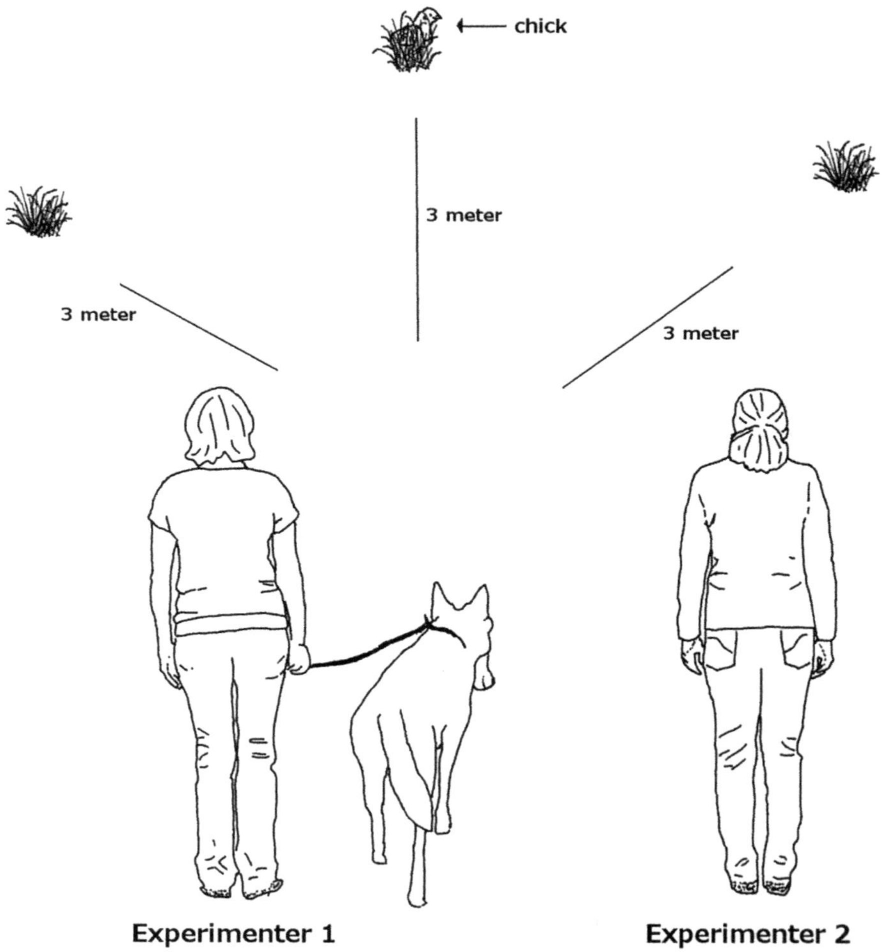

Fig. 8.2 Set-up of the local enhancement task with a dog demonstrator that just hid the chick in the assigned location. Figure adapted from Range and Virányi (2013). https://www.frontiersin.org/files/Articles/65843/fpsyg-04-00868-HTML/image_m/fpsyg-04-00868-g002.jpg; CC BY 3.0; http://creativecommons.org/licenses/by/3.0/

Interestingly, wolves performed significantly better with the human demonstrator hiding the chick than with the dog demonstrator (which was not the case for the dogs), and they successfully differentiated between the two human demonstrations. This surprising finding that the wolves were less successful in finding the chick after the dog demonstration than the human demonstration was partly explained by their lower interest in the former's demonstration, e.g. they paid significantly less attention to dogs than humans. One possible explanation for this may have been that the wolves recognized that the dog demonstrators (in contrast to our pack dogs that are

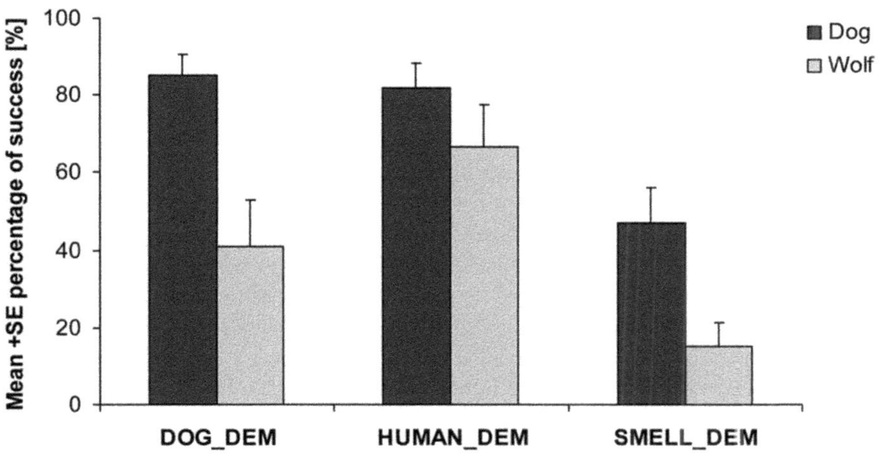

Fig. 8.3 Proportion of success finding the chick for wolves and dogs in the dog demonstrator (DOG_DEM), human demonstrator (HUMAN_DEM), and smell control (SMELL_CON) condition. Figure adapted from Range and Virányi (2013). https://doi.org/10.3389/fpsyg.2013.00868; CC BY 3.0; http://creativecommons.org/licenses/by/3.0/

used to this kind of food) did not like the chicks (we had to train the dogs intensively to convince them to carry the chicks—not because they wanted to eat them but because they wanted to spit them out). If the wolves, like some other animal species (e.g. tamarins) (Snowdon and Boe 2003), are sensitive to a display of disgust, they might have dismissed the value of the food rewards and lost interest in finding it. Alternatively, although both dogs and wolves come to associate humans with food at the Wolf Science Center, since this is often what we use to train and work with them, this link between food and humans in early development may have been stronger for wolves and dogs, leading them to pay more attention to the human than the dog demonstrator.

In another study, we also tested how well the animals learn about hiding locations by observing others. The idea of this experiment was built on the fact that wolves and dogs regularly cache surplus food and retrieve it at a later time. For example, from our own experiences, we know that dogs often cache bones in the garden and recover these stinky titbits sometimes even months later. Caching has also been observed in wolves both in the wild and captivity (Mech 1999; Packard 2003a, b; Peterson and Ciucci 2003; Phillips et al. 1990). The most effective strategy to retrieve the food after an elapsed amount of time is to remember exactly where the cache is located using visual cues. This strategy requires spatial memory, which seems to be the case for most food-caching animals (reviewed in Clayton and Dickinson 1998; Shettleworth 2009). Moreover, apart from having to find their own caches to retrieve the food, food cacher often also face the risk of caches being emptied by other animals. Again, in case one is not dominant, the most effective method is to observe the caching individual and then retrieve the food later when the cacher has left, minimizing the risk of retributions. This ability is called observational spatial memory (OSM).

Our first study had indicated that dogs and also wolves can use information from a demonstrator when locating hidden food (Range and Virányi 2013; see also Heberlein and Turner 2009). Moreover, dogs also seem to be sensitive to potential competitors when retrieving food since they have been shown to approach forbidden food more easily when a human is present in the room and had his/her eyes closed rather than open (Call et al. 2003; Schwab and Huber 2006).

In our study, we aimed to explore if wolves and dogs perform similarly in an observational spatial memory task and how much information they can retain if the number of caches increases. The experiments were conducted in our testing enclosure which contained 131 bricks spread out at a distance of 1 m from each other. Every brick represented a possible hiding location (cache). During 6 sessions the individuals had to retrieve food items (4, 6, or 8 items) either after having seen a human hide the food (total of 18 sessions) or without the benefit of observing the hiding process (total of 18 sessions). All trials were randomized within and between individuals to control for learning effects.

In all sessions, the subject was called into the shifting system. Then, in order to increase the animal's motivation, the experimenter threw the number of chicks to be cached one by one to the floor in front of the animal, on the other side of the fence. The experimenter then entered the test enclosure and picked up the chicks one by one while the animal was watching and hid them in full view of the animal. In control sessions, chicks were hidden before the start of the trial in the absence of the subjects, and a visual barrier was installed on the shifting system to prevent the animals from looking into the test enclosure. The animals were then led into the shifting system, and the experimenter threw the respective number of chicks over the fence, which was still visible for the subject. The experimenter then entered the test enclosure and sat still for 2 min hiding the chicks that had been thrown over the fence in a bag. We refrained from the experimenter moving around in the test enclosure and hiding the chicks in the presence of the animals, since the subjects could potentially still hear the experimenter walking around. The experimenter then exited the test enclosure while the animal was released back into the part of the tunnel system without a visual barrier, and then released into the test enclosure, free to search for the caches.

The animals, especially the wolves, absolutely loved this experiment as judged by their eagerness to participate. One time, I saw the wolf Kaspar jumping over his pack mates Aragorn and Shima just to reach the slide first when called to participate in the study!

Observing the hiding process helped both wolves and dogs, since they found more caches in the test compared to the control condition and they were also faster. Wolves however found more caches than dogs in both conditions and were faster if eight chicks were hidden—which may again be an effect of persistency with the wolves not giving up easily (see Chap. 9) and/or might also be an effect of the wolves' better ability to represent numbers (see Chap. 10). Another aspect that is quite interesting is that they optimized their search pattern, travelling less in the later than in the earlier trials to retrieve the food rewards.

8.1.2 Imitation

In the third experiment, we were interested in whether the wolves and dogs would imitate an action demonstrated by a familiar pet dog, in order to open a puzzle box. We tested our animals at the age of 6 months since at this time wolves start to go hunting with their parents. The pet dogs we used as demonstrators have been present during the hand-raising of our wolf and dog study populations and were dominant over all subjects at the age of 6 months. Half of our wolves and dogs saw a familiar pet dog opening a novel box by pushing down a lever with her mouth, while the other half saw a familiar pet dog opening the same box by pressing the lever down using his paw (Fig. 8.4).

After each demonstration, the observers were led to the box on the leash and could retrieve a piece of food from the box opened by the demonstrator. When the animals had observed a total of six demonstrations, they were released, and they could attempt to access the closed box themselves. Interestingly, all of the 12 wolves, but only 4 out of the 15 dogs tested, managed to open the box at least once (Fig. 8.5). Moreover, only 2 of the 4 dogs that opened the box managed more than once, whereas 10 of the 12 wolves opened the box more than once. This would suggest that, differently from dogs, the wolves had learned something causal about opening the box the first time. Geronimo, one of the wolves, only opened the box once and then showed no further interest in it, keeping busy doing other things in the room. The second wolf that opened the box only once, Wamblee, got visibly quite scared when the lid of the box popped open, and he no longer even approached the box.

Most importantly, we found that the wolves significantly matched the demonstrated action when manipulating the box for the first time, whereas dogs used the two methods pretty much at random. Moreover, when actually opening the box the first time, 9 of the 12 wolves matched the observed method, while none of

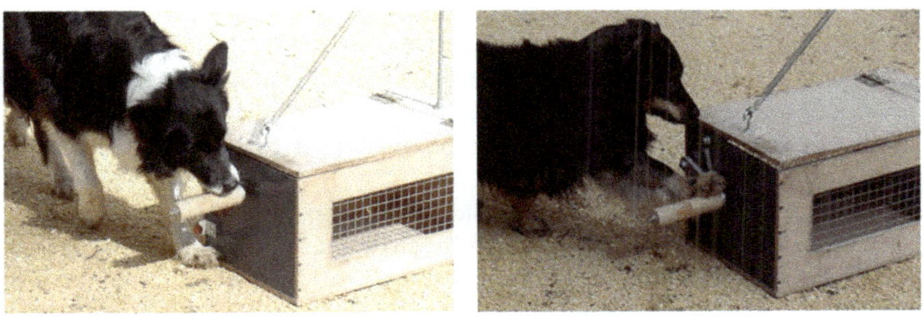

Fig. 8.4 The two dog demonstrators opening the box either with the mouth or with the paw. Figure adapted from Range and Virányi (2014). https://doi.org/10.1371/journal.pone.0086559; CC BY 4.0 https://creativecommons.org/licenses/by/4.0/

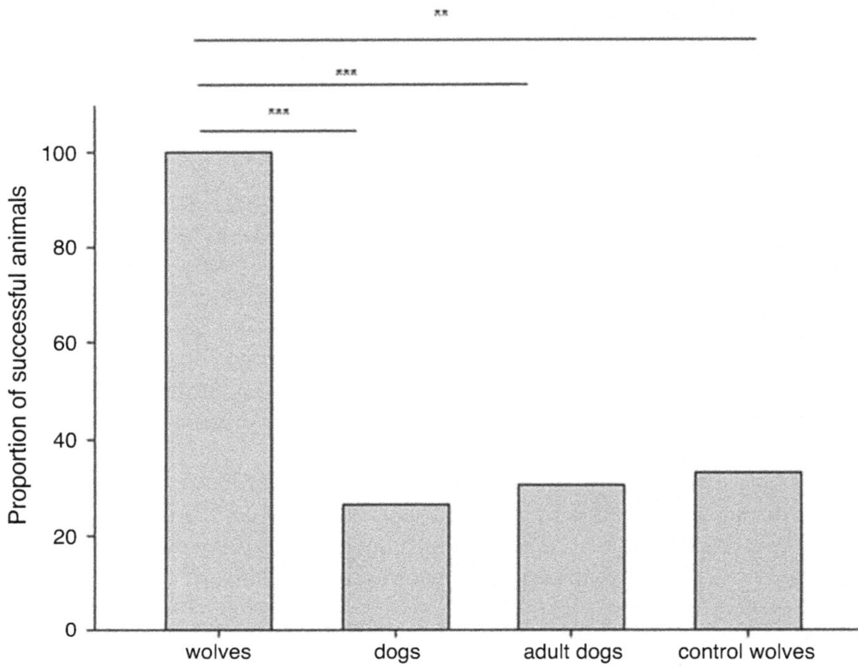

Fig. 8.5 The graph depicts the proportion of animals that successfully opened the box in their first trials in the four test groups (wolf and dog test groups, adult dog control group, wolf control group). Figure adapted from Range and Virányi (2014). https://doi.org/10.1371/journal.pone.0086559; CC BY 4.0 https://creativecommons.org/licenses/by/4.0/

the dogs did. These results suggest that while the wolves imitated the demonstrators' actions in this task, dogs did not.

There are several potential explanations for these results, some of which we have tried to rule out. First, based on the higher manipulative tendency and persistence of wolves (see Chap. 9), wolves might be more motivated to interact with the box leading them to be also more successful (although this would still not explain why they matched the demonstrated actions more than the dogs). However, we found no indication that the dogs were less motivated to do something with the box since they approached and started to manipulate it as quickly as the wolves and actually most wolves had opened the box already at a time when the dogs were still manipulating the box.

Second, we ruled out that the difference between wolves and dogs arose due to differential developmental pathways (e.g. wolves develop faster in terms of their sensory and motor abilities than dogs; see Lord 2013) by retesting 13 of our dogs when they were at least 9 months older. In fact, at this age, the dogs did not perform any better than when they were younger, and again only four opened the box with only one being successful in the first and second round (this was Binti, who was successful only once at 6 months). An independent study with pet dogs conducted at the Clever Dog Lab further confirmed that even older dogs are rather unsuccessful in opening the box after having seen a conspecific demonstration (Range et al. 2009a).

A third possibility is that the wolves might have a better causal understanding (Lampe et al. 2017; Chap. 10), which could explain their higher success rate in this task. To evaluate this possibility, we tested an additional six task-naïve wolves (three 6-month-old animals and three adult animals). These animals were also allowed to take food from the box after it was opened by a familiar pet dog, but crucially, they did not see the actions the dog performed on the box. We achieved this by putting a barrier between the observer and the demonstrator dog so that the wolf saw the demonstrator dog moving towards the box, then heard some sounds of opening the box, and was then allowed to go himself to retrieve the food from the box by going around the barrier. Only one of the six wolves opened the box by using the lever, which clearly shows that seeing the full demonstration considerably helped the wolves to solve this problem in the first attempt (Fig. 8.5). However, it is still possible that with the success and then additional attempts to solve the problem, the wolves would also achieve some causal understanding of the opening mechanism.

Fourth, it has been proposed that instead of trying to solve problems by themselves, dogs often turn towards humans by approaching them or looking back at them to enlist their help (see Chap. 7). Interestingly, several of the dogs approached the human who was filming the experiment, and several of them also went to the dog demonstrator, greeting them submissively, suggesting they may be seeking help or support from the individuals in the room. However, no matter why they approached the human or demonstrator in that situation, they did so only after they had tried for quite some time to open the box by themselves, and actually the wolves managed to open the box before the dogs even approached the experimenter or dog demonstrator.

Finally, given our results regarding the wolf-dog difference in their tolerance of conspecifics in a food context, it is possible that the dogs perceived the box as 'belonging' to the demonstrator and thus avoided or were more reluctant than the wolves in interacting with the box leading to a less successful performance. The behaviour of some of the dogs—approaching the demonstrator in a very submissive way—might support this notion. However, it is worth noting that the dogs all went to the box *before* interacting with the demonstrator and the latency to interact with the box did not differ between wolves and dogs. Nevertheless, running another imitation study where the demonstrator and the humans are removed from the room before the subject can try to get access to the hidden reward would provide further insights.

8.1.3 Summary Social Learning

To sum up, the three studies comparing wolves and dogs in regard to social learning showed that both wolves and dogs perform somewhat similarly in the local enhancement and the observational memory task. But when it comes to paying attention to the exact details of a demonstration and to imitate an action, wolves clearly outperform dogs. Whether these differences indeed stem from differences in cognitive abilities or rather from differences in general purpose mechanism like attention, working memory, inhibition, or motivation needs still to be explored.

8.2 Cooperation

Whereas in social learning, animals profit from watching others solving various challenges, in cooperative interactions the animals need to team up to work together to solve the task at hand. However, for both abilities, the animals need to pay close attention to the 'partner' and need to carefully choose whom to copy or who to cooperate with.

The frequency and complexity of cooperating with others is a hallmark of human society and likely exceptional, if not unique (Melis et al. 2006a, b). We strongly depend on the ability to flexibly cooperate with others, even non-kin, to achieve goals and to attain resources (Melis and Semmann 2010). For example, scientists cooperate with other scientists they have never seen before on the other side of the globe, nations build spaceships together, or we cooperate to take care of each other's children. While complex forms of cooperation can also be observed in non-human organisms (see Melis and Semmann 2010), the phylogenetic origin of cooperation is unclear and remains one of the greatest challenges in evolutionary and cognitive research.

Cooperation is defined as an interaction between at least two individuals that results in net benefits for all the individuals involved (Bergmuller et al. 2007). However, in order to fully explain a certain behaviour, one has to investigate four distinguishable but complementary levels: phylogeny, ontogeny, mechanism, and adaptive value (Tinbergen 1963). For example, the evolution of cooperation is functionally explained in terms of mutualism, kin selection, or reciprocal altruism (Hamilton 1964; Trivers 1971). While in mutualism the behaviour that benefits others also benefits the actor so that he/she can potentially learn about these benefits (e.g. cooperative hunting), in regard to kin selection and reciprocal altruism, the return benefits, i.e. positive consequences for the performer and/or its kin, may be quite obscure or delayed (de Waal and Suchak 2010). Consequently, it is improbable that return benefits play a motivational role in triggering and maintaining cooperative behaviour, which is likely to be mediated by other proximate processes. These may include mental dispositions which rely on cooperation activating brain reward systems. In other words, ultimate level explanations are needed to explain the selection pressures which led to general 'cooperative attitudes' (proximate mechanisms). This is a nice example that it needs the interaction of both ultimate and proximate levels to explain what is going on. Only in the last decades have researchers started investigating the ontogeny of cooperative skills and the proximate mechanisms (including mental) they require.

To date, mainly cognitive requirements, such as numerical discrimination and score keeping, have been studied in animals when investigating proximate mechanism of cooperation (Stevens and Hauser 2004). However, the cognitive processes of cooperation are functionally embedded within the emotional systems and thus should be investigated together. Several studies have revealed that in humans, intrinsic reinforcement and affects are associated with triggering, as well as maintaining and regulating, cooperation (Frank 1988; Rachlin 2002). These and other results have led to the suggestion that humans' exceptional success as a species may be based not only on our unique cognitive abilities but also emotional attitudes that ensure cooperative motivation (Tomasello and Carpenter 2007; Tomasello et al. 2005).

In the last decade, we have investigated several of the proximate mechanisms that might allow wolves and dogs to successfully cooperate with each other and, potentially, with humans. Prosocial behaviours (voluntary, 'friendly' actions that benefit another) are thought to be important in triggering cooperation (Burkart et al. 2014; Gintis et al. 2005): often an individual has to make the first step to cooperate while it does not yet know if the other one will return the favour. On the other hand, sensitivity towards inequitable outcomes after a cooperation event is thought to stabilize cooperation by allowing partners to monitor whether their own outcome is acceptable in relation to the partner's, or whether finding a new cooperation partner is required. Moreover, cooperation with others requires some coordination and potentially the understanding that you need a partner to be successful and what the individual roles in this cooperation should be.

8.2.1 Prosocial Behaviour

Prosocial behaviour is defined as any behaviour performed by one individual to alleviate another's need or improve its welfare (Silk 2007). Typically, the distinction between an altruistic and a prosocial behaviour has been based on the fact that whereas prosociality involves no cost to the actor, altruism does. For instance, 'helping' behaviour, which implies a cost to the actor, would be considered not just prosocial (in that it alleviates another's need or improves its welfare) but also an altruistic act.

Prosocial behaviours are considered to be the foundation of human large-scale cooperation (Burkart et al. 2014; Gintis et al. 2005). At the proximate level, two main mechanisms have been suggested to be involved in prosocial behaviour: empathy, i.e. the capacity to recognize, understand, and share the others' emotional state, and sympathy, i.e. the ability to be concerned for others (Batson et al. 1981; Eisenberg and Miller 1987). That prosocial behaviour is indeed an important basis for cooperation is supported by a few studies. First, it has been shown that if empathy is experimentally increased by nasal administration of oxytocin, people are more generous (Zak et al. 2005). Second, in humans, people feel really good after helping others (Steger et al. 2008), which is supported by an activation of reward-related brain areas (Harbaugh et al. 2007). So it is emotionally rewarding to actually help others, which is likely based on empathy-related mechanisms, since it has been shown in brain imaging studies that there is an overlap between the substrates of cognitive empathy of personal feeling states and the situation of the other (Preston et al. 2007; see also Carr et al. 2003). Accordingly, supporting early theoretical considerations (Trivers 1971), several lines of empirical research suggest that simple forms of empathy are important to trigger prosociality especially between non-kin (Axelrod 1984), particularly in contexts which cannot be understood in terms of long-term self-interest or anticipated reciprocity (Batson and Ahmad 2001; Batson et al. 1995).

Indeed a number of authors have suggested that the same mechanisms underlie prosocial behaviours in non-human species as well (de Waal 2008; Preston and de Waal 2002). It is thought that empathy evolved through kin selection and reciprocal altruism as the proximate mechanism that ensures helping vulnerable offspring within a social group (Goetz et al. 2010). Accordingly, prosocial tendencies are strongest towards familiar and bonded partners such as kin and group mates, in both human (Maner and Gailliot 2007; O'Gorman et al. 2005) and non-human primates (Preston and de Waal 2002) as well as in rodents (Langford et al. 2006; Rice and Gainer 1962) and dogs (Quervel-Chaumette et al. 2015). However, some authors have argued that what may appear as prosocial behaviours are simply a product of a higher motivation for sociality (or social tolerance; Yamamoto et al. 2009), or a strategy adopted to avoid harassment from conspecifics (Gilby 2006; Stevens 2004). Moreover, especially proactive prosociality should be supported by sophisticated cognitive capacities such as cultural learning, theory of mind, perspective taking, and moral judgements, which are thought to be present only in humans (e.g. Silk et al. 2005).

Two experimental paradigms that were devised to investigate the potential presence, prevalence, and the underlying mechanisms of active prosocial tendencies in non-human primates have been adapted to dogs and wolves:

1) The so-called prosocial choice test (PCT) in which animals have a choice to either benefit just themselves or both themselves and a partner or, in some simpler versions, decide whether they would give food to their partner or not, in neither case receiving any benefits themselves.
2) The more naturalistic paradigm, the 'food-sharing' paradigm, in which an individual is given the possibility of sharing food with his/her conspecifics. While the prosocial choice tests can be considered to be showing active prosocial behaviour, allowing another one to share the resource would be considered as more passive prosocial behaviour and can be driven rather by a high social tolerance than prosocial attitudes.

8.2.1.1 The Prosocial Choice Test (PCT)

In this test, subjects are typically given a choice between two reward combinations, one of which delivers a food item to the subject and their partner (prosocial choice) and the other which rewards only the subject (selfish choice; Colman et al. 1969). However, a previous pilot experiment at the Clever Dog Lab in Vienna revealed that this set-up is probably too complicated at least for dogs, since they did not learn to differentiate between the two options even via an intensive training phase.

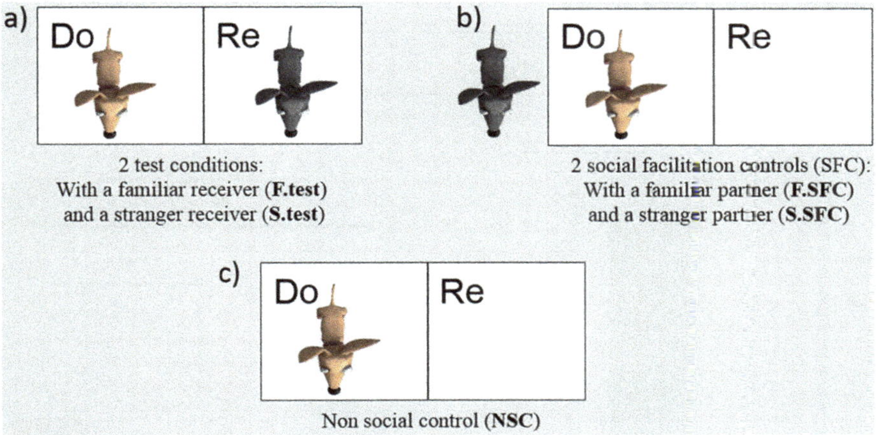

Fig. 8.6 Location of the donor (Do) and receiver (Re) dog in each condition a) social partner test, b) social facilitation control, and c) non-social control. Figure adapted from Quervel-Chaumette et al. (2015). https://www.nature.com/articles/srep18102/; CC BY 4.0; http://creativecommons.org/licenses/by/4.0/

Accordingly, we simplified the procedure, giving the pet dogs the choice between choosing to give a reward to the partner only or not at all. They themselves never received a reward (see also Burkart et al. 2007). Pet dogs were then successfully tested using a token choice task (Dale et al. 2016) and bar-pulling (Quervel-Chaumette et al. 2015) version of this test with a familiar conspecific partner and an unknown conspecific partner. In contrast to previous studies, we tested the animals not just in the basic control condition, where no partner was present, but also in social facilitation controls. This latter control condition is crucial for such experiments since it controls for the effect that the mere presence of a conspecific may have on the individual's motivation to engage in certain behaviours, either motivating or inhibiting them, rather than a prosocial attitude of the animal (Marshall-Pescini et al. 2016). This effect has been termed 'social facilitation'. In our view, this social facilitation control is more relevant than a non-social control condition, where no partner is present during the test (Fig. 8.6)

We found that the pet dogs gave more food in the social partner condition than in the control conditions in the bar-pulling experiment—but only with the familiar partner (Fig. 8.7; Quervel-Chaumette et al. 2015). In the stranger condition, the pet dogs actually withheld food and gave more food in the control condition than in the social partner test condition! However, in the token choice experiment, while the dogs gave more food to the familiar partner, and less food to the stranger in the social partner test conditions than in the no-social control condition, there was no significant difference in comparison with the social facilitation controls, indicating that the presence of the partner was sufficient to trigger or inhibit pressing the giving token (Dale et al. 2016).

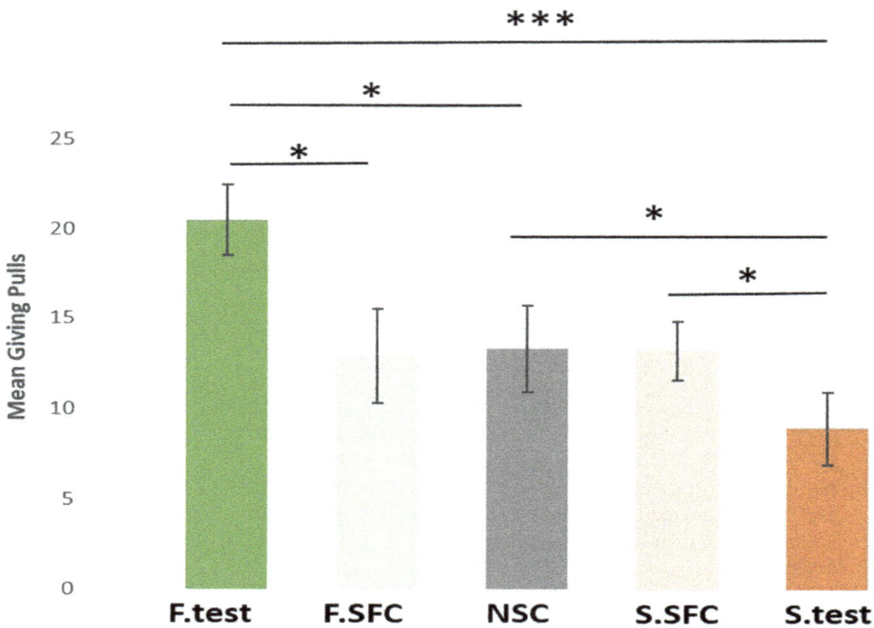

Fig. 8.7 Mean number of giving pulls (mean ± s.e.m., $n = 16$ individuals) in which donors pulled the giving (0/1) tray across conditions (*$p < 0.05$, **$p < 0.01$, ***$p < 0.001$). F. test = familiar partner test condition; F.SFC = familiar social control condition; NSC = non-social control condition; S.SFC = stranger social control condition; S. test = stranger partner test condition. Figure adapted from Quervel-Chaumette et al. (2015). https://www.nature.com/articles/srep18102/;

To compare prosocial behaviours between wolves and dogs and to investigate whether the responses are consistent across different tasks, we used three different prosocial choice paradigms at the Wolf Science Center: a touchscreen version, a location choice paradigm, and a group-service paradigm.

The Touchscreen Version

In the current task, we used two adjacent rooms with a transparent door in between them. In one room the subject could work on the touchscreen, whereas in the other, the food was delivered. This 'receiver' room was either empty or housed a conspecific partner, depending on the condition (Fig. 8.8).

Subjects were first trained to press a 'giving' symbol with their nose on the touchscreen, which delivered a reward to the adjacent receiver enclosure, over a 'control' symbol, which provided no reward. The wolves and dogs were familiar with the use of a touchscreen apparatus prior to the study, and thus the training did not pose any difficulty. During this training phase, the transparent door to the receiver enclosure was open, allowing the animals to go over and collect the reward after a 'giving' choice. During the test conditions, a pack member or a non-pack

Fig. 8.8 Set-up of the test condition. The subject could choose on the touchscreen whether to give a reward or not (**a**). After a 'giving' choice by the subject on the touchscreen (right), the partner (left) was rewarded (**b**). A transparent door between the two enclosures allowed the subject to see whether a partner was rewarded for their actions or not, as well as the identity of the partner. Figure adapted from Dale et al. (2019). https://doi.org/10.1371/journal.pone.0215444; CC BY 4.0; http://creativecommons.org/licenses/by/4.0/

member was positioned in the receiver enclosure, and the see-through door was closed so that the subjects no longer had access to the rewards they produced, but instead the respective partner would be able to eat it. In the control conditions, instead of being placed in the 'receiver room', the potential receiver was placed in another room (still next to the subject's touchscreen room), where they no longer had access to the reward but could be seen by the individual operating the touchscreen. In all four conditions, we then measured in how many trials the subjects were pressing the 'giving' symbol on the touchscreen given that the rewards were devoured by an affiliative partner (pack mate) or an outsider.

Since the subjects themselves were never rewarded during these tests, between each condition, we ran *motivation sessions* in order to ensure that the motivation to work on the touchscreen was similar prior to each condition. These sessions were identical to the training sessions, whereby the subject had access to the receiver enclosure, allowing them to retrieve generated rewards. Moreover, to ensure that on any given day the subjects were similarly motivated to work, after each test, we presented them with six *knowledge-probe trials*, where touching the giving symbol now delivered food to their own room while the partner remained in their original locations.

It turned out that overall the wolves were quite willing to give some food to their pack mates, but not to their non-pack members, and that this response was not

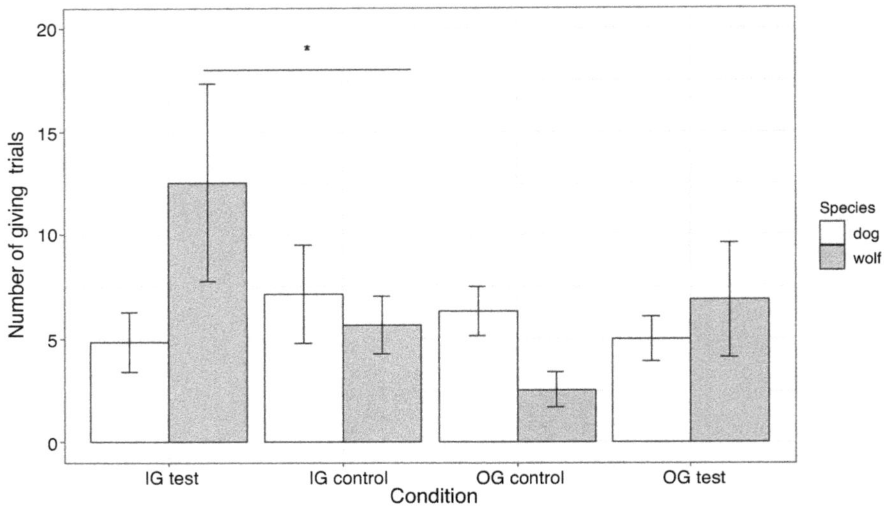

Fig. 8.9 Number of giving trials performed by wolves and dogs in the test and respective social facilitation control conditions IG = in-group, OG = out-group. Test = partner had access to the reward, control = partner was present but had no access to the reward. Figure adapted from Dale et al. (2019). https://doi.org/10.1371/journal.pone.0215444; CC BY 4.0; http://creativecommons.org/licenses/by/4.0/

triggered by social facilitation. In contrast to the wolves, the dogs showed a session effect, giving less and less food over sessions, independent of the partner or condition, thus showing no prosocial behaviour (see Fig. 8.9). Importantly for the comparison of the wolves and the dogs, all subjects performed all six knowledge-probe trials at the end of each session demonstrating that the subjects did not stop working on the touchscreen due to distraction or stress from the presence of a partner, and they were willing to resume working immediately when they could receive the rewards themselves. Additionally, the dogs performed well in the motivation sessions—suggesting that they were willing to keep working when the rewards were for themselves (Fig. 8.9).

The performance of the Wolf Science Center dogs differs from the results found with pet dogs at the Clever Dog Lab, who were quite willing to give food to their familiar partner but not to a stranger (Dale et al. 2016), similarly to our wolves. Various explanations could account for this difference including (1) the different socialization/training history of the animals and (2) the paradigms we used which differed somewhat. The pet dogs were studied using a bar-pull paradigm, where the animals had to pull in a table to provide food for the partner, and a token choice paradigm, where dogs had the option to touch a wooden token with their nose that delivered a reward to an adjacent receiver enclosure (Dale et al. 2016). The latter task was actually very similar to the touchscreen task we used at the Wolf Science Center. Interestingly, while the pet dogs gave more food to a familiar partner in both tasks, they differed in the two tasks in regard to their behaviour in the social facilitation conditions as explained above. These differences in the results show that even minor changes in the test paradigms might result in somewhat different results even if the same population of animals is used. On one hand, this is a concern, since the responses of the animals seem often very specific to certain test situations; to really understand on what base they make their decisions, we need to ask the same question using different methodological approaches. On the other hand, such sensitivity to the exact test conditions gives us the opportunity to ask what differed between the experimental set-ups and come up with different hypotheses that can be tested. In the end, this should allow us to understand when animals make which decisions and what the possible constraints they face are.

In regard to the studies mentioned above, we think that the token choice task might have been cognitively more challenging than the simpler bar pull. More specifically, in the token choice task, the food was not visible during the choice phase, requiring the animals to keep out-of-sight food in mind, whereas in the bar-pulling task, being able to track the food from the moment it was placed on the tray to the mouth of the partner (or the empty enclosure) may have helped dogs to discriminate between the conditions in which the partner obtained the treat or not. Moreover, in the token choice task, the dogs had to learn an abstract association between a symbol and the food reward (as opposed to the direct connection between food and shelf in the bar pull). This greater cognitive demand was also apparent in the training phase, where the dogs learned to choose the 'giving' option faster in the bar pull task than the token choice task. Given that the token exchange was quite similar in these aspects to the touchscreen task that we conducted at the Wolf Science Center, these aspects might also explain the differences between wolves and dogs, given that wolves seem to be better at tracking food quantities when not in sight (Range et al. 2014) and have better causal understanding (Lampe et al. 2017; see also Chap. 10).

To investigate whether, indeed, the wolves are more prosocial than the dogs and this is not just an artefact of this one task, we ran two more studies using different methodologies.

Location Choice

Hernandez-Lallement et al. (2015) designed this task to study prosocial behaviour in rats. They devised a double T-maze, where the rats could choose between two compartments, one in which they themselves and their partner received a reward and one where only they themselves received a reward (Fig. 8.10). In the experiment, the rats made prosocial choices on average 55% of the time in this test condition. As a control, an inanimate toy rat replaced the living partner, and the actors chose the prosocial compartment significantly more in the partner than in the toy condition, indicating that the actors understood the difference between these two conditions.

The general procedure we used for the wolves and dogs remained mostly the same as with the rats. Ahead of every test (whether for the experimental or control condition), the animals started with two exposure sessions allowing the actor to

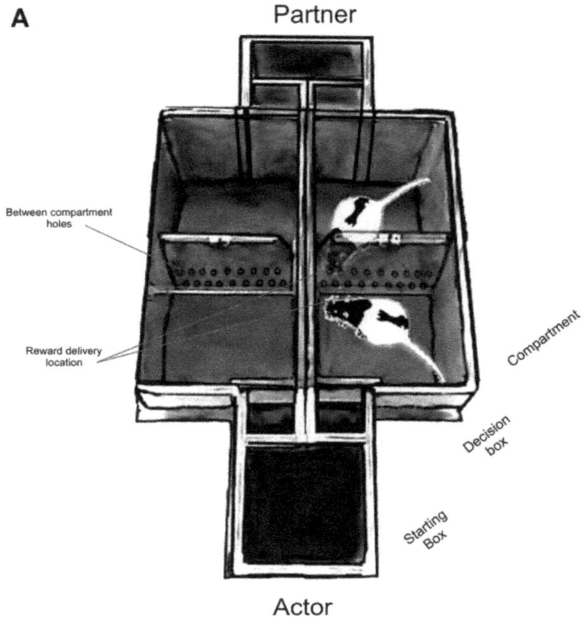

Fig. 8.10 Experimental set-up. Figure adapted from Hernandez-Lallement et al. (2015). https://doi.org/10.3389/fnins.2014.00443; CC BY 4.0

become familiar with task contingencies, i.e. which compartment delivered the prosocial reward distribution and which resulted in a non-prosocial choice. In these trials, actors were 'forced' to enter both compartments in a pseudo-randomized order by opening the door of only one compartment per trial. The procedure of the test condition was the same as that of the exposure sessions, except that both compartments were simultaneously opened before each trial so that the actor could decide to go into either the prosocial compartment or the selfish compartment.

With the wolves and dogs being much bigger than the rats, we had to use a much larger dimension of a T-maze. This also meant that we had to reduce the number of trials, since our animals could not be asked to run so much in each session. Accordingly, instead of 105–150 free-choice trials per condition for the rats, we ran only 72 free-choice trials per condition. Another change we made is that in the control condition, instead of using a toy rat, in our study the partner was present but was never rewarded, even when the actor entered the prosocial compartment. Moreover, instead of 48 rats, we could only test 7 animals as subjects (4 dogs and 3 wolves). We started off by testing more animals, but actually the first subjects based their choice of compartment on the movement of the experimental helper that sequentially opened the slides (doors) before the subjects were released (even though the experimental helpers used our pulley system to open the doors and we placed covers along the fence between the animals and humans, and thus were hardly seen). Thus, we had to disregard these animals and start anew, slightly changing the procedure by including two helpers who opened both compartments simultaneously.

In the end, we did not find that our animals made prosocial choices in this task except one wolf, Yukon. It is interesting that it was Yukon who showed prosocial behaviour as in the touchscreen task she was also prosocial, but only towards the out-group member, suggesting that she can be prosocial if she chooses to be.

Although we had thought that this is a rather simple task for the wolves and dogs it again turned out to be rather challenging for them. One problem in comparison with the rat study was that we just could not run as many trials as they did, which might have constrained the ability of our animals to learn the contingencies of the task.

Group-Service Paradigm

Finally, we used a similar group-service seesaw paradigm as implemented by Horn et al. (2016) to assess prosocial behaviour in our animals.

The seesaw construction is positioned in- and outside the animals' enclosure, so that if an animal steps on the platform inside the enclosure (Fig. 8.11), the apparatus tips down and food is delivered through a tube so that another group member can eat it (Aragorn is doing that in the back of the picture). If Shima steps off the platform, the seesaw tips back up and the food, if not eaten, rolls back into the tube—thus Shima cannot step on the apparatus and then move over to get the food!

This paradigm has previously been used for primates and magpies (Burkart and van Schaik 2013; Horn et al. 2016) and allows testing prosocial behaviours in a group setting. As in the previous tasks, the animal has a choice of delivering a treat to another group member, while the donor itself does not get any reward for itself. In a

Fig. 8.11 Seesaw apparatus with Shima and Aragorn (photo taken by Rooobert Bayer)

control condition, access to the treats is blocked, allowing for testing for the animal's comprehension of the task, e.g. the animals should stop using the apparatus when no more food is delivered. Additionally, the apparatus can be fixed in such a way that animals can access the food without help from their pack members, which has been used to calculate the evenness of the distribution of taken treats within each group (Pielou 1977; Burkart and van Schaik 2013) as a measure of tolerance.

Burkart and van Schaik (2013) invented this paradigm to assess social tolerance and prosocial behaviour in three non-human primate species that differ in their breeding system: in Japanese macaque females raise their infant alone, in Capuchin monkey females are sometimes supported by the fathers, and common marmosets are cooperative breeders. They found high social tolerance in the Capuchin monkeys and common marmosets, but a lower value for the independently breeding Japanese macaques. Only the common marmosets showed prosocial behaviour by delivering food to group members in 60.9% of all trials with all group members operating the apparatus (Burkart and van Schaik 2013). The same paradigm with a few species-specific adjustments has been applied in two groups of the cooperatively breeding azure-winged magpie (Horn et al. 2016). As expected, both magpie groups showed high levels of social tolerance, comparable to the cooperatively breeding common marmosets (Horn et al. 2016).

The advantage of this apparatus is that it is relatively easy to understand and not much manipulation is necessary (big plus for the dogs!). We were really excited about this study, and still are—now you probably wonder what we found... well, well, we have not yet analysed the data! So look out for our future studies!

8.2.1.2 Food Sharing

Food sharing is universal among human cultures and has received considerable attention as a model for the evolution of altruistic behaviour (Gurven 2004). It is also a relatively common form of prosocial behaviour among animals; however, the frequency and modality of food sharing varies considerably among species (see Stevens and Gilby 2004, for a review).

Food sharing has been defined as the transfer of a defendable food item from one food-motivated individual to another. The level of prosociality during food sharing may be measured by the reactions of the food owner to approaches by others and the way in which food is transferred. Overall, positive reactions to approaches and tolerant sharing would reflect prosociality, whereas negative reactions and forced transfers would not. In line with this reasoning, a number of different behavioural categories of food sharing have been identified in the literature, the main ones being:

a) Unsolicited/spontaneous giving, i.e. active transfers (hand to hand; mouth to mouth; hand to mouth or vice versa) initiated by the owner in the absence of any form of request or begging by the recipient
b) Solicited sharing, i.e. active transfer of food by the owner following a request/ begging gesture from the recipient
c) Tolerated theft/scrounging, i.e. a passive transfer in which the owner allows the recipient to take food directly from his/her hand/mouth

d) Co-feeding, i.e. a passive transfer in which the owner allows the other to take food that is in direct proximity and could be monopolized

The tests we have carried out in terms of food sharing would fall into the categories of co-feeding and tolerated theft/scrounging, since the dominant animals have the possibility of monopolizing the food resource, but they may also allow others to co-feed.

Our various tolerance tests carried out to date either at the dyad or group level (Dale et al. 2017b; Range et al. 2015; see Chap. 5) have all revealed a similar picture: wolves were more tolerant than dogs, allowing subdominant animals to co-feed, and they escalated less often into severe aggression than dogs. Moreover, during the breeding season, both species and both sexes exhibit decreased aggression in a food-sharing context, and in dogs, females spend more time in proximity to the male s food and start eating sooner when in heat (Dale et al. 2017a). This latter effect was absent in wolves, which might be explained by the fact that wolves are generally more tolerant with conspecifics, allowing them to be closer to the food than dogs independently from the breeding status. This difference is in line with the hypothesis that promiscuity may promote a greater sensitivity to food-for-sex effects (Gwynne 1984; Jaeggi and Van Schaik 2011).

8.2.1.3 Summary Prosocial Studies

Overall the results from these different paradigms suggest that wolves are more prosocial than dogs when raised in a comparable way. While the wolves provided food towards pack mates in the touchscreen and group-service paradigms, the dogs did not show any prosocial tendencies. These results are supported by the more naturalistic food-sharing situations, where dogs are a lot less tolerant towards each other than the wolves.

8.2.2 Inequity Aversion

Inequity aversion is defined as partners resisting inequitable outcomes, and in humans it seems to be based on the simultaneous evaluation of costs and gains in comparison with those of a partner. The prevailing hypothesis regarding the ultimate function of inequity aversion suggests it as a mechanism that stabilizes cooperation among unrelated individuals (Brosnan 2011; Fehr and Schmidt 1999; Fehr and Schmidt 2006). Thus, if an individual realizes that it has not received the same outcome for the same effort in a cooperative interaction, it should react to this knowledge, either by finding a new partner or by punishment of non-cooperators. Some evidence from experiments with humans indicates that active, costly punishment, or the threat thereof, suffices to achieve full cooperation (Fehr and Gächter 2000, 2002; Fowler 2005; Henrich et al. 2006; Sääksvuori et al. 2011). Inequity aversion functions as the underlying motivation driving such punishment (Dawes et al. 2007; Raihani and McAuliffe 2012).

In humans, reactions to inequity are linked to negative emotions such as anger (E. Fehr and Gachter 2002), which in turn may trigger the willingness of individuals to punish those not contributing properly to a cooperative interaction, even if it's costly. Moreover, in humans, shame, in the case of the individual, who is contributing less, may increase the individual's contribution to cooperative actions in the future (Bowles and Gintis 2003).

Several experiments carried out with non-human animals suggest that reacting negatively to an unequal reward distribution is not a uniquely human trait; non-human primates (Brosnan and de Waal 2003; Brosnan et al. 2006; Cronin and Snowdon 2008; Fletcher 2008; Melis et al. 2009; Neiworth et al. 2009), domestic dogs (see McGetrick and Range 2018, for a review), and rats (Oberliessen et al. 2016) all prefer an equitable distribution of reward.

Most studies have investigated inequity aversion using different versions of the so-called exchange task. In such a task, the experimenter asks for a certain action (e.g. giving back a token), first from the subject and then the same action from the partner. Depending on the condition, the experimenter rewards the animals either equally (e.g. with the same quantity and quality of food) or unequally (e.g. with different food types or quantities). In dogs, most studies have been conducted using the paw task, where pairs of dogs from the same household were seated next to each other while a human experimenter knelt before them asking the dogs, alternately, to give their paw (Fig. 8.12). Upon successfully performing, depending on the condition, either they both received the same reward (one piece of sausage each, *equity*

Fig. 8.12 Set-up of the paw task. The red bowl in front of the experimenter contains both reward types. The owner is standing behind the dogs. Figure adapted from Brucks et al. (2016). https://doi.org/10.1371/journal.pone.0153799; CC BY 4.0; http://creativecommons.org/licenses/by/4.0/)

condition) or one received a low-quality reward while the other received a high-quality reward (piece of sausage vs. kibble, *quality inequity condition*) or one received a reward while the other received nothing (sausage vs. nothing, *quantity inequity condition*). In addition to these social conditions, the animals were also tested individually, to see whether they would continue to give the paw in the absence of the partner even if they received nothing (*no-reward control*). This condition controls for the possibility that the animals just stop participating in the experiment if they do not receive any food themselves rather than reacting to the fact that their partner received some food while they did not (inequity). In all of these conditions, it was then recorded for how long the animals would continue to participate in the task and how often a command had to be repeated before they would comply as a measure of their willingness to cooperate.

To investigate if wolves and dogs at the Wolf Science Center react similarly, we started to run the same task with one dyad of wolves, Kaspar and Aragorn, as all of our wolves are well trained in various 'obedience' commands, including 'paw' from puppyhood on. We put up a small fence, fastened the leashes of the two wolves to two hooks in the wall behind them, and then had a trainer ask the wolves for their paws alternately. The trainer was sitting on a chair in front of the animals; the reward bowl was between her knees. We started with the *equity condition* with both animals receiving the same kind of food, and it went reasonably well, with both animals willing to give their paw to receive the rewards. However, already in this condition, where both were treated equally, the animals were very excited and moving around a lot, jumping up at the fence to look over (despite the fact that it was just a wire mesh fence). Moreover, having considerably larger paws than dogs and responding very enthusiastically, the trainer had quite sore arms after the test. Nevertheless, we also attempted the *quality inequity condition*, with the dominant wolf (Kaspar) receiving the lower-quality reward. It was clear very quickly that he was not happy with this reward distribution at all, and after threatening the trainer two times, we decided to stop the experiment there and then for evident safety reasons.

After this first attempt, we redesigned the experiment. Instead of asking the animals to give the paw to the trainer, we asked them to press a buzzer, with the trainer standing on the other side of a fence and requesting the press by pointing at the buzzer (Fig. 8.13). Importantly, we also put the two animals in separate enclosures, each of which was equipped with their own 'buzzer' apparatus, thereby avoiding the possibility of aggression over competition for the device between the two animals (Essler et al. 2017).

Similarly to the prior dog studies (see McGetrick and Range 2018, for review), wolves and dogs very quickly refused to continue to work in the presence of a rewarded partner if they themselves received nothing. This refusal was likely due to the presence of the partner, not the lack of reward, as they continued to work for longer when they were alone but still did not receive a reward. Moreover, in contrast to previous studies on pet dogs (Brucks et al. 2016; Range et al. 2009b), both wolves and dogs also completed fewer trials in the *quality inequity condition* (high- vs. low-quality reward) than in the *equity condition*, suggesting that the animals were also sensitive to the difference in type of rewards received (Fig. 8.14).

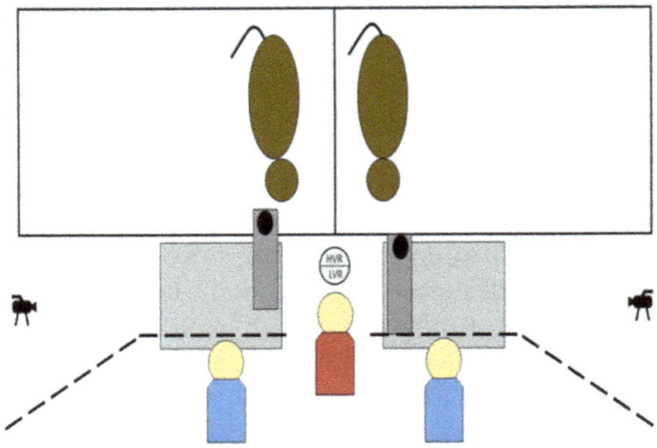

Fig. 8.13 Overhead view of the buzzer task set-up. Figure adapted from Essler et al. (2017). Copyright license number 5053080727612; reprinted with permission from Current Biology

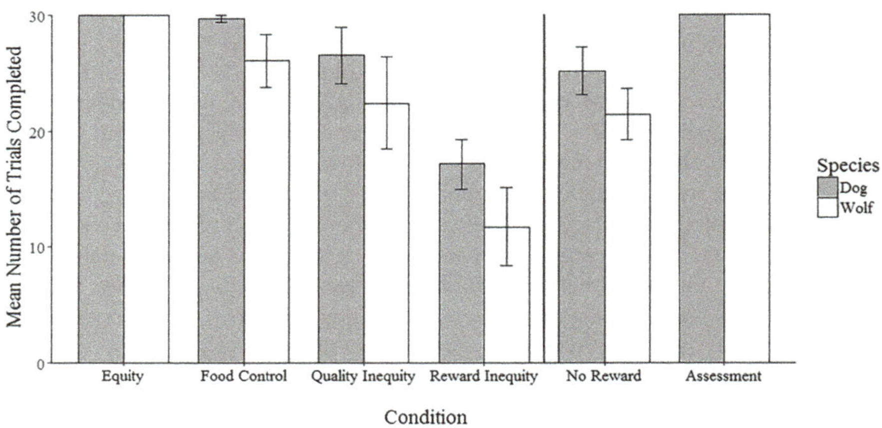

Fig. 8.14 Results of the inequity aversion study at the Wolf Science Center. Figure adapted from Essler et al. (2017). Copyright license number 5053080727612; reprinted with permission from Current Biology

Interestingly, in terms of the number of commands that needed to be given in the quality inequity condition compared to the *equity condition*, wolves and dogs differed. While the wolves needed to be prompted much more often to press the buzzer if their partner received a higher-quality reward than if both received the same, dogs did not, suggesting that wolves were more sensitive to the *unequal reward distribution* than the dogs. Moreover, dominant animals reacted more strongly to being treated worse than their subordinate partners than vice versa. This might be partly due to the fact that during normal pack life, it is the dominant individual that has the priority of access to the resources and can secure the better quality food (Clutton-Brock 1995; Mech and Boitani 2003), leading to a violation of expectation in the dominant partner if these roles are reversed.

Moreover, to investigate whether the inequity experienced by the subject directly affected subsequent interactions with its partner, as would be expected if inequity aversion does function to maintain cooperative partnerships, an 'interaction test' was presented to the animals immediately after specific test sessions (see Brucks et al. 2016, for this procedure used with pet dogs). In this follow-up test, both the subject and the partner were released in a large enclosure together and observed in their interactions with each other right after the social conditions. Indeed, like pet dogs, both wolves and pack-living dogs spent less time with their partner after the *reward inequity condition* than after the *equity condition*, suggesting that their experience with their partner influences future interactions. Moreover, in both the wolves and dogs, we found that rank influenced how fast they would approach their partner after the test. If the dominant animal was the subject in the *inequity condition*, it approached the subdominant faster after being treated unequally than when treated equally. This effect was absent if the subdominant animal was the subject. This again supports our results that dominant animals seem to react more strongly to being treated badly than their subordinate partners and that this might have consequences for further interactions.

This latter notion is further supported by results from the pet dogs (Brucks et al. 2016), where a food-sharing experiment was conducted right after the social conditions of the exchange task. The animals' experience in the exchange task influenced the co-feeding in the subsequent situation, with dogs being more likely to share if they were previously tested in the equity condition than any of the *unequal conditions* (i.e. reward inequity and quality inequity conditions). In addition, the duration of co-feeding was linked to the amount of experienced inequity, i.e. how long they continued to give the paw even if they did not get any rewards for doing it (some dogs stopped complying after only a few trials, whereas others continued for many more trials): dogs showed less co-feeding when they experienced the inequity over a longer time period.

Another context in which inequity has been discussed is play behaviour, which is often regarded as a cooperative interaction. Also, it has been suggested that individuals refrain from playing others who do not play in a 'symmetrical' way, i.e. take the loser or winning role equally (Bekoff 1974, 1977, 2001; see Chap. 5). However, so far there is little evidence that dogs adhere to the so-called '50:50' rule (Bauer and Smuts 2007; Ward et al. 2008). In a study carried out with 3–5-month-old wolf puppies and between mixed dyads of juvenile and adult wolves at the Wolf Science Center, we also found no evidence for the 50:50 rule at any time during our study period. However, dyads consisting of two puppies showed significantly more equal play than dyads consisting of one puppy and one adult. These results suggest that wolf puppies are more likely to play on equal terms with similarly aged partners. However, regardless of age pairings, roles in play largely mirrored those outside of the play context in terms of the dominance relationship (Essler et al. 2016; for an in-depth discussion of play and these studies, please see Chap. 5).

8.2.2.1 Summary

Our results suggest that wolves and dogs are sensitive towards unequal treatment. Moreover, being treated unequally appears to have consequences for subsequent interactions between partners, even more so when the unequal treatment is directed towards the more dominant of the two individuals. This latter point is theoretically significant since it means that being treated unequally may not carry the same weight for all animals in a social group, but rather an animal may perceive a situation as more or less acceptably inequitable depending on the relationship with a specific partner. Interestingly, in the play context, dogs and wolves do not seem to react negatively if they are mainly in the losing role, which could be due to the fact that their roles during play mirror their normal dominance relationship, without bearing the consequences of non-play interactions. Puppies, which may not have established a clear dominance relationship yet (see also Chap. 5 for the development of dominance), seem to keep the roles more at equal terms and may use play to establish dominance relationships.

8.2.3 Coordination in a Cooperative Problem-Solving Task

Probably the most popular method to test if, and to what degree, animals coordinate with each other in cooperative problem-solving tasks is the ‘loose string task’, in which two individuals have to simultaneously pull on two ends of a rope to pull a baited tray within reach. If only one end of the string is pulled, the other end becomes inaccessible, and the tray cannot be moved forward anymore, rendering the trial unsuccessful. Since the rope is clearly visible, the animals may, at least theoretically—if they have some means-end understanding—see and understand the effect their actions have when they pull too much on one end, which makes it a good task for a variety of species. This experiment has been used with a wide variety of species including chimpanzees (Hirata and Fuwa 2007; Melis et al., 2006a, b; Suchak et al. 2014); macaques (Molesti and Majolo 2016); ravens (Massen et al. 2015); rooks (Seed et al. 2008); elephants (Plotnik et al. 2011); grey parrots (Péron et al. 2011, b); and kea (Heaney et al. 2017), with many succeeding in solving the task after being initially trained individually to pull the tray forward by pulling both ends of the rope together. The beauty of the string-pulling task is that one can design all different kind of variants of the basic design, allowing us to explore the limits of the animals’ understanding of the ‘cooperative’ nature of the task. For example, a ‘two-apparatus’ condition (Drea and Carter, 2009) has been devised where the two animals need to coordinate in both ‘time and space’ to first go to one apparatus together and then move to the other one in order to get all rewards. Another condition that has been used in many different species is the ‘delay condition’, where the subject is released before the partner with the rationale being that if the former understands that it needs the partner to succeed, they should wait to pull the rope until its arrival at the apparatus. Some researchers have also used two different apparatuses, one of which needs a partner to be solved and the other one can be solved alone. This set-up allows for testing whether the animals realize that they need the partner in one

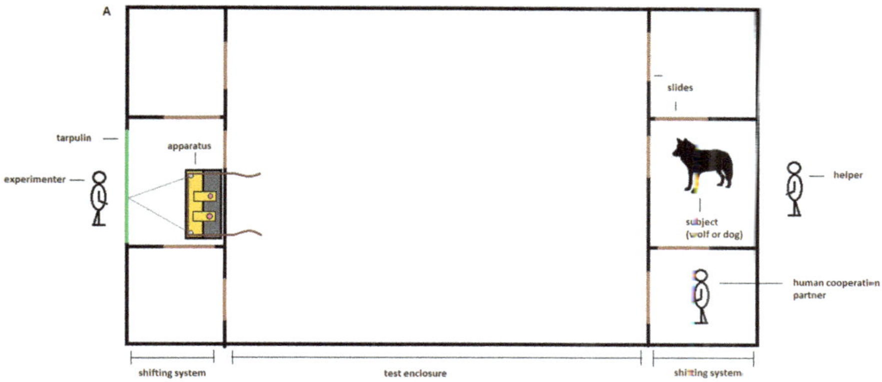

Fig. 8.15 Experimental set-up of the string-pulling studies (sketch by Corinna Kratz)

condition but not the other and whether they would go as far as to recruit a partner if they needed help.

The loose string paradigm seemed a very good opportunity to test wolves' and dogs' cooperative skills. While this paradigm had not been used with wolves, pet dogs had already been tested before we started with our studies. In that study, Ostojjic and Clayton (2014) tested working dogs together with a conspecific as well as a human partner. Like most other species, the pet dogs were all highly trained individually to pull both ends of the rope together, before being tested with separate ropes and conspecific partners. Actually, of the 29 dogs that started the training, only 11 dogs completed the tests—the others dropped out either because the dogs did not advance in the training sessions or because the owners did not have enough time. In the conspecific experiment, the dogs were successful in the *simultaneous condition* and waited up to 2 sec if the partner was delayed. With a human partner, while successful if the human arrived at the same time, the dogs did not wait the 15 s required for the human to arrive in the *delay condition*.

With these results in mind, we started our more than 4-year journey to investigate the coordination abilities of our animals in the loose string paradigm. All tests were performed in our test enclosures with the starting position of the animals/humans in the small compartments and the table(s) being set up at the other side of the enclosure so that the ropes were within reach but the apparatuses were placed behind the fence (Fig. 8.15). By pulling on the rope ends simultaneously, the tray would move forward, and the food would then become within reach of the animals' snout.

8.2.3.1 Animal-Animal Coordination

Since we were interested in whether our animals could also spontaneously figure out how to work together, we abandoned the training procedure in our first experiment and just confronted different dyads with the loose string apparatus (Marshall-Pescini et al. 2017). Seven wolf and eight dog dyads were compared in this version of the *simultaneous condition*, where both animals are released at the same time. The wolf

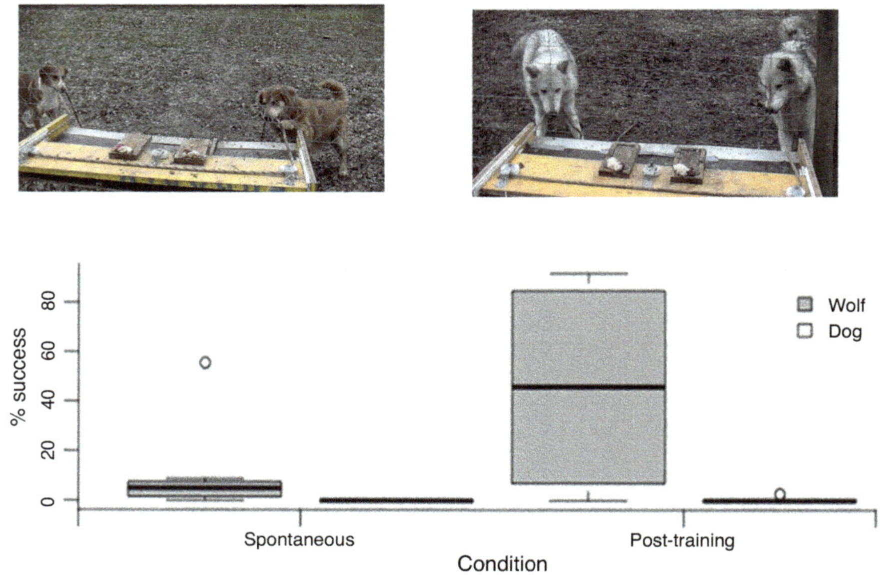

Fig. 8.16 Pictures: dogs and wolves working on the loose string paradigm (pictures by Camille Basin). Graph: success of wolves and dogs in the string-pulling task with conspecifics without training and after the classical training. Figure adapted from Marshall-Pescini et al. (2017)

and dog dyads tested were comparable in their level of experience and were either (a) completely task-naïve and had not been exposed to the apparatus beforehand or (b) one of the two animals had been tested previously with another pack member but had failed to solve the task. All animals had previous experience of pulling a string to obtain an attached piece of food from experiments during puppyhood. Five of the seven wolf dyads succeeded in at least one trial (success rate per dyad: 3–56% of trials), while none of the dog dyads succeeded (Fig. 8.16). Obviously, we were rather surprised by these results and decided that we would now go back to the classical pre-training procedure to see if that would help the unsuccessful animals to solve the task.

Four wolf dyads (three of which had been tested in the *simultaneous condition* without training and less than a 10% success rate) and six dog dyads (four of which had been tested in the spontaneous condition) were submitted to the classical training procedure. During the 2–3 months of training, they learned, using a shaping procedure, to solve the string-pulling task alone by taking both ends of the rope in their mouth and pulling. While in the beginning of the training the two ends of the ropes were intertwined, as the animals succeeded, the ropes were moved farther and farther apart up to a distance of 10 cm. Once they had successfully solved this training step by taking first one rope and then the other into their mouth and pulling, they were (re-)tested in the simultaneous condition together with their partner. This individual training apparently helped the wolves to solve the task, since three of the four wolf dyads succeeded in 14–92% of trials (Fig. 8.16). However, it only slightly increased the performance of our dogs: just two of the six dog dyads succeeded—but still only in a single trial each (3% of the trials)—suggesting that even with the training the dogs would not cooperate with each other in this paradigm despite the fact that they knew how to pull the rope.

We then continued to test just the wolves in more complex versions of the task to explore whether they would coordinate in time and space and whether they would wait for a delayed partner. All animals were tested with all others within theirs packs to see whether the identity of the partner would make a difference in their ability to coordinate their actions. Animals that had not been successful with one partner, but were then successful with another, were later retested with their original partner to see if the bad performance was just due to them not understanding the task, or whether it was rather affected by the combination of the two animals.

Twelve dyads were tested in the *two-apparatus condition*, where they first had to solve one and then the other string-pulling apparatus together. All animals succeeded to solve both apparatuses in 74% of the trials (range 20–100%). Interestingly, it mattered with whom they were paired: animals with a better affiliative relationship and a closer rank position outperformed the other dyads. In the *delay condition*, where one partner was released 10 sec earlier and had to wait for the partner to pull the rope, all animals except Shima performed rather well with 55–94% of successful trials if the partner was delayed by 10 sec. In these trials, the wolves often first ran over to the apparatus, but then looked back to their partner, who was still in the small compartment, clearly demonstrating that they realized that the partner had to be present so that they could solve the task. Some of the animals even ran back to the compartment, waiting there until the partner was released and then running over together to the apparatus. Shima instead just pulled the rope out or in later trials explored the testing enclosure.

Overall, we found these results exciting, as they show that wolves coordinate their actions in this loose string paradigm remarkably well. Behavioural observations supported the impression of complex coordination: Tala, the younger of two females, had taken over the dominant position in their pack from Shima when she was only 1 year old—well ahead of the loose string experiments. Although Shima came to accept her new position in the dominance hierarchy, the two of them never had a relaxed relationship after the turnover. When the two of them were paired to solve the task, they had a problem since both of them initially strongly preferred the same side of the apparatus to pull the rope. During the first two sessions, they were unsuccessful because Tala always replaced Shima on the preferred side, but Shima still was never willing to move to the other side of the table. Inevitably, Tala pulled the rope out and the trial was terminated. Then, suddenly, in the third session, after looking at Shima standing on the preferred side, Tala evidently gave in, switched sides, and pulled the other side, which she continued to do from then on. This was remarkable, since actually it was the dominant animal that gave up her preference to allow the dyad to succeed. Another telling anecdote was that Kaspar, the dominant animal in the dyad, once 'punished' Aragorn (second ranking in the pack) vigorously by pinning him to the ground after Aragorn messed up a trial in the delay condition where he arrived first and pulled the rope out. In the trials after the punishment, Aragorn run to the rope, took it into his mouth, and waited motionless for Kaspar to arrive, and together they nicely pulled again and were successful. These anecdotes suggest that the wolves might have quite a high potential to negotiate with each other, but possibly also use punishment in order to achieve cooperation.

Aside from the wolves' success in the *simultaneous condition*, they also showed high levels of success when the task became more complex, and they had to coordinate to first go to one apparatus and then the other one to solve both (*two-apparatus* condition). Moreover, they seemed to have at least a basic understanding that they needed the partner present at the table to solve the task (delay), with the possible exception of Shima, who only waited in 8% of the trials for her partner—not really a shining performance! However, the performance of our dogs was much worse and rather perplexing—especially given the previous results on the pet dogs (Ostojic and Clayton, 2014)—so we decided to try to figure out why our dogs did not seem to be able to solve the problem.

First, we analysed the difference in behaviour between the wolves and dogs in their first session of the *simultaneous condition* (including those before and after the individual training) to see if it would provide us with any clues. In order to be successful, animals would need to pull on the end of a rope, and they would have to do so at the same time. While the dogs pulled less than the wolves in the spontaneous *simultaneous condition* (without prior training), they did not differ from each other after the training. So, by itself, while this could explain why wolves outperformed the dogs in the *simultaneous condition* without training, it could not explain why wolves were also much better after the training.

Since we knew from all our previous tolerance tests (see Chap. 5) that dogs might try to avoid each other around a resource, we investigated if the dogs were less likely than the wolves to be at the apparatus at the same time. The analyses of the data showed that this was not the case when considering proximity (both animals within a one body length radius of the tray); however, compared to wolves the dogs were less likely to manipulate the apparatus. So typically, what we would see is one dog pulling the rope a bit or pawing at the tray and then taking a step back, and then the other dog would come forward and interact with the tray for a bit. The fact that they so rarely interacted with the tray at the same time made it basically impossible for them to discover the solution, which required them to simultaneously pull the two rope ends.

Thus, it seemed that the failure of the dogs to solve the task was due to how they avoid each other and maintain a distance when in the presence of a food source. But what would happen if at least one of the two dogs would understand that they could only solve the task if they work together? This idea was actually based on a video from one of the wolf dyads.

Kaspar, after being successful with two other partners, had been paired with Tala, a female pack member, who had never really been successful before (a trial here and there). When they had their first trial, Kaspar arrived at the apparatus first and took the rope into his mouth, but did not pull. When Tala arrived, she first looked at the other end of the rope but then came over to Kaspar and looked at his end of the rope.

He then pulled a little bit, but stopped again. Tala then tried to take his rope, but he nicely told her (by showing a stiff posture, raising his hackles and his tail) that this end of the rope was his. She then moved over to the other end of the rope, and as soon as she started to pull, Kaspar synchronized his behaviour with hers, finally pulling together and getting their reward. From this first successful trial, they were highly successful on all subsequent trials. While we probably will never know what exactly was on Kaspar's mind in this situation, his patience and the fact that he did not jump on her and told her off when she tried to take his end of the rope definitely made it possible for them to solve the task together. Accordingly, we wondered what would happen if we trained up the dominant dogs of each pack with a stooge (a pet dog of our trainers who had been extensively trained on the task) until they really understood that they need the partner. Would this realization make them more tolerant and hence more willing to let the subordinate individuals of their packs interact with the apparatus at the same time? So that is exactly what we did.

After spending another 2–3 months training up the most dominant animal in two of our dog packs (Nuru and Meru) with a pet dog and making them wait up to 1 min for the partner to really make sure they understood the need for a second dog to solve the task, we tested them with task-inexperienced dogs from their respective packs. One of these dyads (Nuru-Pepeo) was very successful, allowing us to test them also in the *two-apparatus condition* and *delay condition*. Similar to wolves, the successful, non-trained dog was then defined as an expert and tested with the other pack members as well.

At the end, we had six mixed-experience dog dyads (one expert and one naïve partner) that we could compare with eight inexperienced dog dyads from the previous experiment. We found that in contrast to the mean success of inexperienced dog dyads (just 0.3% of trials), mixed-experience dog dyads were successful on an average of 19% of trials. However, although the performance of the dogs had increased dramatically, in comparison with similarly mixed-experience wolf dyads ($N = 5$; mean success rate: 63% of trials), the performance was still inferior. Although the sample size was small ($N = 6$ dyads), we nevertheless found that as for wolves, the affiliative bond and the rank distance affected their likelihood of success. These results support the idea that the dogs' difficulties in solving this task are largely linked to their social dynamics in a resource situation where a potential for conflict exists (see also a discussion about these aspects in Chap. 5).

As a next step in our developing project, we decided to test the wolves and the dogs with a human partner. The idea was that if, indeed, tolerance is the main reason why the dogs cannot solve this task together, they should perform much better with a human, where this is not an issue.

8.2.3.2 Animal-Human Coordination

Similarly, to the animal-animal tests, we first started with the simultaneous condition, and then, if successful, the animals were subsequently tested in the *two-apparatus* and the *delay conditions* (Fig. 8.17) (Range et al. 2019). Since we knew that humans are able to coordinate with each other, we put the emphasis on testing the animals. So, in the first, *simultaneous condition*, we released the human and

Fig. 8.17 A wolf (**a**) and a dog (**b**) working with a human cooperation partner (photographs by Rooobert Bayer). Figure adapted from Friederike Range et al. (2019). https://www.nature.com/articles/s41598-019-40468-y/; CC BY 4.0; http://creativecommons.org/licenses/by/4.0/

animal partners in such a way that in half of the trials the animal and in the other half the human would arrive slightly (3 sec) ahead of the partner and thus could choose which end of the rope to pull. Since most of our animals had a rather strong side preference in this task, the human was told to always choose the preferred side of the respective animals. Thus, to be successful, the animal really had to adjust its behaviour to that of the human.

The human experimenters in this study were our animal trainers or a hand-raiser, who had good relationships with the animals. They received a set of instructions how to behave in the various situations in order to standardize their behaviour as much as possible. Overall, the instructions were chosen in such a way that there was very little communication going on from the human to the animal to avoid the human guiding the cooperation. Thus, the humans were not allowed to talk to or look at the animal and had defined paths of where to walk to get to the apparatus(es) and exact instruction on how long to wait before starting to pull the rope, as well as how fast they pull the rope. Importantly the rope was pulled independent of the behaviour of the animal partner, e.g. even if the dog or wolf were not there, the human partner started pulling the rope after 3 sec.

To control for the fact that all our animals had had different experiences with the loose string paradigm by now, we included two variables in all statistical analyses to see if it affected their performance: (a) whether an animal had received any training to solve the task alone (see above) and (b) their previous success levels in the task, i.e. the average percentage of trays successfully solved across testing in all conditions with all previous conspecific partners.

In the *simultaneous condition*, statistically wolves and dogs performed similarly (mean success rate: wolves, 61.5% of trials; dogs, 49% of trials). Animals got better over sessions and previous experience mattered only to a minor degree. Not surprisingly, the dyads were more successful if the human took the non-preferred side of the animal than if the animal had to adjust (wolves, mean animal first 88% vs. human first 50%; dogs, mean animal first 70% vs. human first 40%). Interestingly, when the human arrived first, 14 of the 15 wolves and also 2 of the 12 dogs grabbed the rope from the human in an attempt to pull the rope on their preferred side. Since the human partners were instructed not to get into conflicts with the animals (after all we are working with wolves), these trials failed since the human stepped back and did not engage in pulling the string anymore. Over time these animals actually decreased in their rope stealing habits, suggesting that they realized that this was not helpful for the successful solving of the task.

Nine wolves and 7 dogs also participated in the *two-apparatus condition* (Range et al. 2019). Here the human was always released first so that the animal had to adjust its' behaviour based on the human's decision to approach one or the other of the trays. But once they had reached the tray, we allowed the animal the choice of which side they preferred to pull the rope on. After successfully solving the first apparatus (or failing if the rope was pulled out or stolen by the animal partners), the human cooperation partner either (1) followed the animal, if it moved towards the second tray, or (2) waited for 5 sec before approaching the second tray. Again, we found no differences in how successful the animals were in terms of solving both trays in a trial (wolves, 76% of trials; dogs, 67% of trials). We then investigated the behaviour of the wolves and dogs in more detail when moving from the first to the second apparatus and found that wolves were more likely than dogs to take the lead in moving from the first to the second tray (wolves in 60% and dogs in 35% of trials). This difference in behaviour was unaffected by wolves' prior experience with the trays. Finally, in trials in which the human led from the first to the second tray ($N = 50$ for wolves and $N = 82$ for dogs), we found that the dogs were more likely to follow the human partner (doing so in 70% of trials) than wolves (which followed in 42% of trials). Again, the difference between wolves and dogs was independent of previous success. These results suggest that while wolves and dogs readily cooperate with humans, they might do so with a different mindset. Wolves seemed to be overall more assertive on their role, stealing the rope and leading the way, whereas dogs rather stayed back and followed the human's lead.

Finally, like in the previous animal-animal interactions, we tested the animals in the *delay condition*, releasing the human partner 10 sec after the subject had been released. Animals were successful on average in 88% of trials across the two sessions, and we found no difference between wolves and dogs. This was again surprising since in the pet dogs' study of Ostojjic and Clayton (2014), the dogs did not wait the 15 s required for the human to arrive in the *delay condition*. Thus, our results suggest that the dogs have the cognitive capacity to understand or learn over the course of the experiment that they need a partner in such cooperative tasks.

Moreover, after testing our dogs with the human partner and seeing that they were actually successful, we wanted to know if their failure with the conspecific was maybe due to the limited experience they have had at this time or rather about their relationship. Thus, we retested eight dogs that cooperated very successfully with their human partners with their original conspecific partners. Although we only had four dyads, the results were clear since none of the dog dyads solved a single trial together despite all the successful experience they had had in the meanwhile with the human partners (unpublished data). This suggests that the failure with their conspecific partners was not due to a lack of cognitive skills (i.e. understanding of the task) but rather due to social relationship factors.

8.2.3.3 The Role of the Partner

Compared to other species, we humans show extreme flexibility in our ability to cooperate with unrelated individuals. This is potentially due to a number of factors: humans recognize the gains to be made from cooperation over years or decades, even when there are initial short-term costs, and they can assess the intentions of the other party, can use language to formulate agreements that limit the potential costs, and are also aware of the potential consequences of not cooperating (e.g. punishment, loss of allies). However, successful cooperation may not require any such understanding of complex contexts and consequences. For example, many non-human species reach functionally cooperative outcomes despite a lack or limited understanding of the partner's intentions and the consequences of their and their partners' decisions during the interaction. By studying especially this latter aspect—the understanding of different roles during cooperation—we can start to understand the decisions that govern animals in cooperative interactions and can further understand what exactly is, or is not, unique about human cooperation, and how humans may have exploited their unique cooperative abilities to achieve our astounding degree of interdependence.

So, continuing with the loose string paradigm, we started to investigate a bit further whether our animals 'understood' the need for a partner in the task and, more importantly, if they really understood it enabling them to flexibly use this knowledge or if they had learned simple rules based on the contingencies over repeated exposures. Part of the answer to this question we had already received in the delay condition of the previous experiment: wolves and also dogs (if cooperating with a human partner) could (and would) wait for the partner's arrival for at least 10 sec. This suggests that the animals had at least learned that a partner needed to be present at the tray in order to solve the task. However, they would also go one step further and recognize in which situations they needed a partner, and as a consequence appropriately adjust their behaviour—which would go beyond the simple learning of contingencies.

One approach to answer this question is to compare the animal's behaviour in a cooperative vs. solo condition. While, in the cooperative condition, the subject needs the partner to solve the task, in the *solo condition*, the string-pulling apparatus is altered by tying the rope to the centre of the tray and having a single dangling rope end in the enclosure. Pulling this rope alone will move the tray forward and allow the animals to obtain the reward, i.e. no partner is necessary to solve the task.

Although various animal species have been tested for their understanding of the need for a partner in the string-pulling task (e.g. chimpanzees: Melis et al. 2006a, b; elephants: Plotnik et al. 2011; grey parrots, and keas: Heaney et al. 2017; Péron et al. 2011, b; but see rooks: Seed et al. 2008), it is important to note that only in the chimpanzee study (Melis et al. 2006a, b) and in a study with groupers and moray eels that used a somewhat different design (Vail et al. 2014) did the animals have to actively recruit a partner (by, e.g. opening a door for them). Active recruitment is arguably more challenging than simply going to the 'cooperative' apparatus when the partner is present and to the solo apparatus when alone (keas, grey parrots) or only inhibiting the pulling behaviour in the cooperative condition (elephants); first, because the animals need to move away from the visibly baited apparatus to recruit the partner, which has been shown to be difficult for many species due to limited inhibitory control abilities (Amici et al. 2008; MacLean et al., 2014). Second, when the animal inspects the apparatus, it has to decide in the absence of the partner whether to recruit it or not. Considering their extensive previous experience with the apparatus, animals may have acquired a simple rule of thumb: 'if the partner is present go to the cooperative apparatus'. This rule, however, is not sufficient when the partner has to be actively recruited.

Accordingly, we set out to train our animals to step on a wooden sign (a star). If they stepped on it, a sliding door of one of the small compartments of the testing enclosure would open. Once they had learned this association, they were given a pre-exposure session with an apparatus that only had one rope attached in the middle and thus could be solved alone. Then, the animals were exposed to test sessions in which they were confronted with four solo and four cooperative trials intermixed. As before, the test apparatus was set up on one side of the test enclosure while the animal was positioned on the other side in the middle compartment. A potential human collaborator was positioned in the compartment next to the animal—either to the left or right side. While they could solve the string-pulling apparatus by themselves in the solo condition, in the cooperative condition, they would have to go back to the other side of the test enclosure and release the human cooperation partner by stepping on the wooden star next to her compartment, so that she could come out and help in solving the task. To make life a bit more difficult for the animals, two stars were present—one in front of the enclosure with the human partner and one in front of the empty compartment on the other side of the compartment from where the animal started. This was to allow us to differentiate between the possibilities that the animals simply exhibited the last trained action—stepping on the star—instead of actually trying to recruit a partner.

Not surprisingly, the wolves and dogs were really good in the solo trials, solving them in 72.5% of the trials. In the cooperation trials, they recruited a partner in 56% of the trials. Interestingly, in 71% of the trials, the animals first went to inspect the table demonstrating that they did not rely on the small differences of the baiting procedure that would have allowed them to differentiate between the two conditions. Instead they went to the table and pulled the rope, but just a little bit, and they did this in 57% of the trials in the cooperation condition and 73% of the trials in the solo condition. Pulling the rope provides feedback on whether or not you need a partner: if the rope moves, but the tray does not, you need a partner. Interestingly, the animals only recruited a partner in about half of the cooperation trials. But when they did step on a marker, they were more likely to step on the marker next to the human compartment than the marker next to the empty compartment. So why did they not recruit the partner more often?

One idea based on observing the animals during the task was that they really had to walk a lot in our sizable test enclosure (25 m). They first had to cross the testing enclosure to walk over to the apparatus after being released, and then, once they realized that they were facing a cooperation trial and needed a partner, they had to walk all the way back to release the human partner and then walk once again over to the other side. Anyhow, to see whether the travel distance required made a difference, we changed the spatial configuration and used the two compartments next to the one where the apparatus was set up. Now, the distance that had to be covered to recruit the partner was a much shorter. While this change increased the stepping on the marker next to the human from the last session of the previous set-up (58%) to the new set-up (70%), it did not really increase the success rate once the partner had been recruited because they actually had been already rather good in the last session with the original set-up (70%). Moreover, the animals' success rate with this new set-up was no longer significantly affected by the solo vs. cooperation conditions indicating that both conditions were similarly 'easy' for the animals to solve.

8.2.3.4 Summary

The results from these various loose string paradigm studies suggest that both wolves and dogs are able to coordinate their actions with certain partners in order to solve cooperative tasks, even the more complex versions. They also understand, or learn to understand, over the course of the experiments, that a partner needs to be present in order to be successful. Interestingly, while the wolves cooperated successfully with conspecifics as well as with a human partner, dogs only succeeded with the humans. This suggests that the dogs' failure to cooperate with conspecifics in this paradigm is not due to some cognitive limitations but rather supported our intuition that the problem was really about social relationships, notably tolerance.

Finally, our detailed analyses of the wolves' and dogs' behaviour with the humans revealed some remarkable differences. While both were successful in cooperating with us, wolves seem to be less shy in trying to grab the rope, if the

Table 8.1 Summary of studies on cooperation

Test	Task	Effect
Prosociality	Token choice (touchscreen)	Yes (with pack members): Wolves No: Dogs
	Location choice	Neither wolves nor dogs
	Group-service paradigm	Preliminary: Yes (with pack members): Wolves No: Dogs
	Food sharing (Chap. 5)	Wolves > dogs
Inequity aversion	Buzzer task	Yes: Wolves = dogs
	Play	No: Wolves = dogs
Coordination (string pulling)	With conspecifics	Wolves >> > dogs
	With humans	Wolves = dogs, but wolves lead, and dogs follow
	Understanding the role of a partner	Yes: Wolves = dogs

human happens to be on their preferred side. And in the two-apparatus condition, it became clear that wolves would rather lead (and perhaps just expect humans to follow), whereas dogs were more inclined to wait for the human to take the initiative and then follow their lead. These differences in behaviour may go some way in explaining why dogs, despite their predatory ancestral grandfathers, now make such wonderful compliant pets, whereas wolves remain the highly cooperative, yet also confident, strong-willed creatures that demand a certain dose of respect from us 'puny' humans.

Furthermore, we found that both wolves and dogs could flexibly adjust their behaviour to whether or not they required a partner to solve a string-pulling task, recruiting a partner only when needed. This suggests that they have at least a basic understanding of the role of a partner in a task requiring cooperation, inhibitory control of inappropriate responses, and a precise coordination of actions.

8.2.3.5 What Factors Influence the Success of Cooperation?

Having conducted a number of studies on the animals' skills thought to be important for cooperation, and having observational data on the animals' social relationships (see Chap. 5), we investigated how social and non-social factors influence the wolves' interactions in coordination, prosocial, and inequity aversion tasks. We concentrated at the wolf results, given that the dogs' performance in two of the tasks discussed above (prosociality and coordination) was not very impressive and, hence, could not be included in the analyses. The overall results are summarized in Fig. 8.18. In essence, social factors such as dominance rank and affiliation mainly influenced how the animals interacted with each other (see also Chap. 5), while the non-social factors had only a limited effect, if any (Dale et al. 2020). Accordingly, in the end, most of these aspects of cooperation in wolves, like in humans, seem to be closely linked to aspects of social mechanisms, notably emotions.

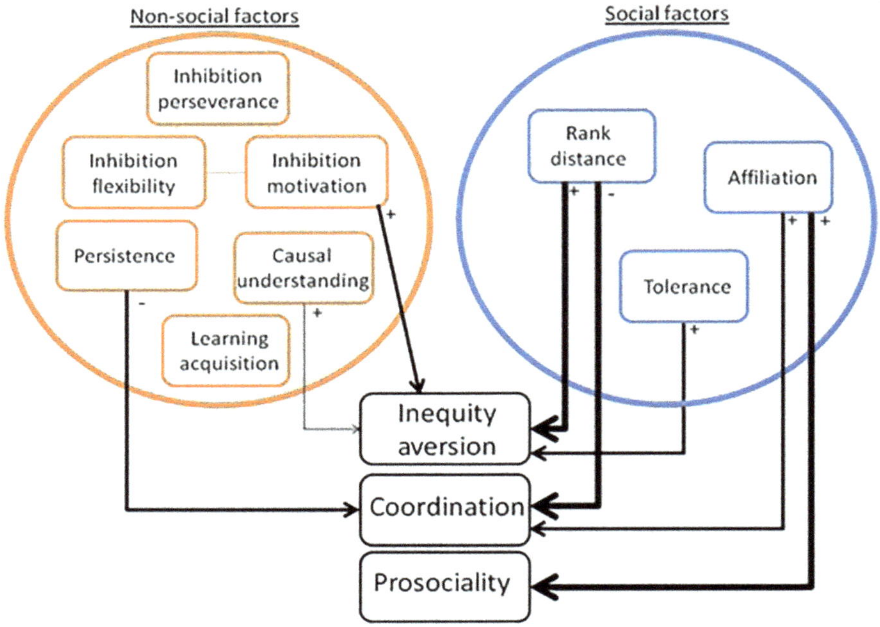

Fig. 8.18 An infographic to illustrate the results of the analyses showing that social factors have a stronger influence on the performance of the animals in cooperative tasks (coordination, prosociality, and inequity aversion) than the non-social factors. The thickness of the lines depicts the significance of the effect ($p < 0.05$, 0.01, or 0.0001). Figure adapted from Dale et al. 2020. https://www.nature.com/articles/s41598-020-68734-4/; CC BY 4.0; http://creativecommons.org/licenses/by/4.0/

8.2.4 Summary Cooperation

The studies reviewed in this chapter all demonstrate that wolves and dogs share with us at least some of the abilities that seem to be important also for human cooperation: they—at least wolves and pet dogs—seem to have some prosocial tendencies, are inequity averse, can coordinate their actions, and have a basic understanding of the role of their partner in cooperative interactions.

As pointed out, cooperation is closely linked to underlying mental mechanisms such as emotions in humans. In a number of species including wolves, the social relationships between partners (in terms of dominance, affiliative bonds, and tolerance) have been shown to strongly affect who will cooperative with whom and how successful such cooperative interactions are, suggesting that, also in non-human animals, emotions are strongly involved in governing such behaviours. In addition to social relationships, non-social factors provide the mechanistic base of cooperation, for example, inhibitory control, i.e. *blocking an aversive impulse in favour of a more appropriate option when it is advantageous to do so. But also other factors such as persistence, learning, and causal understanding are all known to influence problem solving skills (see* Chap. 9*).*

References

Amici F, Aureli F, Call J (2008) Fission-fusion dynamics, behavioral flexibility, and inhibitory control in primates. Curr Biol 18(18):1415–1419. https://doi.org/10.1016/j.cub.2008.08.020

Axelrod RM (1984) The evolution of cooperation. Basic Books, New York

Batson CD, Ahmad N (2001) Empathy-induced altruism in a prisoner's dilemma II: what if the target of empathy has defected? Eur J Soc Psychol 31:25–36. https://doi.org/10.1002/ejsp.26

Batson CD, Batson JG, Todd RM, Brummett BH, Shaw LL, Aldeguer CM (1995) Empathy and the collective good: caring for one of the others in a social dilemma. J Pers Soc Psychol 68:619–631

Batson CD, Duncan BD, Ackerman P, Buckley T, Birch K (1981) Is empathic emotion a source of altruistic motivation? J Pers Soc Psychol 40(2):290

Bauer EB, Smuts BB (2007) Cooperation and competition during dyadic play in domestic dogs, *Canis familiaris*. Anim Behav 73:489–499. https://doi.org/10.1016/j.anbehav.2006.09.006

Bekoff M (1974) Social play in coyotes, wolves and dogs. Bioscience 24(4):225–230. https://doi.org/10.2307/1296803

Bekoff M (1977) Mammalian dispersal and the ontogeny of individual behavioral phenotypes. Am Nat 111(980):715–732

Bekoff M (2001) Social play behaviour. Cooperation, fairness, trust, and the evolution of morality. J Conscious Stud 8(2):81–90

Bergmuller R, Johnstone RA, Russell AF, Bshary R (2007) Integrating cooperative breeding into theoretical concepts of cooperation. Behav Process 76(2):61–72

Bowles S, Gintis H (2003) The origins of human cooperation. In: Hammerstein P (ed) Genetic and cultural evolution of cooperation. MIT Press, Cambridge

Brosnan SF (2011) A hypothesis of the co-evolution of cooperation and responses to inequity. Front Neurosci 5:43

Brosnan SF, de Waal FBM (2003) Monkeys reject unequal pay. Nature 425(6955):297–299

Brosnan SF, Freeman C, De Waal FBM (2006) Partner's behavior, not reward distribution, determines success in an unequal cooperative task in capuchin monkeys Am J Primatol 68(7):713–724

Brucks D, Essler JL, Marshall-Pescini S, Range F (2016) Inequity aversion negatively affects tolerance and contact-seeking Behaviours towards partner and experimenter. PLoS One 11(4): e0153799. https://doi.org/10.1371/journal.pone.0153799

Burkart JM, van Schaik C (2013) Group service in Macaques (*Macaca fuscata*), Capuchins (*Cebus apella*) and Marmosets (*Callithrix jacchus*): a comparative approach to identifying proactive prosocial motivations. J Comp Psychol 127(2):212–225

Burkart JM, Allon O, Amici F, Fichtel C, Finkenwirth C, Heschl A et al (2014) The evolutionary origin of human hyper-cooperation. Nat Commun 5:4747

Burkart JM, Fehr E, Efferson C, van Schaik CP (2007) Other-regarding preferences in a non-human primate: common marmosets provision food altruistically. Proc Natl Acad Sci U S A 104(50): 19762–19766. https://doi.org/10.1073/pnas.0710310104

Call J, Bräuer J, Kaminski J, Tomasello M (2003) Domestic dogs (*Canis familiaris*) are sensitive to the attentional state of humans. J Comp Psychol 117(3):257–263. https://doi.org/10.1037/0735-7036.117.3.257

Carr L, Iacoboni M, Dubeau M-C, Mazziotta JC, Lenzi GL (2003) Neural mechanisms of empathy in humans: a relay from neural systems for imitation to limbic areas. Proceedings of the National Academy of Science 100:5497–5502

Clayton NS, Dickinson A (1998) Episodic-like memory during cache recovery by scrub jays. Nature 395(6699):272–274. https://doi.org/10.1038/26216

Clutton-Brock J (1995) Origins of the dog: domestication and early history. The domestic dog: Its evolution, behaviour and interactions with people, 7–20

Colman AD, Liebold KE, Boren JJ (1969) A method for studying altruism in monkeys. Psychol Rec 19(3):401–405. https://doi.org/10.1007/BF03393866

Cronin KA, Snowdon CT (2008) The effects of unequal reward distributions on cooperative problem solving by cottontop tamarins, *Saguinus oedipus*. Anim Behav 75:245–257

Dale R, Marshall-Pescini S, Range F (2017a) Do females use their sexual status to gain resource access? Investigating food-for-sex in wolves and dogs. Current Zoology 63(3):323–330. https://doi.org/10.1093/cz/zow111

Dale R, Marshall-Pescini S, Range F (2020) What matters for cooperation? The importance of social relationship over cognition. Sci Rep 10(1):1–14

Dale R, Palma-Jacinto S, Marshall-Pescini S, Range F (2019) Wolves, but not dogs, are prosocial in a touch screen task. PLoS One 14(5):e0215444. https://doi.org/10.1371/journal.pone.0215444

Dale R, Quervel-Chaumette M, Huber L, Range F, Marshall-Pescini S (2016) Task differences and Prosociality; investigating pet dogs' prosocial preferences in a token choice paradigm. PLoS One 11(12):e0167750. https://doi.org/10.1371/journal.pone.0167750

Dale R, Range F, Stott L, Kotrschal K, Marshall-Pescini S (2017b) The influence of social relationship on food tolerance in wolves and dogs. Behav Ecolol Sociobiol 71:107–131. https://doi.org/10.1007/s00265-017-2339-8

Dawes CT, Fowler JH, Johnson T, McElreath R, Smirnov O (2007) Egalitarian motives in humans. Nature 446(7137):794

de Waal FBM (2008) Putting the altruism back into altruism: the evolution of empathy. Annu Rev Psychol 59:279–300. https://doi.org/10.1146/annurev.psych.59.103006.093625

de Waal FBM, Suchak M (2010) Prosocial primates: selfish and unselfish motivations. Phil Trans R Soc B 365:2711–2722. https://doi.org/10.1098/rstb.2010.0119

Drea CM, Carter AN (2009) Cooperative problem solving in a social carnivore. Anim Behav 78(4): 967–977. https://doi.org/10.1016/j.anbehav.2009.06.030

Eisenberg N, Miller PA (1987) The relation of empathy to prosocial and related behaviors. Psychol Bull 101(1):91

Essler JL, Cafazzo S, Marshall-Pescini S, Virányi Z, Kotrschal K, Range F (2016) Play behavior in wolves: using the '50:50' rule to test for egalitarian play styles. PLoS One 11(5):e0154150. https://doi.org/10.1371/journal.pone.0154150

Essler JL, Marshall-Pescini S, Range F (2017) Domestication does not explain the presence of inequity aversion in dogs. Curr Biol 27(12):1861–1865.e1863. https://doi.org/10.1016/j.cub.2017.05.061

Fehr E, Gachter S (2002) Altruistic punishment in humans. Nature 415(6868):137–140

Fehr E, Gächter S (2000) Cooperation and punishment in public goods experiments. Am Econ Rev 90(4):980–994. https://doi.org/10.1257/aer.90.4.980

Fehr E, Schmidt KM (1999) A theory of fairness, competition, and cooperation. Q J Econ 114(3): 817–868

Fehr E, Schmidt KM (2006) The economics of fairness, reciprocity and altruism–experimental evidence and new theories. Handbook of the economics of giving, altruism and reciprocity 1: 615–691

Fletcher GE (2008) Attending to the outcome of others: disadvantageous inequity aversion in male capuchin monkeys (*Cebus apella*). Am J Primatol 70(9):901–905. https://doi.org/10.1002/ajp.20576

Fowler JH (2005) Altruistic punishment and the origin of cooperation. Proc Natl Acad Sci U S A 102(19):7047–7049. https://doi.org/10.1073/pnas.0500938102

Frank RH (1988) Passions within reason: the strategic role of the emotions. W.W. Norton, New York

Galef BG, Giraldeau LA (2001) Social influences on foraging in vertebrates: causal mechanisms and adaptive functions. Anim Behav 61:3–15

Gilby IC (2006) Meat sharing among the Gombe chimpanzees: harassment and reciprocal exchange. Anim Behav 71(4):953–963. https://doi.org/10.1016/j.anbehav.2005.09.009

Gintis H, Bowles S, Boyd RT, Fehr E (2005) Moral sentiments and material interests: the foundations of cooperation in economic life, vol Vol. 6. MIT press

Giraldeau L-A (1997) The ecology of information use. In: Krebs JR, Davies NB (eds) Behavioural ecology an evolutionary approach, 4th edn. Blackwell Scientific Publications, pp 42–68

Goetz JL, Keltner D, Simon-Thomas E (2010) Compassion: an evolutionary analysis and empirical review. Psychol Bull 136(3):351–374. https://doi.org/10.1037/a0018807

Gurven M (2004) To give and to give not: the behavioral ecology of human food transfers. Behav Brain Sci 27(4):543–559. https://doi.org/10.1017/S0140525X04000123

Gwynne DT (1984) Courtship feeding increases female reproductive success in bushcrickets. Nature 307(5949):361–363

Hamilton WD (1964) The genetical evolution of social behaviour. J Theor Biol 7:1–52

Harbaugh W, Mayr U, Burghart. (2007) Neural responses to taxation and voluntary giving reveal motives for charitable donations. Science 325:1622–1625

Heaney M, Gray RD, Taylor AH (2017) Keas perform similarly to chimpanzees and elephants when solving collaborative tasks. PLoS One 12(2):e0169799

Heberlein M, Turner D (2009) Dogs, *Canis familiaris*, find hidden food by observing and interacting with a conspecific. Anim Behav 78(2):385–391. https://doi.org/10.1016/j.anbehav.2009.05.012

Henrich J, McElreath R, Barr A, Ensminger J, Barrett C, Bolyanatz A et al (2006) Costly punishment across human societies. Science 312(5781):1767–1770

Hernandez-Lallement J, van Wingerden M, Marx C, Srejic M, Kalenscher T (2015) Rats prefer mutual rewards in a prosocial choice task. Front Neurosci 8:443. https://doi.org/10.3389/fnins.2014.00443

Heyes CM (1994) Social learning in animals: categories and mechanisms. Biol Rev 69:207–231

Hirata S, Fuwa K (2007) Chimpanzees (Pan troglodytes) learn to act with other individuals in a cooperative task. Primates 48:13–21

Horn L, Scheer C, Bugnyar T, Massen JJ (2016) Proactive prosociality in a cooperatively breeding corvid, the azure-winged magpie (*Cyanopica cyana*). Biol Lett 12(10):20160649

Jaeggi AV, Van Schaik CP (2011) The evolution of food sharing in primates. Behav Ecol Sociobiol 65(11):2125–2140

Laland KN (2004) Social learning strategies. Learning & Behavior 32(1):4–14

Lampe M, Bräuer J, Kaminski J, Virányi Z (2017) The effects of domestication and ontogeny on cognition in dogs and wolves. Sci Rep 7(1):11690. https://doi.org/10.1038/s41598-017-12055-6

Langford DJ, Crager SE, Shehzad Z, Smith SB, Sotocinal SG, Levenstadt JS et al (2006) Social modulation of pain as evidence for empathy in mice. Science 312(5782):1967–1970. https://doi.org/10.1126/science.1128322

Lord K (2013) A comparison of the sensory development of wolves (*Canis lupus lupus*) and dogs (*Canis lupus familiaris*). Ethology 119(2):110–120. https://doi.org/10.1111/eth.12044

MacLean EL, Hare B, Nunn CL, Addessi E, Amici F, Anderson RC et al (2014) The evolution of self-control. Proc Natl Acad Sci 111(20):E2140–E2148. https://doi.org/10.1073/pnas.1323533111

Maner JK, Gailliot MT (2007) Altruism and egoism: prosocial motivations for helping depend on relationship context. Eur J Soc Psychol 37(2):347–358. https://doi.org/10.1002/ejsp.364

Marshall-Pescini S, Dale R, Quervel-Chaumette M, Range F (2016) Critical issues in experimental studies of prosociality in non-human species. Anim Cogn 19(4):679–705. https://doi.org/10.1007/s10071-016-0973-6

Marshall-Pescini S, Schwarz JFL, Kostelnik I, Virányi Z, Range F (2017) Importance of a species' socioecology: wolves outperform dogs in a conspecific cooperation task. Proc Natl Acad Sci 114(44):11793–11798. https://doi.org/10.1073/pnas.1709027114

Massen J, Ritter C, Bugnyar T (2015) Tolerance and reward equity predict cooperation in ravens (*Corvus corax*). Sci Rep 5(15021):1–11. https://doi.org/10.1038/srep15021

McGetrick J, Range F (2018) Inequity aversion in dogs: a review. Learn Behav:1–22

Mech LD (1999) Alpha status, dominance, and division of labor in wolf packs. Canadian Journal of Zoology-Revue Canadienne De Zoologie 77(8):1196–1203

Mech LD, Boitani L (2003) Wolf social ecology. In: Mech LD, Boitani L (eds) Wolves: behavior, ecology, and conservation. The University of Chcago Press, Chicago, London, pp 1–35

Melis AP, Hare B, Tomasello M (2006a) Chimpanzees recruit the best collaborators. Science 311: 1297–1300

Melis AP, Hare B, Tomasello M (2006b) Engineering cooperation in chimpanzees: tolerance constraints on cooperation. Anim Behav 72:275–286

Melis AP, Hare B, Tomasello M (2009) Chimpanzees coordinate in a negotiation game. Evol Hum Behav 30(6):381–392. https://doi.org/10.1016/j.evolhumbehav.2009.05.003

Melis AP, Semmann D (2010) How is human cooperation different? Philosophical Transactions Of The Royal Society B-Biological Sciences 365(1553):2663–2674. https://doi.org/10.1098/rstb.2010.0157

Molesti S, Majolo B (2016) Cooperation in wild barbary macaques: factors affecting free partner choice. Anim Cogn 19(1):133–146. https://doi.org/10.1007/s10071-015-0919-4

Neiworth JJ, Johnson ET, Whillock K, Greenberg J, Brown V (2009) Is a sense of inequity an ancestral primate trait? Testing social inequity in cotton top tamarins (*Saguinus oedipus*). J Comp Psychol 123:10–17

O'Gorman R, Wilson DS, Miller RR (2005) Altruistic punishing and helping differ in sensitivity to relatedness, friendship, and future interactions. Evol Hum Behav 26(5):375–387. https://doi.org/10.1016/j.evolhumbehav.2004.12.006

Oberliessen L, Hernandez-Lallement J, Schäble S, van Wingerden M, Seinstra M, Kalenscher T (2016) Inequity aversion in rats, *Rattus norvegicus*. Anim Behav 115:157–166. https://doi.org/10.1016/j.anbehav.2016.03.007

Ostojic L, Clayton N (2014) Behavioural coordination of dogs in a cooperative problem-solving task with a conspecific and a human partner. Anim Cogn 17(2):445–459. https://doi.org/10.1007/s10071-013-0676-1

Packard JM (2003a) Wolf behavior: reproductive, social and intelligent. In: Mech LD, Boitani L (eds) Wolves: behavior, ecology and conservation. University of Chicago Press, Chicago, pp 35–65

Packard JM (2003b) Wolf behavior: reproductive, social and intelligent. In: Mech LD, Boitani L (eds) Wolves: behaviour, ecology and conservation. University of Chicago Press, Chicago, pp 35–65

Péron F, Rat-Fischer L, Lalot M, Nagle L, Bovet D (2011) Cooperative problem solving in African grey parrots (*Psittacus erithacus*). Anim Cogn 14:545–553

Peterson RO, Ciucci P (2003) The wolf as a carnivore. In: Mech LD, Boitani L (eds) Wolves: behaviour, ecology and conservation. The University of Chicago Press, Chicago, London, pp 104–130

Phillips DP, Danilchuk W, Ryon J, Fentress JC (1990) Food-caching in timber wolves, and the question of rules of action syntax. Behav Brain Res 38(1):1–6. https://doi.org/10.1016/0166-4328(90)90018-A

Pielou EC (1977) Mathematical ecology. Wiley, New York

Plotnik JM, Lair R, Suphachoksahakun W, Waal FBM (2011) Elephants know when they need a helping trunk in a cooperative task. Proceedings of the National Acadademy of Sciences USA 108:5116–5121

Preston SD, Bechara A, Damasio H, Grabowski TJ, Stansfield RB, Mehta S, Damasio AR (2007) The neural substrates of cognitive empathy. Soc Neurosci 2(3–4):254–275. https://doi.org/10.1080/17470910701376902

Preston SD, de Waal FBM (2002) Empathy: its ultimate and proximate bases. Behav Brain Sci 25(1):1–20

Quervel-Chaumette M, Dale R, Marshall-Pescini S, Range F (2015) Familiarity affects other-regarding preferences in pet dogs. Sci Rep 5:18102. https://doi.org/10.1038/srep18102

Rachlin H (2002) Altruism and selfishness. Behav Brain Sci 25:239–296

Raihani NJ, McAuliffe K (2012) Human punishment is motivated by inequity aversion, not a desire for reciprocity. Biol Lett 8(5):802–804

Range F, Heucke SL, Gruber C, Konz A, Huber L, Virányi Z (2009a) The effect of ostensive cues on dogs' performance in a manipulative social learning task. Appl Anim Behav Sci 120(3–4): 170–178. https://doi.org/10.1016/j.applanim.2009.05.012
Range F, Horn L, Virányi Z, Huber L (2009b) The absence of reward induces inequity aversion in dogs. Proc Natl Acad Sci U S A 106(1):340–345. https://doi.org/10.1073/pnas.0810957105
Range F, Jenikejew J, Schröder I, Virányi Z (2014) Difference in quantity discrimination in dogs and wolves. Front Psychol 5:1299
Range F, Marshall-Pescini S, Kratz C, Virányi Z (2019) Wolves lead and dogs follow, but they both cooperate with humans. Sci Rep 9(1):3796. https://doi.org/10.1038/s41598-019-40468-y
Range F, Ritter C, Virányi Z (2015) Testing the myth: tolerant dogs and aggressive wolves. Procedings of Royal Society London 282(1807). https://doi.org/10.1098/rspb.2015.0220
Range F, Virányi Z (2013) Social learning from humans or conspecifics: differences and similarities between wolves and dogs. Front Psychol 4:868. https://doi.org/10.3389/fpsyg.2013.00868
Range F, Virányi Z (2014) Wolves are better imitators of conspecifics than dogs. PLoS One 9(1): e86559. https://doi.org/10.1371/journal.pone.0086559
Renevey N, Bshary R, van de Waal E (2013) Philopatric vervet monkey females are the focus of social attention rather independently of rank. Behaviour 150(6):599–615. https://doi.org/10.1163/1568539X-00003072
Rice GEJ, Gainer P (1962) "Altruism" in the albino rat. J Comp Physiol Psychol 55:123–125
Sääksvuori L, Mappes T, Puurtinen M (2011) Costly punishment prevails in intergroup conflict. Proc R Soc B Biol Sci 278(1723):3428–3436
Schwab C, Huber L (2006) Obey or not obey? Dogs (*Canis familiaris*) behave differently in response to attentional states of their owners. J Comp Psychol 120:169–175
Seed AM, Clayton NS, Emery NJ (2008) Cooperative problem solving in rooks (*Corvus frugilegus*). Proceedings of the Royal Society B-Biological Sciences 275:1421–1429
Shettleworth SJ (2009) Cognition, evolution, and behavior. Oxford University Press, Oxford
Silk JB (2007) Empathy, sympathy, and prosocial preferences in primates. In: Dunbar RIM, Barrett L (eds) Oxford handbook of evolutionary psychology. Oxford University Press, Oxford, pp 115–126
Silk JB, Brosnan SF, Vonk J, Henrich J, Povinelli DJ, Richardson AS et al (2005) Chimpanzees are indifferent to the welfare of unrelated group members. Nature 437(7063):1357–1359
Snowdon CT, Boe CY (2003) Social communication about unpalatable foods in tamarins (*Saguinus oedipus*). J Comp Psychol 117(2):142–148. https://doi.org/10.1037/0735-7036.117.2.142
Spence KW (1996) The nature of discrimination learning in animals. Psychol Rev 43:427–449
Steger M, Kashdan T, Oishi S (2008) Being good by doing good: daily eudaimonic activity and Well-being. J Res Person 42:22–42
Stevens JR (2004) The selfish nature of generosity: harassment and food sharing in primates. Proceedings of the Royal Society of London Series B-Biological Sciences 271(1538):451–456
Stevens JR, Gilby IC (2004) A conceptual, framework for nonkin food sharing: timing and currency of benefits. Anim Behav 67:603–614. https://doi.org/10.1016/j.anbehav.2003.04.012
Stevens JR, Hauser MD (2004) Why be nice? Psychological constraints on the evolution of cooperation. Trends Cogn Sci 8(2):60–65
Suchak M, Eppley TM, Campbell MW, de Waal FB (2014) Ape duos and trios: spontaneous cooperation with free partner choice in chimpanzees. PeerJ 2:e417
Thorndike EL (1898) Animal intelligence: an experimental study of the associative processes in animals. Psychol Rev Monogr Suppl 2(8)
Thorpe WH (1956) Learning and instinct in animals. Methuen, London
Tinbergen N (1963) On aims and methods of ethology. Zeitschrift fur Tierpsychologie 20:410–433
Tomasello M, Carpenter M (2007) Shared intentionality. Dev Sci 10(1):121–125. https://doi.org/10.1111/j.1467-7687.2007.00573.x
Tomasello M, Carpenter M, Call J, Behne T, Moll H (2005) Understanding and sharing intentions: the origins of cultural cognition. Behav Brain Sci 28:675–735
Trivers RL (1971) The evolution of reciprocal altruism. Q Rev Biol 46:35–57

Vail AL, Manica A, Bshary R (2014) Fish choose appropriately when and with whom to collaborate. Curr Biol 24(17):R791–R793. https://doi.org/10.1016/j.cub.2014.07.033
Ward C, Bauer EB, Smuts BB (2008) Partner preferences and asymmetries in social play among domestic dog, *Canis lupus familiaris*, littermates. Anim Behav 76(4):1187–1199
Whiten A, Ham R (1992) On the nature and evolution of imitation in the animal kingdom: reappraisal of a century of research. Adv Study Behav 21:239–283
Yamamoto S, Humle T, Tanaka M (2009) Chimpanzees help each other upon request. PLoS One 4(10):e7416. https://doi.org/10.1371/journal.pone.0007416
Zak PJ, Kurzban R, Matzner WT (2005) Oxytocin is associated with human trustworthiness. Horm Behav 48(5):522–527. https://doi.org/10.1016/j.yhbeh.2005.07.009
Zentall TR (2004) Action imitation in birds. Learning & Behavior 32(1):15–23

9 Physical Cognition Part I: Problem Solving and Its Correlates

On a daily basis, animals are exposed to many challenges in the social and physical domains. The problem-solving strategies they employ are usually a mix of adaptation, their ability to generalize, and system constraints (e.g. the limits of the brain). In the physical environment, predator avoidance and foraging pose the largest challenges, whereas the latter is likely more relevant for wolves and dogs. When moving through the environment, animals may also encounter novel opportunities which pose challenges. They first need to approach the unknown and then using either already existing behaviour patterns to exploit them or develop new ones. These challenges may change depending on the environment and the nature of the problems (see Lefebvre 2011 for a review on foraging innovations). Being able to come up with solutions to new challenges is likely to have fitness advantages since it allows access to resources (see Lefebvre and Sol 2008). Not surprisingly, understanding how individuals solve problems related to their physical environment has received at least as much attention as how animals solve the challenges in the social domain (Chap. 7), and researchers have argued about which of these selective pressures drive the evolution of the brain (see Holekamp and Benson-Amram 2017 for a nice review on the various hypotheses).

In any case, the most common paradigm to test problem-solving behaviour in the physical domain is to present animals with puzzle boxes that contain one or more food rewards. To get access to the reward(s), the animals have to open the box by finding a solution to unlock one or more mechanisms. These paradigms allow scientists to measure if the animals solve the task by insight (e.g. understanding the properties of the locks) or trial-and-error learning. It also allows to measure their persistence in trying to solve a certain problem and how flexible they are (e.g. if another approach is not successful). One of the simplest versions that has been used with carnivores is a metal cage containing a chunk of meat, enclosed by a spring-loaded door that can be opened by moving a sliding latch (Benson-Amram et al. 2013, 2016; Benson-Amram and Holekamp 2012; Borrego and Gaines 2016).

The first study that compared wolves' and dogs' independent problem-solving abilities using puzzle boxes was carried out by the Frank's (Frank and Frank 1985).

F. Range, S. Marshall-Pescini, *Wolves and Dogs*, Fascinating Life Sciences,
https://doi.org/10.1007/978-3-030-98411-3_9

They tested the four wolf and four malamute pups at the age of 10 weeks in a series of manipulation tasks involving food-containing puzzle boxes of varying complexity. The basic problem the pups faced was having to pull a food bowl out from under a box. This could be achieved either directly or only after removing an obstacle. For example, they had to pull on a string, overturn the box, or push a plunger that would slide the food bowl from under the box. In total, there were six different problem-solving tasks. The animals were tested in eight trials to solve the various problems and were given only 2 min per trial. According to the authors, this time limitation gave a certain advantage to those animals that planned their actions beforehand and 'performed their trial-and-error experiments internally' as well as to animals that had better generalization skills, i.e. able to adapt previously successful solutions to the new problem and animals that remembered previous trials. Some of the manipulations also required some serial organization (e.g. first pulling the container out with a string before they could manipulate the container directly), which would likely be facilitated by means-end understanding, e.g. understanding that a series of progressing steps are the means to achieve a certain goal. All wolves that participated in a given trial solved all problems except the last one (the plunger), which was only solved by one individual. The malamutes performed significantly worse than the wolves, with only a few animals solving some of the problems. The Frank's describe the behaviour of the animals in their article as wolves 'attacking each puzzle immediately' (pp. 271) and persisting 'until the problem was solved or time ran out' (pp. 271), while the dog pups tended to rather quickly seek human attention and then lay down until the end of the test once they found out that the food was not easily accessible (Frank and Frank 1985). This behaviour pattern has been reviewed in detail in Chap. 7. The Frank's also note that the wolves might have performed even better if it weren't for the fact that the last test was inconveniently scheduled on the fourth of July—America's Independence Day—which is traditionally accompanied by lots of rather noisy fireworks. This distracted the animals considerably, and even when retested 2 days later, they appeared nervous and jumpy, paying little attention to the food and the box.

This study nicely demonstrates that variations in problem-solving skills between species and individuals may be due to the motivational level (reactions to novelty, perseverance) (Griffin and Guez 2017; Teschke et al. 2011, 2013), differences in attention, memory and inhibitory control (general-purpose mechanisms) (Griffin and Guez 2017; Tebbich et al. 2016), or their associative learning and causal reasoning/inference skills (cognitive processes) (Auersperg et al. 2011; Huber and Gajdon 2006; Taylor et al. 2009, 2010).

In the current chapter, we will thus summarize our knowledge about the motivational factors and general-purpose mechanisms of wolves and dogs, while in Chap. 10 we will review studies on learning and reasoning that might help animals to solve problems in the physical realm.

9.1 Motivational Factors

Motivational factors found to influence problem-solving skills across taxa include neophilia and neophobia (attraction to and the fear of new situations or objects, respectively) (Greenberg and Mettke-Hofmann 2001; Mettke-Hofmann et al. 2002) as well as persistence (i.e. task-directed motivation; the time an animal spends tackling a task). Persistence predicts success in almost every study where it has been considered (e.g. Borrego and Gaines 2016; Griffin et al. 2014; Huebner and Fichtel 2015; Morand-Ferron et al. 2011, but see Griffin and Diquelou 2015). However, persistence may be detrimental if not paired with behavioural flexibility (using a number of different object-directed manipulative behaviours), since animals could just persist in an ineffectual behaviour (Benson-Amram and Holekamp 2012). It is important to note that while each of these correlates contributes to problem-solving success, they are not independent of each other (Rowe and Healy 2014). Moreover, Sih and Del Giudice (2012) proposed that persistence, neophilia/neophobia, and motivation may form parts of a behavioural syndrome. If these are indeed personality traits, they would likely be correlated with each other and be stable over time and context (Réale et al. 2007).

9.1.1 Neophobia and Exploration

Motivators for an action could be intrinsic, where an individual pursues an activity 'for the sake of it', such as exploring an environment and manipulating objects or extrinsic, where an animal performs an activity in pursuit of an end goal such as food, water, or mating opportunities (Barto 2013). Several comparisons between wolves and dogs suggest that wolves and dogs differ in their intrinsic rather than their extrinsic motivation.

In terms of intrinsic motivation, early studies comparing wolves and dogs led most authors to agree that wolves, while more neophobic than dogs even if intensely socialized by humans (Fentress 1967; Klinghammer and Goodmann 1987; Zimen 1987), are also more explorative. Wolves start to explore their environment and objects already at the age of 2 weeks (Frank and Frank 1982; Packard 2003), while dogs do so only at the age of 4 weeks (Fox 1964; Rheingold 1963; Scott and Fuller 1965). Interestingly, a recent study revealed that this is not due to wolves developing faster than dogs, but, while wolves already explore when they are still blind and deaf, dogs explore only once all their senses (olfaction, audition, and vision) are operational (Lord 2013).

That wolves and dogs also differ in terms of neophobia and exploration when they grow older becomes apparent from several more recent studies. A study of 18 dog and 17 wolf puppies' explorative and neophobic behaviours in a novel environment and with a novel object was carried out at the Wolf Science Center (Marshall-Pescini et al. 2017b). A novel environment test was conducted at 5 weeks of age, and a novel object test was carried out twice: at 6 and at 8 weeks of age. The novel objects were a remote-controlled toy dog that moved around barking and

Fig. 9.1 Novel objects in the puppy tests at the Wolf Science Center. Figure reproduced from Marshall-Pescini et al. 2017b. https://doi.org/10.3389/fpsyg.2017.00180; CC BY 4.0; http://creativecommons.org/licenses/by/4.0/

wagging the tail and a remote-controlled car moving around in the room, either with a cardboard/plastic box on top of it or not (Fig. 9.1). The toy dog and the two different versions of the car were presented to all puppies once at each age. The likelihood of making contact or not with the object during the trial was somewhat higher for wolves than dogs. Indeed, on average at 6 weeks, 56% of dogs and 82% of wolves and at 8 weeks 65% of dogs and 88% of wolves contacted the object. Results showed that there was no difference between wolves and dogs in the latency to get in contact with the objects, but wolves were more persistent in exploring both the environment and the objects, and this was the case at all ages. There were no differences in the frequency of fear-related behaviours and time spent in proximity to humans. Stress-related behaviours were similarly expressed at 5 and 6 weeks, although wolves showed a higher frequency of such behaviours at 8 weeks, maybe indicating an onset of higher neophobia at this time. Another study conducted at 4 months of age with the Budapest group found that wolf pups spent more time exploring an unfamiliar outdoor kennel than dog pups (Topál et al. 2005).

In a recent longitudinal study comparing two litters of Alaskan huskies ($N = 12$) and three litters of European grey wolves ($N = 13$), which were hand-raised under identical conditions, Hansen Wheat et al. (2019) found similar results in terms of neophobia, but different ones in terms of exploration. The animals were tested repeatedly with different novel objects (rolled-up mattress, wheelbarrow (up-side down), stuffed wolverine toy, etc.) at the ages of 6, 10, 14, 18, 22, and 26 weeks in a familiar room. The puppies were taken into the room by a caregiver, who then quickly left again. The test was observed via video cameras and lasted 10 min, unless terminated earlier if the objects were destroyed ($N = 11$ trials). The authors found that while there was no significant difference between wolves and dogs in their latencies to approach the novel object at 6 weeks of age, at 26 weeks of age, the dogs were significantly faster, suggesting that while the neophobic response of the wolves did not change with time, the dogs' fear reaction decreased. Moreover, the dogs also seemed to investigate the object less with age than the wolves, for which the age effect was less pronounced. However, there were no differences between the two

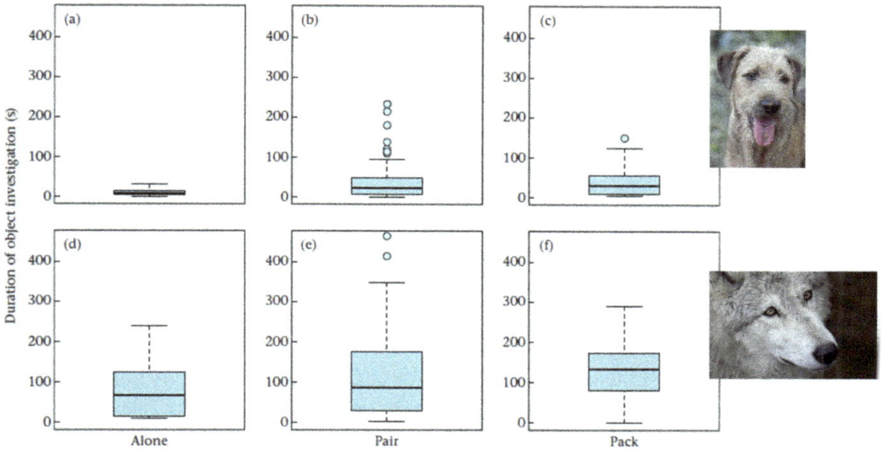

Fig. 9.2 The graphs show how long adult wolves and dogs investigated the novel objects in the alone pair and pack condition. Figure adapted from Moretti et al. (2015) (copyright licence number 5053081493048), reprinted by permission from Animal Behaviour; pictures taken by Daniel Zupanc

species in terms of getting into contact with the objects or manipulating them, nor did age have an effect on these two measures.

Finally, Moretti et al. (2015) confronted adult wolves and dogs with novel objects (bicycle, book, teddy bear, balloons, helmets, baby buggy, garden dwarf, etc.) for 15 min in a familiar environment at the Wolf Science Center in three different conditions: alone (twice), with a partner (twice with each of the other pack members), and in the pack (twice), always with novel objects. We found that although wolves were slower at contacting a novel object in a familiar environment and fled from it more often than dogs, dogs were less likely to approach the novel object than wolves (7 of the 13 dogs did not approach the object in at least one of the tests, while all 11 wolves always approached). Moreover, wolves were more persistent in their investigation of the objects than dogs. Both wolves and dogs manipulated objects for longer when paired or in the pack than when alone. Hence, it would seem that although wolves take longer to approach the objects, once they do, they explore more (see Chap. 5 for a discussion of the effect of social relationships on the animal's explorative behaviours) (Fig. 9.2).

Given these overall results, it appears that dogs and wolves may follow a different developmental trajectory in relation to their explorative tendencies, with wolves becoming more explorative than dogs with age, while at the same time being more neophobic. This pattern might be adaptive considering that young wolves should stay within the den or close to it and not venture into unknown territories, but then need to explore new areas when migrating and new potential food resources when they become older. At the same time, they need to be careful when encountering new objects—especially everything that is connected to humans.

The costs and benefits of exploration are influenced by different ecological factors. Neophilia is thought to be driven by the benefits of exploration, for instance the discovery of new resources, finding resources, or developing new means of obtaining familiar resources. On the other hand, neophobia may be driven by the costs of exploration, e.g. risk of predation (Greenberg and Mettke-Hofmann 2001). For wolves both the potential gains and also the costs of exploration (i.e. risk of being attacked by neighbouring groups) are high. In contrast, selection pressure is released on both factors for dogs, which could explain their higher neophobia and neophilia. In line with this idea, empirical studies on birds and non-human primates have confirmed the link between explorative behaviour and feeding ecology (Gilby and Wrangham 2007; Haun et al. 2011; Heilbronner et al. 2008; Rosati et al. 2007; van Oers et al. 2004) and the motivation to take risks to get access to high-quality rewards. Considering that wolves have higher explorative tendencies and a reliance on hunting behaviour but dogs rather scavenge on distributed and more stable food sources, it can be hypothesized that in the feeding context, wolves would be willing to take higher risks than dogs to get access to high-value rewards. To test the risk preference of wolves and dogs, we adapted a method previously used with apes (Heilbronner et al. 2008; Rosati and Hare 2012, 2013) in which the animals are presented with a two-way choice task (Fig. 9.2). One choice results in a 'safe outcome' where animals always obtain a less preferred food item (i.e. a dry food pellet). The alternative choice represents the 'risky outcome', where on 50% of trials, subjects obtain a preferred food item (i.e. a piece of meat), but in the remaining 50%, they obtained a non-edible item (a stone). Before each trial, subjects are shown the potential outcome of each choice, so they are aware of the 'odds' of the risky vs. safe outcome (Fig. 9.3).

We found that, indeed, wolves consistently chose the risky option significantly more than dogs in all test sessions, making between 70% and 95% risky choices compared to dogs that choose the risky option in 38%–76% (Marshall-Pescini et al. 2016). However, it is worth noting that in the beginning, dogs seemed to have some trouble understanding the contingencies of the task, as was revealed in comprehension sessions carried out before and after the testing.

These results could be driven by a higher motivation of wolves to get access to high-quality resources and thus potentially take high risks. However, it is also possible that due to their risky way of life (e.g. hunting is dangerous; see Chap. 3), wolves are in general more risk-prone than dogs.

To better understand the influence of food quality and to validate the results of all other experiments, where we use food as an *extrinsic motivator*, we set out to test food preferences of wolves and dogs. As hinted at above, animals may have preferences for certain kinds of food over others depending on their feeding niche and could thus differ in their levels of motivation when attempting to obtain them (e.g. Auersperg et al. 2013; Fontenot et al. 2007; Hillemann et al. 2014; Wascher et al. 2012). For example, considering their reliance on hunting, wolves may have evolved strong preferences for specific foods (such as meat), while dogs, being more generalist, may have lost some of these preferences. Moreover, another crucial aspect to consider is the effect of hunger, which may affect preference patterns in

Fig. 9.3 Set-up of the risk-taking task, which was rather complicated not just for the animals but also for the humans. Importantly, the animals could not see the experimenter's face but only her hands, so that no cueing was possible (picture taken by Ingo Besserdich)

both dogs and wolves. The subjective feeling of hunger is linked to motivation (Berridge 2004; Day et al. 1998), as it may lead to animals consuming novel foods (Ramsey et al. 2007) and modify their foraging styles, e.g. feed at locations that they might otherwise avoid (Croy and Hughes 1991). We generally attempt to rule out differences in motivation as caused by specific food preferences and/or hunger by (1) conducting preference tests prior to the experimental sessions and (2) controlling the hunger status in experiments involving different food types (e.g. in the risk-taking experiment described above and the inequity aversion study; Chap. 8). However, a general difference in food preference would be a major problem for our comparative approach. Accordingly, in a recent study, we set out to test our animals' food preferences in two different experimental set-ups: a classic two-choice paradigm, which has been used for testing food preferences in several taxa (Bradshaw et al. 2000; Fontenot et al. 2007; Hillemann et al. 2014; Laska et al. 2000; Wascher et al. 2012), and a cafeteria task, where five options were available and the animals could choose three (Rao et al. 2018b). There were two main reasons for conducting two different food preference tests. Firstly, it allowed us to assess the consistency in the animals' preferences between tests (i.e. while the two-choice paradigm provides information about which food an animal prefers from a pair, it is difficult to say whether the animal would choose similarly when presented with multiple choices). Secondly, it allowed us to avoid some of the shortcomings of the two-choice paradigm (i.e. task contingencies and experience with other, similar two-choice tasks may affect the animals' behaviours, e.g. side biases). In the two-choice task, animals could choose between the different combinations of 1-day old (dead) chicks, cow meat, sausage, and dry food. In the cafeteria task, also tofu was available. All animals were tested in two conditions: in one condition they were fed approximately 15 hours prior (i.e. the previous evening) to the test session, which is in line with the normal feeding schedule at the Wolf Science Center. In the second condition, the wolves were not fed the evening before testing, and the dogs were fed less than half their regular measures of dry food approximately 15 hours prior to the test session, since for medical reasons we could not skip a meal for some of the dogs.

Luckily for all our various tests, we found only minor differences in dogs' and wolves' food preferences, suggesting that overall, they should have had a similar motivation to get to the food reward in the different tests. Both strongly preferred meat over dry food in both testing paradigms. The only observed difference is related to the choice diversity, where wolves were less diverse (or more 'selective') than dogs when not hungry. The three kinds of food that were chosen most by wolves and dogs in both paradigms were chicks, meat, and sausage. While choosing the protein-rich chicks is a good choice and in line with work on macronutrient selection, not choosing tofu or dry food, which have comparable or higher calorific value, is probably due to the sensory qualities (taste and smell)—after all, wolves and dogs are not nutrition scientists. Another possible explanation is that the animals choose partly based on the fat content (meat and sausage had the highest) as has also been observed in humans and several other animals (Drewnowski and Greenwood 1983).

9.1.2 Persistence and a Little Bit of Motor Diversity

While dogs and wolves do not seem to differ in their food preferences, they definitely differ in their persistence. Persistence, in the context of problem solving, has generally been quantified as the time an individual spends engaged with an object to obtain a reward (Benson-Amram and Holekamp 2012; Borrego and Gaines 2016; Griffin and Guez 2014; Thornton and Samson 2012). As mentioned above, it is one of the most powerful indicators of problem-solving performance, predicting success in problem-solving tasks in studies with carnivores such as lions, spotted hyenas, leopards, tigers (Borrego and Gaines 2016), hyenas (Benson-Amram and Holekamp 2012; Benson-Amram et al. 2013), and meerkats (Thornton and Samson 2012).

The first tasks that suggested differences in the persistence of wolves and dogs was the experiment carried out by the Frank's reviewed above showing that while wolves were persistent, dogs instead looked at the human.

The wolves' higher persistence and dogs' higher inclination to look back at the experimenter have been observed in other experiments as well (Brubaker et al. 2017; Marshall-Pescini et al. 2017a; Miklósi et al. 2003; Udell 2015; see Chap. 7), and researchers have argued that the difference between wolves and dogs could be caused by one of two reasons. First, it is possible that dogs' increased looking behaviour is due to a process of selection for such a trait and/or because of a prior history of reinforcement of this behaviour (see Chap. 7). In the human-dominated niche of dogs, humans often provide support in all important domains, including access to resources such as food (Marshall-Pescini et al. 2017a). Hence, dogs might come to expect that humans will solve problems for them and thus turn to humans for help without trying hard to solve problems by themselves. However, a recent study by our group on pet dogs and free-ranging dogs (Lazzaroni et al., submitted, Chap. 7) rather supports that differences in persistence are adaptations to the change in the feeding ecology during the domestication process (Fleming et al. 2017), with differences in 'looking back' rather being a by-product of such differences in persistence.

Wolves are primarily hunters (between 10 and 50%; Mech et al. 2015) and thus need to be highly persistent to survive. Dogs, however, are primarily scavengers (Chap. 4) and may not need to be as persistent. As described in Chap. 7 a number of studies found wolves to be more persistent than dogs (Marshall-Pescini et al. 2017a; Miklósi et al. 2003; Udell 2015). However, humans were always present as a confounding variable as the aim of these studies was human communication rather than persistence per se. Furthermore, different raising of some of the tested populations makes interpretation of the results difficult.

Rao and Bernasconi et al. (2018a) set out to specifically test the persistence of wolves and dogs with the same life experiences, in a task that was unsolvable, and with no humans present to influence results. We conducted the study in the animals' home enclosures since it is where the animals least expect a human to be present. Indeed, at the Wolf Science Center, behavioural tests are normally conducted in specific 'testing enclosures', and humans (including trainers) only visit the animals in the home enclosures in very specific contexts (i.e. pack visits, animal care, and short training demonstrations during public-guided tours). Moreover, after enclosure cleaning, trainers usually hide food rewards in the home enclosure as a form of enrichment before the animals are released back in. Since this enrichment programme happens once per week, the animals immediately start searching for food rewards upon release back into the home enclosure and also manipulate plants/trees to get at the hidden treasures. Accordingly, this set-up is perfect to test if wolves and dogs differ in their persistence when confronted with a food reward hidden in an inaccessible object.

Rao and Bernasconi presented our dogs and wolves with two different unsolvable tasks on two separate occasions. Each task consisted of an object baited with food that was inaccessible and unsolvable to the animal. One object was a 'lion feeder' (an enrichment object used in some zoos), and the other was a modified, perforated PVC sewage pipe (Fig. 9.4). Subjects were tested early in the morning, having not been fed the evening before and with a high-quality food reward hidden in the object to assure similarly high motivation in wolves and dogs.

Supporting our previous results on neophobia (Moretti et al. 2015), while only 14 out of 17 dogs manipulated the ball and 10 out of 15 dogs manipulated the pipe, all wolves manipulated both objects, indicating again a lower interest by the dogs in interacting with unfamiliar objects. Not surprisingly, the wolves as a group were more persistent than dogs in their manipulation of both objects. Interestingly however, there were two male dogs that were highly persistent, manipulating especially the pipe for almost as long as some of the wolves. Moreover, contact latency (the

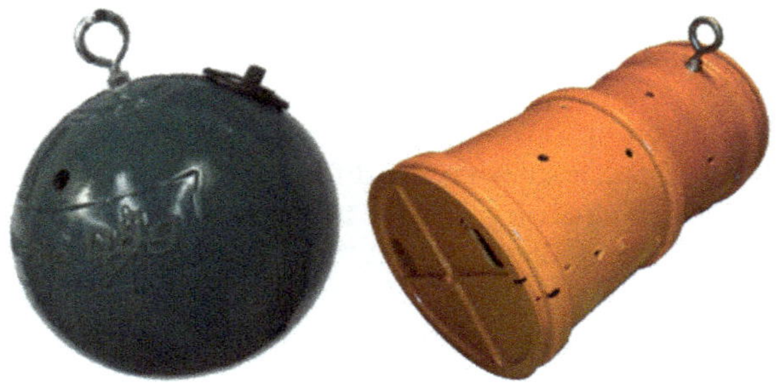

Fig. 9.4 The two objects used for the persistence study. Figure reproduced from Rao, Bernasconi et al. 2018a. https://doi.org/10.7717/peerj.5944; CC BY 4.0; http://creativecommons.org/licenses/by/4.0/

time it took the animals from a two-meter radius to make physical contact with the object) decreased with persistence in both dogs and wolves and with age in the dogs, but not the wolves. Finally, motor diversity (the number of different object-directed manipulative behaviours exhibited) increased with persistence in both dogs and wolves. It was higher in wolves than in dogs when interacting with the ball, but not when interacting with the pipe (Fig. 9.5).

To investigate whether this lower persistence of our dogs is a general trait of dogs or specific to our population of the WSC dogs (they might have learned to rely on human support rather than trying to solve problems themselves), we used the same test to investigate persistence in pet dogs in Vienna and free-ranging dogs at our study site in Morocco. The dogs were tested with the lion feeder ball, and additionally the free-ranging dogs and pet dogs were also tested with a plastic bottle (Lazzaroni et al. 2019), an object which is often found in the piles of rubbish the dogs scavenge in. All animals were tested in a familiar environment in the absence of conspecifics and humans.

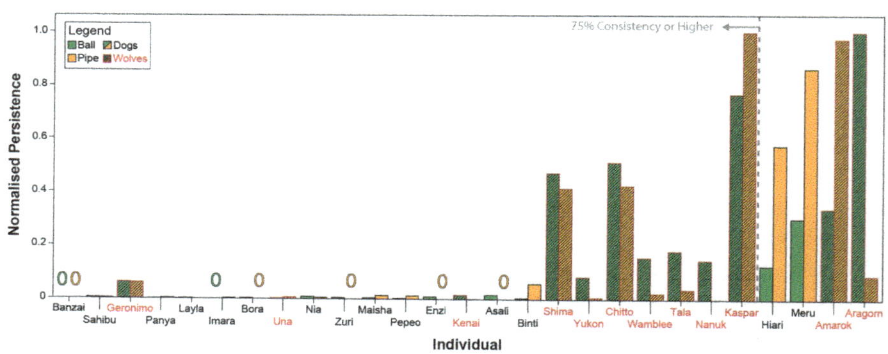

Fig. 9.5 Every individual's persistence in both tasks, rescaled from 0 to 1 for comparability. Green bars indicate persistence with the ball; orange bars indicate persistence with the pipe. Zeros indicate that the individual did not manipulate the object at all. Individuals with red names and hashed bars are wolves; individuals with black names and non-hashed bars are dogs. Individuals are arranged from left to right in descending order of consistency in persistence across tasks. Figure reproduced from Rao, Bernasconi et al. 2018a. https://doi.org/10.7717/peerj.5944; CC BY 4.0; http://creativecommons.org/licenses/by/4.0/

We found that while the different populations did not differ in contact latency and exploration (i.e. the duration of sniffing), only three free-ranging dogs actually spent time manipulating the ball. Thus, overall free-ranging dogs were less likely to manipulate the ball than pet dogs and WSC dogs. In terms of persistence, only the pet dogs and the WSC dogs were compared and no difference emerged. We were a bit surprised about these results, and so Martina also had the idea to test the dogs with plastic bottles that she baited with food. The wolves would have made a short-lived experiment out of the bottle, so we did not test them and thus also not the pack dogs. Interestingly, when comparing the performance of the free-ranging dogs with the two objects (the ball and the plastic bottle), we found that there was no effect of neophobia, since they approached both objects equally quickly. However, the free-ranging dogs spent a lot longer manipulating the bottle than the ball.

This study and others with free-ranging dogs in India, which have shown free-ranging dogs to be as persistent as pet dogs (Brubaker et al. 2017), strengthen the conclusion that as a species, dogs are less persistent than wolves when manipulating objects but also that dog populations differ depending on their experience. However, to really complete our dataset, we would need to test also free-ranging wolves.

Finally, Hiestand (2011) compared 7 6-month-old and two adult wolves, as well as 40 adult German shepherds in their ability to pull ropes from a ceiling to get access to food. The difficulty of the tests was increased incrementally by adding more ropes (medium and/or long) and including novel locations for the ropes. Moreover, the various tasks tested different aspects of problem solving including:

1) Exploration, e.g. the animals had to figure out that they needed to pull the ropes from the ceiling
2) A sequential learning component, since in some tasks the ropes had to be pulled in a certain sequence
3) Flexibility, since in one task the previously learned order had to be reversed
4) Persistence in attempting to solve the task despite repeated failure (for instance when the animals could not figure out the correct order in which to pull the ropes)

During all tasks, a human was present encouraging the animals when deemed necessary. Hiestand found that without prior training, the adult wolves readily pulled the ropes already in the first trial (with a single rope), whereas four of seven juvenile wolves were successful from trial three onwards. In contrast, only 2 dogs solved the tasks by themselves, while 35 required shaping to pull the rope and 3 dogs had to be withdrawn from the study. Adding new ropes in new locations did not constitute a problem for the wolves, but it did for the dogs. Especially when 2 long ropes were added in test 2, only 15 of the 29 dogs tested were successful in at least one trial. In test 3, where the 2 ropes had to be pulled in the opposite order, neither the 2 adult wolves nor the 12 dogs that were tested had problems solving the task. In test four, yet another rope (short, stopped over the head) was positioned in a new place, and the ropes had to be pulled in order. While the 2 adult wolves solved the task, for the 13 dogs and 5 juvenile wolves that were tested, the short rope needed to be moved to the familiar location, and the order requirement had to be dropped for them to be successful at all. Nevertheless, both groups were successful only in 20% of the trials. In test five the three ropes had to be pulled in a certain order, which was then reversed in test six. The two adult wolves and four dogs participating in test five solved the problem easily, while in test six, the dogs outperformed the wolves (see Chap. 10 for a discussion of these results). Interestingly, 'one of the wolves, after nine unsuccessful trials, in which she received no reward, she had a perfect trial, pulling the three ropes in the correct order' (p. 851). Hiestand (2011) concluded that adult wolves are better problem-solvers than dogs, probably due to higher persistence in manipulating the objects and eventually hitting on the correct solution. Dogs simply seemed to give up a lot earlier, which is not really surprising after all the new studies that we just reviewed above showing that dogs are a lot less persistent than wolves.

9.1.3 Summary: Motivational Factors

To sum up, the combined results suggest that wolves are more likely to approach novel objects, even though it may take them longer to do so; they are more persistent manipulating them than dogs. Wolves are potentially also more motivated and take higher risks in interacting with objects in their environment. Since persistence is one of the most robust predictors of problem solving, it is important to take this difference into account when designing experiments aimed at comparing the two species. For example, in the cooperative string-pulling experiments reviewed in Chap. 8, we found that wolves were initially more active in pulling the ropes, which might have explained their higher success compared to the dogs in the first sessions without training. Only after training the individual animals to solve the task and then retesting them with a partner revealed that persistency is at least not the only factor driving this difference. Instead tolerance around food resources may have been more crucial. However, experiments about behavioural flexibility, a rather important factor in how effective your interaction with the environment is, would be needed to better understand how effective the higher persistence of wolves compared to dogs really is (Table 9.1).

9.2 Attention and Working Memory

Selective attention is the ability to focus on task-relevant information while inhibiting signals from irrelevant stimuli (Olsen 2018), and this ability seems to be closely intertwined with working memory capacity (WMC). Both mechanisms and inhibitory control (see below) are thought to be crucial when facing problem-solving tasks. Engle (2002) describes *working memory* as a system with two components: a temporary memory storage that keeps a small amount of information active (i.e. available for processing) and an executive component that regulates the contents

Table 9.1 Summary of results of the various studies on motivation

Parameters	Age	Species	References
Neophobia			
Environment	5 w	W = D	Marshall-Pescini et al. (2017b)
Objects			
Latency contact	6 > 8 w	W = D	
Fear behaviours	6 w	W = D	
	8 w	W > D	
Objects			Hansen Wheat et al. (2019)
Latency approach	6, 10, 14, 18, 26 w	W = D	
	22 w	W > D	
Latency contact	6 . . .26 w	W = D	
Exploration			
Environment	5 w	W > D	Marshall-Pescini et al. (2017b)
Objects	6 & 8 w	W > D	
Objects			Hansen Wheat et al. (2019)
Investigating	6, 10, 14, 18, 26 w	W = D	
	22 w	W > D	
Manipulation	6 . . .26 w	W = D	
Environment	4 m	W > D	Topál et al. (2005)
Objects	Adults	W > D	Moretti et al. (2015)
Persistence			
Puzzle boxes	10 w	W > D	Frank and Frank (1985)
String pulling (impossible task)	4 m	W > D	Miklósi et al. (2003)
Container (impossible task)	Adults	W > D	Marshall-Pescini et al. (2017a)
Rope pulling	Adult W & pet D + juvenile wolves	W > D	Hiestand (2011)
Container	Adult W, pet D, shelter D	W > D	Brubaker et al. (2017); Udell (2015)
Ball/pipe	Adult W and D	W > D	Rao et al. (2018a)
Behavioural flexibility			
Ball/pipe	Adult W and D	W = D	Rao et al. (2018a)

w = weeks, m = months, W = wolves, D = dogs

of the former. Working memory capacity (WMC) is the amount of information that an individual is able to hold in an active state and is a good predictor of performance on various tasks in humans, such as reading comprehension (Daneman and Carpenter 1980) and multitasking (Hambrick et al. 2010). Attention and WMC are linked, which may be related to proactive interference (i.e. the interference of older memories with the retrieval of newer ones). It is thought that individuals with higher WMC are able to use attentional control to combat the effects of proactive interference under normal circumstances, but the limited nature of attention makes it so that this effect gets dampened when distractions are strong enough (Engle 2002). Thus, individuals with high WMC seem to be better at focusing on relevant information than those with low WMC (Kane and Engle 2000), which in turn is thought to aid performance when problem-solving.

To our knowledge no study has specifically tested *attention and/or working memory* capacities in wolves compared to dogs yet. From some of the studies reviewed so far, it seems that wolves might pay somewhat more attention to certain aspects of their environment. For example, they performed better in the imitation task (Chap. 8), the quantity discrimination task (Chap. 10), and the risk aversion task (this chapter)—each of these paradigms required the animals to pay attention to the details of the procedure, hinting towards wolves being more attentive and/or having better working memory capacities. Also, the fact that they made decisions faster in the learning and reversal task, but were equally successful (Chap. 10), suggests that they may have higher working memory capacities. However, whether this is true or not needs to be studied in more detail.

9.3 Inhibitory Control

Inhibitory control is defined as blocking an impulsive or prepotent response in favour of a more appropriate alternative when it is advantageous to do so. Being able to withhold such actions is highly important for all kinds of social interactions and decision-making processes for the following reasons: (1) it can reduce distraction caused by irrelevant but salient cues and/or increase the ability of an individual to take the complete set-up of the task into account as their attention is not captured by a single salient stimulus (such as a food reward) (Diamond 1990), and (2) it can help animals enhance their behavioural flexibility, increasing their ability to adapt to the current situation. This may be especially true in an unstable or rapidly changing environment, since in these contexts a once advantageous behaviour can quickly become useless.

Levels of inhibitory control have been linked to a variety of cognitive performance measures in children and adolescents (Carlson and Moses 2001; Duckworth and Seligman 2005; Mischel et al. 1988, 1989; Tangney et al. 2004) as well as general intelligence in chimpanzees (Beran and Hopkins 2018). However, previous studies have also demonstrated that inhibitory control is not a unitary construct but rather context-specific (e.g. Tsukayama et al. 2012) and individual performances seem to differ between inhibitory control tasks (Addessi et al. 2013; Amici et al. 2008; Bray et al. 2014; e.g. Brucks et al. 2017a; Van den Bergh et al. 2006). Accordingly, inhibitory control has been categorized into:

1) Cognitive inhibition, referring to low-level actions not relevant to the task, e.g. the Stroop task, when there is a mismatch between the name of a colour and the colour it is printed in (i.e., the word ‘red’ is printed in blue ink instead of red ink), people take longer and are more prone to errors when asked to read the word than when the colour of the ink matches the name of the colour.

2) Motor inhibition, defined as blocking a prepotent motor response (e.g. going around a barrier instead of trying to go through it).
3) Self-control, usually measured in delay of gratification tasks, in which individuals are required to make a temporal decision between an immediate low-quality reward and a delayed high-quality reward (e.g. would you rather have a dry piece of bread now or a delicious steak in 15 min?) (Mischel et al. 1989).

Most of the studies to date comparing wolves and dogs have focused on motor and cognitive inhibition, whereas only one so far has compared the animals' behaviour in a delay of gratification task.

9.3.1 Cognitive Inhibition

Cognitive inhibition has only been looked at directly in the context of two reversal-learning studies until now. In this task, the animals first learned an association between the choice of a specific object and a reward (S+), and in a second step, these object contingencies were reversed in such a way that the previously negative object (S-) was now being rewarded.

In order to succeed, this task requires that the animal changes its previously successful strategy; thus, it needs to be flexible and at the same time also be able to inhibit choosing the previously rewarded object after the reward contingencies are reversed. Accordingly, reversal-learning tasks have frequently been used in the literature across a range of different species (e.g. pigeons: Laude et al. 2016; dogs: Tapp et al. 2003; mice: Pattij et al. 2003; tamarins: Hauser et al. 2002; corvids: Bond et al. 2007; chimpanzees: Wobber and Hare 2009), not only to assess learning capacities but also to test for behavioural flexibility and inhibitory control.

The first such study was conducted by the Frank's when the wolves and dogs were 15 weeks old, first with the mother-raised animals and later with the human-socialized animals (see Chap. 10 for the details). While the malamute pups performed much better than mother-raised wolf pups during the reversal, the human-socialized wolves outperformed both groups (Frank and Frank 1987), suggesting better cognitive inhibition abilities in the wolves compared to the dogs. However, a second study with adult animals at the Wolf Science Center found no differences in the reversal-learning performance of the wolves and dogs (Brucks et al. 2019, supplementary material). Accordingly, further studies will be necessary to understand whether wolves and dogs differ in regard to cognitive inhibition.

9.3.2 Motor Inhibition

In contrast to cognitive inhibition, quite a few studies have tested the motor inhibition skills of wolves and dogs. Again, the first studies were conducted by the Frank's, who used different paradigms to test passive and active motor inhibition. In the passive task (based on Scott et al. 1967), the 7-week-old animals were required to stay on a squared (1 x 1 m) platform placed in one corner of the testing enclosure for 15 sec (Frank 2011; Frank and Frank 1987). The experiment was repeated on 10 continuous days and included some training. Basically, the animals were wearing a choke chain that was connected to a string. The string ran through a pulley system above the platform to a tinted observation window, where the experimenter could then give a sharp tug to the choke chain via the string. On the first day, the animals were held in position on the platform for 5 sec with the help of the string and the choke chain. On the second day, the animal received the same 'preliminary trial' as on day one and then two training trials in which the subject was placed on the platform by one experimenter and punished by a sharp tug on the choke chain by an experimenter sitting behind the observation window if it tried to leave the platform before 15 sec was up. In the remaining days of the experiment, the animals always received a preliminary trial and then five training trials with the only difference that, according to performance, the time required to stay on the platform increased from 15 sec to 120 sec. No more than five corrections were administered per trial. The number of correction trials and how long the animals would spend on the platform before any correction trial was administered were measured.

Not surprisingly, the malamutes received fewer correction and stayed three times longer on the platform than the wolves (12 vs. 4 min, total intra-trial best performance times) (Frank and Frank 1987). However, since this first group of wolves might have been nervous due to their missing socialization with humans, Frank repeated the study with four wolf pups from the highly human-socialized wolves (Frank 2011). Moreover, they subjected an additional three highly socialized wolf pups to the same procedure but with the difference that the punishment was not administered by the choke chain, but rather by a sniper, who shot pellets at the animals to punish them. The results suggested that while species difference still emerged if human-socialized wolves and malamutes were compared when receiving the choke-chain punishment, the difference disappeared if the human-socialized wolves with sniper punishment were compared to the malamutes with choke-chain punishment. However, since the dogs were not tested with the sniper punishment and the sample size was really low, comparisons have to be treated with a grain of salt. Moreover, wolves and likely also other animals are sensitive to such stressors, making any conclusions about their cognitive abilities in such extreme tasks difficult.

However, in regard to staying in one place and wait, we have found that it is *very* hard if not impossible for wolves to learn. We usually train (with positive reinforcement) our animals to stay on a spot while we move away. Not surprisingly, dogs do rather well there as we all probably know also from training pet dogs. The funniest story I have heard in that respect was from an animal trainer, who had asked her Border collie to wait on the agility course, but she forgot him there, while she went to have a drink in the clubhouse. After 2 hours the dog walked by, looked in, and gave his owner a dirty look, before continuing to the car where he jumped into his box! Well, this ain't ever going to happen with a wolf. I tried over a year to train Kaspar to stay on a spot while moving away and then also moving around him. Staying for longer than maybe 10–20 sec was difficult and moving around him? The best I managed was half a circle! And I tried. . . hard. He never got the point why on earth he should stay in one place and not turn with me, come to me, or explore the test enclosure or room or home enclosure.

But let's come back to the Frank's. They also compared their animals in an active inhibition task, where the mother-raised animals had to learn to properly walk on a leash without jumping, biting the leash, or tugging, but rather walk at the trainer's side with no tension on the leash. The leash walk was 150 m long and incorporated doors, stairs, and gates. The animals were walked on a choke chain and a leather leash. The training started when the animals were 11 weeks old and continued for 10 days. In the first 2 days, the animals were carried to the starting point and then were led to the enclosure gate, where they were picked up again and taken to the kitchen to get a reward. Afterwards they were carried back to the gate and led back to the home barn. On the following days, the animals had to walk through the gate, two doors, up a short flight of stairs and then into the kitchen, where they received a food reward. Afterwards they were led back to the barn where they received a second food reward. The performance of the animals was only measured on the walk to the kitchen, not back. While there were few differences between wolves and dogs on day one, the malamutes nicely improved over sessions, inhibiting the unwanted behaviours, whereas the wolves did not improve and performed significantly worse on the last day. Again, these experiments were repeated with seven properly human-socialized wolves a few years later with four wolves being tested with the choke chain and three wearing a harness during the experiment. As with the mother-raised wolves, the species differences remained significant both in terms of best day performance and improvement, but the harness made a big difference for the wolves, and if the harness-trained socialized wolves were compared with the malamutes on their best day, no difference emerged. Although, as said before, these comparisons are difficult to interpret due to the small sample sizes and the differences in socialization and the restraints used, one valid conclusion for sure is that wolves really do not like choke chains! (Who would blame them? They are fine on a leash if trained nicely; see Chap. 6).

Another study, showing wolves' aversion to being restrained, was a 'pointing study' by Martha Gasci and colleagues (2009). The animals were 3–4 months old when participating in the study (see Chap. 7) and were held by a human hand-raiser while an experimenter tried to establish eye contact to then demonstrate the pointing gesture. Compared to the dogs, the wolves took longer to establish eye contact and struggled more, even trying to bite the hand-raiser. This was interpreted as better inhibitory control in the dogs compared to the wolves.

One test that has been used repeatedly by researchers to test motor inhibition is the barrier or detour task. Actually, our animals at the Wolf Science Center quickly learn that during walks through the forest, they need to return the same way they went into the forest if the leash gets stuck on a tree. Anyhow, more recent tests have used a slightly different approach with a barrier. Here a food reward is put behind a fence, and the animals need to detour the fence, i.e. move away from the food reward that they can see and want to obtain, in order to actually get access to the reward (Fig. 9.6).

Fig. 9.6 Sequence of a detouring task at the WSC during winter (pictures taken by Lisa Wallis)

Frank and Frank (1988) tested the mother-reared wolf and dog pups at the age of 6 weeks on three different days with different barriers. On the first day, the barrier was 2.4 m long, on the second day 7.3 m long, and on the third day, they used a U-shaped barrier. Each animal was subjected to two trials with each barrier with one trial reinforced with food and the second trial reinforced with access to an adult malamute that the pups were eager to interact with (social reward). The wolves committed significantly fewer errors (stopped or changed direction) in this test than the malamute pups.

Another study that used the barrier task was run at the Wolf Science Center, when the animals (14 wolves, 13 dogs) were 10 months old (Marshall-Pescini et al. 2015). We used a V-shaped fence and conducted a total of three trials. In the first trial, while one experimenter held the animal on a collar, the second experimenter showed a dead chick to the subject and then walked to the tip of the V and dropped the chick on the other side of the fence. Once the experimenter was back and positioned behind the animal, the first experimenter released the subject. In the second trial, a small door on one side of the fence was opened so that the animals could just put their head through to get the reward and, theoretically, did not need to walk around the fence to get to the food reward. During the opening of the door, the subject was distracted so that it did not see what was happening. Otherwise the procedure was the same. The third trial was identical to the first.

In the first trial, the wolves detoured the fence faster than the dogs (mean: wolves, 22.4; dogs, 34.1 sec), which was explained by the wolves spending less time at the fence in proximity to the food, i.e. showing more inhibitory control than the dogs. Interestingly, two dogs did not manage to solve the task within 1 min. In the second trial, the wolves and dogs did not differ from each other in the latency to get the food. However, four wolves actually never used the shortcut through the open door but instead sprinted around the fence. In the third trial, the wolves again outperformed the dogs with now three dogs unable to complete the task within 60 sec. Finally, the wolves became faster across the three trials and spent less time in proximity to the food, while the dog's performance did not improve.

Another detour paradigm that was used to test the inhibitory control of our animals when they were between 9 months and 3 years old was the cylinder task (MacLean et al. 2014; Marshall-Pescini et al. 2015). The cylinder was open on both sides and attached to a wooden base for support. In the training trials, the cylinder was opaque (Fig. 9.7), while in test trials it was exactly the same but transparent. Accordingly, in the test trials, the animals could see the food through the cylinder,

Fig. 9.7 Training trial with a pet dog. Figure reproduced from Marshall-Pescini et al. 2015. https://doi.org/10.1371/journal.pone.0118469; CC BY 4.0; http://creativecommons.org/licenses/by/4.0/

introducing a competition between the correct motor response (established during the training trials) and visual input (which could lead to the animal trying to approach the food directly and then bumping into the front of the cylinder). In order to move on to the test trials, the animal was required to make a correct first choice (i.e. not touch the cylinder on the exterior before entering the open end with the snout) in four of five consecutive training trials. Ten test trials were conducted right after the training criterion was reached.

While the two groups performed equally well in the training, dogs had more correct responses (e.g. not manipulating the transparent cylinder first) than the wolves in the test trials. It was interesting that the dogs basically performed at ceiling level in this task. One idea we had was that it might be an effect of dogs being more obedient than the wolves—after all, the test trials were preceded by a training phase where the animals learned a certain behaviour. Accordingly, we went and also tested pet dogs that had received intensive obedience training and those that were rather naïve in that aspect, and one group of pet dogs was tested without the prior training with the opaque cylinder. No differences emerged between all four groups (trained pet dogs, untrained pet dogs, no training phase pet dogs, pack dogs) in the number of correct test trials! Another aspect that we found interesting is that 8 of the 14 wolves tested scored on a par with dogs (9 or 10 out of 10 correct trials), whereas scores for the remaining individuals ranged from 2 to 7 correct trials out of 10. This suggests a much higher variability in the wolf population than in the dogs, hinting to the possibility that there may be multiple factors affecting wolves' inhibitory control.

Despite the fact that these were both detour tasks, the performance of the wolves or dogs did not correlate between tasks. Based on these different results, we decided to run an entire battery of tasks aimed at measuring both motor and cognitive inhibition of our animals, to try and understand if there are at least some common features that underlie the performance of the animals across tasks, as we previously showed in pet dogs (Brucks et al. 2017a). Specifically, we tested our adult animals in the following tasks (Brucks et al. 2019): motor inhibition skills were tested in a detour-reaching test ('box test') and a 'middle cup test'. In the 'box test', a big transparent Plexiglas box with a piece of sausage inside was presented to the animals. The box was open on one side, and the animals were required to find this open side to stick their head in and retrieve the sausage. In each trial, the position of the open side relative to the animal was changed. Since the box was transparent and thus the animals could see the reward, the individuals needed to inhibit the urge to try to get their nose close to the reward while circling around the box looking for the open side. Thus, this test measured the animals' ability to inhibit the motor action to get to the reward directly through the transparent barrier. This is similar to the cylinder task (see above), but since there was no training involved in the box test (except a habituation phase with an opaque box) and the configuration changed in each trial, we hoped that we would not find a similar ceiling effect as before.

The middle cup test consisted of a repeated choice between three transparent cups, set up next to each other (distance 45 cm between cups). In each trial, only two of the cups were baited with food, and they were either next to each other (control trials) or at the two ends (test trials). Importantly, the animals were allowed to make only two choices, so that they had to inhibit knocking over empty cups (loss of a reward). This was especially difficult in the test condition, where they had to skip the middle cup when moving from one side to the other to retrieve the two rewards. In contrast to the box test, this test additionally involved a choice component.

Cognitive inhibition was assessed in a reversal-learning test using an object choice task (see above).

Finally, we implemented a test combining aspects of motor and cognitive inhibition, which we refer to as the 'buzzer test'. Here, the animals were again faced with the problem of gaining access to a reward that was located inside a box, but this time they had to press a buzzer with their paw to open the box. While in the initial training phase, the box was opaque and the buzzer was placed next to the box, in the actual test, the box was transparent and the buzzer was positioned 2.5 m away from the box. Accordingly, in order to gain access to the reward, the individuals needed to inhibit manipulating the box but instead turn away from the reward and go and press the buzzer. This test included components of motor inhibition (i.e. manipulating the box, movement towards the buzzer), but also aspects of cognitive inhibition (i.e. referring to learned procedure (= pressing buzzer) in a tempting situation).

We tried to minimize the influence of the experimenters as much as possible by having the handlers positioned at least 2.5 m away and the experimenter hiding behind an opaque fence where possible (middle cup test, reversal-learning and buzzer task).

We first analysed the different tests separately and found that the control and training conditions were easier to solve for the animals than the test conditions, suggesting that we had indeed captured the animals' inhibitory control abilities. Especially the middle cup task turned out to be difficult and few animals were successful in the test trials. Moreover, for each task, we conducted a principal component analysis which revealed two components describing the behaviour of the animals in each test. Interestingly, of the eight emerging components, wolves and dogs differed in two: wolves exhibited a faster decision time in the middle cup test and likewise showed a shorter choice time in the reversal-learning test, which actually could be an indication of better working memory capacities of wolves compared to dogs (see above).

Secondly, we were interested in whether animals would behave consistently across tasks. However, similarly to previous studies (e.g. Bray et al. 2014; Brubaker et al. 2017), the individual performances in the single tests were not correlated with each other—neither in dogs nor wolves—again supporting the previous findings that inhibitory control is context-specific. We then took a different approach and investigated whether there are common mechanisms underlying the performance of the animals in the different tasks by using an exploratory approach (principal component analyses). Interestingly, three components emerged, which seemed to capture the underlying structure of the animals' behaviour. We referred to these components as 'motivation', 'flexibility', and 'perseveration'. The first component, motivation, was based on the individuals' activity and speed in making a choice, but also their attention. It is important so that an individual engages in problem solving to start with. The 'flexibility' component included measures relating to the animals' adaptability to changing aspects of the test (e.g. open sides, object contingencies), but also their persistence in sticking with their initially learned behaviour. As discussed above, the ability to flexibly switch between options, e.g. inhibit previously learned response patterns, has often been linked to success in changing environments (e.g. Bond et al. 2007). Finally, our 'perseveration' component included measures of persistence from the different tests. These perseverative actions (i.e. repetition of the same behaviours independent of feedback) are thought to occur due to inhibition problems and can prevent individuals from flexibly adjusting to a given situation (see Hauser et al. 1999, for a review). Two interesting aspects are emerging here: first, if we reconsider which factors might influence problem-solving abilities as outlined above, motivational factors, general-purpose mechanisms, and cognitive processes, we realize that our three components really mirror the motivational factors consisting of motivation, persistence, and behavioural flexibility. Second, these components were rather similar to our previous study with pet dogs (Brucks et al. 2017a), in that the same variables grouped together, suggesting that these are rather robust results and that the test battery seems to capture important aspects of not just inhibitory control but maybe even problem-solving performance.

Regarding the species differences, we only found an age—species interaction for the 'motivation' component indicating that younger dogs were less motivated than older dogs or wolves, whereas no such effect could be found for the wolves. Based on previous results, we might have also expected that the wolves and dogs differ at least also in the perseveration component since wolves have been shown to be more persistent than dogs. To reconcile this difference, it is worth noting that here the perseveration referred to continuing to do something despite a better option being available/known, rather than just continuing to try to solve the same problem as is tested in the persistence tests.

9.3.3 Self-Control

One other aspect of inhibitory control that has not been captured in any of these tests described so far is self-control. It is usually measured by applying an exchange paradigm in which individuals need to resist taking an initial low-value reward for a certain amount of time before they are offered the chance of exchanging it for a better reward. In children it has been tested using marshmallows and thus has been dubbed the marshmallow test (Mischel et al. 1989). Basically, in that human version, kids are left alone sitting at a table in an empty room with a marshmallow in front of them. They are told that if the experimenter comes back in 15 min and they have not eaten the marshmallow, they will get two instead of one marshmallow. Anyhow, performance in delay of gratification tasks seems to be affected by several factors including the quality of the reward (e.g. Auersperg et al. 2013; Brucks et al. 2017b; Hillemann et al. 2014; Wascher et al. 2012) as well as the strategy employed to resist the temptation to take the food (Evans and Beran 2007; Koepke et al. 2015). For example, human children use different strategies like playing (Steelandt et al. 2012) or sometimes even dozing to distract themselves. Pepperberg, a US-American scientist, showed footage at a conference once, where her grey parrot Alex displayed some of the same behavioural strategies as the kids including dozing to distract himself from taking the peanut in a delay of gratification task. It was hilarious. Similarly, some dogs have been observed to doze or look away from the reward during the task, and these dogs perform better than animals that focused on the rewards and did not use such strategies (Brucks et al. 2017b).

Obviously, we were interested also how the wolves and our pack dogs would react to such a task. So we set out to test young wolves (<1 year) using both a quantity exchange, where the animals had to exchange a smaller number of food items for a larger number of food items, and a quality exchange, where the animals had to exchange a lower-quality food for a better-quality food item. To better understand possible differences in the performance of wolves and dogs, we investigated the influence of the reward type (sausage vs. meat), the behavioural strategies, and their motivation to participate in the task. First, they learned in a training phase that they had to return a stone placed on a board fixed on rails by pushing it with the nose to the other side of a barrier where the experimenter was sitting (Fig. 9.8). In return, the experimenter pushed out a second board with a reward placed on top to the animal's side of the barrier. Once the animals had learned this basic action, they learned that they needed to wait 2 sec before they could give back the stone. In the next training step, they learned that they could exchange a

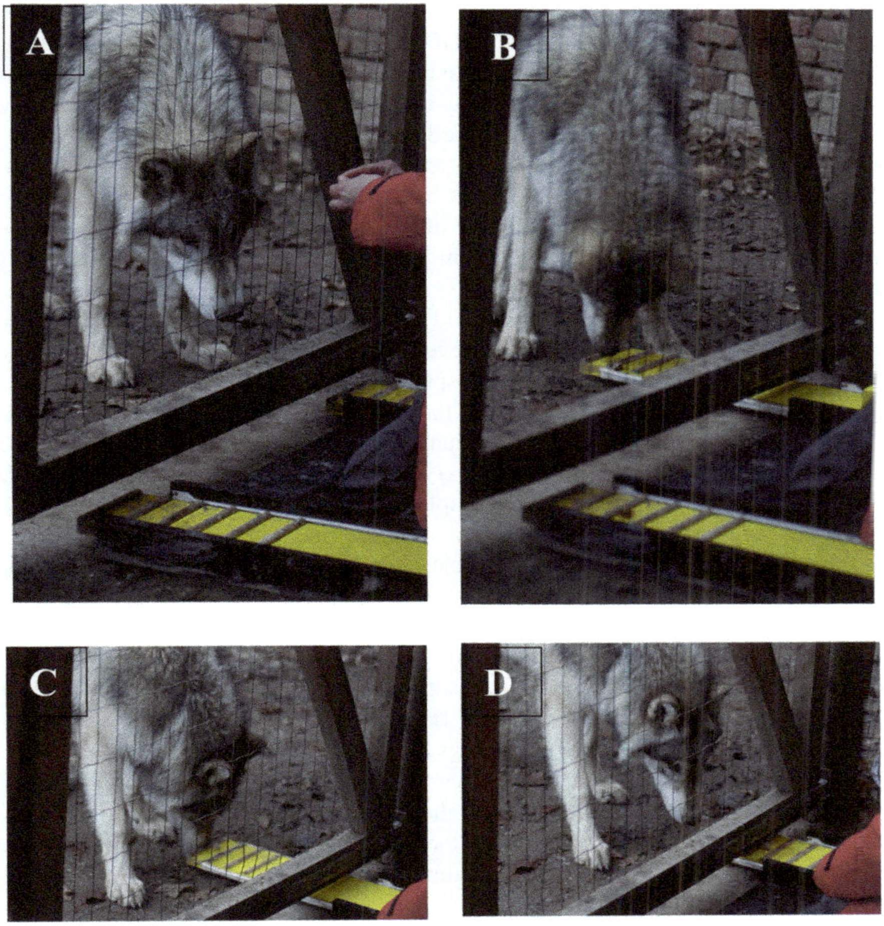

Fig. 9.8 Geronimo doing the exchange task in the training phase with a stone and a piece of dry food. (**a**) The experimenter shows him the potential rewards. She puts a stone on one tray and a piece of dry food on the other tray. Then, she pushes the tray with the stone on it out, so that it becomes available to Geronimo on the other side of the fence. (**b**) Geronimo looks at the stone. (**c**/**d**) He pushes the tray with the stone on it with his nose back on the other side of the fence. Now the experimenter will push the tray with the dry food on it out to the wolf. With training the time delay between the low-value reward becoming available to the animal and being allowed to push it back (the experimenter will hold the tray in place by pushing from the other side) will be increased to 2 sec. Pictures reproduced from Range et al. 2020. https://link.springer.com/article/10.1007/s10071-020-01346-7/; CC BY 4.0; http://creativecommons.org/licenses/by/4.0/

piece of dry food (low-value reward; see above) for a piece of sausage or meat (high-value reward) or a smaller vs. a larger quantity of dry food (1 vs. 4, that wolves and dogs can discriminate easily; see Chap. 10). Once they understood that, they again were trained to wait at least 2 sec before they could return the dry food. In the test phase, two different types of trials were conducted: *motivation trials* with the same delay duration as in the training (2 s) and *delay trials* with increasing delay duration. Motivation trials were included to test that the animals were sufficiently motivated to participate in the experiment. In the delay trials, the delay between the immediately accessible low-quality/low-quantity reward and the delayed high-quality/high-quantity reward was increased in consecutive steps (2 s, 5 s, 10s, 15 s, 20s, 25 s, 35 s, 45 s, 60s, 75 s, 100 s, 150 s, 175 s). Six test sessions were conducted per delay stage before passing on to the next stage. If the subject was successful (= waiting for high-quality reward) in at least one trial within the six sessions, it was tested in the next step with a longer delay.

We started by training six 10-month-old wolves with three trained for the quantity and three for the quality condition. However, two of the three wolves in the quantity group failed to reach training criterion despite several weeks of training, and the third was excluded because he did not relax in the test room. Therefore, we abandoned the quantity condition, pursuing only the quality exchange condition. In the end, we tested eight wolves (age range: 1–6 years) and five dogs (age, 1 years) (Range et al. 2020). Since it took several years to complete the dataset, this test probably took as much patience on our side as on the side of the animals.

However, the results at least were clear: dogs outperformed the wolves with waiting an average of 66 s, while wolves only waited on average 23 s. Also, if we look at the top performance, a dog was clearly the winner: Asali waited 150 s, whereas Aragorn reached the maximum delay of 35 s (see Fig. 9.3). Interestingly, our Wolf Science Center dogs waited almost twice as long (mean) compared to 12 pet dogs tested in a similar set-up (Brucks et al. 2017b). Also, variance in performance was much higher in pet dogs than in wolves. This difference could be due either to the relatively standardized conditions at the Wolf Science Center, both in regard to the training procedures and the life experiences compared, to pet dogs that receive individualized training by their owners and have vastly variable life experiences. Secondly, we tested the animals face to face with one of our animal trainers, while in the study with pet dogs, the experimenter and owner were hidden behind a cover and not visible during the test.

This difference in set-up might be relevant, since we found in other studies that dogs are much easier to socially inhibit than wolves. For example, in our cooperative string-pulling task with humans (Chap. 8), we saw that wolves were more likely to steal the rope from a human partner, while dogs hardly ever did that. Moreover, the dogs were also more likely to wait for the human to initiate an action. Similarly, in the pointing studies (Chap. 7), we showed that the dogs were more likely to follow the pointing gesture than wolves, which might be interpreted as following a command. All these studies suggest that humans can easily inhibit dogs (even if not intending to do so), whereas wolves might care less. In the current experiment, having the human experimenter sitting in front of the animal and facing it while holding up the better reward might have led dogs to perceive the situation as more constrained, waiting for cues from the human experimenter that they could take the food. A similar argument could explain also the patience pet dogs demonstrated in a previous study (Leonardi et al. 2012), where two dogs waited up to 20 min for the

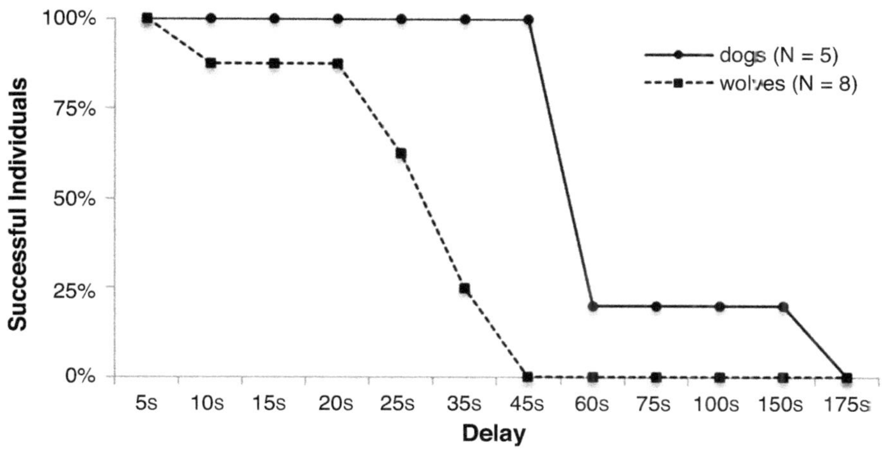

Fig. 9.9 Performance of individuals in the different delays. The figure shows the percentage of successful individuals across delay stages separately for dogs and wolves. Figure reproduced from Range et al. 2020. https://link.springer.com/article/10.1007/s10071-020-01346-7/; CC BY 4.0; http://creativecommons.org/licenses/by/4.0/

better reward. In that study, the human experimenter also faced them and thus might have inhibited the dogs from taking the food—at least the more sensitive ones. Thus, it is possible that our dogs were socially more inhibited by the experimenter at the Wolf Science Center (Fig. 9.9).

Furthermore, food quality (meat/sausage) did not influence how long the animals waited for the better reward. Interestingly, dogs and wolves did vary somewhat in their strategies that they employed to resist the temptation of taking the food. Dogs were more successful if they moved less during longer delays, whereas locomotion did not influence wolves' performance. Instead wolves performed better if they alternated their gaze between the high- and low-value rewards. Dogs and wolves behaved similarly in that they looked away from the food (dogs more so than wolves), as has been reported also for other species including children (Mischel and Ebbesen 1970), chimpanzees (Evans and Beran 2007), a grey parrot (Koepke et al. 2015), and pet dogs (Brucks et al. 2017b).

So, while this study suggests that dogs are much better at self-control, we will need to repeat the study with a new cohort of dogs, testing the animals in the absence of a human experimenter. It will be interesting to see if then the variance between animals increases, though we doubt it will change anything in the species difference we found. But who knows, it would not be the first surprising result.

9.3.4 Summary Inhibition

Studies on inhibitory control abilities draw a mixed picture. While the dogs seem to be better in self-control (or is it social inhibition/control?), every other aspect of inhibitory control is not quite as clear-cut, with some studies showing better performance of the wolves and others of the dogs. Interestingly, in the two studies using barriers, wolves performed much better than the dogs. While of course this could be indeed linked to inhibitory control, it is also possible that the wolves have a better spatial perception/orientation than the dogs, which would allow them to realize that they can reach the food quickly by just doing some additional steps. To date, however, we have no knowledge if wolves and dogs indeed differ in this aspect. Finally, in the components underlying the performance of the animals across a range of inhibition tasks, no real differences emerged except maybe that the wolves are faster in making decisions than the dogs. This might, again, be more related to the overall processing capacity of wolves compared to dogs, rather than inhibition. However, again, we know nothing about processing capacity to date. Future studies should pay attention to using set-ups that exclude humans as potential influencing factors (Table 9.2).

9.4 General Summary

The data available to date suggest that while wolves are more neophobic than dogs, they are also more explorative and manipulative when encountering new objects in their environment. These are important aspects since they suggest that, even if there are no differences in the cognitive abilities between wolves and dogs, wolves would probably be more successful in solving all sorts of physical tasks just because they would have better chances to stumble over the solution during their manipulations. In other words, they would be more successful than dogs irrespective of whether both species approach problems with a trial-and error learning approach rather or trying to figure out the underlying properties of the tasks. In combination with being more sensitive towards details of a situation and maybe better working memory capacities, wolves should be in general better problem-solvers. With regard to inhibitory control, dogs might have an advantage in some situations especially if humans are involved since they might be more socially inhibited to react right away and thus might have more time to understand the set-up.

Table 9.2 Summary of results of the studies focusing on general-purpose mechanisms

Parameters	Task	Age	Species	References
Cognitive inhibition	Reversal learning	15 d	W > D	Frank and Frank (1987)
	Reversal learning	2–9 y	W = D	Brucks et al. (2019)
Motor inhibition	Staying on platform (choke chain) Wolves with harness vs. malamutes with choke chain	7 w	W < D W = D	Frank (2011); Frank and Frank (1987)
	Walking on the leash (choke chain) Wolves with harness vs. malamutes with choke chain	11 w	W < D W = D	Frank (2011); Frank and Frank (1987)
	Restrained during pointing test	3–4 m	W < D	Gácsi et al. (2009)
	Detour task Barrier (U-shaped fence) Detour (V-shaped fence) Cylinder task Box task	 6 w 10 m 3–6 y 2–9 y	 W > D W > D W < D W = D	 Frank and Frank (1988) Marshall-Pescini et al. (2015) Brucks et al. (2019)
Motor + cognitive inhibition	Buzzer task	2–9 y	W = D	Brucks et al. (2019)
Self-control	Delay of gratification	1–6 y	W < D	Range et al. (2020)

D = days, w = weeks, m = months, y = years

References

Addessi E, Paglieri F, Beran MJ, Evans TA, Macchitella L, De Petrillo F, Focaroli V (2013) Delay choice versus delay maintenance: different measures of delayed gratification in capuchin monkeys (*Cebus apella*). J Comp Psychol 127:392–398. https://doi.org/10.1037/a0031869

Amici F, Aureli F, Call J (2008) Fission-fusion dynamics, behavioral flexibility, and inhibitory control in primates. Curr Biol 18:1415–1419. https://doi.org/10.1016/j.cub.2008.08.020

Auersperg AMI, Laumer IB, Bugnyar T (2013) Goffin cockatoos wait for qualitative and quantitative gains but prefer 'better' to 'more'. Biol Lett 9:20121092. https://doi.org/10.1098/rsbl.2012.1092

Auersperg AMI, von Bayern AMP, Gajdon GK, Huber L, Kacelnik A (2011) Flexibility in problem solving and tool use of kea and new Caledonian crows in a multi access box paradigm. PLoS One 6:e20231. https://doi.org/10.1371/journal.pone.0020231

Barto AG (2013) Intrinsic motivation and reinforcement learning. In: Baldassarre G, Mirolli M (eds) Intrinsically motivated learning in natural and artificial systems. Springer, Berlin, Heidelberg, pp 17–47. https://doi.org/10.1007/978-3-642-32375-1_2

Benson-Amram S, Dantzer B, Stricker G, Swanson EM, Holekamp KE (2016) Brain size predicts problem-solving ability in mammalian carnivores. Proc Natl Acad Sci U S A 113:2532–2537. https://doi.org/10.1073/pnas.1505913113

Benson-Amram S, Holekamp KE (2012) Innovative problem solving by wild spotted hyenas. Proc R Soc B 279:4087–4095. https://doi.org/10.1098/rspb.2012.1450

Benson-Amram S, Weldele ML, Holekamp KE (2013) A comparison of innovative problem-solving abilities between wild and captive spotted hyaenas, *Crocuta crocuta*. Anim Behav 85: 349–356. https://doi.org/10.1016/j.anbehav.2012.11.003
Beran MJ, Hopkins WD (2018) Self-control in chimpanzees relates to general intelligence. Curr Biol 28:574–579. https://doi.org/10.1016/j.cub.2017.12.043
Berridge KC (2004) Motivation concepts in behavioral neuroscience. Physiol Behav 81:179–209. https://doi.org/10.1016/j.physbeh.2004.02.004
Bond AB, Kamil AC, Balda RP (2007) Serial reversal learning and the evolution of behavioral flexibility in three species of north American corvids (*Gymnorhinus cyanocephalus, Nucifraga columbiana, Aphelocoma californica*). J Comp Psychol 121:372–379. https://doi.org/10.1037/0735-7036.121.4.372
Borrego N, Gaines M (2016) Social carnivores outperform asocial carnivores on an innovative problem. Anim Behav 114:21–26. https://doi.org/10.1016/j.anbehav.2016.01.013
Bradshaw JWS, Healey LM, Thorne CJ, Macdonald DW, Arden-Clark C (2000) Differences in food preferences between individuals and populations of domestic cats *Felis silvestris catus*. Appl Anim Behav Sci 68. https://doi.org/10.1016/S0168-1591(00)00102-7
Bray EE, MacLean EL, Hare BA (2014) Context specificity of inhibitory control in dogs. Anim Cogn 17:15–31. https://doi.org/10.1007/s10071-013-0633-z
Brubaker L, Dasgupta S, Bhattacharjee D, Bhadra A, Udell MAR (2017) Differences in problem-solving between canid populations: do domestication and lifetime experience affect persistence? Anim Cogn 20:717–723. https://doi.org/10.1007/s10071-017-1093-7
Brucks D, Marshall-Pescini S, Range F (2019) Dogs and wolves do not differ in their inhibitory control abilities in a non-social test battery. Anim Cogn 22:1–15. https://doi.org/10.1007/s10071-018-1216-9
Brucks D, Marshall-Pescini S, Wallis LJ, Huber L, Range F (2017a) Measures of Dogs' inhibitory control abilities do not correlate across tasks. Front Psych 8:849. https://doi.org/10.3389/fpsyg.2017.00849
Brucks D, Soliani M, Range F, Marshall-Pescini S (2017b) Reward type and behavioural patterns predict dogs' success in a delay of gratification paradigm. Sci Rep 7:42459. https://doi.org/10.1038/srep42459
Carlson SM, Moses LJ (2001) Individual differences in inhibitory control and children's theory of mind. Child Dev 72:1032–1053. https://doi.org/10.1111/1467-8624.00333
Croy MI, Hughes RN (1991) Effects of food supply, hunger, danger and competition on choice of foraging location by the fifteen-spined stickleback, *Spinachia spinachia* L. Anim Behav 42: 131–139. https://doi.org/10.1016/S0003-3472(05)80613-X
Daneman M, Carpenter PA (1980) Individual differences in working memory and reading. J Verb Learn Verb Behav 19:450–466. https://doi.org/10.1016/S0022-5371(80)90312-6
Day JEL, Kyriazakis I, Rogers PJ (1998) Food choice and intake: towards a unifying framework of learning and feeding motivation. Nutr Res Rev 11:25–43. https://doi.org/10.1079/Nrr19980004
Diamond A (1990) Developmental time course in human infants and infant monkeys, and the neural bases of, inhibitory control in reaching. In: Diamond A (ed) The development and neural bases of higher cognitive functions. New York Academy of Sciences, New York, pp 637–676
Drewnowski A, Greenwood MRC (1983) Cream and sugar: human preferences for high-fat foods. Physiol Behav 30:629–633. https://doi.org/10.1016/0031-9384(83)90232-9
Duckworth AL, Seligman MEP (2005) Self-discipline outdoes IQ in predicting academic performance of adolescents. Psychol Sci 16:939–944. http://www.jstor.org/stable/40064361
Engle RW (2002) Working memory capacity as executive attention. Curr Dir Psychol Sci 11:19–23. https://doi.org/10.1111/1467-8721.00160
Evans TA, Beran MJ (2007) Chimpanzees use self-distraction to cope with impulsivity. Biol Lett 3: 599–602. https://doi.org/10.1098/rsbl.2007.0399
Fentress JC (1967) Observations on behavioral development of a hand reared male timber wolf. Am Zool 6:339–351. https://doi.org/10.1093/icb/7.2.339

Fleming PJS, Nolan H, Jackson SM, Ballard G-A, Bengsen A, Brown WY, Meek PD, Mifsud G, Pal SK, Sparkes J (2017) Roles for the Canidae in food webs reviewed: where do they fit? Food Webs 12:14–34
Fontenot MB, Watson SL, Roberts KA, Miller RW (2007) Effects of food preferences on token exchange and behavioural responses to inequality in tufted capuchin monkeys, *Cebus apella*. Anim Behav 74:487–496. https://doi.org/10.1016/j.anbehav.2007.01.015
Fox MW (1964) The ontogeny of behaviour and neurologic responses in the dog. Anim Behav 12: 301–310. https://doi.org/10.1016/0003-3472(64)90016-8
Frank H (2011) Wolves, dogs, rearing and reinforcement: complex interactions underlying species differences in training and problem-solving performance. Behav Genet 41:830–839. https://doi.org/10.1007/s10519-011-9454-5
Frank H, Frank M (1987) The University of Michigan canine information-processing project (1979-1981). In: Frank H (ed) Man and Wolf. Advances, Issues, and Problems in Captive Wolf Research. Dr. W. Junk Publishers, NL, Dordrecht, pp 143-167
Frank H, Frank MG (1982) Comparison of problem-solving performance in six-week-old wolves and dogs. Anim Behav 30:95–98. https://doi.org/10.1016/S0003-3472(82)80241-8
Frank H, Frank MG (1985) Comparative manipulation test performance in ten-week-old wolves (*Canis lupus*) and Alaskan malamutes (*Canis familiaris*): a Piagetian interpretation. J Comp Psychol 99:266–274. https://doi.org/10.1037/0735-7036.99.3.266
Frank MG, Frank H (1988) Food reinforcement versus social reinforcement in timber wolf pups. Bull Psychonom Soc 26:467–468. https://doi.org/10.3758/BF03334915
Gácsi M, Győri B, Virányi Z, Kubinyi E, Range F, Belényi B, Miklósi A (2009) Explaining dog wolf differences in utilizing human pointing gestures: selection for synergistic shifts in the development of some social skills. PLoS One 4:e6584. https://doi.org/10.1371/journal.pone.0006584
Gilby IC, Wrangham RW (2007) Risk-prone hunting by chimpanzees (*Pan troglodytes schweinfurthii*) increases during periods of high diet quality. Behav Ecol Sociobiol 61:1771–1779. https://doi.org/10.1007/s00265-007-0410-6
Greenberg R, Mettke-Hofmann C (2001) Ecological aspects of Neophobia and Neophilia in birds. In: Nolan V, Thompson CF (eds) Current ornithology, vol 16. Springer, Boston, MA, pp 119–178. https://doi.org/10.1007/978-1-4615-1211-0_3
Griffin AS, Diquelou M, Perea M (2014) Innovative problem solving in birds: a key role of motor diversity. Anim Behav 92:221–227. https://doi.org/10.1016/j.anbehav.2014.04.009
Griffin AS, Diquelou MC (2015) Innovative problem solving in birds: a cross-species comparison of two highly successful passerines. Anim Behav 100:84–94. https://doi.org/10.1016/j.anbehav.2014.11.012
Griffin AS, Guez D (2014) Innovation and problem solving: a review of common mechanisms. Behav Proc 109:121–134. https://doi.org/10.1016/j.beproc.2014.08.027
Griffin AS, Guez D (2017) Solving foraging problems: top-down and bottom-up perspectives on the role of cognition. In: ten Cate C, Healy SD (eds) Avian Cognition. Cambridge University Press, Cambridge, pp 119–140. https://doi.org/10.1017/9781316135976.008
Hambrick DZ, Oswald FL, Darowski ES, Rench TA, Brou R (2010) Predictors of multitasking performance in a synthetic work paradigm. Appl Cognit Psychol 24:1149–1167. https://doi.org/10.1002/acp.1624
Hansen Wheat C, van der Bijl W, Temrin H (2019) Dogs, but not wolves, lose their sensitivity toward novelty with age. Front Psych 10:2001. https://doi.org/10.3389/fpsyg.2019.02001
Haun DBM, Nawroth C, Call J (2011) Great apes' risk-taking strategies in a decision making task. PLoS One 6:e28801. https://doi.org/10.1371/journal.pone.0028801
Hauser MD, Kralik J, Botto-Mahan C (1999) Problem solving and functional design features: experiments on cotton-top tamarins, *Saguinus oedipus oedipus*. Anim Behav 57:565–582
Hauser MD, Santos LR, Spaepen GM, Pearson HE (2002) Problem solving, inhibition and domain-specific experience: experiments on cottontop tamarins, *Saguinus oedipus*. Anim Behav 64: 387–396. https://doi.org/10.1006/anbe.2002.3066

Heilbronner SR, Rosati AG, Stevens JR, Hare B, Hauser MD (2008) A fruit in the hand or two in the bush? Divergent risk preferences in chimpanzees and bonobos. Biol Lett 4:246–249. https://doi.org/10.1098/rsbl.2008.0081

Hiestand L (2011) A comparison of problem-solving and spatial orientation in the wolf (*Canis lupus*) and dog (*Canis familiaris*). Behav Genet 41:840–857. https://doi.org/10.1007/s10519-011-9455-4

Hillemann F, Bugnyar T, Kotrschal K, Wascher CAF (2014) Waiting for better, not for more: corvids respond to quality in two delay maintenance tasks. Anim Behav 90:1–10. https://doi.org/10.1016/j.anbehav.2014.01.007

Holekamp KE, Benson-Amram S (2017) The evolution of intelligence in mammalian carnivores. Interface Focus 7:20160108. https://doi.org/10.1098/rsfs.2016.0108

Huber L, Gajdon GK (2006) Technical intelligence in animals: the kea model. Anim Cogn 9:295–305. https://doi.org/10.1007/s10071-006-0033-8

Huebner F, Fichtel C (2015) Innovation and behavioral flexibility in wild red-fronted lemurs (*Eulemur rufifrons*). Anim Cogn 18:777–787. https://doi.org/10.1007/s10071-015-0844-6

Kane MJ, Engle RW (2000) Working-memory capacity, proactive interference, and divided attention: limits on long-term memory retrieval. J Exp Psychol-Learn Mem Cogn 26:336. https://doi.org/10.1037/0278-7393.26.2.336

Klinghammer E, Goodmann PA (1987) Socialization and management of wolves in captivity. In: Frank H (ed) Man and Wolf. Advances, Issues, and Problems in Captive Wolf Research. Dr. W. Junk Publishers, NL, Dordrecht, pp 31–59

Koepke AE, Gray SL, Pepperberg IM (2015) Delayed gratification: a Grey parrot (*Psittacus erithacus*) will wait for a better reward. J Comp Psychol 129:339–346. https://doi.org/10.1037/a0039553

Laska M, Hernandez Salazar LT, Rodriguez Luna E (2000) Food preferences and nutrient composition in captive spider monkeys, *Ateles geoffroyi*. Int J Primatol 21:671–683. https://doi.org/10.1023/A:1005517421510

Laude JR, Pattison KF, Rayburn-Reeves RM, Michler DM, Zentall TR (2016) Who are the real bird brains? Qualitative differences in behavioral flexibility between dogs (*Canis familiaris*) and pigeons (*Columba livia*). Anim Cogn 19:163–169. https://doi.org/10.1007/s10071-015-0923-8

Lazzaroni M et al (2019) The role of life experience in affecting persistence: a comparative study between free-ranging dogs, pet dogs and captive pack dogs. PLoS One 14:e0214806. https://doi.org/10.1371/journal.pone.0214806

Lefebvre L (2011) Taxonomic counts of cognition in the wild. Biol Lett 7:631–633. https://doi.org/10.1098/rsbl.2010.0556

Lefebvre L, Sol D (2008) Brains, lifestyles and cognition: are there general trends? Brain Behav Evol 72:135–144. https://doi.org/10.1159/000151473

Leonardi RJ, Vick SJ, Dufour V (2012) Waiting for more: the performance of domestic dogs (*Canis familiaris*) on exchange tasks. Anim Cogn 15:107–120. https://doi.org/10.1007/s10071-011-0437-y

Lord K (2013) A comparison of the sensory development of wolves (*Canis lupus lupus*) and dogs (*Canis lupus familiaris*). Ethology 119:110–120. https://doi.org/10.1111/eth.12044

MacLean EL et al (2014) The evolution of self-control. Proc Natl Acad Sci U S A 111:E2140–E2148. https://doi.org/10.1073/pnas.1323533111

Marshall-Pescini S, Besserdich I, Kratz C, Range F (2016) Exploring differences in dogs' and wolves' preference for risk in a foraging task. Front Psych 7. https://doi.org/10.3389/fpsyg.2016.01241

Marshall-Pescini S, Rao A, Virányi Z, Range F (2017a) The role of domestication and experience in 'looking back' towards humans in an unsolvable task. Sci Rep 7:46636. https://doi.org/10.1038/srep46636

Marshall-Pescini S, Virányi Z, Kubinyi E, Range F (2017b) Motivational factors underlying problem solving: comparing wolf and dog Puppies' explorative and neophobic behaviors at 5, 6, and 8 weeks of age. Front Psych 8. https://doi.org/10.3389/fpsyg.2017.00180

Marshall-Pescini S, Virányi Z, Range F (2015) The effect of domestication on inhibitory control wolves and dogs compared. PLoS One 10:e0118469. https://doi.org/10.1371/journal.pone 0118469

Mech LD, Smith DW, MacNulty DR (2015) Wolves *on the* Hunt. The Behavior of Wolves Hunting Wild Prey. University of Chicago Press, Chicago, London. https://doi.org/10.7208/chicago/9780226255286.001.0001

Mettke-Hofmann C, Winkler H, Leisler B (2002) The significance of ecological factors for exploration and Neophobia in parrots. Ethology 108:249–272. https://doi.org/10.1046/j.1439-0310.2002.00773.x

Miklósi Á, Kubinyi E, Topál J, Gácsi M, Virányi Z, Csányi V (2003) A simple reason for a big difference: wolves do not look Back at humans, but dogs do. Curr Biol 13 763–766. https://doi.org/10.1016/s0960-9822(03)00263-x

Mischel W, Ebbesen EB (1970) Attention in delay of gratification. J Pers Soc Psychol 16:329–337. https://doi.org/10.1037/h0029815

Mischel W, Shoda Y, Peake PK (1988) The nature of adolescent competencies predicted by preschool delay of gratification. J Pers Soc Psychol 54:687–696. https://doi.org/10.1037/0022-3514.54.4.687

Mischel W, Shoda Y, Rodriguez ML (1989) Delay of gratification in children. Science 244:933–938. https://doi.org/10.1126/science.2658056

Morand-Ferron J, Cole EF, Rawles JEC, Quinn JL (2011) Who are the innovators? A field experiment with 2 passerine species. Behav Ecol 22:1241–1248. https://doi.org/10.1093/beheco/arr120

Moretti L, Hentrup M, Kotrschal K, Range F (2015) The influence of relationships on neophobia and exploration in wolves and dogs. Anim Behav 107:159–173. https://doi.org/10.1016/j.anbehav.2015.06.008

Olsen MR (2018) A case for methodological overhaul and increased study of executive function in the domestic dog (*Canis lupus familiaris*). Anim Cogn 21:175–195. https://doi.org/10.1007/s10071-018-1162-6

Packard JM (2003) Wolf behavior: reproductive, social and intelligent. In: Mech LD, Boitani L (eds) Wolves: behaviour, ecology and conservation. University of Chicago Press, Chicago, pp 35–65

Pattij T, Broersen LM, van der Linde J, Groenink L, van der Gugten J, Maes RAA, Olivier B (2003) Operant learning and differential-reinforcement-of-low-rate 36-s responding in 5-HT_{1A} and 5-HT_{1B} receptor knockout mice. Behav Brain Res 141:137–145. https://doi.org/10.1016/S0166-4328(02)00345-5

Ramsey G, Bastian ML, van Schaik C (2007) Animal innovation defined and operationalized. Behav Brain Sci 30:393–437. https://doi.org/10.1017/S0140525x07002373

Range F, Brucks D, Virányi Z (2020) Dogs wait longer for better rewards than wolves in a delay of gratification task: but why? Anim Cogn 23:443–453. https://doi.org/10.1007/s10071-020-01346-7

Rao A, Bernasconi L, Lazzaroni M, Marshall-Pescini S, Range F (2018a) Differences in persistence between dogs and wolves in an unsolvable task in the absence of humans. PeerJ 6:e5944. https://doi.org/10.7717/peerj.5944

Rao A, Range F, Kadletz K, Kotrschal K, Marshall-Pescini S (2018b) Food preferences of similarly raised and kept captive dogs and wolves. PLoS One 13:e0203165. https://doi.org/10.1371/journal.pone.0203165

Réale D, Reader SM, Sol D, McDougall PT, Dingemanse NJ (2007) Integrating animal temperament within ecology and evolution. Biol Rev 82:291–318. https://doi.org/10.1111/j.1469-185X.2007.00010.x

Rheingold HL (ed) (1963) Maternal behavior in mammals. Wiley, New York

Rosati AG, Hare B (2012) Decision making across social contexts: competition increases preferences for risk in chimpanzees and bonobos. Anim Behav 84:869–879. https://doi.org/10.1016/j.anbehav.2012.07.010

Rosati AG, Hare B (2013) Chimpanzees and bonobos exhibit emotional responses to decision outcomes. PLoS One 8:e63058. https://doi.org/10.1371/journal.pone.0063058
Rosati AG, Stevens JR, Hare B, Hauser MD (2007) The evolutionary origins of human patience: temporal preferences in chimpanzees, bonobos, and human adults. Curr Biol 17:1663–1668. https://doi.org/10.1016/j.cub.2007.08.033
Rowe C, Healy SD (2014) Measuring variation in cognition. Behav Ecol 25:1287–1292. https://doi.org/10.1093/beheco/aru090
Scott JP, Fuller JL (1965) Genetics and the social behavior of the dog. The classic study. University of Chicago Press, Chicago, London
Scott JP, Shepard JH, Werboff J (1967) Inhibitory training of dogs: effects of age at training in basenjis and Shetland sheepdogs. J Psychol 66:237–252. https://doi.org/10.1080/00223980.1967.10544902
Sih A, Del Giudice M (2012) Linking behavioural syndromes and cognition: a behavioural ecology perspective. Phil Trans R Soc B 367:2762–2772. https://doi.org/10.1098/rstb.2012.0216
Steelandt S, Thierry B, Broihanne M-H, Dufour V (2012) The ability of children to delay gratification in an exchange task. Cognition 122:416–425. https://doi.org/10.1016/j.cognition.2011.11.009
Tangney JP, Baumeister RF, Boone AL (2004) High self-control predicts good adjustment, less pathology, better grades, and interpersonal success. J Pers 72:271–324. https://doi.org/10.1111/j.0022-3506.2004.00263.x
Tapp PD, Siwak CT, Estrada J, Head E, Muggenburg BA, Cotman CW, Milgram NW (2003) Size and reversal learning in the beagle dog as a measure of executive function and inhibitory control in aging. Learn Mem 10:64–73. https://doi.org/10.1101/lm.54403
Taylor AH, Hunt GR, Medina FS, Gray RD (2009) Do new Caledonian crows solve physical problems through causal reasoning? Proc R Soc B 276:247–254. https://doi.org/10.1098/rspb.2008.1107
Taylor AH, Medina FS, Holzhaider JC, Hearne LJ, Hunt GR, Gray RD (2010) An investigation into the cognition behind spontaneous string pulling in new Caledonian crows. PLoS One 5. https://doi.org/10.1371/journal.pone.0009345
Tebbich S, Griffin AS, Peschl MF, Sterelny K (2016) From mechanisms to function: an integrated framework of animal innovation. Phil Trans R Soc B 371:20150195. https://doi.org/10.1098/rstb.2015.0195
Teschke I, Cartmill EA, Stankewitz S, Tebbich S (2011) Sometimes tool use is not the key: no evidence for cognitive adaptive specializations in tool-using woodpecker finches. Anim Behav 82:945–956. https://doi.org/10.1016/j.anbehav.2011.07.032
Teschke I, Wascher CAF, Scriba MF, von Bayern AMP, Huml V, Siemers B, Tebbich S (2013) Did tool-use evolve with enhanced physical cognitive abilities? Phil Trans R Soc B 368:20120418. https://doi.org/10.1098/rstb.2012.0418
Thornton A, Samson J (2012) Innovative problem solving in wild meerkats. Anim Behav 83:1459–1468. https://doi.org/10.1016/j.anbehav.2012.03.018
Topál J, Gácsi M, Miklósi Á, Virányi Z, Kubinyi E, Csányi V (2005) Attachment to humans: a comparative study on hand-reared wolves and differently socialized dog puppies. Anim Behav 70:1367–1375. https://doi.org/10.1016/j.anbehav.2005.03.025
Tsukayama E, Duckworth AL, Kim B (2012) Resisting everything except temptation: evidence and an explanation for domain-specific impulsivity. Eur J Personal 26:318–334. https://doi.org/10.1002/per.841
Udell MAR (2015) When dogs look back: inhibition of independent problem-solving behaviour in domestic dogs (*Canis lupus familiaris*) compared with wolves (*Canis lupus*). Biol Lett 11: 20150489. https://doi.org/10.1098/rsbl.2015.0489
Van den Bergh F, Spronk M, Ferreira L, Bloemarts E, Groenink L, Olivier B, Oosting R (2006) Relationship of delay aversion and response inhibition to extinction learning, aggression, and sexual behaviour. Behav Brain Res 175:75–81. https://doi.org/10.1016/j.bbr.2006.08.003

van Oers K, Drent PJ, de Goede P, van Noordwijk AJ (2004) Realized heritability and repeatability of risk-taking behaviour in relation to avian personalities. Proc R Soc Lond B 271:65–73. https://doi.org/10.1098/rspb.2003.2518

Wascher CAF, Dufour V, Bugnyar T (2012) Carrion crows cannot overcome impulsive choice in a quantitative exchange task. Front Psych 3:118. https://doi.org/10.3389/fpsyg.2012.00118

Wobber V, Hare B (2009) Testing the social dog hypothesis: are dogs also more skilled than chimpanzees in non-communicative social tasks? Behav Proc 81:423–428. https://doi.org/10.1016/j.beproc.2009.04.003

Zimen E (1987) Ontogeny of approach and flight behavior towards humans in wolves, poodles and wolf-poodle hybrids. In: Frank H (ed) Man and Wolf. Advances, Issues, and Problems in Captive Wolf Research. Dr. W. Junk Publishers, NL, Dordrecht, pp 275–292

Physical Cognition Part II: Learning and Reasoning

10

Aside from the general-purpose mechanisms and motivation discussed in Chap. 9, researchers have also investigated if and what knowledge about the physical world animals possess to use to solve their daily challenges and how they acquire such information, i.e. through learning or reasoning (Shettleworth 2010).

While some researchers have suggested that associative learning (covariation between elements) may be sufficient for animals to negotiate the necessities of their environment, others argue that inferential reasoning is also involved. Whereas learning involves associating events simultaneously occurring in space and time (and hence includes some trial and error), reasoning allows perceived events to be combined with imagined events or events that are not necessarily co-occurring (see also Maier and Schneirla 1935), hence relying on memory, notably in the form of mental representations.

Accordingly, the ability to attribute causality to specific elements of the environment (inferential reasoning) may facilitate animals in sorting out relevant information from the many events that co-vary and allow individuals to respond in a context-adequate and adaptive way when facing new, incomplete, or contradictory information. Moreover, numerical competence, or at the very least a capacity to discriminate between quantities on a perceptual basis, would be highly adaptive when faced with foraging decisions (Gallistel 1990; Stephens and Krebs 1986).

To date, only a few studies have set out to compare learning, inference abilities, and numerical competence in wolves and dogs, despite the fact that already Harry Frank proposed that wolves and dogs might differ specifically in these aspects. In his theoretical model of wolf and dog information processing (Frank 1980) (Chap. 11), he proposes that wolves might have acquired a 'cognitive subsystem, which may have evolved concomitantly with the rise of group hunting, and is characterized by complexity, foresight, a capacity for internal mental representation (imagery), and insight into rudimentary means-ends relations' (cited from Frank and Frank 1985). In contrast to wolves, Harry Frank proposed that environmental effects on feeding and mating are buffered in dogs (and in other domesticated species) by humans and, thus, natural selection on dogs' individual problem-solving might have been relaxed

F. Range, S. Marshall-Pescini, *Wolves and Dogs*, Fascinating Life Sciences,
https://doi.org/10.1007/978-3-030-98411-3_10

(Frank 1980; Hemmer 1990). Accordingly, he predicted that wolves should outperform dogs in problem-solving tasks that require 'discovering' or 'reasoning' out the correct response and are not merely solved by trial-and-error learning. The fact that dogs have relatively smaller brains than wolves (Herre and Röhrs 2013) and that pet dogs often perform suboptimally on tasks where they have to rely on inferences about their environment (see Müller et al. 2011, 2014a, b, 2016; Osthaus et al. 2003, 2005; Riemer et al. 2014, but see Kundey et al. 2010 for positive results) are in line with this hypothesis.

One observation that made us wonder whether wolves indeed might have a better causal understanding than dogs is their reactions towards the electric fence. The wolves would usually get a shock from the electric fence just once, if at all, and consistently avoid it from that point on if the electricity was on. The feeling we got was that a single experience was enough for them to clearly link the fence to the shock. This was especially surprising in one situation, when I (Friederike) wanted to take Shima out on a walk. I joined Shima in her enclosure, put on a collar and the leash, opened the enclosure door while I was holding her collar and, not paying sufficient attention, got an electric shock from the wire. The electric shock wave also reached Shima, who jumped back squeaking. I thought 'Oh no, that's it, now I destroyed our relationship'. Well, I was wrong. Shima just looked at the fence and then happily came out with me, showing no sign that she connected the shock with me. In contrast, pet dogs seem in general more resistant in learning this connection. Thus, these observations could point to wolves having better associative learning abilities (one-shot learning) and/or being better at linking the effect to the potential cause.

In this chapter, we will review studies comparing wolves' and dogs' abilities to learn by association and their inference skills. Finally, we will summarize what is known about their numerical competence.

10.1 Learning

Associative learning is the process by which a person or animal learns an association between two stimuli or events. Learning is an adaptation to a viable environment, allowing animals to face and exploit novel sources of information that a hard-wired mechanism would struggle to capture reliably. Such learning is highly advantageous for all living organisms since it enables an organism to adapt its behaviour during its lifetime, without requiring genetic changes across generations, and, unlike other sources of behavioural plasticity that occur during development, learning can result in rapid rather than gradual behavioural modifications but still is limited by the species-specific reaction norm (Dukas 2013).

The learning speed of non-human animals is often assessed using a two-choice procedure, where two stimuli are presented simultaneously and only the selection of the target stimulus leads to a reward. Over successive trials, the animals learn that they should ignore the irrelevant or unrewarded stimulus and only choose the target stimulus. Binary discriminations can involve spatial, object, or colour cues and have been tested in many different animal species.

While the main measure of interest is often the number of trials necessary to achieve a certain, previously defined criterion such as 80 or 90% correct choices, other measurements can also be interesting. For example, perseverative errors, whereby the animal continues to choose the unrewarded stimulus over and over again, suggest that the individual requires repeated negative feedback in order to learn. Such perseverative errors can be considered a sign of reduced cognitive flexibility (Scott 1962).

Another approach to assess animals' flexibility is reversal learning. In reversal learning tests, two stimuli change their reward contingencies after the initial discrimination learning (acquisition phase). The Franks tested four wolf and four malamute puppies at the age of 15 weeks (Frank and Frank 1987). The animals had to learn to choose either a black or a white wooden block from a total of three blocks with the other two blocks being of the other colour (white or black, respectively). If they choose correctly and removed the correct block, they had access to a reward inside. Olfaction was controlled for by smearing some food also into the grooves underneath the other blocks. During the experiment, the experimenter was not visible to the subject, but could observe it through a one-way mirror.

In the initial study with the mother-reared wolves, the wolves learned the discrimination in the acquisition slightly faster—but not significantly so—than the dogs (mean: 82.5 vs. 93.75 trials), while the malamutes were significantly faster in the reversal than the wolves, with two wolves never reaching criterion at all. However, variance in the wolf performance was rather high, suggesting that maybe some of the animals were too nervous to concentrate on the task. This notion was confirmed when testing four fully human-socialized wolf pups that the Franks raised later (Frank 2011). These four pups were not just significantly faster than the dog pups in both the acquisition and the reversal phase of the experiment, but they were also significantly faster than the mother-reared animals in the reversal phase.

So far, we have only conducted one study investigating associative and reversal learning at the Wolf Science Center, and it was part of a study focusing on inhibitory control (see Chap. 9; Brucks et al. 2019). In this test, the adult wolves and dogs initially learned to discriminate between two plastic objects of different shapes and colours. In the acquisition phase, the objects were pushed out from behind a barrier by an experimenter, while the subject was held 2 meters away by a trainer. The subject was then released and allowed to make a choice by touching the object with their nose. After a correct choice, the experimenter lifted the object, and the subject could take the reward underneath. After an incorrect choice, the subject was shown that the chosen object was not baited and then the rewarded object was quickly lifted so that the subject could see the reward but could not take it. Criterion was reached when the subject chose the correct object on 9/12 trials within 1 session. We found that in the acquisition phases, there were no species differences in the number of sessions needed to reach the learning criterion, but on average wolves were faster (18.77 s) than dogs (26.28 s) in making a choice.

After reaching criterion, the object contingencies were reversed such that the previously rewarded object was now unrewarded and vice versa. Similar to the acquisition phase, wolves and dogs did not differ in the number of correct choices in the first session of the reversal, but again we found a species effect in the speed of choice, with wolves (23.14 s) on average being significantly faster than dogs (38.71 s) in making their choices.

In another study, we trained our wolves and pet dogs to discriminate between two quantities (numerical competence; see below) (Rivas-Blanco et al. 2020). The 9 wolves that reached criterion were faster in reaching criterion than the 16 pet dogs, suggesting that wolves might be better at associative learning if the difficulty increases. We definitely need more studies in that respect!

The Franks also tested the associative learning abilities using T-mazes. In the first study, they tested four 13-week-old wolf puppies on a six-unit T-maze task. To exit the maze, the pups had to adopt either a strategy in which they predominately turned to the left or one in which they predominately turned to the right (Frank and Frank 1987). After 2 days of habituation to the testing arena and the presence of the human, the animals were tested on one trial per day for ten consecutive days. The results of the wolf puppies were compared to a sample of pet dog puppies of various breeds that had been tested by Scott and Fuller (Scott and Fuller 1965). The wolf puppies clearly outperformed the dog puppies in all three measurements: number of total errors, number of errors on their best trial, and number of trials to the best trial.

In another test, the Franks investigated whether four 22-week-old mother-reared wolves and three malamutes would learn to differentiate between a visual (flashing light) and auditory (police whistle) cue telling them which end of the T-maze arm was open, which would allow them to escape the maze faster. After a total of 40 training trials—administered over 4 days—in which only the arm linked to the respective cue (light or whistle) was open, the animals were tested in 16 discrimination trials per day over 8 days. Overall, none of the wolf pups, but two of the malamute pups, reached the set criteria. However, considering the discrepancies seen in associative learning abilities between mother-reared and human-reared pups in the previously described study by the Franks, the results in the T-maze might have been due to the fact that the wolf pups were not properly socialized with humans. They may have been too agitated during the task to pay close attention to their surroundings, and/or their anxiety may have reduced their motivation to work for treats (Frank et al. 1989).

Actually, a lack of food motivation is something we also often observe in our animals (both wolves and dogs), and it manifests mainly in two ways. First, if the animals are too stressed or uncomfortable in a testing situation, they become indifferent to even the best food rewards. While we have seen this happening mainly with the puppies, it has occasionally also happened with adult animals when confronted with new situations. Second, specific animals—especially some of our 'prima donnas'—may develop an insistent (and often incomprehensible to us) dislike of a specific test. At this point, no matter their hunger levels or how amazingly enticing the reward offered, they just refuse to participate. Some of them will play passive-resistant games, just lying down and taking a nap, others (a couple of dogs in particular) will busy themselves with hunting mice around the enclosure, or if they somehow guess the test they dislike is scheduled (e.g. by seeing one of the students with a specific apparatus passing in front of them), they simply refuse to come out of their home enclosures. Interestingly, we have had these 'problems' mainly when our animals were going through adolescence or when they were still quite young—with age, they seem to become more reasonable and in some cases even more eager to 'work'.

Finally, one other study explained in detail in Chap. 9 (Hiestand 2011) compared four adult German Shepherds and two adult wolves in a reversal test. In the acquisition phase, they had to first pull three ropes of different lengths from a ceiling in a certain order. The order in which they had to pull the rope was then reversed. While the wolves and dogs did not differ in the number of trials it took them to learn the initial correct order in the acquisition phase, the dogs outperformed the wolves in the reversal phase. The two wolves got the sequence correct only in 1/10 and 2/5 trials, respectively, within an imposed time limit of 2 minutes, whereas one dog succeeded in 5/5 and another in 4/5 trials. How this is possible remains unclear though. Given that the animals had learned the initial order just beforehand, how would they know that they had to reverse the sequence suddenly? So how can the dog be correct in 5/5 trials? Even 4/5 is really good, given that it is a sequence of three ropes they have to pull. Anyhow, it is also not reported if the other two dogs were actually tested and if so, how they performed. Interestingly though, one of the wolves had a perfect trial—likely by chance—pulling the three ropes in the correct order after nine unsuccessful trials, in which she received no reward. None of the dogs sustained that degree of failure and still continued to try to solve the problem.

10.1.1 Summary: Learning

Thus, to sum up, if relaxed during testing, wolf puppies seem to be better at associative learning tasks than similarly raised dog puppies. In adult animals, however, we found no differences in learning accuracy between wolves and dogs, but wolves were generally much faster in making choices in the acquisition and reversal phases. At present, however, there are only a limited number of studies on these issues.

10.2 Inferential Reasoning

Different types of inferential reasoning have been described. *Inference by exclusion* refers to the ability to infer the characteristic of an unknown option by logically eliminating the other known options (Call 2004), e.g. in a choice between two cups, one of which contains food, based on seeing only the empty cup, the animal is able to correctly select the full cup (Call 2004). In this task, however, an avoid-the-empty-cup strategy could also be used, rather than inference. To exclude this explanation, more recent studies instead presented straight or L-shaped tubes (Fig. 10.1) (Jelbert

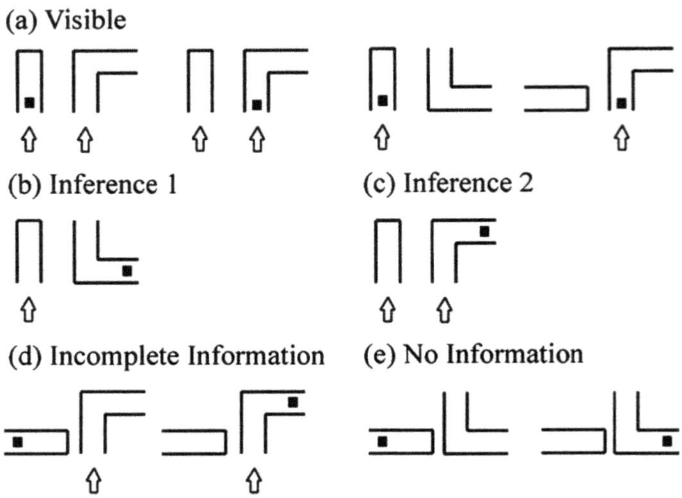

Fig. 10.1 Diagram of the L-shaped tube task. The arrow indicates the tube openings through which the subjects can see. Figure adapted from Jelbert et al. (2015) (copyright license number 5053091037167); reprinted by permission from the *Journal of Comparative Psychology*

et al. 2015). Here, the animals can look inside two tubes, of which one is baited and the other one is not. In the visible condition (Fig. 10.1a), they can see the treat in the baited tube and choose accordingly. In the inference trials, however, they cannot see the treat and thus should choose the tube that they cannot look into or that is obviously bend (Fig. 10.1 b). In the incomplete and no information trials (Fig. 10.1c), they cannot make any informed decision and thus should perform at chance level. Another task is the invisible displacement task, where the target object is moved from one screen to another by using a container so that the animal actually never sees the target object being moved, but needs to infer its location based on the displacement container.

A second type is *diagnostic inference*, i.e. reasoning from the consequent (the effect) to the antecedent (the cause). It can be based on affirming or denying the consequent. An example of the former is that upon hearing a rattling noise (the effect) coming from a cup that is being shaken, compared to a cup that when shaken remains silent, the animal infers the food is in the rattling cup. While chimpanzees have been shown to be able to solve such tasks, dogs have been reported to be rather bad, at least compared to apes (Bräuer et al. 2006). In contrast, inference based on denying the consequent occurs if the animal hears that the shaking does not produce noise and hence selects the alternative cup. The latter type of inferential reasoning has been shown much more rarely in non-human species (see Völter and Call 2017 for a review).

A third type is *predictive inference*, i.e. inferring the consequent (effect) from the antecedent (cause). Predictive inferences are especially relevant for action planning, and thus most tasks require animals to anticipate the effect of an action on the

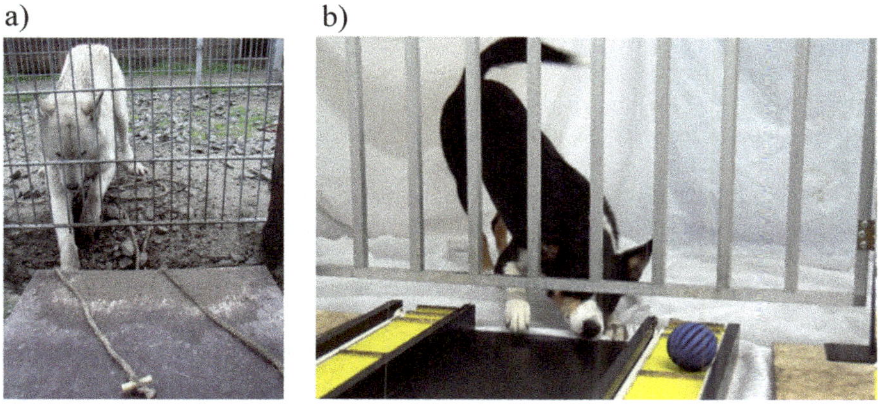

Fig. 10.2 Predictive inference is often tested using the string-pulling task (**a**) or the support problem (**b**). Figure a adapted from Range et al. (2012) (copyright license number 5053041149215); reprinted by permission from Springer

trajectory or status of a food item. Tasks that have been used to test for predictive inference usually involve object-object relations such as the string-pulling task, where the animals need to pull in an out-of-reach object attached to a string (Osthaus et al. 2005) (Fig. 10.2a), or the on-off problem, where an object is placed on a support (a board or cloth) and the animal needs to pull in the support to reach the object (Hauser et al. 1999) (Fig. 10.2b).

Importantly, object-object relations are governed by physical laws, and while animals might benefit from their inherent predictability, the ability to reason by inference might depend on the sensitivity of the animals to specific laws related to the object's properties (e.g. solidity, weight, gravity).

Almost nothing is known about inferential reasoning in wolves and dogs and how it compares. In the following, we will discuss the few studies that have been conducted according to the inference they are likely testing.

10.2.1 Reasoning by Exclusion

Being able to infer the location of objects/animals that have disappeared from view (referred to as object permanence) is highly relevant for canines, especially during hunting. It requires an understanding that objects continue to exist even when they are out of sight, which researchers have referred to as object permanence. Sometimes, however, prey and/or social partners may not stay hidden behind a specific obstacle, but may continue to move to another location while being either in or out of sight of the subject. Moving from behind one obstacle to another while briefly returning in clear view of the subject is referred to as 'visible displacement', whereas moving from one hiding location to the next while remaining out of sight of the subject is referred to as 'invisible displacement'. In the latter case, to find the prey or social partner when it moves to a new location while remaining out of sight, animals must be able to track the displacement of the prey/social partner and infer its position by using perceptual cues, such as footprints left in the grass or the change of the hiding locations. Obviously, being able to track such displacements would have clear advantages for a species that depends on hunting live prey.

Jean Piaget, a French psychologist, studied object permanence in human infants and observed that they progressed through different stages of object permanence during development. He proposed a 6-stage developmental schema (Piaget 1954), which was later extended by Uzgiris and Hunt (1975) to 15 stages. While the first stages are very simple with objects/food only being partially hidden behind occluders, the tasks become more and more complex in the later stages. For example, as described in Chap. 7, in the A-not-B task, the subject first has to repeatedly retrieve an object/reward hidden behind the same occluder before the object/reward is visibly moved to another location. As we saw, human infants as well as many animal species commit the so-called A-not-B error in this task by continuing to search behind the first occluder (A) even when the object has been visibly moved behind or under a second occluder (B, visible displacement). The most complex version of the object permanence task is the 'invisible displacement', in which the target object, after being hidden in the first location, is placed into an opaque displacement device and then, still hidden inside the device, is moved behind a second occluder and left there. During both the visible and invisible displacements, the complexities can be increased by adding sequential displacements, e.g. where an object disappears first behind one location and then reappears to disappear again behind a second and third location and so on. Human children are able to develop a mental image, hold it in mind, and manipulate it to solve problems, including object permanence problems that are not based solely on perception at roughly 2 years of age (between 18 and 24 months). Then they can reason about where the object may be when invisible displacement occurs. Object permanence has been studied in many different species and has proven useful to compare this kind of cognitive performance between different taxa (Pepperberg 2002).

Dogs have been tested by different researchers in object permanence tasks, with highly contradictory results. Some early studies suggest that they can solve invisible displacement tasks not only as adults but already at the age of 3 months (Gagnon and Doré 1992, 1994). However, follow-up studies including more controls showed that dogs are unable to track objects in invisible displacement tasks (Doré et al. 1996; see also Fiset and Plourde 2013; Miller et al. 2009; Rooijakkers et al. 2009) and rather use the visible location of the displacement device as a cue to make their choices (Collier-Baker et al. 2004; Fiset and LeBlanc 2007). To date, object permanence in wolves and dogs has been compared in just one study.

Fiset and Plourde (2013) tested wolves from Wolf Park Indiana and pet dogs in several visible and invisible displacement tasks. Since the wolves at Wolf Park are not trained to be held on a leash, the studies were conducted in a dynamic fashion: The experimenter called the attention of the animal by showing a treat. Once the animal looked at and started to move towards the experimenter, the experimenter carried out the displacements and then stepped back.

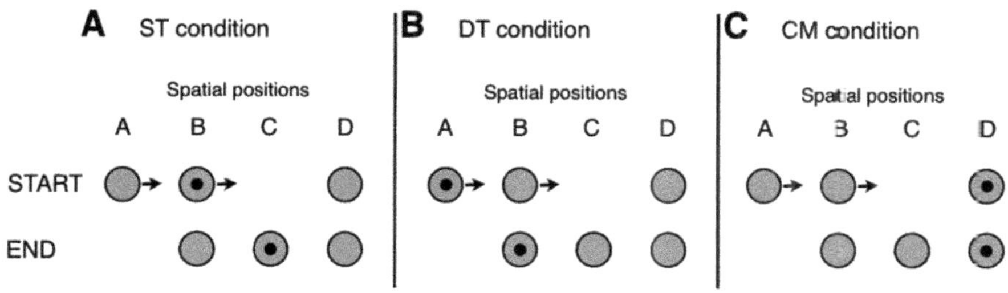

Fig. 10.3 Transposition conditions. A: substitution transposition; B: double transposition; C: movement control. Figure adapted from Fiset and Plourde (2013) (copyright license number 5053100190457); reprinted by permission from the *Journal of Comparative Psychology*

In the visible displacement tasks, they tested 9 wolves and 22 pet dogs first in the A-not-B task, then a single visible displacement problem, and finally a double visible displacement problem. They had three blue opaque semicircular screens (1–3) set up. In the A-not-B task, a bowl with a treat inside was first hidden for five consecutive trials behind Screen 1 (called A trials). On the sixth trial, however, it was hidden behind Screen 3 (called B trial). Only two wolves and three dogs made mistakes in the B trials, searching behind the wrong screen, suggesting that neither wolves nor dogs committed the A-not-B error.

Next, the animals were tested in a single visible displacement problem, where the experimenter displaced the bowl from the start position in full view of the subject behind one of the three screens. In the second task, they were tested in a double visible displacement problem, where the experimenter first displaced the bowl from the start position in full view of the subject behind one of the three screens and then immediately removed it from the screen and visibly moved it behind a second but adjacent screen. Both wolves and dogs performed above chance in both tasks, but were slightly worse in the double visible displacement than both the A-not-B task and the single displacement test. The main error they committed was to search the first screen behind which the object disappeared.

In the invisible displacement tasks, they tested 5 wolves and 19 dogs. However, instead of using the original tasks designed by Piaget, they used spatial transposition problems that are thought to be easier for animals to solve. In these tasks, after the reward is hidden in a target container, the container is moved to an adjacent spatial position, and the animal's task is to track the displacement of the target container to locate the food reward that has invisibly changed its location (see Fig. 10.3). The animals were tested in three conditions with three containers present in each condition: In the substitution transposition condition, after the target container was moved to a previous empty location, an identical but empty container was moved to the initial position of the target container. In the double transposition condition, the initial position of the target container was left empty. And in the control condition, the target container remained at its initial position. While the wolves and dogs did not differ, both performing above chance in both the control and the double transposition condition, they performed below chance in the substitution transposition condition. Interestingly, they mainly searched for the food in the non-target container that was moved to the initial position of the target container. These data are in line with previous studies (Doré et al. 1996; Rooijakkers et al. 2009) and suggest that dogs and wolves are unable to fully represent invisible displacements (Collier-Baker et al. 2006; Fiset and LeBlanc 2007).

10.2.2 Diagnostic Inference

At the Wolf Science Center, we investigated the animals' ability for diagnostic inferencing in two tasks (see Fig. 10.4 for the setup). In the first one, animals were presented with two opaque containers: one contained stones and food; the other was empty. With the help of a fishing line, the experimenter shook first one and then the other container. In the second task, two wooden shapes were presented to the animal: one was inclined and hiding the food, while the other was lying flat on the surface.

The wolves ($N = 12$) outperformed the pack ($N = 14$) and pet dogs ($N = 12$) in both of these tasks (Lampe et al. 2017), demonstrating that they could use the cues provided to make the correct inferences. These results of the dogs are in line with a previous study, where pet dogs performed badly and significantly worse than chimpanzees (Bräuer et al. 2006). However, in both studies, a number of crucial

Fig. 10.4 'Choice table'. The subject will touch the desired object with its nose. The experimenter will manipulate the objects using thongs and will stay out of sight behind a curtain. Figure reproduced from Lampe et al. (2017). https://www.nature.com/articles/s41598-017-12055-6/; CC BY 4.0; http://creativecommons.org/licenses/by/4.0/

control conditions were missing to differentiate whether causal relationships or perceptual cues were used to solve the task (e.g. the noisy cup is simply more salient).

On a side note, the dogs were attentive towards the table for more than 80% of the time, while the wolves, which were more successful, spent only 40% of their time looking at the table, since they also had to engage in 'essential' activities such as howling and exploring the area. This reminds us somewhat of the results in the learning tasks (Chap. 9) where the wolves made their decisions faster than the dogs, but were equally successful suggesting that maybe the wolves process information faster than the dogs.

10.2.3 Predictive Inference

There has been only one study comparing wolves and dogs on predictive inference, and it did so by using a means-end understanding task. Means-end understanding in animals is usually investigated by offering a choice of two options, one of which is in physical connection to a desired out-of-reach object, e.g. a piece of food (e.g. Hauser et al. 1999; Heinrich 1995; Heinrich and Bugnyar 2005; Köhler 1925; Pepperberg 2004). To solve the problem, the subjects need to understand that the food can be obtained by, for example, pulling the string which is connected to the food rather than the unconnected string. The assumption is that subjects that do not recognize that pulling the string is the 'means' for achieving the desired 'ends' of bringing the food within reach will only succeed if repeated exposure allows for associative learning (Thorndike 1898).

Both dogs and wolves have been shown to retrieve out-of-reach food by means of pulling an attached string (Miklósi et al. 2003; Osthaus et al. 2005). However, when tested in a two-choice paradigm where one string was attached to the food and the other not, dogs usually pull the string closest to the food reward independent of whether the string is actually connected to the food or not, i.e. they commit the so-called proximity error (Osthaus et al. 2005). The proximity bias, which has also been found in other experiments with dogs (Osthaus et al. 2005), could be the result of a latent predisposition to concentrate on the closest available food.

To investigate whether wolves and dogs differ in their means-end understanding and in their way of committing mistakes, we ran a string-pulling task with wolves and dogs that had somewhat similar, but not exactly the same, experiences, since the study was conducted at the very beginning of the Wolf Science Center and we had yet to raise our own dogs and a sufficient number of wolves (Range et al. 2012). Nevertheless, all the wolves were hand-raised animals; six belonged to Wolf Park, USA, and three belonged to the WSC. The ten dogs were Siberian Huskies of the 'Mountain Wolf Farm' in Austria and were kept in small groups in kennels. They were sled dogs that had been raised by their mothers and transferred to the kennels at the age of 3 months.

In our experiment, we first provided the animals with only one string attached to a food reward that was positioned behind the fence, to see if the animals would

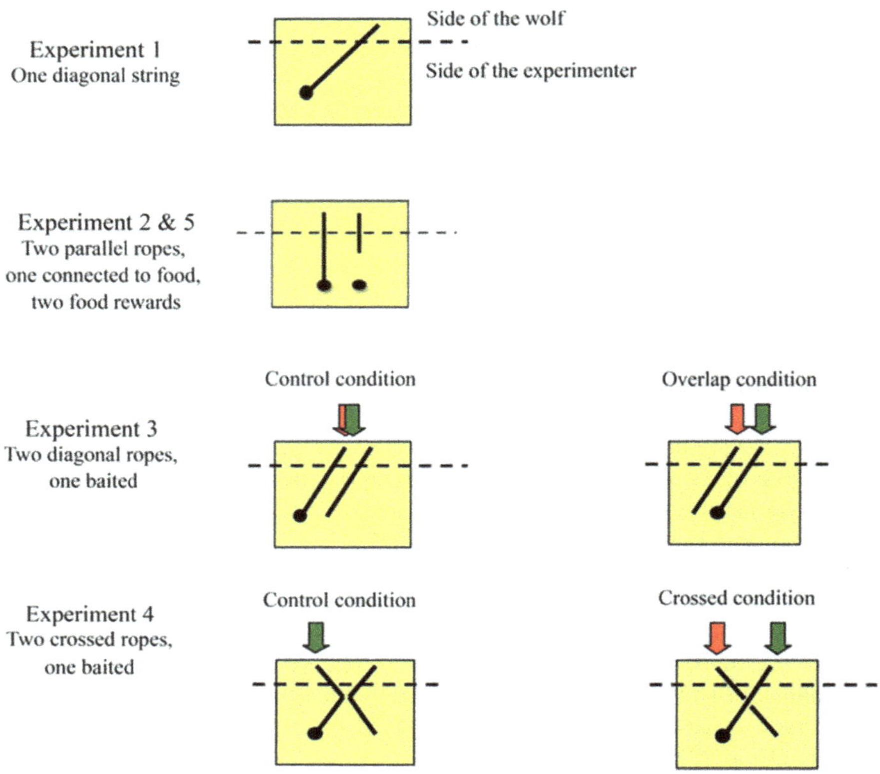

Fig. 10.5 String-pulling conditions. Figure adapted from Range et al. (2012) (copyright license number 5053041149215); reprinted by permission from *Animal Cognition*

spontaneously use the string to get access to the food or whether training would be required (Fig. 10.5). The string was laid out diagonally, and while all animals solved the task easily, many of them first pawed at the fence location closest to the food.

Next, we used two strings, laid out parallel to each other, each with a reward at the end. While one was connected to a piece of food, the other was not, and there was a 10 cm gap between the end of the string and the food reward. In this setup, proximity is not a confounding factor, since the distance from the food to the end of each rope is the same (Fig. 10.5). Surprisingly, the wolves and dogs learned something different across the 20 trials we conducted. While the wolves learned to pull the correct string and performed significantly above chance level in the last ten trials, the dogs were significantly *below* chance in the last ten trials. Apparently, they had developed a preference for the shorter, unconnected, and not-rewarded string. Why this preference developed is completely unclear, and we have not come up with any good ideas to explain this counterproductive learning. Furthermore, at the individual level, one wolf successfully solved the first six trials, and another was significantly successful (14 of 20 trials). None of the dogs was successful at the individual level.

After these conditions, we exposed the animals to two further experiments: In experiment 3 (Fig. 10.5), two ropes, one with food attached and one without, were laid in parallel at an acute angle to the fence resulting in either the 'overlap' or the 'control' condition. In the overlap condition, if both ropes were laid out with a tilt to the right, and the food attached to the left rope, the end of the non-baited rope was in line with the food. If the food was attached to the right rope, no overlap between the baited rope and the accessible end of the non-baited rope would occur, and the food was closest to the correct string ('control' condition). In experiment 4 (see Fig. 10.5), the strings were crossed, and thus choosing based on proximity would lead to failure. In the control condition, perceptually, the setup was similar to the crossed condition, but each rope was bent into a < shape, and the two ropes touched at the tip (><) so a choice based on proximity would lead to success. We ran 24 trials in each experiment with 12 control and 12 experimental trials intermixed. The results of these two experiments showed that wolves and dogs performed above chance in the control, but at chance in the test conditions. Interestingly, at the individual level, 2 wolves and 2 dogs significantly performed above chance or at least approached significance in the last 12 trials of experiment 3, suggesting some learning across trials. Similarly, in experiment 4, while most dogs chose based on proximity and most wolves developed a side preference, 2 wolves got better over trials with 1 of them significantly performing above chance in the last 12 trials. These individual data are interesting since they demonstrate that some animals have some basic means-end understanding and/or learn relatively quickly to pay attention to the relevant features.

However, solving these tasks might not necessarily imply an understanding of the underlying physical concept of connectivity, but rather solving such tasks may be based on perceptual features such as the physical contact between the 'means' and the 'end' (Tomasello and Call 1997). A study on New Caledonian crows, for example, found that spontaneous string pulling did not depend on means-end understanding but on learning mediated by a perceptual-motor feedback cycle, e.g. pulling the baited string moves the meat closer to the individual (Taylor et al. 2010; but see Schuck-Paim et al. 2009). Accordingly, several authors have suggested that non-human animals do not have a representation of unobservable physical forces linking elements of our environment (Povinelli 2000; Tomasello and Call 1997), but rather acquire simple abstract rules based on observable features of their environment and come to rely on these to solve specific problems. This general conclusion is in line with the few studies conducted on pet dogs, suggesting that they have a rather poor understanding of the invisible forces defining the relationship between the elements within our environment (Müller et al. 2014a, 2016).

10.2.4 Summary: Inference

*Taken together, the results above hint at wolves showing a more efficient associative learning mechanism (*Frank 2011*;* Range et al. 2012*) and potentially a better diagnostic inferencing ability (noisy cup test and slanting board;* Lampe et al. 2017*). However, considering the missing control conditions in regard to the latter* *and the fact that we have no comparative studies on the other inference types nor the animals' understanding of the basic physical laws (solidity, weight, gravity), no conclusion can be currently drawn in regard to whether wolves indeed have a better understanding than dogs of the invisible forces governing their environment.*

10.3 Numerical Competence

Being able to discriminate between different quantities yields evident advantages for the fitness and survival (Byrne 1995). For example, animals could optimize their energy intake based on such assessments by choosing the spot with more food items. Indeed, several studies both in the laboratory (e.g. great apes: Hanus and Call 2007) and in the wild (e.g. monkeys: Hauser et al. 2000) have shown that given the choice between two quantities, animals will show a preference for the larger one. Furthermore, quantity judgements seem to play an important role also in regard to intergroup conflicts over food, mates, and territory (Parker 1974). Studies on lions, chimpanzees, wolves, and dogs (Bonanni et al. 2011; Harrington and Mech 1979; McComb et al. 1994; Sillero-Zubiri and Macdonald 1998; Wilson et al. 2001) revealed that animals withhold their participation in intergroup conflicts over resources if the number of animals in the rival group is larger than in their own group. Given that intergroup conflicts, at least in wolves, are one of the major causes of death (see Chap. 3), being able to discriminate between quantities can make a huge a difference.

In such contexts, animals can base their decisions on the number of items (a quantity assessment) or some correlated perceptual features. Numbers refer to counting (one by one) and are accurate, while quantities assessments are estimates that are only an approximation (Gallistel and Gelman 2000). Moreover, for small numbers (up to six), animals could rely on subsidizing (e.g. a perceptual low-level process of pattern recognition (Fig. 10.6)) (Gallistel and Gelman 2000). Finally,

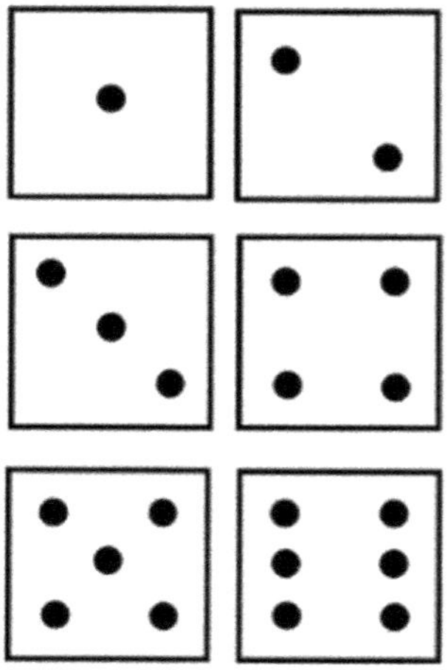

Fig. 10.6 Subsidizing: We do not need to count to know the numbers displayed on a dice

animals could just base their decision on perceptual features such as the size of a food pile or the surface area covered by it.

The main experimental setup that has been used to investigate numerical competence in non-human animals and to understand what information the animals rely on when making their decisions is a two-choice paradigm. In this paradigm, animals are encouraged to choose between two different food quantities, which can be presented either simultaneously or sequentially and can be visible or invisible at the time of choice. In cases where the two sets are presented at the same time, correlated perceptual features such as surface area, brightness, etc. could drive the decision of the animals unless properly controlled for (e.g. Brannon and Terrace 2000). This problem is diminished if the items of each set are presented sequentially and are invisible during the choice, so that the subject never sees the entire content of either set at any time during the presentation. However, this requires that the animal modifies the representation of each set's content as one item is added after the other and then mentally/internally compare the two representations to choose the larger set. While this latter version of the task allows the exclusion of a perceptually based choice, the animals could still rely on numerical or quantity information to make their choice since the number of food items is perfectly correlated with the amount of food.

So far, we've tested wolves and dogs on two tasks at the Wolf Science Center to investigate their ability to discriminate between different quantities and to understand whether they rely on perceptual features or whether they have an internal representation of (small) quantities.

In the first task, we used a very small set (up to four), but presented the different quantities sequentially by inserting food items (cheese) into opaque tubes, so that the animals never saw the entire sets of food items at the same time (Range et al. 2014; Utrata et al. 2012). After the animals observed the insertion of both sets, they were released and could make a choice. During the entire experiment, only the hands of the experimenter were visible, and they were retracted behind a screen during the choice. A total of 13 adult dogs and 11 adult wolves at the Wolf Science Center participated in this study. We first trained the animals on the general procedure, and once they had successfully discriminated between 1 and 4, we tested the following combinations: 1 vs. 4, 1 vs. 3, 1 vs. 2, 2 vs. 4, 2 vs. 3, and 3 vs. 4. Overall, all animals selected the larger quantity more often than the smaller quantity, and no difference between wolves and dogs emerged. These results remained robust also when excluding the combination 1 vs. 4, the pair known from the training phase. However, when analysing each numerical pair separately, wolves were significantly above chance at all tested ratios, while dogs performed at chance level when asked to discriminate 2 vs. 3 and 3 vs. 4.

To investigate if the wolves and dogs used numerical information and not alternative strategies such as paying attention to the amount of time it takes to insert the food items into the tubes (which of course is shorter the fewer items are inserted), we conducted a control condition, where we equalled the number of items inserted into each tube by adding stones to the smaller quantity of cheese pieces ('time control', Fig. 10.7a). We tested the animals in three sets of different ratios. Overall,

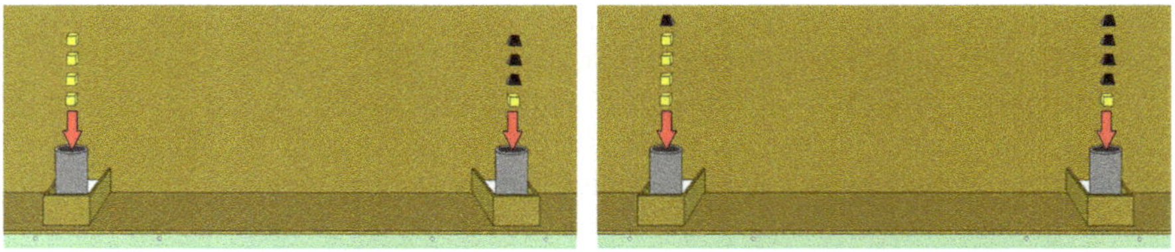

Fig. 10.7 Illustrations of the time (**a**) and stone (**b**) control conditions. Figure adapted from Utrata et al. (2012). https://www.frontiersin.org/articles/10.3389/fpsyg.2012.00505/full; CC BY 3.0; http://creativecommons.org/licenses/by/3.0/

while the wolves still selected the larger amount above chance level, dogs' performance did not differ from chance. More precisely, the dogs' performance was above chance with a ratio of 0.25 (1:4), below chance with a ratio of 0.5 (1:2), and at chance level with a ratio of 0.67 (2 vs. 3). The wolves performed above chance at all three ratios.

However, since the stones were always added last, making a 'clink' sound, the animals could have just avoided the tube where the sound was made instead of relying on the numerical information. Thus, we added yet another control condition, adding a stone also to the bigger amount of cheese and a number of stones to the smaller cheese amount so that the number of items were equal on both sides ('stone control', Fig. 10.7b). Similar to the first control, the wolves outperformed the dogs and choose the larger quantity more often than the dogs, performing above chance level while the dogs did not. Wolves performed above chance level at 0.25 and 0.67, but only approached significance at 0.5, while the dogs performed at chance level at all three ratios.

A more recent study from Wolf Park presented five pet dogs and five human-socialized wolves with two quantities of food items (range 1–4) simultaneously and allowed them to choose only one of the two visible options (Petrazzini and Wynne 2017). While both wolves and dogs choose the larger quantity significantly above chance, dogs selected the set with more food items only when presented with 1 vs. 3 and 2 vs. 4 items but not when presented with 2 vs. 3 or 3 vs. 4 items. These results show that dogs' accuracy was affected by numerical ratio, while no ratio effect was observed in wolves, supporting our conclusions even when a simpler setup was used.

In contrast to these current studies, previous studies showed that dogs could discriminate between two small quantities of one to five items (Ward and Smuts 2007; West and Young 2002). However, in line with our observations, the dogs performed worse when the presented quantities were invisible during the choice (Ward and Smuts 2007; see Baker et al. 2011 for similar results on coyotes), suggesting that dogs (and coyotes) chose based on alternative strategies such as surface area/volume of the presented food items or pattern recognition rather than making a numerical judgment. Moreover, dogs' and coyotes' performances decreased with increasing ratio between two presented quantities, while the wolves' performance was not affected by ratio.

However, all of the above studies—also the ones on pet dogs—are limited in terms of the largest numbers tested and their ability to control for some possibly confounding factors. So far, dogs have only been tested with numbers up to 9 and wolves up to 4, which may dampen the effect of the ratio on the subjects' performance. Testing the animals with bigger numbers would likely not be feasible when using sequential paradigms, as that would require them to memorize the quantities of two large sets of items, which could make it harder for them to draw comparisons. Moreover, all of the studies used food as the item to be counted, which may be distracting. Also related to the presence of food, although unlikely, previous studies do not conclusively rule out olfaction as a potential non-numerical strategy the animals may be using to select the larger amount. Accordingly, one of our students in the Clever Dog Lab designed a second study using a touchscreen apparatus (Rivas-Blanco et al. 2020). The touchscreen technology, the touchscreens, as well as the programming of the software were donated by Dietmar Schinnerl, who contacted me (Friederike) at one point with this idea of designing touchscreens for pet dogs to play games. He was happy that we tested his touchscreens by conducting a numerical competence study. Using a touchscreen conferred many advantages, such as allowing for greater standardization in the presentation of stimuli, avoidance of potential experimenter cues, removal of the possibly confounding effect of the presence of food during the choice, and the ability to control non-numerical cues in great detail. Thus, unlike in previous studies, the current study controlled for cumulative surface area, spacing, and shape of the array of stimuli.

Pohl set out to first test pet dogs for her master thesis at the Clever Dog Lab. In a second step, we then tested our wolves at the Wolf Science Center. Unfortunately, the screens broke, and Schinnerl had moved on with his life, so we never managed to test also our pack dogs. And it is important to note here that the procedure somewhat differed between the pet dogs and the wolves, for example, in terms of how often they were tested per day and how many days were between test sessions (pet dogs usually had several sessions after each other and mostly only 1 week between test days, whereas the wolves hardly ever had more than one session per day and sometimes weeks or even months between test sessions). Thus, the results cannot be directly compared—at least one should only do so very carefully.

The touchscreen consisted of a screen fixed on a metal plate with two separate acrylic glass panes placed in front of it: one covering the left side and the other one the right side. After an initial training of how to use the touchscreens, the animals were first trained to select either the larger or smaller quantity of dots (up to eight) presented simultaneously on the touchscreen on the two sides (Fig. 10.8). The animals could then choose either the right or left side by pressing the glass pans with their nose, which activated the pressure-measuring elements to which the panes were connected. If they choose the correct stimuli (either the larger or smaller quantity depending on the group they were assigned to), a high-pitched sound was played, and they received a reward from a treat dispenser ('Manners Minder' by Premier).

During the training phase, only certain 'training' combinations (e.g. 1:6, 1:5, 2:6; 2:7; 3:8) were shown so that the animals learned what was expected of them. Half of

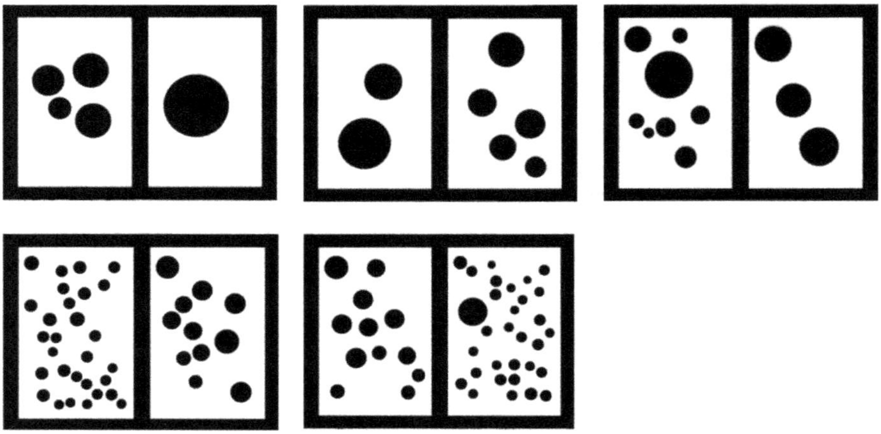

Fig. 10.8 Example stimuli of the touchscreen studies. Figure adapted from Rivas-Blanco et al. (2020). https://doi.org/10.3389/fpsyg.2020.573317; CC BY 4.0; http://creativecommons.org/licenses/by/4.0/

the animals had to choose the larger quantity (group A); the other half had to choose the smaller quantity (group B). Once they reliably choose the correct quantity (larger or smaller depending on the group), they were submitted to the first test. Here, they were tested on both familiar training and novel test pairs of quantities ranging from 1 to 8.

Sixteen of 18 pet dogs successfully completed the training phase (626.47 ± 335.58 trials) and proceeded to the first test phase, in which novel 'test' combinations were inserted between the familiar training trials. In the test phase, the animals were not rewarded for the test trials (e.g. they received no feedback whether they choose correctly or not) avoiding that the animals learn the new combinations during the testing. Overall, the dogs performed above chance with all familiar training pairs, and interestingly, they also performed significantly above chance in all novel combination of quantities except 6:7 and 7:8, which constitute the highest ratios of 0.86 and 0.88.

After this first test, 8 pet dogs continued with the second test with quantities up to 32. Again, we first had a training phase now including stimuli with higher quantities (9:27; 11:29; 9:32), which the pet dogs completed with an average of 873.00 ± 690.91 trials. In the test phase, they successfully discriminated between all presented training combinations. In the test trials, we only presented some combinations, and all pairs of numbers representing a ratio below 0.63 were selected above chance; above that, only 14:21, 17:24, 21:28, and 12:15 were successfully discriminated (Fig. 10.9).

At the Wolf Science Center, we trained 11 wolves. Geronimo and Yukon did not reach criterion; for whatever reason, those two did never really figure out the touchscreen. They would participate in some easier touchscreen studies, but the ones including long and extensive training just never worked with them despite all

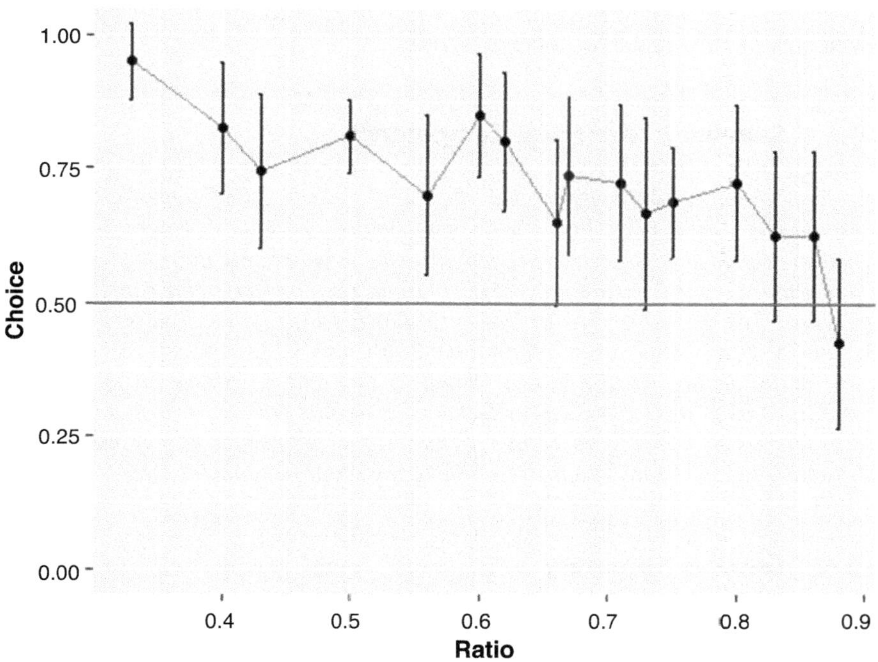

Fig. 10.9 Effect of the ratio on the probability of success in the test trials with higher quantities (error bars set at the 95% confidence interval; the red line represents probability of success by chance). Figure adapted from Rivas-Blanco et al. (2020) https://doi.org/10.3389/fpsyg.2020.573317; CC BY 4.0; http://creativecommons.org/licenses/by/4.0/

our efforts and various approaches to teach them what to do. The other nine wolves needed on average 486.86 ± 426.77 trials to reach criterion. In the test phase, the wolves performed significantly above chance in all training combinations and were also quite successful with the novel test trials, where they significantly discriminated all combinations up to a ratio of 0.80 with the exceptions of 2:3, 5:7, and 6:8, which means they were also not successful with 5:6, 6:7, and 7:8.

Six of the eight wolves also participated in the second test with the larger quantities; however, we had big problems here with the touchscreens, which had more and more problems and data management issues, so we need to be careful especially in comparing the results with those of the pet dogs. The six wolves needed on average 2702.00 ± 2092.97 trials to proceed to the test phase. In the test phase, they solved all the trained combinations successfully; however, they performed above chance only in six combinations: 10:30; 12:30, 9:18, 13:26, 15:25, and 19:29.

In addition to these methods and results described here, we also conducted additional analyses and control trials, to investigate whether non-numerical factors influenced the performance of our animals (Rivas-Blanco et al. 2020). These controls indicated that both wolves and dogs may have used continuous, non-numerical cues, such as size and shape of the stimuli, in conjunction with the numerical information to solve this task, which might explain why dogs performed much better in this task than in previous ones.

10.3.1 Summary: Numerical Competence

Together these results suggest that dogs and wolves differ somewhat in their quantitative competences. In the tasks with smaller quantities, wolves outperform dogs, and their success in the sequential tasks may suggest that they base their choices at least partly on numerical information. The dogs might base their quantity judgements partly on perceptual features, allowing them to succeed in some tasks but not in those where this information is not available. Overall, both species successfully distinguished between stimuli of different quantities up to 32 items and ratios up to 0.60–0.80, and with larger quantities, their performance was in accordance with Weber's law (which predicts worse performances at higher ratios), suggesting that previous results were due to the small magnitudes tested.

10.4 General Summary

The studies outlined above suggest that domestication may have partially negatively affected some of the dogs' cognitive abilities, possibly due to a relaxation of selective pressures on such skills (Frank 1980; *Frank and Frank* 1982*). While wolves appear*

Table 10.1 Summary of studies investigating learning and reasoning

	Wolves	Dogs	References
Associative learning and reversal			
Colour discrimination (puppies)	Wolves > dogs		Frank and Frank (1987)
Object discrimination and reversal	Wolves = dogs		Brucks et al. (2019)
Sequence learning	Dogs > wolves		Hiestand (2011)
Inference by exclusion			
Visible displacements	Yes	Yes	Fiset and Plourde (2013)
Invisible displacements	No	No	
Diagnostic inferences	Yes	No	Lampe et al. (2017)
Predictive inference (means-end understanding)	(No)	No	Range et al. (2012)
Numerical competence			
Simultaneous presentation	Yes	No	Petrazzini and Wynne (2017)
Simultaneous presentation (touchscreen)		*Pet dogs*	
1–8	Yes	Yes	Rivas-Blanco et al. (2020)
1–32	Partly	Ratio < 0.63	
Sequential presentation	Yes	No	Range et al. (2014)

to be better than dogs in certain tasks (e.g. *numerical competence and potentially inferential reasoning and associative learning), they show an equally moderate performance in other areas such as the invisible displacements in the object permanence tasks or the means-end string-pulling task. To address contexts and causalities of these aspects, it will be necessary to investigate their learning and inference skills in much more detail, designing multiple paradigms aimed at uncovering the ability to use the same underlying mechanism* (e.g. *connectivity*).

References

Baker JM, Shivik J, Jordan KE (2011) Tracking of food quantity by coyotes (*Canis latrans*). Behav Proc 88:72–75. https://doi.org/10.1016/j.beproc.2011.08.006

Bonanni R, Natoli E, Cafazzo S, Valsecchi P (2011) Free-ranging dogs assess the quantity of opponents in intergroup conflicts. Anim Cogn 14:103–115. https://doi.org/10.1007/s10071-010-0348-3

Brannon EM, Terrace HS (2000) Representation of the Numerosities 1-9 by rhesus macaques (*Macaca mulatta*). J Exp Psychol Anim Behav Process 26:31–49. https://doi.org/10.1037/0097-7403.26.1.31

Bräuer J, Kaminski J, Riedel J, Call J, Tomasello M (2006) Making inferences about the location of hidden food: social dog, causal ape. J Comp Psychol 120:38–47. https://doi.org/10.1037/0735-7036.120.1.38

Brucks D, Marshall-Pescini S, Range F (2019) Dogs and wolves do not differ in their inhibitory control abilities in a non-social test battery. Anim Cogn 22:1–15. https://doi.org/10.1007/s10071-018-1216-9

Byrne RW (1995) Primate cognition: comparing problems and skills. Am J Primatol 37:127–141. https://doi.org/10.1002/ajp.1350370206

Call J (2004) Inferences about the location of food in the great apes (*Pan paniscus*, *Pan troglodytes*, *Gorilla gorilla*, and *Pongo pygmaeus*). J Comp Psychol 118:232–241. https://doi.org/10.1037/0735-7036.118.2.232

Collier-Baker E, Davis JM, Nielsen M, Suddendorf T (2006) Do chimpanzees (*Pan troglodytes*) understand single invisible displacement? Anim Cogn 9:55–61. https://doi.org/10.1007/s10071-005-0004-5

Collier-Baker E, Davis JM, Suddendorf T (2004) Do dogs (*Canis familiaris*) understand invisible displacement? J Comp Psychol 118:421–433. https://doi.org/10.1037/0735-7036.118.4.421

Doré FY, Fiset S, Goulet S, Dumas M-C, Gagnon S (1996) Search behavior in cats and dogs: interspecific differences in working memory and spatial cognition. Anim Learn Behav 24:142–149. https://doi.org/10.3758/BF03198962

Dukas R (2013) Effects of learning on evolution: robustness, innovation and speciation. Anim Behav 85:1023–1030. https://doi.org/10.1016/j.anbehav.2012.12.030

Fiset S, LeBlanc V (2007) Invisible displacement understanding in domestic dogs (*Canis familiaris*): the role of visual cues in search behavior. Anim Cogn 10:211–224. https://doi.org/10.1007/s10071-005-0255-1

Fiset S, Plourde V (2013) Object permanence in domestic dogs (*Canis lupus familiaris*) and Gray wolves (*Canis lupus*). J Comp Psychol 127:115–127. https://doi.org/10.1037/a0030595

Frank H (1980) Evolution of canine information processing under conditions of natural and artificial selection. Z Tierpsychol 53:389–399. https://doi.org/10.1111/j.1439-0310.1980.tb01059.x

Frank H (2011) Wolves, dogs, rearing and reinforcement: complex interactions underlying species differences in training and problem-solving performance. Behav Genet 41:830–839. https://doi.org/10.1007/s10519-011-9454-5

Frank H, Frank M (1987) The University of Michigan canine information-processing project (1979-1981). In: Frank H (ed) Man and Wolf. Advances, Issues, and Problems in Captive Wolf Research. Dr. W. Junk Publishers, NL, Dordrecht, pp 143–167

Frank H, Frank MG (1982) On the effects of domestication on canine social development and behavior. Appl Animal Ethol 8:507–525. https://doi.org/10.1016/0304-3762(82)90215-2

Frank H, Frank MG (1985) Comparative manipulation test performance in ten-week-old wolves (*Canis lupus*) and Alaskan malamutes (*Canis familiaris*): a Piagetian interpretation. J Comp Psychol 99:266–274. https://doi.org/10.1037/0735-7036.99.3.266

Frank H, Frank MG, Hasselbach LM, Littleton DM (1989) Motivation and insight in wolf (*Canis lupus*) and Alaskan malamute (*Canis familiaris*): visual discrimination learning. Bull Psychonom Soc 27:455–458. https://doi.org/10.3758/BF03334654

Gagnon S, Doré FY (1992) Search behavior in various breeds of adult dogs (*Canis familiaris*): object permanence and olfactory cues. J Comp Psychol 106:58–68. https://doi.org/10.1037/0735-7036.106.1.58

Gagnon S, Doré FY (1994) Cross-sectional study of object permanence in domestic puppies (*Canis familiaris*). J Comp Psychol 108:220–232. https://doi.org/10.1037/0735-7036.108.3.220

Gallistel CR (1990) The Organization of Learning. Learning, development, and conceptual change. MIT Press

Gallistel CR, Gelman R (2000) Non-verbal numerical cognition: from reals to integers. Trends Cogn Sci 4:59–65. https://doi.org/10.1016/s1364-6613(99)01424-2

Hanus D, Call J (2007) Discrete quantity judgments in the great apes (*Pan paniscus, Pan troglodytes, Gorilla gorilla, Pongo pygmaeus*): the effect of presenting whole sets versus item-by-item. J Comp Psychol 121:241–249. https://doi.org/10.1037/0735-7036.121.3.241

Harrington FH, Mech LB (1979) Wolf howling and its role in territory maintenance. Behaviour 68: 207–249. https://doi.org/10.1163/156853979X00322

Hauser MD, Carey S, Hauser LB (2000) Spontaneous number representation in semi-free-ranging rhesus monkeys. Proc R Soc B 267:829–833. https://doi.org/10.1098/rspb.2000.1078

Hauser MD, Kralik J, Botto-Mahan C (1999) Problem solving and functional design features: experiments on cotton-top tamarins, *Saguinus oedipus oedipus*. Anim Behav 57:565–582. https://doi.org/10.1006/anbe.1998.1032

Heinrich B (1995) An experimental investigation of insight in common ravens (*Corvus corax*). Auk 112:994–1003. https://doi.org/10.2307/4089030

Heinrich B, Bugnyar T (2005) Testing problem solving in ravens: string-pulling to reach food. Ethology 111:962–976. https://doi.org/10.1111/j.1439-0310.2005.01133.x

Hemmer H (1990) Domestication. The decline of environmental appreciation. Cambridge University Press, Cambridge

Herre W, Röhrs M (2013) Haustiere-Zoologisch Gesehen. Springer Spektrum, Berlin

Hiestand L (2011) A comparison of problem-solving and spatial orientation in the wolf (*Canis lupus*) and dog (*Canis familiaris*). Behav Genet 41:840–857. https://doi.org/10.1007/s10519-011-9455-4

Jelbert SA, Taylor AH, Gray RD (2015) Reasoning by exclusion in new Caledonian crows (*Corvus moneduloides*) cannot be explained by avoidance of empty containers. J Comp Psychol 129: 283–290. https://doi.org/10.1037/a0039313

Köhler W (1925) The mentality of apes. Harcourt, Brace & Company, New York, London

Kundey SM, De Los RA, Taglang C, Baruch A, German R (2010) Domesticated dogs' (*Canis familiaris*) use of the solidity principle. Anim Cogn 13:497–505. https://doi.org/10.1007/s10071-009-0300-6

Lampe M, Bräuer J, Kaminski J, Virányi Z (2017) The effects of domestication and ontogeny on cognition in dogs and wolves. Sci Rep 7:11690. https://doi.org/10.1038/s41598-017-12055-6

Maier NRF, Schneirla TC (1935) Principles of animal psychology. In: Principles of animal psychology. McGraw-Hill, New York, NY, US

McComb K, Packer C, Pusey A (1994) Roaring and numerical assessment in contests between groups of female lions, *Panthera leo*. Anim Behav 47:379–387. https://doi.org/10.1006/anbe.1994.1052

Miklósi Á, Kubinyi E, Topál J, Gácsi M, Virányi Z, Csányi V (2003) A simple reason for a big difference: wolves do not look Back at humans, but dogs do. Curr Biol 13:763–766. https://doi.org/10.1016/s0960-9822(03)00263-x

Miller HC, Gipson CD, Vaughan A, Rayburn-Reeves R, Zentall TR (2009) Object permanence in dogs: invisible displacement in a rotation task. Psychon Bull Rev 16:150–155. https://doi.org/10.3758/PBR.16.1.150

Müller CA, Mayer C, Dörrenberg S, Huber L, Range F (2011) Female but not male dogs respond to a size constancy violation. Biol Lett 7:689–691. https://doi.org/10.1098/rsbl.2011.0287

Müller CA, Riemer S, Range F, Huber L (2014a) Dogs' use of the solidity principle: revisited. Anim Cogn 17:821–825. https://doi.org/10.1007/s10071-013-0709-9

Müller CA, Riemer S, Virányi Z, Huber L, Range F (2014b) Dogs learn to solve the support problem based on perceptual cues. Anim Cogn 17:1071–1080. https://doi.org/10.1007/s10071-014-0739-y

Müller CA, Riemer S, Virányi Z, Huber L, Range F (2016) Inhibitory control, but not prolonged object-related experience appears to affect physical problem-solving performance of pet dogs. PLoS One 11. https://doi.org/10.1371/journal.pone.0147753

Osthaus B, Lea SEG, Slater AM (2005) Dogs (*Canis lupus familiaris*) fail to show understanding of means-end connections in a string-pulling task. Anim Cogn 8:37–47. https://doi.org/10.1007/s10071-004-0230-2

Osthaus B, Slater AM, Lea SEG (2003) Can dogs defy gravity? A comparison with the human infant and a non-human primate. Dev Sci 6:489–497. https://doi.org/10.1111/1467-7687.00306

Parker GA (1974) Assessment strategy and the evolution of fighting behaviour. J Theor Biol 47: 223–243. https://doi.org/10.1016/0022-5193(74)90111-8

Pepperberg IM (2002) The value of the Piagetian framework for comparative cognitive studies. Anim Cogn 5:177–182. https://doi.org/10.1007/s10071-002-0148-5

Pepperberg IM (2004) "insightful" string-pulling in Grey parrots (*Psittacus erithacus*) is affected by vocal competence. Anim Cogn 7:263–266. https://doi.org/10.1007/s10071-004-0218-y

Petrazzini MEM, Wynne CDL (2017) Quantity discrimination in canids: dogs (*Canis familiaris*) and wolves (*Canis lupus*) compared. Behav Proc 144:89–92. https://doi.org/10.1016/j.beproc.2017.09.003

Piaget J (1954) The development of object concept (M. Cook, Trans.). In: Piaget J, Cook M (Trans.) The construction of reality in the child. Basic Books, pp 3–96. https://doi.org/10.1037/11168-001

Povinelli DJ (2000) Folk physics for apes. Oxford University Press, New York

Range F, Jenikejew J, Schröder I, Virányi Z (2014) Difference in quantity discrimination in dogs and wolves. Front Psych 5:1299. https://doi.org/10.3389/fpsyg.2014.01299

Range F, Möslinger H, Virányi Z (2012) Domestication has not affected the understanding of means-end connections in dogs. Anim Cogn 15:597–607. https://doi.org/10.1007/s10071-012-0488-8

Riemer S, Müller C, Range F, Huber L (2014) Dogs (*Canis familiaris*) can learn to attend to connectivity in string pulling tasks. J Comp Psychol 128:31–39. https://doi.org/10.1037/a0033202

Rivas-Blanco D, Pohl I-M, Dale R, Heberlein MTE, Range F (2020) Wolves and dogs may rely on non-numerical cues in quantity discrimination tasks when given the choice. Front Psych 11. https://doi.org/10.3389/fpsyg.2020.573317

Rooijakkers EF, Kaminski J, Call J (2009) Comparing dogs and great apes in their ability to visually track object transpositions. Anim Cogn 12:789–796. https://doi.org/10.1007/s10071-009-0238-8

Schuck-Paim C, Borsari A, Ottoni EB (2009) Means to an end: Neotropical parrots manage to pull strings to meet their goals. Anim Cogn 12:287–301. https://doi.org/10.1007/s10071-008-0190-z

Scott JP, Fuller JL (1965) Genetics and the social behavior of the dog. The classic study. University of Chicago Press, Chicago, London

Scott WA (1962) Cognitive complexity and cognitive flexibility. Sociometry 25:405–414. https://doi.org/10.2307/2785779

Shettleworth SJ (2010) Cognition, evolution, and behavior, 2nd edn. Oxford University Press, New York, Oxford

Sillero-Zubiri C, Macdonald DW (1998) Scent-marking and territorial behaviour of Ethiopian wolves *Canis simensis*. J Zool (Lond) 245:351–361. https://doi.org/10.1111/j.1469-7998.1998.tb00110.x

Stephens DW, Krebs JR (1986) Foraging theory. Monographs in Behavior and Ecology. Princeton Univ. Press, Princeton, N.J

Taylor AH, Medina FS, Holzhaider JC, Hearne LJ, Hunt GR, Gray RD (2010) An investigation into the cognition behind spontaneous string pulling in new Caledonian crows. PLoS One 5. https://doi.org/10.1371/journal.pone.0009345

Thorndike EL (1898) Animal intelligence. An experimental study of the associative processes in animals. Psychol Rev Monogr Suppl 2:i–109. https://doi.org/10.1037/h0092987

Tomasello M, Call J (1997) Primate Cognition. Oxford University Press

Utrata E, Viranyi Z, Range F (2012) Quantity discrimination in wolves (*Canis lupus*). Front Psych 3:505. https://doi.org/10.3389/fpsyg.2012.00505

Uzgiris IC, Hunt JM (1975) Assessment in infancy: ordinal scales of psychological development. University of Illinois Press, Urbana

Völter C, Call J (2017) Causal and inferential reasoning in animals. In: Call J, Burghardt GM, Pepperberg IM, Snowdon CT, Zentall T (eds) APA handbook of Comparative Psychology: Perception, learning, and cognition. APA Handbooks in Psychology, pp 643–671. https://doi.org/10.1037/0000012-029

Ward C, Smuts BB (2007) Quantity-based judgments in the domestic dog (*Canis lupus familiaris*). Anim Behav 10:71–80. https://doi.org/10.1007/s10071-006-0042-7

West RE, Young RJ (2002) Do domestic dogs show any evidence of being able to count? Anim Cogn 5:183–186. https://doi.org/10.1007/s10071-002-0140-0

Wilson ML, Hauser MD, Wrangham RW (2001) Does participation in intergroup conflict depend on numerical assessment, range location, or rank for wild chimpanzees? Anim Behav 61:1203–1216. https://doi.org/10.1006/anbe.2000.1706

11 Domestication Hypotheses Relating to Behaviour and Cognition: Which Are Supported by the Current Data?

The great tragedy of science—the slaying of a beautiful hypothesis by an ugly fact.

Thomas Huxley

In Chap. 1, we started off the book by presenting some of the ideas relating to *how* domestication from wolf to dogs may have happened and to what extent the evidence supports 'self-domestication' in contrast to a 'human-selection' scenario (i.e. adopting puppies) in the initial stages of this process. Here we present concepts relating to *what changed* as a result of the domestication process, with a particular emphasis on behaviour and on the potential *selective pressures*.

The 'how' question remains largely unanswerable without the fabled time machine. The best shot we have may come from the closest approximation of science to such a time machine, i.e. the concerted efforts of geneticist, neurologists, archaeologists, anthropologists, and animal behaviour scientists (e.g. see Larson et al. 2014). The 'what changed' question can, in our opinion, be addressed by the careful comparison of the wolves and dogs we have today. Still a number of important points, which we already raised in Chap. 1, need to be briefly re-iterated, before evaluating results from all our own and other people's research.

First, just to highlight once again, it is important to note that we are comparing wolves and dogs *today*, with at least 30,000 years of separate histories shaping their cognition, morphology, genetics, and behaviour. And as we saw in Chap. 1, we do not know to what extent the wolves we have today are like the ancestral progenitor of dogs considering the growing genomic evidence that dog's ancestral wolf population is now extinct (Freedman et al. 2013; Bergström et al. 2020). How representative today's wolves are of the dogs' ancestor remains an open question. For example, due to the lack of strong, century-long hunting pressure by humans, the ancestral wolves have been probably less shy and, therefore, more approachable for people than the contemporary wolves.

However, the best way of addressing such a comparison at the moment is by looking at the currently living subspecies of wolves and assessing their diversity. In terms of the kind of ecological niche they fill, although there are variations (e.g. more

F. Range, S. Marshall-Pescini, *Wolves and Dogs*, Fascinating Life Sciences,
https://doi.org/10.1007/978-3-030-98411-3_11

nomadic vs. territorial populations; see Musiani et al. 2007), they mostly are cursorial hunters cooperating to bring down medium to large prey. The social structure is also remarkably similar, characterized by pair-bonding, family units with alloparental care from all pack members, and occasional inclusion of non-related individuals in the packs. It is also important to note that a number of these traits (pair-bonding being perhaps the strongest one) are also shared with the rest of the *Canis* genus, except domestic dogs (Lord et al. 2013). Thus, based on the current evidence, it would seem reasonable (and more conservative) to assume that an extinct branch of the wolf family would still share the same characteristics currently displayed by the rest of the genus *Canis*, in particular, group hunting, monogamy, and likely alloparental care. These aspects are those we consider of fundamental importance when analysing the potential forces/selective pressures shaping dogs' cognition and behaviour.

Second, it is important to remember that during the ensuing more than 30,000 years of wolf-dog histories, significant events occurred to both dogs and wolves. This included severe persecution of wolves (evident in genetic bottlenecks—see Pilot et al. 2014; Fan et al. 2016) and an outstanding worldwide population expansion for dogs. These opposite trends may have led to inadvertent as well as conscious selection for different traits in the two species.

Third, genetic admixture with dogs has characterized wolves' past and is an ongoing phenomenon to this day (Bergström et al. 2020). Some wolf populations show much higher levels of dog admixture than others (e.g. European vs. American populations) (Pilot et al. 2014; Fan et al. 2016), and this aspect also needs to be considered in evaluating results of the behaviour research.

Fourth, even if we compare wolves and dogs that have had the same experiences throughout their lifetimes, we cannot assume that they learn the same things from these experiences. They may perceive situations very differently (Kortekaas et al. 2013) (e.g. more stressful, more competitive) and thus extract different information from the same experiences, which will then influence their behaviour in the future. However, if we can ensure that they have been exposed to the same learning opportunities, such differential epigenetic processes are based on their different genetic make-ups, which then hints at an evolutionary difference between the two species. So, we are confident that our experiments indeed measure what we intend to measure, namely, differences due to domestication and not experience.

In sum, the differences between todays' wolves and dogs must be considered as the result of a potentially meandering domestication *process* (as selection pressures on dogs vary due to culture and societal conditions in the human host populations), i.e. all the past and ongoing changes that have occurred and continue to occur to these two species. Accordingly, when studying these animals, we need to keep in mind the origin of the populations we study as well as their experiences.

11.1 A Comprehensive Theory for All Domesticated Animals: 'The Domestication Syndrome'

Domestic animals differ from their wild ancestors in physiological, morphological, and behavioural traits (Darwin 1868; Belyaev 1979; Trut et al. 2009). At the behavioural level, domestic animals are thought to be less fearful of humans and more socially tolerant with conspecifics (e.g. less aggressive). Other traits include reduced brain size, depigmentation, shorter skulls and smaller teeth, altered size and shape of ears and tails, hypersexuality and altered estrous cycles, as well as changes in concentrations of adrenocorticotropic hormones and neurotransmitters. These changes are considered to be *inextricably linked* and are referred to as 'domestication syndrome' (Price 1984, 1999, 2002). Based on the selective breeding experiment on the silver form of the red fox, it has been suggested that *tameness*, i.e. reduction in fear and aggression, is the trait that was under direct selection during the domestication process (Belyaev 1979; Trut et al. 2004, 2009), which was then responsible for all the other changes in domestic animals compared to their wild type (Chap. 1; see Box 1.1).

In 2014, Wilkins and colleagues (Wilkins et al. 2014) suggested that these apparently unrelated traits defining the domestication syndrome are the result of mild alteration of neural crest cells during embryonic development (*neural crest cell hypothesis*). Neural crest cells that develop at the edge of the neural tube during the vertebrate embryonic development migrate throughout the body to give rise to various tissue types and organs (including the skull, sympathetic ganglia, adrenal medulla, and melanoblasts) and influence a range of neurobiological and physiological functions, including the size and activity of the adrenal glands, which play a key role in the stress response (Wilkins et al. 2014).

Although the idea of one mechanism explaining all the different changes observed in domestic animals is compelling, at the moment, there are still many open questions, and rigorous tests of predictions from this hypothesis are still scarce (Sánchez-Villagra et al. 2016; Lord et al. 2020). In particular, the following questions have been raised: (1) whether indeed all domestic animals show the proposed changes compared to the wild type, (2) whether the proposed changes are in fact the result of a selection for *a single* trait and whether that trait is indeed tameness (Lord et al. 2020), and (3) if the proposed changes are linked/correlated to each other (hence can be considered a 'syndrome') and whether they do in fact have the same genetic basis (Wright 2015; Sánchez-Villagra et al. 2016; Wright et al. 2020). Below, we briefly review the evidence for each point, with a particular focus on the behavioural and to some extent hormonal aspects (see Wright 2015; Sánchez-Villagra et al. 2016; and Wright et al. 2020 for more in-depth analyses of morphological traits).

11.1.1 Evidence For and Against the Domestication Syndrome

Regarding the *differences between domestic and wild form (Point 1)*, two aspects have been considered: The first aspect is whether there is consistency in the morphological traits that changed from wild-type to domestic forms across a variety of species, as would be predicted by the domestication syndrome (DS). Recent reviews suggest that in fact, depending on the species, the number of morphological traits affected can vary distinctly between wild and domestic type and the consistency across species is not high (Sánchez-Villagra et al. 2016; Lord et al. 2020; but see Zeder 2020).

The second aspect is whether in fact the predicted behavioural changes do hold up to closer scrutiny. A number of studies have compared these two forms raised under similar conditions to allow for valid comparisons. As predicted by the domestication syndrome, domestic guinea pigs are less aggressive and show more socio-positive behaviours, including higher frequency of courtship behaviours; they have a reduced reactivity of the hypothalamic-pituitary-adrenal (HPA) axis but higher basal testosterone levels and are less explorative/attentive towards their environment than their wild counterparts (Kaiser et al. 2015). Similarly, domesticated chickens are also less aggressive (Wirén et al. 2009); less fearful of novel objects, environments, and humans; and less explorative and active than red junglefowl, the ancestor of domestic chicken (Wright et al. 2012). Cortisol responses following restraining were shown to be higher in red junglefowl than domestic chicken, although their recovery rate was also faster (Ericsson et al. 2014) and early in life, contrary to the domestication syndrome predictions, domestic chicks show stronger HPA reactivity than junglefowl chicks (Ericsson and Jensen 2016). Wolf-dog comparisons are consistent with the prediction of decreased neophobia in dogs, but wolves are more explorative and more persistent in manipulating new objects than dogs (Moretti et al. 2015, Marshall-Pescini et al. 2017a, b, c, d, Rao et al. 2018; Chap. 9). However, wolves are more tolerant of conspecifics (group members) around a feeding source than dogs, and preliminary data suggest no species difference in frequencies of aggressive and affiliative exchanges with conspecific pack members or maybe even more severe aggression in dogs than wolves (Chap. 5). Finally, dogs' basal cortisol levels are consistently higher than wolves (Vasconcellos et al. 2016; Wirobski et al. 2020).

With regard to selection for a single trait—*tameness*—being responsible for such changes in domestic animals *(Point 2)*, a selective breeding experiment on a farmed silver form of the red fox provided the inspiration for the domestication syndrome hypothesis (see Chap. 1, Box 1.1). Selection for increased 'tameness' (i.e. reduction of intrinsic fear and aggression towards caregivers) in the foxes resulted—within a few generations—in a population displaying dog-like behaviour (Trut et al. 2009) including being more vocal. These individuals also exhibited morphological and physiological traits characteristic of domesticated mammals, such as floppy ears, curly or short tails, modified coat coloration and an altered estrous cycle, and a reduced basal and stress-induced cortisol response (Trut et al. 2009). However, like in other species, the domesticated foxes showed superior competitive abilities over a food source (potentially linked to higher aggression rates) as compared with unselected control foxes (Harri et al. 2003). Recently, criticism has been levelled against the farm fox experiment based on the fact that both the tame-selected populations and the controls originated from captive populations that have already been subjected to artificial selection and inbreeding, thereby only minimally mirroring the actual domestication process (Lord et al. 2020; but see Trut et al. 2020). Furthermore, in the farm fox experiment, the scientists also selected a line of foxes for 'increased aggression', and interestingly, a number of traits associated with the domestication syndrome slowly started appearing in this line as well (Trut et al. 2017). Indeed, a recent study showed parallel changes to the brain anatomy of both the tame and aggressive selected lines (Hecht et al. 2021) and none of the expected changes to skull morphology in the tame-selected vs. control foxes, but significant differences compared to wild-born animals (Kistner et al. 2021). Such striking results may indicate that selection along the shy-bold axis in any direction (or even mere inbreeding) may affect the neural crest cells in a similar way. These results highlight the need for a more concerted effort in the comparison between aggressive and tame-selected foxes, but including also populations that have not undergone inadvertent selection for captive living.

In a parallel program, selecting for tameness in rats, similar adrenal hypofunction and changes in coat colouration were found in rats selected for reduced flight distance from humans compared to controls (Albert et al. 2009). However, contrary to the domestication syndrome predictions, coat colouration in rats was not associated with the same gene loci that affected tameness (Albert et al. 2009).

Finally, red junglefowl selected for reduced fear of humans for five to eight generations were heavier at hatch, showed less fear in an open-field and novel-object tests, spent more time exploring and foraging in a food-choice test, and had smaller brain and testes size compared to controls (Agnvall et al. 2015, 2017; Katajamaa et al. 2018). Selected birds were also more dominant, outcompeting the control group over a food resource (Agnvall et al. 2014), which might be explained by domestication shifting animals from being shy to being bold, which also implies a greater disposition to be dominant in the social domain. However, contrary to the domestication syndrome predictions, no differences between strains emerged in their corticosterone response to physical restraint (Agnvall et al. 2015).

The *covariation of behavioural and morphological traits (Point 3)*, which would indicate that indeed these traits form a 'syndrome', has been scrutinized considerably less than the other two points. The domestication syndrome hypothesis predicts correlations between specific (1) morphological traits (e.g. positive correlation between coat colouration and floppy ears) and (2) behavioural traits (e.g. positive correlation between human-directed tameness and conspecific sociability, but a negative correlation between such prosocial behaviours and aggression and neophobia) and (3) positive correlations between morphological and behavioural traits (e.g. the more morphological traits present, the stronger the behavioural effects).

In pure-bred domestic dogs, correlations between *behavioural traits* do emerge more strongly in breeds originating from East Asia, considered as 'ancient' compared to more 'modern' European breeds (Wheat et al. 2019a, b). Furthermore, in both breed groups, neophobic behaviour is strongly positively correlated with a tendency towards aggression (Wright et al. 2012; Fadel et al. 2016). In a number of species, some evidence has emerged for *morphology-behaviour trait associations*. For example, in captive mink, the darker coat colour was associated with higher aggression levels (Ducrest et al. 2008), and depigmentation was associated with less defensive behaviour towards humans (Trapezov et al. 2008). However, in breed dogs, no positive correlations emerge between *domestication syndrome-related morphological traits*, and they appear to have no predictive value in relation to behavioural patterns associated with the domestication syndrome (Sánchez-Villagra et al. 2016; Wheat et al. 2019a, b). In fact, some domestic animals, e.g. cats, lack morphological traits such as curly tails or floppy ears unless there were very stringent selection efforts (as seen in its occurrence in specific breeds), but display behavioural traits associated with the domestication syndrome. Furthermore, some morphological traits thought to be domestication syndrome constituents, such as white colouration, could have been directly selected for by humans (Linderholm and Larson 2013). Such decoupling of domestication syndrome-related morphological and behavioural traits has led some authors to question the existence of a common mechanism for them (Sánchez-Villagra et al. 2016, Wheat et al. 2019a, b).

A number of hypotheses have been put forward in relation to the genetic basis of domestication and more specifically: the *destabilizing selection for tameness hypothesis* (Belyaev 1979) (see Chap. 1, Box 1.1), the *neural crest hypothesis* (Wilkins et al. 2014), and the *modularity hypothesis* (Wright 2015). All these hypotheses take as a starting point that the domestication syndrome-related traits do occur and differ between the domestic and wild type, but they differ in regard to the genetic mechanisms underlying these traits.

The 'alteration' of neural crest cells that has been suggested as the developmental mechanism for the domestication syndrome has a complex, multiple pleiotropic genetic basis because of the large number of genes involved in the formation and migration of the neural crest cells and complex interactions among them (Wilkins et al. 2014, Simões-Costa and Bronner 2015). Several studies have shown that different genes are associated with the regulation of the neural crest cell migration across species (cats: Montague et al. 2014; horses: Librado et al. 2017; dogs: Pendleton et al. 2018; reindeer: Lin et al. 2019; silver fox: Wang et al. 2018). Evidence for the pleiotropic effect of genes associated with the neural crest cell function comes from a study on *Williams-Beuren syndrome* in humans. This syndrome is associated with the lack of social inhibition and excessive friendliness towards strangers, reduced reactive aggression, and craniofacial dysmorphism and results from a hemi-deletion of a chromosomal region including the *BAZ1B* gene—a chromatin-dependent transcription factor (Kaplan et al. 2001). This gene was shown to play a key role in the neural crest cell migration and in the regulation of neural crest-specific transcriptional circuits as well as in shaping human craniofacial features (Zanella et al. 2019). This finding provides evidence for the pleiotropic effect of the transcriptional regulation of the neural crest cell migration on specific morphological and behavioural traits that are affected by the domestication process, such as craniofacial features, sociability, and aggression levels.

The alternative proposed, the *clustered-linkage* (Wright 2015) (or *modularity*; Wright et al. 2020) *hypothesis*, rather suggests that linked clusters of genes/quantitative trait loci played a particularly large role in the domestication process. Such gene clusters have been identified in the genome of multiple domestic plant and animal species and provide an alternative mechanism through which the appearance of the phenotypic array predicted by the domestication syndrome may have emerged (Wright 2015). Future studies combining genetics with behavioural studies across domesticated species need to test the different hypotheses.

11.2 What Brought About the Behavioural Wolf-Dog Differences We See Today and Which Hypotheses Are Supported by the Current Data?

The domestication syndrome proposed that the selection for a single trait, tameness, brought about the multitude of morphological, hormonal, and behavioural changes we observe in domestic animals. Researchers focusing on dog and wolf behaviour and cognition have also suggested that during domestication, selection for a number of different traits may have occurred, including most prominently attachment, sociality, attention, inhibition, and cooperativeness. In most cases, these traits are suggested to have been selected in combination with tameness (reduced fear and aggression). A number of other researchers have instead considered the changes observed in dogs as the product of a relaxation on specific selective pressures still active in wolves (potentially also acting together with a selection for tameness). In the end, likely, all these processes contributed to the domestication of the wolf.

As the field has expanded, so has the number of hypotheses. It is important to note that these hypotheses mostly emerged from studies on different (narrow) aspects of the wolf-dog comparison (see Fig. 11.1 and Table 11.1), in particular from studies comparing (1) wolf-dog differences in human-directed behaviours and understanding of human communication, (2) wolf-dog social behaviours with conspecifics, or (3) wolf-dog differences in interactions with their environment (e.g. problem-solving). Thus, here we present the hypotheses according to this grouping (Fig. 11.1).

Having presented the major hypotheses relating to dog domestication in each group, we will shortly review the data in support of each, drawing from our own and other people's studies.

11.2.1 Interactions with Humans

11.2.1.1 Part I: Domestication Hypotheses

Based on more recent wolf-dog comparisons focusing on human-related behaviours, a number of hypotheses have been proposed, which suggest that life in the human environment was associated with selective pressures for specific traits in dogs. Some of these hypotheses propose direct selection on traits related to cognition (social

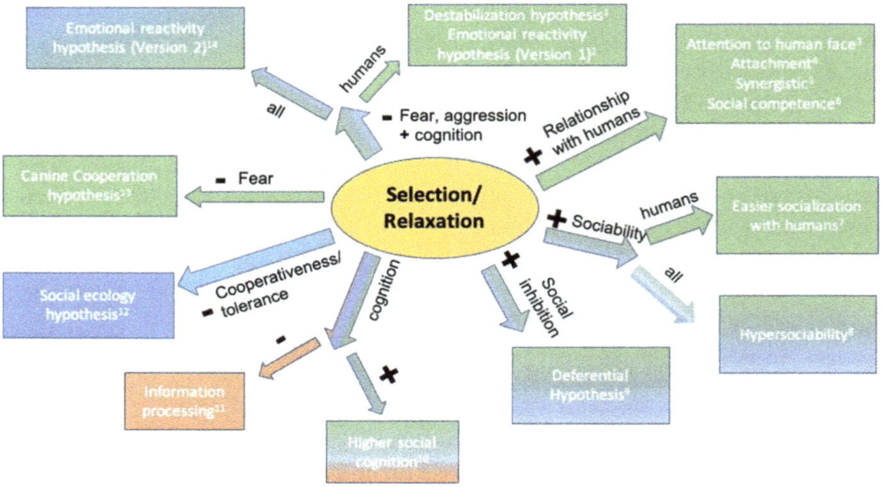

Fig. 11.1 Summary of the various domestication hypotheses predicting changes in cognition (orange), human-directed behaviour (green), or conspecific interaction (blue).[1](Belyaev 1979), [2](Hare and Tomasello 2005a, b),[3](Miklósi et al. 2003), [4](Topál et al. 2005a, b), [5](Gácsi et al. 2009), [6](Miklósi and Topál 2013),[7](Klinghammer and Goodmann 1987), [8](Bentosela et al. 2016; Vonholdt et al. 2017a, b), [9](Range et al. 2019), [10] (Hare et al. 2002),[11] (Frank 1980), [12] (Marshall-Pescini et al. 2017a, b, c, d), [13] (Range and Virányi 2015),[14] (Hare et al. 2012; Hare 2017)

cognition, attention, inhibition), while others propose changes in temperament (fearfulness, aggressiveness, sociability), which might or might not enhance cognitive skills. Some of these earlier hypotheses have also then been combined to form more embracing hypotheses.

Selection for Temperament Traits

Tameness (i.e. Reduced Fear and Aggression of Humans)

Picking up on Belyaev's idea of selection for tameness (see Chap. 1, Box 1.1) and the success of tame foxes and domestic dogs in pointing studies compared to wolves, Hare and Tomasello (2005a, b) formulated their *emotional reactivity hypothesis*, which proposes that in a first step, as a by-product of the selection for reduced fear and aggression of humans, dogs were as comfortable with humans as with conspecifics, allowing them to use their species-specific cooperative and communicative skills with humans. After this initial change, during a second step, selection may have acted on the existing variations of these traits further increasing dogs' flexibility and cognitive adaptation to human interaction. Accordingly, dogs that are socialized with both conspecifics and humans should cooperate and communicate with humans as well as they do with conspecifics, and indeed, as well as wolves cooperate with each other. However, dogs should cooperate with humans better than wolves (even if wolves are socialized with humans and hence the fear element is removed), since, according to the hypothesis, humans selected for specific traits in

Table 11.1 Summary of hypotheses with the respective predictions and in historical order

Hypothesis	Based on	Evolutionary effect on dogs	Predictions on		
			Problem-solving skills	Interactions with conspecifics	Interactions with humans
Destabilizing selection for tameness hypothesis (Belyaev, 1979)	Silver fox experiment	Selection against fear & aggression (Stress hormone levels: D < W)			D > W
? (Feddersen-Petersen, 2004)	Comparisons of packs of wolves and different dog breeds	Relaxation of selection for conspecific cooperation/conflict/aggression		Aggression: D > W Cooperation: W > D	
Information processing hypothesis (Frank, 1980)	Problem solving studies in wolves and dogs	Relaxation of natural selection	W > D	ritualized aggression: W > D	sensitivity to wide band of stimuli (e.g. human cues, learning of associations) & plasticity: D > W
Human Socialization Hypothesis (Klinghammer & Goodmann, 1987; Scott & Fuller, 1965).	Raising of wolf and dog pups	Selection for an easier and broader/more generalizable form of socialization			Socialization: D easier than W

(continued)

Table 11.1 (continued)

Hypothesis	Based on	Evolutionary effect on dogs	Predictions on		
			Problem-solving skills	Interactions with conspecifics	Interactions with humans
Enhanced social cognition hypothesis (Hare, Brown, Williamson, & Tomasello, 2002)	Pointing studies of (non-socialized) wolves and dogs	?			Social communicative skills: D > W
Attention towards the human face (Miklósi et al., 2003)	Pointing studies & looking back study	Selection on willingness to establish eye contact			Eye contact: D > W Communication: D > W Obedience: D > W
Emotional Reactivity Hypothesis (Version1) (Hare & Tomasello, 2005)	Pointing studies of silver foxes, (non-socialized) wolves and dogs	Selection against fear & aggression		Tolerance & Cooperation: D = W	Tolerance & Cooperation: D = W
Emotional Reactivity Hypothesis (Version 2)	Pointing studies of silver foxes, (non-socialized) wolves and dogs	Selection against fear & aggression		Tolerance & Cooperation: D = W	Tolerance & Cooperation: D = W

(continued)

Table 11.1 (continued)

Hypothesis	Based on	Evolutionary effect on dogs	Predictions on		
			Problem-solving skills	Interactions with conspecifics	Interactions with humans
Attachment hypothesis (Topál et al., 2005)	Attachment studies wolf/dog puppies	Selection for the capacity to form attachments with humans			Attachment: only D Reliance on humans: D > W Communication: D > W Synchronized Collaboration: D > W
Synergistic hypothesis (Gácsi et al., 2009)	Pointing and attachment studies	Selection for reduced fear and aggression & better recognition of humans as social partners	Inhibitory control: D > W		Eye contact: D > W Communication: D > W
Evolutionary Social Competence Hypothesis (Miklósi & Topál, 2013)(see also (Topál et al., 2009)	Pointing and attachment studies	Convergent evolution of dogs and humans			Bond: D > W Prosocial behavior; social competence: D > W
Two Stage hypothesis (Udell, Dorey & Wynne, 2010)	Pointing studies	Environment is more important than genetical differences		socio-cognitive skills: W = D if same experience	

(continued)

Table 11.1 (continued)

Hypothesis	Based on	Evolutionary effect on dogs	Predictions on		
			Problem-solving skills	Interactions with conspecifics	Interactions with humans
Hypersociability hypothesis (Bentosela, Wynne, D'Orazio, Elgier, & Udell, 2016; Vonholdt, Udell, & Sinsheimer, 2017)	Social preference (humans) and problem-solving studies	Selection on structural variants of two genes linked to Williams-Beuren syndrome	W > D	Motivation to seek social contact (dependent on socialization): D > W	
Mechanistic: OT underlying the social bond with humans (Nagasawa et al., 2015)	Social interaction study with wolves/dogs (different socialization)	Selection on OT system (OT: D >W)			Eye contact: D > W Communication: D > W Bond: D > W

(continued)

Table 11.1 (continued)

Hypothesis	Based on	Evolutionary effect on dogs	Predictions on		
			Problem-solving skills	Interactions with conspecifics	Interactions with humans
Deferential Behaviour Hypothesis (Range, Marshall-Pescini, Kratz, & Virányi, 2019)	Various experiments carried out at the WSC in the last ten years	Selection for social inhibition	Confidence: W > D => better problem solving skills?	Tolerance: W > D Cooperativeness: W > D Reconciliation: W > D	Obedience: D > W Strong influence of human presence during experiments
Canine Cooperation hypothesis		Selection for cooperation in wolves as a basis for the human-dog relationship		W – W Interactions = D – H interactions	
Social ecology hypothesis		Adaptation to different feeding niches	No general predictions but dependent on the specific situation		

Predictions in relation to changes in cognition (orange), in human-directed behaviour (green), conspecific interaction (blue) or a mix of those (yellow)
D = dog; W = wolf; H = Human

dogs that would facilitate such cooperation (second step). Similar to the Budapest group (Kubinyi et al. 2007; Topál et al. 2009a, b), they propose that these initial changes in temperament traits (which then allowed changes in the cognitive domain) may mirror the evolution of social and cognitive abilities in humans (see related articles on human self-domestication hypotheses; Hare 2017; Sánchez-Villagra and Schaik 2019; Shilton et al. 2020).

Later on, Hare et al. (2012) (see also Hare 2017), expanding on the human self-domestication aspect of the hypothesis, also imply that selection for tameness actually increased dogs' tolerance and cooperativeness (and related cognitive skills) not just towards humans but also towards other dogs. Accordingly, this version of the emotional reactivity hypothesis would predict that dogs should outperform wolves in cooperative and communicative interactions with conspecifics as well as with humans (see below interactions with conspecifics).

Sociability and Attachment

To date, several studies have suggested that domestication might have altered dogs' sociability towards humans. The first hypothesis is the *human socialization hypothesis*, suggesting that dogs were selected for an easier and broader/more generalizable form of socialization with humans compared to wolves, while socialization of wolves with human needs to start much earlier to achieve a trustful relationship and remains largely focused on individual humans (Scott and Fuller 1965; Klinghammer and Goodmann 1987).

In 2005, Topal and colleagues (Topál et al. 2005a, b) proposed that it is not just the case that dogs can be more easily socialized with humans but also that dogs might have evolved an evolutionary novel 'capacity for attachment to humans that is functionally analogous to that present in human infants' (pp. 1373) during domestication: an ability absent in wolf pups (*attachment hypothesis*). Moreover, Miklósi and Topál suggest that the attachment bond also plays an important role in allowing complex interspecific communication to emerge and, thus, facilitate synchronized collaboration between dogs and humans (Miklósi and Topál 2013) (see also below the *attention hypothesis*). However, it remains unexplained, whether these enhanced socio-cognitive abilities are based on specific changes in the cognitive skills or whether they are just the consequence of staying close to the human and thus having better opportunities to learn from humans and/or perceive human action's as more relevant due to the attachment.

The more recent *hypersociability hypothesis* suggests that compared to wolves, dogs show an increased motivation to seek affiliative social contact with humans independent of familiarity (Bentosela et al. 2016; Vonholdt et al. 2017a, b). Thus, this hypothesis includes at least two components: (1) better acceptance of unfamiliar partners and (2) social interest in general ('exaggerated motivation to seek social contact'). Compared to the previous hypothesis, it suggests that the observed genetic changes from wolves to dogs were not specifically selected to create a social bond between dogs and humans. Rather, dogs show genetic predispositions to hypersociability towards any companion. Accordingly, experience, in the sense of the degree of socialization with humans (or potentially any other species), still plays a crucial role in determining dogs' social responses. In other words, if wolves and dogs are both socialized with humans and conspecifics, dogs would be predicted to be more social (independent of familiarity) than wolves towards humans and conspecifics, e.g. spend more time with them. Interestingly, human-directed social behaviours in both wolves and dogs were linked to structural variants of two genes previously implicated in the Williams-Beuren syndrome (WBS) in humans, providing a possible genetic basis on which directional selection might have been targeted (Vonholdt et al. 2017a, b). This syndrome is characterized by delayed development, cognitive impairment, behavioural abnormalities, and hypersociability (vonHoldt et al. 2010). However, it is also not quite clear whether the kind of sociability proposed by the hypersociability hypothesis is not just a consequence of dogs being easier to socialize than wolves (*human socialization hypothesis*).

Selection Related to Socio-cognitive Skills

Socio-cognitive Skills (General)

One of the first hypotheses related to an enhancement of cognitive skills in dogs compared to wolves was proposed by Hare and colleagues (2002). The authors suggest that through exposure to the cognitively challenging human social environment, dogs evolved functionally human-analogue social-communicative skills. The basis for this hypothesis was dogs' superior abilities in following the human pointing gesture compared to (zoo-kept) wolves. However, the hypothesis is very vague in terms of which abilities were selected for and why.

Social Cooperative Skills

In a review presenting why dogs may be a good species to study social cognition, Miklósi et al. (2004) suggest that domestication may have enhanced dogs' social skills allowing them to develop complex cooperative interactions with humans.

Attention Towards the Human Face

Based on dog puppy's enhanced willingness to establish eye contact and their better performance in pointing tasks compared to wolf pups, Miklósi et al. (2003) proposed that the main difference between wolves and dogs in their interactions with humans is in the dogs' inclination to look at the human's face. Whereas in many animal species direct, staring eye contact is considered a threat, in humans, this behaviour is considered the first step in initializing and maintaining communication with others and hence a precursor of numerous socio-cognitive abilities (see Chap. 7). It does not only serve to transmit the intent to communicate but also serves to clarify who the recipient of the communicative interaction is. Thus, as a consequence of 'positive feedback processes' (both evolutionary and ontogenetic), Miklósi and colleagues suggest that 'dogs' readiness to look at the human face has led to complex forms of dog-human communication that cannot be achieved in wolves even after extended socialization' (Miklósi et al. 2003, p. 763). They further argue that the susceptibility of dogs towards human ostensive-referential signals activates a 'ready-to-obey' attitude in dogs (see also Topál, Gácsi et al. 2005; Topál et al. 2005a, b), which results in following the human instruction without necessarily understanding the causal structure of the interaction.

In line with this idea, based on their study on oxytocin levels following gazing behaviour between wolf-owner and dog-owner dyads, Nagasawa et al. (2015) suggest that an oxytocin-mediated positive feedback loop mediated by gaze lies at the heart of the dog-human bond.

Inhibition

In 2009, Gásci et al. (2009) offered their *synergistic hypothesis* suggesting that through the selection for reduced fear and aggression, dogs are predisposed to develop better action inhibition than wolves (see also Gácsi et al. 2005), allowing for better cooperation with humans. In combination with selection for faster and easier recognition of humans as social partners, which improves dogs' sensitivity to salient human communication, dogs look more readily at the human face providing the basis for complex forms of dog-human communication.

Deferential Behaviour

Years of working with wolves and dogs and running various experiments showed us (Virányi and Range 2014) that indeed the human-directed behaviours of wolves and dogs differ even when socialized similarly. While it is very clear that it is much easier to socialize a dog than a wolf, and dogs are much more relaxed in regard to who interacts with them than wolves, wolves can be very attentive, communicative, cooperative, and even tolerant *if* they choose to. And we guess this, the *if*, is the key word here. Wolves are rather self-confident—in general more so than dogs—and primarily follow their own preferences. But if you have a very good relationship with them, they do make concessions. I (Friederike) have been in situations, where I kept a wolf from chasing someone else out of his enclosure and in no situation did he attack me keeping our relationship intact (this does not mean that it might always be the case). So based on such experiences, observations, and also experiments, we started to get an idea how wolves tick.

Studying the social interactions of dogs and wolves, we have found consistent differences both in intra- and interspecific contexts indicating that dogs tend to avoid conflicts and follow the human's lead, suggesting that at least since the Neolithic Revolution dogs were (consciously or unconsciously) selected for increased submissive inclinations (*deferential behaviour hypothesis*; Range et al. 2019). In other words, through selection for higher sensitivity towards social inhibition, dogs lost some of their 'self-confidence', allowing humans, but also higher-ranked partners, to take over the leading role without being contested. In terms of human-animal cooperation, it means that dogs cooperate in a deferential and compliant manner rather than as equal partners.

Which elements during dog domestication selected for these characteristics is an open question. The new foraging niche with smaller dispersed itemized foods rather than larger carcasses (requiring sharing) may have selected for individuals that avoid costly conflicts and use distance maintenance as a conflict management strategy. The 'respect' for another's resource ownership, without engaging in conflict, may have also increased humans' acceptance of dogs staying in proximity to them, especially in situations related to food (carrying food, cooking). It is also very likely that humans (unconsciously and maybe also consciously) selected those animals that were less inclined to steal food and more likely to accept the inhibitory and leading role of the humans. Accordingly, we proposed that after an initial selection against fear during the domestication process, dogs were selected for increased deference in order to minimize conflicts over resources, to ensure safe co-habitation and co-working in a way that humans lead and dogs follow (Range et al. 2019).

Composite Hypotheses

In 2009, Topál et al. (2009a, b) proposed a *system-theoretical approach* suggesting that the similarities between human and dog social behaviour are not the result of a fundamental change in one single trait but rather may be the result of small genetic changes, which affected various aspects (components) of sociality in concert (attachment, synchronization, etc.). They emphasize the analogy between the evolution of dog's social skills with some aspects of the hominization process over the past six million years, resulting in increased sociality, cooperativeness, and communicative skills in humans as compared to chimpanzees. Due to a kind of convergent evolution in the dogs, they highlight that while dogs possess a set of functionally analogous skills corresponding to that of humans (derived from homologous traits of the wolf), the underlying cognitive capacities and physiological mechanisms controlling these skills may differ between dogs and humans. The idea of convergent evolution being responsible for similarities in the dog-wolf and human-chimpanzee comparisons was also presented by Hare et al. (2002) and expanded on in later papers (Hare and Tomasello 2005a, b; Hare et al. 2012; Hare 2017) (Fig 11.2).

Trying to reconcile previous domestication hypotheses, Miklosi and Topal (2013) introduced the notion of *evolutionary social competence* (ESC), i.e. a suite of behaviours relating to dogs' interspecific capabilities. Again, the idea is that, since humans' social competence is biased towards prosocial behaviour and the fact that humans are an integral part of dogs' social environment, similar changes in dogs' social competence are promoted by natural and artificial selection. Importantly, the result of this process is an *evolutionarily novel*, *interspecific* social competence in dogs allowing them to establish a variety of social relationships with humans. This evolutionary novel competence is thought to be brought about by (genetic) changes in the set point(s) and the degree of behavioural (cognitive) plasticity that led to a different cognitive functioning in dogs as compared to wolves. Consequently, rearing wolves and dogs in the same environment will not be sufficient to achieve dog-like social competence in wolves; however, intensely human-socialized wolves may outperform shelter dogs, who may have been hindered by a lack of human exposure in their social competence skills. Finally, linking back to a previous hypothesis, they propose that the ability of dogs to form attachment bonds with humans is at the core of their social competence with other secondary 'modules' emerging based on the specific human environment the animals are exposed to.

Finally, the *two-stage hypothesis* (Udell et al. 2010) proposes that dogs and wolves have similar socio-cognitive skills and their differences stem from their

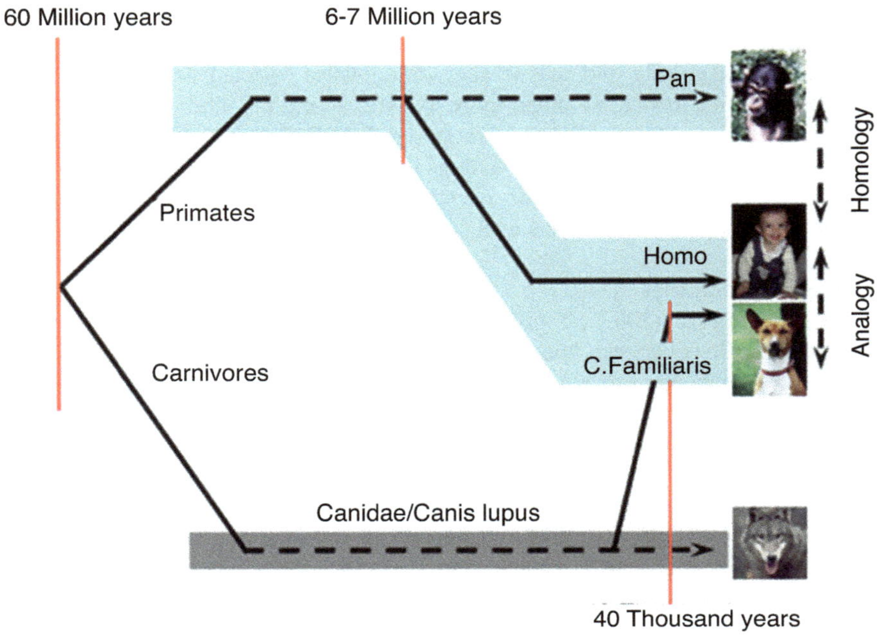

Fig. 11.2 Illustration of the domestication process in comparison to human evolution. Figure adapted from Kubinyi et al. (2007); Copyright © 2007 Kubinyi, Virányi, Miklósi; reprinted with permission from Kubinyi

ontogeny; dogs are highly cooperative with humans because they (1) accept humans as social companions, which is acquired early in life through socialization, and (2) learn from humans to follow their gestures and actions through conditioning, whereas wolves do not typically live with humans. Interestingly, similar to the hypersociability hypothesis (see above), the two-stage hypothesis also does not claim that this process requires that the social companion is human, but can be any other animal. This view suggests that the evolutionary adaptation to the human environment is of negligible importance in the emergence of dog social communication skills. Thus, any specific genetic changes and/or changes in behaviour-organizing mechanisms is neither necessary nor sufficient to account for 'human-like' social skills in dogs.

11.2.1.2 Part II: Data in Support of These Hypotheses

Sociability and Attachment

Several of the earliest studies comparing wolves and dogs suggest that it is easier to socialize dogs than wolves and that the socialization window is longer in dogs (*human socialization hypothesis*). Although specific studies on this aspect have not been carried out in recent years, there is little doubt that this observation holds true based on our and other's experience on the considerable 7/24 effort that needs to go into socializing wolf pups. However, beyond this basic difference, is there evidence for further changes in sociability with humans in dogs brought about by domestication?

Reviewing the studies looking at potential differences in social behaviour towards caregivers and strangers in wolves and dogs (Chap. 5) indicates that both have a clear capacity to distinguish between them already at an early age (Gácsi et al. 2005; Topál et al. 2005a, b). Furthermore, although early studies, in line with a number of hypotheses (*attachment hypothesis*; *social competence hypothesis*), suggested that in contrast to dogs, wolves would not form 'attachment bonds' with humans (Gácsi et al. 2005, Topál et al. 2005a, b), more previous results showed that wolves show all the behavioural indication of an attachment bond with their caregivers (Hall et al. 2015; Lenkei et al. 2020; Wright et al. 2020). Interestingly, at an early age, wolves appear to show a greater curiosity than dogs towards strangers (Gácsi et al. 2005, Topál et al. 2005a, b), spending more time with them, which may have masked their bond with the caregiver in the relevant test settings. This greater curiosity towards novelty as compared to dogs is also apparent towards inanimate objects (Chap. 9) (Moretti et al. 2015; Marshall-Pescini et al. 2017a, b, c, d).

However, in studies at the Wolf Science Center and elsewhere, when testing adults, dogs approach and stay in physical contact with humans (bonded, familiar, or strangers) more than wolves (in line with the *hypersociability hypothesis*) (Bentosela et al. 2016; Vonholdt et al. 2017a, b; Wirobski et al. 2021). Interestingly, in this respect, the variability in the wolves is often much higher than in the dogs, with some animals seeming to really look for physical contact even with 'just' familiar people, and other wolves that are disinterested or avoidant also when offered positive physical interaction with their closest human partners (Wirobski et al. 2021). However, there are a number of important aspects in relation to these results. In some cases, where wolf-dog differences have been found, these may be better explained by the different social experiences of the populations compared (Bentosela et al. 2016; Vonholdt et al. 2017a, b). The bond that a pet dog establishes with its owner is quite different from that of a pack living dog with its caregivers. This is highlighted by the fact that whereas we found no association between social interactions and oxytocin levels in our animals at the Wolf Science Center regardless of relationship type, we did find a positive association between time spent being cuddled by the owner and oxytocin levels in pet dogs (Wirobski et al. 2021). Furthermore, equally in both wolves and dogs, the time in physical contact with 'just familiar' people was positively correlated with cortisol levels (Wirobski et al. 2021), whereas the proximity and interactions with bonded humans in a training context were associated with a decrease in their cortisol levels (Vasconcellos et al. 2016). Such results highlight the limited meaningfulness when comparing populations of animals that are not raised in the same way, especially when evaluating the relationships the animals form with humans.

A second important aspect is that when testing for sociality towards humans, we need to take into account the other known differences between dogs and wolves in terms of how they interact with their environment. For example, as a number of studies have shown that wolves are more explorative than dogs (see Chap. 9; Udell 2015; Marshall-Pescini et al. 2017a, b, c, d; Rao et al. 2018), it may be necessary to account for these differences in experimental settings by giving wolves sufficient time to explore the testing arena, before introducing the human element for them to interact with. Designing tests that isolate the single variable of interest, while accounting for other differences between the two species, will be the next important challenge for us.

Leadership, Conflict Management, and Aggression

Although in a number of studies dogs spent more time in proximity and physical contact to humans than wolves, an important aspect which needs attention is whether they differ in their motivation to do so. Two studies that investigated this question by (1) giving the animals a choice between food and social contact with a bonded person (Lazzaroni et al. 2020) or (2) investigating whether food or social contact is more rewarding (Feuerbacher and Wynne 2012) found no difference between wolves and dogs.

However, several studies revealed that during interactions with humans, dogs more frequently exhibit self-directed behaviours thought to be related to arousal or stress (lips licking, yawning, etc.) and tail wagging (Gácsi et al. 2013). Furthermore, pilot studies at the Wolf Science Center suggest that dogs also show more greeting and submissive behaviours towards humans than wolves. With 'just familiar' people, both wolves and dogs seem to find physical contact somewhat stressful (cortisol increases), but wolves in most cases just avoid the situation altogether by keeping their distance (Wirobski et al. 2021). But dogs go there anyway! The fact that the dogs approach and allow physical contact with the human even if they are not quite sure about it could be a consequence of selection for compliance. Across several studies, it was found that wolves are less willing to accept humans, when 1. physically restraining them (Chap. 6: Zimen descriptions, when being held as pups during the pointing tasks) (Zimen 1971), 2. forbidding something (Chap. 6: calling away from food in puppy tests, stealing the rope in the cooperative string-pulling task, persistence in getting the meat on the chair, getting the toy back after a playful interaction in adult animals) (Gácsi et al. 2013; Range et al. 2019; Ujfalussy et al. 2020), and 3. taking the leading role (Chap. 8: waiting for humans to initiate action and then following them in the cooperative string-pulling) (Range et al. 2019). Such results point at the possibility that the hierarchical elements of the relationship with humans may differ in meaning for wolves or dogs. Dogs accept the leading role of the human, allowing themselves to be subjected to their handling and trying to do what is expected of them, including coming close and accepting manipulation even if this makes them feel a bit uncomfortable. And they accept these behaviours in general also from unfamiliar people. Wolves are willing to subject themselves to potentially unpleasant situations *if* the relationship with the human is strong and the human asks in a respectful way. But if these two elements are lacking, they seem to see no reason to do so and may even react with agonistic behaviours (in line with Zimen observations, but also Wold Science Center general life).

These results do not support the *emotional reactivity hypothesis*, since while there was selection against fear, compliance might more consistently explain the behaviour of the dogs with both conspecifics and humans in terms of accepting the leading role of higher-ranking individuals (independently if human or conspecific) and trying to avoid any, potentially dangerous conflicts (*deferential behaviour hypothesis*). This interpretation also accounts for the fact that human-socialized wolves are cooperative with bonded humans in various situations (training, cooperative string-pulling, daily handling including things like blood taking), but more on eye level than dogs.

Furthermore, the differences observed in sociability with humans between wolves and dogs (not mirrored in differences in sociability with conspecifics—see below) might not be driven by a motivation to be social per se (*hypersociability hypothesis*), but rather may derive from a difference in (1) how dogs perceive human cues as 'commands' in contrast to requests, (2) how compliant they are in situations when humans are making decisions for them, and (3) how they generalize this compliance to humans also if they have no specific bond with them (*deferential behaviour hypothesis*).

Socio-cognitive Abilities in Relation to Humans

We found no clear support for a selection for increased social attention to humans (*attention hypothesis*) in dogs, since once wolves and dogs are equally socialized to people, wolves are in most cases as attentive to humans as dogs (local enhancement task: Range and Virányi 2013; Chap. 8; training situations: Vasconcellos et al. 2016; Chap. 6; and sensitivity to attentional states Udell and Wynne 2011; Chap. 7).

What about differences between wolves and dogs in their socio-cognitive skills? The most used tasks towards this question are those relating to the animal's understanding of human referential cues. There, dogs are more prone than wolves to establish eye contact with humans and to follow the pointing gesture already at 3–4 months of age (Miklósi et al. 2003; Gácsi et al. 2009), in contrast to wolves starting to follow pointing only with training and/or age (Udell et al. 2008; Virányi et al. 2008; Lampe et al. 2017). However, what do these results mean? Wolves have no trouble following human gaze cues into distance space and around barriers (Range and Virányi 2011), and they can also follow conspecifics' gaze in a two-choice task (Chap. 7). This suggests that wolves have the cognitive abilities to use the directional cues provided by a social partner. However, their performance with a human partner in the two-choice task is very poor. As we discussed in Chap. 7, this might be partly due to the fact that the animals are restrained during the observation phase (*synergistic hypothesis*; Gácsi et al. 2009; Ujfalussy et al. 2020)—which is not the case in the gaze into distance space and gaze around barriers tasks (Range and Virányi 2011). Being restrained might be more distracting for the wolves than the dogs—at least if a human and not a conspecific is providing the cue. With age, maybe the trust and the acceptance of humans restraining them increase, allowing them to succeed in the two-choice pointing task even without specific training (Virányi et al. 2008, Lampe et al. 2017).

Dogs on the other hand may be ok with the two-choice setup, but interpret pointing—at least how we present it (direct eye contact, no accompanying head turn, relatively abrupt, momentary pointing gesture)—as a command rather than a hint. Maybe, as suggested by the *deferential behaviour hypothesis* (Range et al. 2019), dogs are more sensitive to the dominance status of others and acceptant of leadership, which results in different reactions to specific human gestures and behaviours than wolves. Adult pet dogs, for example, may interpret the human gaze cue in the two-choice task as a sign of ownership and therefore avoid it altogether. While the human gaze might not be sufficiently salient for wolves (especially if being distracted by being restrained), if instead a dominant conspecific is giving the cue, they will retrieve the food, even if taking the food from a dominant conspecific might lead to conflicts. Dogs on the other hand, as with the human, will try to avoid any conflict over the resource (Range et al. 2015; Dale et al. 2017a, b). So, in the end, there may not be any difference in their understanding of referential communicative cues, but rather in how the wolves and dogs perceive the testing situations and interpret the cues given.

Further support for this impression comes from the studies where researchers investigated how wolves and dogs communicate with humans. Both wolves and dogs nicely communicate where food is hidden in the showing paradigm Heberlein et al. 2016; Chap. 7). But while the looking back behaviour in the unsolvable task was used to support the idea of a difference between wolves and dogs in their preference to communicate with humans (Miklósi et al. 2003), it rather seems to be indicative of differences in their object-directed persistence (Udell 2015, Marshall-Pescini et al. 2017a, b, c, d; Chap. 9). Thus, at the moment, the data we have do not support the idea that dogs have an overall better understanding of human communicative abilities than wolves nor a more striking and somewhat miraculous ability in communicating with them (*social competence hypothesis*, *emotional reactivity hypothesis*). Rather, results paint a more complex and patchy picture; in some cases, it seems the paradigms used were not carefully enough designed to isolate the single variable of interest, and in other cases, there are still questions whether the animals feel comfortable enough or interpret the setup as the researcher's intended.

Although, as suggested by the *two-stage hypothesis*, across many studies, the importance of the human social experience in affecting results on their communicative abilities has emerged again and again, wolves and dogs, even when raised and kept in similar ways, do show a number of crucial differences in their interaction with humans (see references above). Therefore, differential experience alone cannot explain results, and a wolf, even if raised as a dog, is still very much a wolf.

Instead, the results lend some support for the *canine cooperation hypothesis* and the *deferential behaviour hypothesis*. However, studies (1) to better understand if wolves and dogs differently 'interpret' human cues, which may convey command and/or dominance status, and (2) to compare wolves and dogs in situations of 'disagreement' (or mild conflict) with humans would be helpful to better assess the merits of the latter hypotheses.

11.2.2 Social Behaviour with Conspecifics

11.2.2.1 Part I: Domestication Hypotheses

Based on the comparisons between wolf and dog packs of different breeds, and work with the mixed packs of wolves and poodles, Feddersen-Petersen (2007) noted that in contrast to wolves, 'some dogs breeds are unable to cooperate in a very basic manner and compete within groups, reflected in difficulties in establishing and maintaining the rank order' (page 111); furthermore, 'tactical variants of conflict solving (to appease, animate, or inhibit opponents) do not exist in groups of several dog breeds'. This led her to suggest that 'during domestication dogs have lost essential capabilities necessary to adapt to living exclusively with conspecifics in groups', although some breeds, namely, genetically wolf-like ones (such as Alaskan malamutes, Siberian huskies, and Samoyed) were exceptions. Dogs also show reduced diversity of social expressions compared to wolves, probably also due to their altered morphology, which hinder their social communication with conspecifics leading Feddersen-Petersen to conclude that 'modern dogs seem to be better adapted to communication with humans than with other dogs' (page 117) (Feddersen-Petersen 2007).

The Franks (1982) similarly outline the possibility that, because dogs no longer live and function as pack animals, and hence no longer need to independently determine access to resources such as food and mating partners (because humans do it for them), the selection for ritualized dominance behaviours has been relaxed.

We will refer to these observations as the hypothesis for 'reduced intraspecific social competence'.

When we (Zsofia Viranyi, Kurt Kotrschal, and me, Friederike) started to think about setting up the Wolf Science Center and the kind of scientific questions we were interested in, one aspect that we were rather clear about is that we really wanted to investigate not just the interactions of wolves and dogs with humans but also their interactions with conspecifics. We felt that it was a potentially very important, missing aspect in the domestication story. Consequently, we set up the Wold Science Center in a way that would allow us to pursue this aspect (e.g. raising the animals in peer groups, socializing our pet dogs with the young animals to have good 'conspecific' demonstrators for later on, and keeping the animals in conspecific packs; see Chap. 2). With our first results—the social learning and the gaze following studies—we realized that the picture that was, probably inadvertently, drawn by many domestication hypotheses of wolves being less attentive, more aggressive, and less cooperative than dogs was not quite adequate. We had a lot of discussions about our results and how to frame them also with Sarah Marshall-Pescini and Simona Cafazzo, whom we visited in Rome to see the free-ranging dogs. Rome was probably the birthplace of our hypotheses.

The first one, the *canine cooperation hypothesis* (CCH), is basically about wolves, arguing that due to their dependence on conspecific cooperation in their natural setting (e.g. group hunting, alloparental care), wolves should be very attentive, cooperative, and tolerant with conspecifics. In our view, the wolves' intraspecific cooperativeness (see Chap. 3) was the foundation on which dogs' cooperativeness with humans evolved (Range and Virányi 2014; Virányi and Range 2014; Range and Virányi 2015). In combination with selection for easier socialization with humans (Scott and Fuller 1965; Klinghammer and Goodmann 1987), the dogs were then able to use their wolf-derived social skills also with humans. This aspect of the hypothesis aligns with the emotional reactivity hypothesis (Hare and Tomasello 2005a, b). However, different from it, it suggests that if wolves are given the possibility of seeing humans as social partners (albeit through a more intensive socialization process—see Chap. 2), wolves will be as cooperative with humans as dogs. This also partly explains why wolves and dogs got in contact in the palaeolithic and developed such a strong relationship with each other. Thus, the CCH does not embrace the convergent evolution idea that through a similarity in social environment, dogs have evolved novel, more sophisticated, and more human-like skills (Hare and Tomasello 2005a, b).

However, when formulating the CCH, we already realized that this might not be the whole story, since we also found lower tolerance levels between dogs than between wolves in feeding contexts (Range et al. 2015; Range and Virányi 2015), and similar to descriptions by Feddersen-Petersen, we observed high-intensity aggression more often in our dog packs than in our wolf packs.

Over the next few years, we intensified the number of observations and experiments testing intraspecific behaviour. I (Sarah) joined the Wolf Science Center specifically to work on aspects of intraspecific cooperation, and Simona Cafazzo joined at the same time to look at conflict management behaviours. In general, in the many discussions involving prosecco and good food, we all agreed that we needed to look at wolves' and dogs' different 'lifestyles' to better understand what was going on. In the same way that the CCH was embedded in the observations of the wolves' socio-ecology, i.e. their monogamous mating, group hunting, and parental behaviours, we started thinking that this is where we needed to look for 'inspiration' and maybe even preliminary answers for dogs' behaviour as well. Simona's experience with the free-ranging dogs in Rome opened this door for us. Although when we then started digging deeper in the literature on free-ranging dogs, it became quite clear that there were still a lot of unanswered questions.

In effect, what we aimed to do was what is mostly done when comparing any two species: i.e. characterize their similarities and differences in terms of their socio-ecology and see what predictions would emerge in terms of their behaviours and cognitive abilities. In our thinking, we were heavily affected by some of the observations the Coppingers made in their books (Coppinger and Coppinger 2001, 2016). For example, their observations of free-ranging dogs (see Chaps. 1 and 4) led them to the important consideration that dogs cannot just be seen as a 'human-made' product, but rather should be viewed as a species adapted to a specific environment in which humans are without doubt a highly predominant but not the only element. Comparing the socio-ecology of wolves and dogs as outlined in Chaps. 3 and 4, a few aspects seem important. In contrast to wolves, free-ranging dogs have, in most locations, access to fairly constant, predictable food resources that are also not so dangerous to obtain. They do not need partners to hunt it down, nor do they need any particular persistence to do so—after all, food leftovers usually do not run away, and

Table 11.2 Socio-ecology of wolves and dogs compared

	Wolves	*Dogs*
Predominant food acquisition	Hunting	Scavenging
Food resources	Predominantly meat	High component of grain/starch, human faeces
Distribution	One large monopolizable	Small, distributed items
Predictability	Low: seasonal fluctuation/ changing location	High: seasonally constant/fixed location
Danger/risk	High: prey can be dangerous	None
Reliance on conspecifics	High for large prey	None
Human resource dependence	None	High
Group composition	Family packs	Multi-male/multi-female
Mating system	Pair bonding	Mostly promiscuous
	Seasonal	Non-seasonal
Parental care	By all pack members	Mostly maternal (some cases of paternal)
Pup dependence on adults	High-prolonged	Low-short

also, they come in such small portions that it makes more sense to eat them all rather than share it. These changes in the feeding ecology likely affected many other aspects of their social life as well. Moreover, dogs no longer pair-bond but instead are promiscuous, and group cohesion seems to be diminished. This reduced cooperativeness/cohesiveness could be due to the availability of easily harvested food items allowing all individuals including lactating females as well as relatively young puppies to fend for themselves. This diminishes the need for alloparental investment and the need to be fed by other group members (see also Lord et al. 2013). As a consequence, dogs' reliance on pack members for both foraging and pup-rearing is reduced compared to wolves (Table 11.2).

These differences in both wolves' and dogs' feeding ecology and their social organization likely shaped many different behaviours including how they interact with their surrounding environment (e.g. persistence, neophobia, and risk-taking), but also how they interact with conspecifics (e.g. tolerance around a food source, communication, conflict management, and cooperativeness). Accordingly, we proposed the *social ecology hypothesis* (Marshall-Pescini et al. 2017a, b, c, d) suggesting that, although human selection for specific favourable traits (e.g. tameness, human sociality, attention) likely played a major role in dog domestication, changes in their social ecology (i.e. both their feeding niche and social organization) may also have had a significant impact in moulding dogs' behavioural and cognitive profile, and thus these should be more closely considered both when making predictions and when interpreting wolf-dog differences (Fig. 11.3).

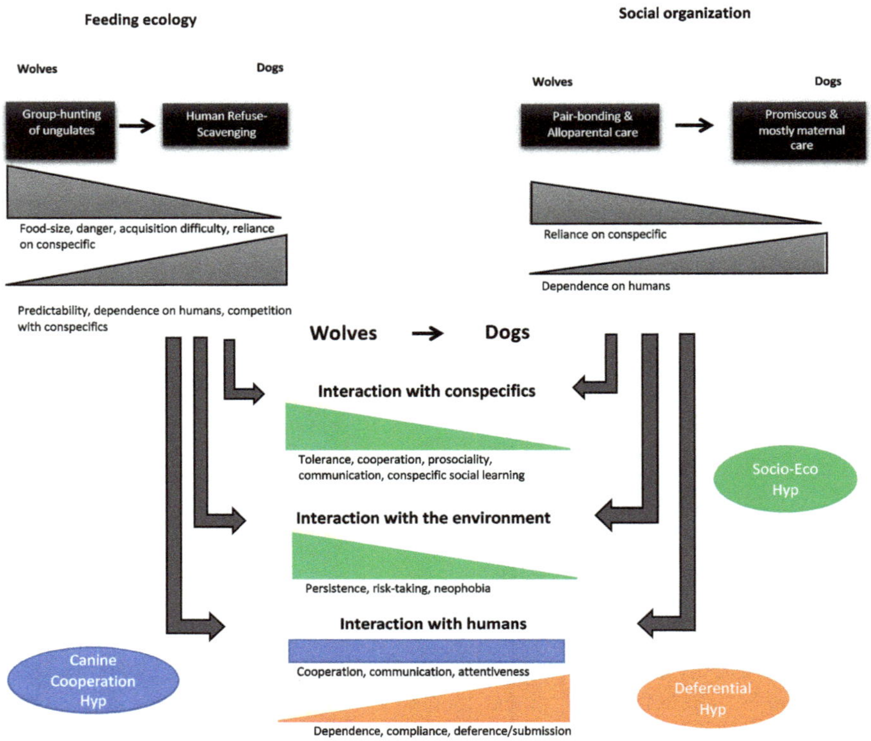

Fig. 11.3 Flowchart summarizing the three hypotheses (canine cooperation hypothesis, social ecology hypothesis, deferential behaviour hypothesis) and their basis suggested by the team of the Wolf Science Center. Figure adapted from Marshall-Pescini et al. 2017. https://doi.org/10.1016/j.cobeha.2017.05.002; http://creativecommons.org/licenses/by/4.0/

11.2.2.2 Part II: Data in Support of These Hypotheses

Both the *hypersociability hypothesis* and the *emotional reactivity hypothesis* predict that also with each other, dogs are more social than wolves. Support for higher sociability in dogs compared to wolves comes from early behavioural observations by Zimen that poodles spend more time close together, when both sleeping and moving around, than wolves (Zimen 1971). However, based on our observations at the Wold Science Center, we found no difference in the time spent in proximity to each other or the exchange of affiliative behaviours in wolf and dog packs (Chap. 5).

As regards aggression in contexts where no food/resource is present, early observations by Zimen, Feddersen-Petersen, and the Franks point to wolves showing more 'ritualized' aggression and dogs showing a greater propensity to escalate into higher-intensity aggressive interactions (Frank and Frank 1982, 1987; Feddersen-Petersen 2004). This is also in line with some of our own data on dogs showing more frequent high-intensity aggression than wolves (Chap. 5; Cafazzo et al. 2018). Based on the long-term observational dataset, the frequency of aggressive behaviours did not differ between wolves and dogs (Chap. 5), but, as discussed, these results need confirmation and a more detailed analysis of the behaviours exhibited by wolves and dogs. Nevertheless, based on current data, we have no support for a reduction in aggressive interactions within dog packs compared to wolf packs (*emotional reactivity hypothesis*) nor increased sociality in terms of higher rates of affiliative, play, or proximity seeking between dogs and wolves (*hypersociability hypothesis*).

Tolerance and aggression around larger pieces of food do vary considerably between wolves and dogs (Range et al. 2015; Dale et al. 2017a, b; Chap. 5). Although access to resources in both is a matter of rank position, in dogs, the most dominant individual monopolizes a large junk of food, and the subordinates do not even approach it. In wolves however, although the higher-ranking animal may be first at the food, it does not monopolize it, resulting in more communal food-sharing in wolves as compared to dogs. Because low-ranking dogs tend to keep away from the common resource, we hardly see agonistic interactions (communication) between the animals, different from wolves, where a continuous 'conversation' about who is going to get what food goes on from both subordinate and dominant animals. In line with these results, in experimental tasks of prosociality, where animals are given the possibility of providing food to their pack mates, wolves consistently show more 'giving' behaviours than dogs (Dale et al. 2019; Chap. 8). Dogs' avoidance of conflicts by using distance-maintenance strategies (Cafazzo et al. 2018) appears to severely constrain their potential to cooperate with each other to gain a common resource, as evident in their poor performance in the string-pulling task (Marshall-Pescini et al. 2017a, b, c, d; Marshall-Pescini et al. 2018; Chap. 8). In contrast, subordinate wolves are not inhibited in approaching and handling a resource at the same time as the dominant animal, which allows them to cooperate to obtain it, notwithstanding potential rank differences (Marshall-Pescini et al. 2017a, b, c, d).

This behaviour with conspecifics corresponds with the evidence-based impression that dogs accept the leading roles of another more easily than wolves. In contrast to dominant dogs usually monopolizing the resources without the subdominant trying to steal it, in wolves, subdominant animals will protest vehemently and successfully to get access to food (Chap. 6; Zimen 1971), and dominant animals are tolerant towards those subdominants (Range et al. 2015; Dale et al. 2017a, b)—in line with the family structure of wolf packs. These results suggest that leadership might be perceived slightly different in dogs and wolves and that these differences are mirrored in how they interact with humans (Range et al. 2019).

Considering these results, in the context of resource sharing, dogs appear to be less tolerant or even prosocial towards pack mates and less able to cooperate with conspecifics (in contrast to what is suggested by the *emotional reactivity hypothesis*) than dogs. Our results are rather in line with the two species' socio-ecology, considering wolves' dependence on group hunting and food-sharing and dogs' shift to a more solitary foraging activity (*social ecology hypothesis*). In line with dogs' potential loss in cooperative and communicative skills in interactions with conspecifics (as hypothesized by Zimen, the Franks, and Feddersen-Petersen and also the *social ecology hypothesis*), we found that after a conflict wolves engage in affiliative approaches towards the opponent to reconcile, whereas dogs rather keep their distance (Chap. 5; Cafazzo et al. 2018). Furthermore, dogs did not appear to be able to learn the subtle details of a behaviour to open an apparatus from observation of a conspecific, which wolves were in fact able to do (Range and Virányi 2014; Chap. 8).

Still, dogs, like wolves, establish rank relationships and different affiliative bonds with conspecifics, which regulate their behaviours in many different contexts, e.g. when encountering novel object (Moretti et al. 2015; Chap. 9) or scary situations (Cimarelli et al. 2021; Chap. 5), in cooperative interactions (Marshall-Pescini et al. 2017a, b, c, d, 2018; Chap. 8) and food-sharing (Dale et al. 2017a, b, 2019; Chaps. 5 and 8), as well as in their reactions to inequity (Essler et al. 2017; Chap. 8). Thus, it is important to note that although the change in dogs' social ecology due to domestication may have altered some of the social skills and strategies used with conspecifics (e.g. conflict management and behaviours in feeding contexts), dogs are still highly social animals, as evidenced by these results and their capacity to live in packs in free-ranging contexts—though seemingly less focused in this respect than wolves.

With regard to aggression towards outgroup members, conflicts between packs appear to be more severe in wolves than dogs (Chaps. 3 and 4). This is reflected in Zimen's observations of dogs' showing lower levels of aggression (i.e. mostly barking) compared to wolves when unfamiliar dogs approached their enclosures (Zimen 1971). However, a more systematic investigation with both captive and free-ranging animals would be needed to better assess the similarities and differences in outgroup behaviour of wolves and dogs; at the moment, such data is very limited.

To conclude, the more recent hypotheses claiming that the conspecific relationships in dogs may have become in general more prosocial during the process of domestication are currently not supported by the data we have. In fact, the classical observing ethologists had a good eye in suggesting that the dogs may have lost some social skills or social motivations, which may be due to the lack of selection pressures to keep the cooperative behaviours functional. The *social ecology hypothesis* picks this up and is currently best supported by the available data.

11.2.3 Interaction with the Environment

11.2.3.1 Part I: Domestication Hypotheses

Based on their comparison of wolf and malamute pups on tasks such as associative learning, problem-solving, and performance in spatial tasks, Frank and Frank (1985) proposed the *information processing hypothesis* (IPH). Here, they suggest that wolves have (1) a repertoire of closed behaviour patterns characterized by being elicited by specific stimulus configurations and exhibited with little plasticity enabling the animals to make correct responses to a set of external stimuli without learning (e.g. recognition of potential mates, care and feeding of neonates, protection of food) and (2) a complex cognitive system, which 'may have evolved concomitantly with the rise of group hunting, and is characterized by complexity, foresight, a capacity for internal mental representation (imagery), and insight into rudimentary means-ends relations' (cited from Frank and Frank 1985). In contrast, natural selection on individual problem-solving has been relaxed in dogs since environmental effects on feeding and mating are buffered by humans (Frank 1980; Frank and Frank 1985; Hemmer 1990). However, to be better suited for interacting with humans, dogs were selected for tractability, i.e. 'sensitivity to a wide band of stimuli including all sorts of verbal cues and physical manipulations that humans use to communicate with animals and behavioural plasticity allowing to shape desired responses and increase their frequency'. Their hypothesis predicts that dogs outperform wolves in tasks where a response is acquired by repetition rather than insight, e.g. if the to-be learned behaviour has no functional connection with the reinforcement. Thus, Frank's IPH pits 'insight' (higher in wolves) against 'associative learning' (better in dogs), which might not be quite accurate (Shettleworth 2012), since, to be able to have insight, animals likely first have to learn certain contingencies by associative learning to be able to transfer this knowledge to a new problem, which they then can solve with 'insight'.

The *social ecology hypothesis* also makes predictions concerning interactions with the environment. Since wolves depend mostly on hunting prey, spanning large territories (Chap. 3), they need to be explorative, persistent, and risk-prone in order to find and bring down potentially dangerous prey. Furthermore, the persecution by humans probably selected for animals that were highly neophobic towards all human-related stimuli. On the other hand, dogs' ease in being close to humans, a greater acceptance (less fear) of change/novel objects in the environment, and a higher behavioural flexibility are useful traits to allow exploitation of the human provided feeding niche.

11.2.3.2 Part II: Data in Support of These Hypotheses

The motivation of wolves and dogs to investigate their environment differs with wolves being more explorative and persistent from a young age onwards in most studies (Chap. 9). This is especially interesting since with age, wolves become more neophobic than dogs, but they still show more interest in novel objects and are more likely to take risks, when obtaining food is at stake (Gácsi et al. 2005; Moretti et al. 2015; Marshall-Pescini et al. 2016, 2017a, b, c, d; Hansen Wheat et al. 2019). This interest in novelty might not just be restricted to objects and the environment but also appears to influence their interactions with humans during puppyhood, when they do show a much interest in strangers (see above, Chap. 6). While several hypotheses suggest that dogs were selected for reduced fear (*destabilizing hypothesis*, *emotional reactivity hypothesis*, *human socialization hypothesis*), predicting lower neophobia in dogs compared to wolves, only the *social ecology hypothesis* predicts the observed differences in exploration and persistence.

The Coppingers suggested that there might have been a rather high selection pressure on dogs to 'wait' or inhibit trying to obtain a resource held by a human (Coppinger and Coppinger 2001, 2016). However, hunting has been suggested to also require considerable inhibitory control abilities. Our current results suggest that inhibitory control abilities might vary considerably depending on context (Gácsi et al. 2009; MacLean et al. 2014; Marshall-Pescini et al. 2015; Brucks et al. 2019; Chap. 9). The dogs clearly outperform the wolves if the human is an integral part of the experimental setup by facing the animals, while they have to control their actions or if some training is required before the test (MacLean et al. 2014; Marshall-Pescini et al. 2015; Range et al. 2020). However, wolves are on *a par* with dogs in other tasks if the human is hidden or absent (inhibition battery) and might even outperform them in tasks where they have to inhibit going directly to a food resource (detour task). The differences in the performance depending on the situation are in line with several studies in various animal species revealing that inhibitory control is highly context specific (Miyake et al. 2000). As researchers, we need to be much more careful when defining our questions and, consequently, designing our experimental setups. In this specific case, we need to design contexts which better reflect the different socio-ecologies of the two species.

In their *information processing hypothesis*, the Franks predict a better sophisticated understanding of their environment, for example, in relation to foresight, means-end understanding, and internal mental representations in wolves than dogs. Similarly, the *social ecology hypothesis* also suggests that inferential reasoning might be highly relevant for wolves in attempting to infer relevant characteristics of their prey, as would be advanced object permanence skills to help track the movement of potential prey or a hunting partner when they shortly disappear behind trees or bushes. In dogs, these abilities may have been retained, or, alternatively, dependence on stable anthropogenic food resources may have relaxed the cognitive processes necessary for the maintenance of such cognitive skills. To date, we only have very few studies comparing these cognitive abilities in wolves and dogs. The summary of the studies in Chap. 10 suggests that both wolves and dogs perform poorly in more advanced object permanence tasks (invisible displacement) and means-end understanding (Range et al. 2012; Fiset and Plourde 2013).

However, dogs seem to perform more poorly than wolves in diagnostic inference tasks (understanding visual and auditory cues indicating the presence of food in one vs. another object) (Lampe et al. 2017) and potentially numerical competence, when an internal representation of quantity is necessary to solve the task (Range et al. 2014). But again, we need to be careful here with far-reaching conclusions, since at least for the diagnostic inference task, only a single study was conducted and some crucial control conditions were missing to allow us to decide whether the wolves based their responses on perceptual cues or on the understanding of causality. In regard to numerical competence, we have just three studies. However, while the first two studies suggested that wolves are better than dogs because numerical ratio did not affect the performance (Range et al. 2014; Petrazzini and Wynne 2017), the latest study found no clear difference between the two species, which might be partly explained by the pre-training involved, allowing the dogs to better understand the contingencies of the task (Rivas-Blanco et al. 2020).

In terms of attentiveness and working memory (Chap. 9), there is circumstantial evidence that wolves might pay more attention to the details of their environment (e.g. in the imitation and risk assessment tasks) (Range and Virányi 2014; Marshall-

Pescini et al. 2016) and may also have a better working memory, in view of their faster decision-making in associative and reversal learning tasks (Brucks et al. 2019). Faster decision-making in wolves compared to dogs was also proposed by the *information processing hypothesis*; however, Frank suggested that the faster decisions were based on closed behaviour repertoires rather than working memory capacity. Faster reactions in decision-making contexts may be under selection as a skill-set required for successful cooperative hunting.

Lastly, one rather important aspect of cognition is associative learning (Chap. 10). Here, the *information processing hypothesis* predicts that dogs are better at it, especially, if there is no causal link between the aspect that needs to be learned and the reinforcement. Unfortunately, again, we have tested associative learning abilities in wolves and dogs in only three studies, which produced mixed results with either wolves being better (visual discrimination learning), dogs being better (sequential rope pulling; Hiestand 2011), or both of them performing on *a par* (association and reversal learning task.

In conclusion, the only two hypotheses that make any predictions about how wolves and dogs may differ in their interaction with the environment are the *information processing hypothesis* and the *social ecology hypothesis*, and based on the current (scant) evidence, these are somewhat supported. But, there are still many, many holes, and in particular, attention, associative and reversal learning, and inference skills require more in-depth analyses.

A final important aspect to consider is that similarities and differences in how the wolves and dogs interact with the environment may also have important implications for all the other aspects we are considering. For example, the wolves' higher persistence in trying to solve a puzzle box was shown to affect their latency to look back towards humans compared to dogs (Marshall-Pescini et al. 2017a, b, c, d Chap. 7). The study was set up to examine human-directed looking behaviour, but in fact persistence played a larger role in explaining results. In contrast, in the cooperative string-pulling task, we carefully excluded the possibility that the species' different persistence levels in manipulating objects was responsible for the differences observed (Marshall-Pescini et al. 2017a, b, c, d). In follow-up studies manipulating the social motivation of the dogs, we were able to confirm that these were driving the results (Range et al. 2019; Chap. 8). Also in regard to the wolves, when running an overall analysis including all cooperative tasks (cooperative string-pulling, inequity aversion, and prosociality), the relationship quality between partners had the largest effect on animal's performance, rather than their inhibitory control, learning speed, causal understanding, or persistence (Dale et al. 2020; Chap. 8).

The challenge for the future will be to more carefully evaluate whether the main differences between wolves and dogs lie in their motivational processes, cognitive abilities, or aspects of their social dispositions.

11.3 Summary and Overall Conclusions

Fuelled by copious amounts of prosciutto and prosecco, we have come to the end of this book. When we started, we thought that it would be relatively easy and fast to write: 'There are not that many studies on wolf-dog comparisons', we said. And this is true to some extent, but still... getting it all 'black on white', and organizing the material so that it made sense to us and (hopefully) you, has been a challenge. In the end, it took more than 2 years!

Considering the evidence so far, results seem to suggest that domestication has had complex effects on dogs' behaviour and cognition that are likely the result of multiple selective pressures. Dogs definitely emerge as less fearful of humans and potentially also less aggressive towards conspecific outgroup members (although on this latter point, there is still much to do) compared to wolves. Dogs' ingroup social dynamics, particularly around food resources, have also changed compared to those of wolves, but the changes are more intricate. Dominant dogs are less tolerant than dominant wolves in the feeding context, and subordinate dogs (but not subordinate wolves) avoid potential conflicts where possible. To some extent, a similar pattern emerges with humans where dogs appear to be more easily inhibited and more likely to comply with the demands we make on them, whereas wolves tend to be more independent and more selective in their choice of when to comply with our demands and when not to. Dogs are also more likely to engage in positive social contact with less familiar people than wolves, potentially mirroring the changes to their more fluid conspecific outgroup dynamics. Currently, there is no evidence for dogs' superior communicative or cooperative abilities when compared to wolves. Indeed, if anything, the opposite is the case, with wolves being more cooperative and prosocial towards their pack members than dogs. Together these results, we think, reflect changes to the socio-ecology of the two species that go hand in hand with the likely human-enacted selective pressures favouring animals that can be easily inhibited and controlled, allowing humans to exploit dogs' skills for their own use.

Nevertheless, many questions remain unanswered (see Box 11.1 *for a summary), and we hope that what we have written inspires even more juicy questions in the future, so that the next generation of researchers can take up the challenge of figuring out wolves and dogs. As always, the scientific process requires us to be open to the possibility that all that we think today may tomorrow be turned on its head!*

Box 11.1 A Few Unanswered Questions

- Human-animal conflict management (what ignites conflict, how they react during conflict and after it), especially in relation to the bond/relationship with the person.
- Do wolves and dogs perceive us differently? Do they see us as social partners at the same eye level or rather someone to whom they look up to? And how much does that depend on specific experiences?

(continued)

Box 11.1 (continued)

- Do dogs (but not wolves?) preferentially use specific communicative cues with humans? And what 'meaning' do these cues have?
- Has domestication changed conspecific communication? Have dogs become more 'subtle' in their communication compared to the extrovert boisterous communication of wolves? And if so, what selective pressure brought this about? And what are the consequences of this change?
- How varied is the social system of both wolves and dogs in the 'wild'? And can such variability be exclusively explained by the respective ecological conditions? For example, would feral dogs living off hunting of prey show a social structure comparable to that of wolves (e.g. dingoes are able to cooperatively hunt large prey such as kangaroos (Allen et al. 2000) and have been observed sharing carcasses (Thomson 1992). Or are there consistent differences between wolves' and dogs' social organization, which cannot be explained by the ecological pressures they live in?
- How do wolves and dogs differ in their learning abilities, causal understanding, and foraging strategies?

References

Agnvall B, Ali A, Olby S, Jensen P (2014) Red Junglefowl (*Gallus gallus*) selected for low fear of humans are larger, more dominant and produce larger offspring. Animal 8(9):1498–1505

Agnvall B, Bélteky J, Jensen P (2017) Brain size is reduced by selection for tameness in red Junglefowl–correlated effects in vital organs. Sci Rep 7(1):1–7

Agnvall B, Katajamaa R, Altimiras J, Jensen P (2015) Is domestication driven by reduced fear of humans? Boldness, metabolism and serotonin levels in divergently selected red junglefowl (*Gallus gallus*). Biol Lett 11(9):20150509

Albert FW, Carlborg O, Plyusnina I, Besnier F, Hedwig D, Lautenschläger S, Lorenz D, McIntosh J, Neumann C, Richter H (2009) Genetic architecture of tameness in a rat model of animal domestication. Genetics 182(2):541–554

Allen L, Thomson P, Lisle A (2000) Pack size and prey behaviour affects prey selection and the predation of livestock by dingoes. Bureau of Rural Sciences, Australian Government, Canberra

Belyaev DK (1979) Destabilizing selection as a factor in domestication. J Hered 70(5):301–308

Bentosela M, Wynne CDL, D'Orazio M, Elgier A, Udell MAR (2016) Sociability and gazing toward humans in dogs and wolves: simple behaviors with broad implications. J Exp Anal Behav 105(1):68–75

Bergström A, Frantz L, Schmidt R, Ersmark E, Lebrasseur O, Girdland-Flink L, Lin AT, Storå J, Sjögren K-G, Anthony D (2020) Origins and genetic legacy of prehistoric dogs. Science 370(6516):557–564

Brucks D, Marshall-Pescini S, Range F (2019) Dogs and wolves do not differ in their inhibitory control abilities in a non-social test battery. Anim Cogn 22(1):1–15

Cafazzo S, Marshall-Pescini S, Lazzaroni M, Virányi Z, Range F (2018) The effect of domestication on post-conflict management: wolves reconcile while dogs avoid each other. R Soc Open Sci 5(7):171553

Cimarelli G, Marshall-Pescini S, Range F, Berghänel A, Virányi Z (2021) Relationship quality affects social stress buffering in dogs and wolves. Anim Behav 178:127–140

Coppinger R, Coppinger L (2001) Dogs. A new understanding of canine origin, behavior and evolution. The University of Chicago Press, Chicago

Coppinger R, Coppinger L (2016) What is a dog? The University of Chicago Press, Chicago

Dale R, Marshall-Pescini S, Range F (2017a) Do females use their sexual status to gain resource access? Investigating food-for-sex in wolves and dogs. Current Zoology 63(3):323–330

Dale R, Marshall-Pescini S, Range F (2020) What matters for cooperation? The importance of social relationship over cognition. Sci Rep 10(1):1–14

Dale R, Palma-Jacinto S, Marshall-Pescini S, Range F (2019) Wolves, but not dogs, are prosocial in a touch screen task. PLoS One 14(5):e0215444

Dale R, Range F, Stott L, Kotrschal K, Marshall-Pescini S (2017b) The influence of social relationship on food tolerance in wolves and dogs. Behav Ecol Sociobiol 71:107–131

Darwin C (1868) The variation of animals and plants under domestication. In: The variation of animals and plants under domestication, 1st edn. John Murray, London

Ducrest A-L, Keller L, Roulin A (2008) Pleiotropy in the melanocortin system, coloration and behavioural syndromes. Trends Ecol Evol 23(9):502–510

Ericsson M, Jensen P (2016) Domestication and ontogeny effects on the stress response in young chickens (Gallus gallus). Sci Rep 6(1):1–7

Ericsson M, Fallahsharoudi A, Bergquist J, Kushnir MM, Jensen P (2014) Domestication effects on behavioural and hormonal responses to acute stress in chickens. Physiol Behav 133:161–169

Essler JL, Marshall-Pescini S, Range F (2017) Domestication does not explain the presence of inequity aversion in dogs. Curr Biol 27(12):1861–1865.e1863

Fadel FR, Driscoll P, Pilot M, Wright H, Zulch H, Mills D (2016) Differences in trait impulsivity indicate diversification of dog breeds into working and show lines. Sci Rep 6(1):1–10

Fan Z, Silva P, Gronau I, Wang S, Armero AS, Schweizer RM, Ramirez O, Pollinger J, Galaverni M, Del-Vecchyo DO (2016) Worldwide patterns of genomic variation and admixture in gray wolves. Genome Res 26(2):163–173

Feddersen-Petersen DU (2004) Hundepsychologie. Kosmos, Stuttgart

Feddersen-Petersen DU (2007) Social behaviour of dogs and related canids. In: Jensen P (ed) The behavioural biology of dogs. Press, Trowbridge, Cromwell, pp 105–119

Feuerbacher EN, Wynne CDL (2012) Relative efficacy of human social interaction and food as reinforcers for domestic dogs and hand-reared wolves. J Exp Anal Behav 98(1):105–129

Fiset S, Plourde V (2013) Object permanence in domestic dogs (*Canis lupus familiaris*) and gray wolves (Canis lupus). J Comp Psychol 127(2):115–127

Frank H (1980) Evolution of canine information processing under conditions of natural and artificial selection. Z Tierpsychol 53(4):389–399

Frank H Frank M (1987) The University of Michigan canine information-processing project (1979–1981). Man and Wolf H. Frank. Dordrecht, Dr W. Junk: 143–167

Frank H, Frank MG (1982) On the effects of domestication on canine social development and behavior. Appl Anim Ethol 8(6):507–525

Frank H, Frank MG (1985) Comparative manipulation test performance in 10-week-old wolves (*Canis lupus*) and Alaskan malamutes (*Canis familiaris*): a Piagetian interpretation. J Comp Psychol 99(3):266–274

Freedman AH, Schweizer RM, Gronau I, Han E, Vecchyo DO-D, Silva PM, Galaverni M, Fan Z, Marx P, Lorente-Galdos B (2013) Genome sequencing highlights genes under selection and the dynamic early history of dogs. PLoS Genet 10(1):e1004016

Gácsi M, Győri B, Miklósi Á, Virányi Z, Kubinyi E, Topál J, Csányi V (2005) Species-specific differences and similarities in the behavior of hand-raised dog and wolf pups in social situations with humans. Dev Psychobiol 47(2):111–122

Gácsi M, Győri B, Virányi Z, Kubinyi E, Range F, Belényi B, Miklósi A (2009) Explaining dog wolf differences in utilizing human pointing gestures: selection for synergistic shifts in the development of some social skills. PLoS One 4(8):e6584

Gácsi M, Vas J, Topál J, Miklósi Á (2013) Wolves do not join the dance: sophisticated aggression control by adjusting to human social signals in dogs. Appl Anim Behav Sci 145(3–4):109–122

Hall NJ, Lord K, Arnold AMK, Wynne CDL, Udell MAR (2015) Assessment of attachment behaviour to human caregivers in wolf pups (*Canis lupus lupus*). Behav Process 110:15–21

Hansen Wheat C, van der Bijl W, Temrin H (2019) Dogs, but not wolves, lose their sensitivity toward novelty with age. Front Psychol 10(2001)

Hare B (2017) Survival of the friendliest: *Homo sapiens* evolved via selection for prosociality. Annu Rev Psychol 68:155–186

Hare B, Brown M, Williamson C, Tomasello M (2002) The domestication of social cognition in dogs. Science 298:1634–1636

Hare B, Tomasello M (2005a) The emotional reactivity hypothesis and cognitive evolution. Trends Cogn Sci 9(10):464–465

Hare B, Tomasello M (2005b) Human-like social skills in dogs? Trends Cogn Sci 9(9):439–444

Hare B, Wobber V, Wrangham R (2012) The self-domestication hypothesis: evolution of bonobo psychology is due to selection against aggression. Anim Behav 83(3):573–585

Harri M, Mononen J, Ahola L, Plyusnina I, Rekilä T (2003) Behavioural and physiological differences between silver foxes selected and not selected for domestic behaviour. Anim Welf 12(3):305–314

Heberlein MTE, Turner DC, Range F, Virányi Z (2016) A comparison between wolves, *Canis lupus*, and dogs, *Canis familiaris*, in showing behaviour towards humans. Anim Behav 122:59–66

Hecht EE, Kukekova AV, Gutman DA, Acland GM, Preuss TM, Trut LN (2021) Neuromorphological changes following selection for tameness and aggression in the Russian farm-fox experiment. J Neurosci 41(28):6144–6156

Hemmer H (1990) Domestication. The decline of environmental appreciation. Cambridge University Press, Cambridge

Hiestand L (2011) A comparison of problem-solving and spatial orientation in the wolf (*Canis lupus*) and dog (*Canis familiaris*). Behav Genet 41(6):840–857

Kaiser S, Hennessy MB, Sachser N (2015) Domestication affects the structure, development and stability of biobehavioural profiles. Front Zool 12(1):S19

Kaplan P, Wang PP, Francke U (2001) Williams (Williams Beuren) syndrome: a distinct neurobehavioral disorder. J Child Neurol 16(3):177–190

Katajamaa R, Larsson LH, Lundberg P, Sörensen I, Jensen P (2018) Activity, social and sexual behaviour in red Junglefowl selected for divergent levels of fear of humans. PLoS One 13(9): e0204303

Kistner TM, Zink KD, Worthington S, Lieberman DE (2021) Geometric morphometric investigation of craniofacial morphological change in domesticated silver foxes. Sci Rep 11(1):1–10

Klinghammer E, Goodmann PA (1987) Socialization and management of wolves in captivity. Man and Wolf. H. Frank. Dordrecht, Dr. W. Junk Publishers, NL, pp 31–59

Kortekaas K, Range F, Virányi Z, Kotrschal K (2013) Heart rate and hear rate variability of dogs (*Canis lupus familiaris*) during physical and mental activities. J Vet Behav 8(4):e45

Kubinyi E, Virányi Z, Miklósi A (2007) Comparative social cognition: from wolf and dog to humans. Comparative Cognition Behav Rev 2:26–46

Lampe M, Bräuer J, Kaminski J, Virányi Z (2017) The effects of domestication and ontogeny on cognition in dogs and wolves. Sci Rep 7:11690

Larson G, Piperno DR, Allaby RG, Purugganan MD, Andersson L, Arroyo-Kalin M, Barton L, Vigueira CC, Denham T, Dobney K (2014) Current perspectives and the future of domestication studies. Proc Natl Acad Sci 111(17):6139–6146

Lazzaroni M, Range F, Backes J, Portele K, Scheck K, Marshall-Pescini S (2020) The effect of domestication and experience on the social interaction of dogs and wolves with a human companion. Front Psychol 11(785)

Lenkei R, Újváry D, Bakos V, Faragó T (2020) Adult, intensively socialized wolves show features of attachment behaviour to their handler. Sci Rep 10(1):1–8

Librado P, Gamba C, Gaunitz C, Der Sarkissian C, Pruvost M, Albrechtsen A, Fages A, Khan N, Schubert M, Jagannathan V (2017) Ancient genomic changes associated with domestication of the horse. Science 356(6336):442–445

Lin Z, Chen L, Chen X, Zhong Y, Yang Y, Xia W, Liu C, Zhu W, Wang H, Yan B (2019) Biological adaptations in the Arctic cervid, the reindeer (*Rangifer tarandus*). Science 364(6446)

Linderholm A, Larson G (2013) The role of humans in facilitating and sustaining coat colour variation in domestic animals. Seminars in cell & developmental biology. Elsevier

Lord K, Feinstein M, Smith B, Coppinger R (2013) Variation in reproductive traits of members of the genus Canis with special attention to the domestic dog (*Canis familiaris*). Behav Process 92: 131–142

Lord KA, Larson G, Coppinger RP, Karlsson EK (2020) The history of farm foxes undermines the animal domestication syndrome. Trends Ecol Evol 35(2):125–136

MacLean EL, Hare B, Nun CL, Addess E, Amic F, Anderson RC, Aureli F, Baker JM, Bania AE, Barnard AM, Boogert NJ, Brannon EM, Bray EE, Bray J, Brent LJN, Burkart JM, Call J, Cantlo JF, Chek LG, Clayton NS, Delgado MM, DiVincenti LJ, Fujita K, Herrman E, Hiramatsu C, Jacobs LF, Jordan KE, Laude JR, Leimgruber KL, Messer EJE, Mour LO, Picard A, Platt ML, Plotnik JM, Range F, Reader SM, Reddy RB, Sandel AA, Santos LR, Schuman K, See AM, Sewal KB, Sha RC, Slocomb KE, Su Y, Takimot A, Tan J, Tao R, van Schaik CP, Virányi Z, Visalberghi E, Wad JC, Watanab A, Widnes J, Young JK, Zental TR, Zhao Y (2014) The evolution of self-control. Proc Natl Acad Sci U S A 111(20):E2140–E2148

Marshall-Pescini S, Basin C, Range F (2018) A task-experienced partner does not help dogs be as successful as wolves in a cooperative string-pulling task. Sci Rep 8(1):16049

Marshall-Pescini S, Besserdich I, Kratz C, Range F (2016) Exploring differences in dogs' and wolves' preference for risk in a foraging task. Front Psychol 7(1241)

Marshall-Pescini S, Cafazzo S, Virányi Z, Range F (2017a) Integrating social ecology in explanations of wolf–dog behavioral differences. Curr Opin Behav Sci 16:80–86

Marshall-Pescini S, Rao A, Virányi Z, Range F (2017b) The role of domestication and experience in 'looking back' towards humans in an unsolvable task. Sci Rep 7:46636

Marshall-Pescini S, Schwarz JFL, Kostelnik I, Virányi Z, Range F (2017c) Importance of a species' socioecology: wolves outperform dogs in a conspecific cooperation task. Proc Natl Acad Sci U S A 114(44):11793–11798

Marshall-Pescini S, Virányi Z, Kubinyi E, Range F (2017d) Motivational factors underlying problem solving: comparing wolf and dog Puppies' explorative and neophobic behaviors at 5, 6, and 8 weeks of age. Front Psychol 8(180)

Marshall-Pescini S, Virányi Z, Range F (2015) The effect of domestication on inhibitory control: wolves and dogs compared. PLoS One 10(2):e0118469

Miklósi Á, Kubinyi E, Topál J, Gácsi M, Virányi Z, Csányi V (2003) A simple reason for a big difference: wolves do not look Back at humans, but dogs do. Curr Biol 13(9):763–766

Miklósi Á, Topál J (2013) What does it take to become 'best friends'? Evolutionary changes in canine social competence. Trends Cogn Sci 17:287–294

Miklósi Á, Topál J, Csányi V (2004) Comparative social cognition: what can dogs teach us? Anim Behav 67(6):995–1004

Miyake A, Friedman NP, Emerson MJ, Witzki AH, Howerter A, Wager TD (2000) The unity and diversity of executive functions and their contributions to complex "frontal lobe" tasks: a latent variable analysis. Cogn Psychol 41(1):49–100

Montague MJ, Li G, Gandolfi B, Khan R, Aken BL, Searle SM, Minx P, Hillier LW, Koboldt DC, Davis BW (2014) Comparative analysis of the domestic cat genome reveals genetic signatures underlying feline biology and domestication. Proc Natl Acad Sci 111(48):17230–17235

Moretti L, Hentrup M, Kotrschal K, Range F (2015) The influence of relationships on neophobia and exploration in wolves and dogs. Anim Behav 107:159–173

Musiani M, Leonard JA, Cluff HD, Gates CC, Mariani S, Paquet PC, Vila C, Wayne RK (2007) Differentiation of tundra/taiga and boreal coniferous forest wolves: genetics, coat colour and association with migratory caribou. Mol Ecol 16(19):4149–4170

Nagasawa M, Mitsui S, En S, Ohtani N, Ohta M, Sakuma Y, Onaka T, Mogi K, Kikusui T (2015) Oxytocin-gaze positive loop and the coevolution of human-dog bonds. Science 348(6232): 333–336

Pendleton AL, Shen F, Taravella AM, Emery S, Veeramah KR, Boyko AR, Kidd JM (2018) Comparison of village dog and wolf genomes highlights the role of the neural crest in dog domestication. BMC Biol 16(1):1–21

Petrazzini MEM, Wynne CD (2017) Quantity discrimination in canids: dogs (*Canis familiaris*) and wolves (*Canis lupus*) compared. Behav Process 144:89–92

Pilot M, Greco C, vonHoldt BM, Jędrzejewska B, Randi E, Jędrzejewski W, Sidorovich VE, Ostrander EA, Wayne RK (2014) Genome-wide signatures of population bottlenecks and diversifying selection in European wolves. Heredity 112(4):428–442

Price EO (1984) Behavioral aspects of animal domestication. Q Rev Biol 59(1):1–32

Price EO (1999) Behavioral development in animals undergoing domestication. Appl Anim Behav Sci 65(3):245–271

Price, E. O. (2002). Animal domestication and behavior., CABI

Range F, Brucks D, Virányi Z (2020) Dogs wait longer for better rewards than wolves in a delay of gratification task: but why? Anim Cogn

Range F, Jenikejew J, Schröder I, Virányi Z (2014) Difference in quantity discrimination in dogs and wolves. Front Psychol 5:1299

Range F, Marshall-Pescini S, Kratz C, Virányi Z (2019) Wolves lead and dogs follow, but they both cooperate with humans. Sci Rep 9(1):3796

Range F, Möslinger H, Virányi Z (2012) Domestication has not affected the understanding of means-end connections in dogs. Anim Cogn 15(4):597–607

Range F, Ritter C, Virányi Z (2015) Testing the myth: tolerant dogs and aggressive wolves. Proc R Soc B 282(1807):20150220

Range F, Virányi Z (2011) Development of gaze following abilities in wolves (*Canis lupus*). PLoS One 6(2):e16888

Range F, Virányi Z (2013) Social learning from humans or conspecifics: differences and similarities between wolves and dogs. Front Psychol 4:868

Range F, Virányi Z (2014) Wolves are better imitators of conspecifics than dogs. PLoS One 9(1): e86559

Range F, Virányi Z (2015) Tracking the evolutionary origins of dog-human cooperation: the "canine cooperation hypothesis". Front Psychol 5

Rao A, Bernasconi L, Lazzaroni M, Marshall-Pescini S, Range F (2018) Differences in persistence between dogs and wolves in an unsolvable task in the absence of humans. PeerJ 5:e5944

Rivas-Blanco D, Pohl I-M, Dale R, Heberlein MTE, Range F (2020) Wolves and dogs may rely on non-numerical cues in quantity discrimination tasks when given the choice. Front Psychol 11(2412)

Sánchez-Villagra MR, Geiger M, Schneider RA (2016) The taming of the neural crest: a developmental perspective on the origins of morphological covariation in domesticated mammals. R Soc Open Sci 3(6):160107

Sánchez-Villagra MR, Schaik CP (2019) Evaluating the self-domestication hypothesis of human evolution. Evol Anthropol 28:1–11

Scott JP, Fuller JL (1965) Genetics and the social behavior of the dog. University of Chicago Press, Chicago

Shettleworth SJ (2012) Do animals have insight, and what is insight anyway? Can J Exp Psychol 66(4):217

Shilton D, Breski M, Dor D, Jablonka E (2020) Human social evolution: self-domestication or self-control? Front Psychol 11:134

Simões-Costa M, Bronner ME (2015) Establishing neural crest identity: a gene regulatory recipe. Development 142(2):242–257

Thomson P (1992) The behavioural ecology of dingoes in northwestern Australia. III. Hunting and feeding behaviour, and diet. Wildl Res 19(5):531–541

Topál J, Gácsi M, Miklósi Á, Virányi Z, Kubinyi E, Csányi V (2005a) Attachment to humans: a comparative study on hand-reared wolves and differently socialized dog puppies. Anim Behav 70:1367–1375

Topál J, Gergely G, Erdőhegyi Á, Csibra G, Miklósi Á (2009a) Differential sensitivity to human communication in dogs, wolves, and human infants. Science 325(5945):1269–1272

Topál J, Kubinyi E, Gácsi M, Miklósi Á (2005b) Obeying social rules: a comparative study on dogs and humans. J Cult Evol Psychol 3(3–4):213–237

Topál J, Miklósi Á, Gácsi M, Dóka A, Pongrácz P, Kubinyi E, Virányi Z, Csányi V (2009b) The dog as a model for understanding human social behavior. Advances in the Study of Behavior, Academic Press 39:71–116

Trapezov O, Trapezova L, Sergeev E (2008) Effect of coat color mutations on behavioral polymorphism in farm populations of American minks (*Mustela vison* Schreber, 1777) and sables (*Martes zibellina* Linnaeus, 1758). Russ J Genet 44(4):444–450

Trut L, Kharlamova A, Vladimirova A, Gerbek Yu E (2017) Ob otbore lisits na agressivnost'i ego korrelirovannykh posledstviyakh [on selection of foxes for enhanced aggressiveness and its correlated implications]. Vavilovskiy zhurnal genetiki i selektsii 21(4):392–401

Trut L, Oskina I, Kharlamova A (2009) Animal evolution during domestication: the domesticated fox as a model. BioEssays 31:349–360

Trut LN, Kharlamova AV, Herbeck YE (2020) Belyaev's and PEI's foxes: a far cry. Trends Ecol Evol

Trut LN, Plyusnina I, Oskina I (2004) An experiment on fox domestication and debatable issues of evolution of the dog. Russ J Genet 40(6):644–655

Udell MAR (2015) When dogs look back: inhibition of independent problem-solving behaviour in domestic dogs (*Canis lupus familiaris*) compared with wolves (*Canis lupus*). Biol Lett 11(9): 20150489

Udell MAR, Dorey NR, Wynne CDL (2010) What did domestication do to dogs? A new account of dogs' sensitivity to human actions. Biol Rev 85(2):327–345

Udell MAR, Giglio RF, Wynne CDL (2008) Domestic dogs (*Canis familiaris*) use human gestures but not nonhuman tokens to find hidden food. J Comp Psychol 122(1):84–93

Udell MAR, Wynne CDL (2011) Reevaluating canine perspective-taking behavior. Learn Behav 39(4):318–323

Ujfalussy DJ, Virányi Z, Gácsi M, Faragó T, Pogány Á, Bereczky BM, Miklósi Á, Kubinyi E (2020) Comparing the tractability of young hand-raised wolves (*Canis lupus*) and dogs (*Canis familiaris*). Sci Rep 10(1):1–14

Vasconcellos A, Virányi Z, Range F, Ades C, Scheidegger JK, Möstl E, Kotrschal K (2016) Training reduces stress in human-socialised wolves to the same degree as in dogs. PLoS One 11(9):e0162389

Virányi Z, Gácsi M, Kubinyi E, Topál J, Belényi B, Ujfalussy D, Miklósi Á (2008) Comprehension of human pointing gestures in young human-reared wolves (*Canis lupus*) and dogs (*Canis familiaris*). Anim Cogn 11(3):373–387

Virányi Z Range F (2014) On the way to a better understanding of dog domestication: aggression and cooperativeness in dogs and wolves. The Social Dog Behaviour and Cognition. J. Kaminski and S. Marshall-Pescini Amsterdam Elsevier: 35–62

Vonholdt B, Udell M and Sinsheimer J (2017a). Simple Genomic Test of Canine Genes Associated With Williams-Beuren Syndrome Could Predict Social Behavior in Domesticated Dogs.

vonHoldt BM, Pollinger JP, Lohmueller KE, Han E, Parker HG, Quignon P, Degenhardt JD, Boyko AR, Earl DA, Auton A, Reynolds A, Bryc K, Brisbin A, Knowles JC, Mosher DS, Spady TC, Elkahloun A, Geffen E, Pilot M, Jedrzejewski W, Greco C, Randi E, Bannasch D, Wilton A, Shearman J, Musiani M, Cargill M, Jones PG, Qian Z, Huang W, Ding ZL, Zhang YP, Bustamante CD, Ostrander EA, Novembre J, Wayne RK (2010) Genome-wide SNP and haplotype analyses reveal a rich history underlying dog domestication. Nature 464(7290): 898–902 vonHoldt BM, Shuldiner E, Koch IJ, Kartzinel RY, Hogan A, Brubaker L, Wanser S, Stahler D, Wynne CDL, Ostrander EA, Sinsheimer JS, Udell MAR (2017b) Structural variants in genes associated with human Williams-Beuren syndrome underlie stereotypical hypersociability in domestic dogs. Sci Adv 3(7):e1700398

Wang X, Pipes L, Trut LN, Herbeck Y, Vladimirova AV, Gulevich RG, Kharlamova AV, Johnson JL, Acland GM, Kukekova AV (2018) Genomic responses to selection for tame/aggressive behaviors in the silver fox (*Vulpes vulpes*). Proc Natl Acad Sci 115(41):10398–10403

Wheat CH, Fitzpatrick JL, Rogell B, Temrin H (2019a) Behavioural correlations of the domestication syndrome are decoupled in modern dog breeds. Nat Commun 10(1):1–9

Wheat CH, van der Bijl W, Wheat CW (2019b) Dogs defy the domestication syndrome: morphology does not covary with predicted behavioural correlations within breeds. bioRxiv 660829

Wilkins AS, Wrangham RW, Fitch WT (2014) The "domestication syndrome" in mammals: a unified explanation based on neural crest cell behavior and genetics. Genetics 197(3):795–808

Wirén A, Gunnarsson U, Andersson L, Jensen P (2009) Domestication-related genetic effects on social behavior in chickens–effects of genotype at a major growth quantitative trait locus. Poult Sci 88(6):1162–1166

Wirobski G, Range F, Schaebs F, Palme R, Deschner T, Marshall-Pescini S (2020) Endocrine changes related to dog domestication: comparing urinary cortisol and oxytocin in hand-raised, pack-living dogs and wolves. Horm Behav 128:104901

Wirobski G, Range F, Schaebs FS, Palme R, Deschner T, Marshall-Pescini S (2021) Life experience rather than domestication accounts for dogs' increased oxytocin release during social contact with humans. Sci Rep 11(1):1–12

Wright D (2015) Article commentary: the genetic architecture of domestication in animals. Bioinformatics and biology insights 9:BBI. S28902

Wright D, Henriksen R, Johnsson M (2020) Defining the domestication syndrome: comment on Lord et al. 2020. Trends in ecology & evolution S0169–5347(0120):30224–3022X

Wright HF, Mills DS, Pollux PMJ (2012) Behavioural and physiological correlates of impulsivity in the domestic dog (*Canis familiaris*). Physiol Behav 105(3):676–682

Zanella M, Vitriolo A, Andirko A, Martins PT, Sturm S, O'Rourke T, Laugsch M, Malerba N, Skaros A, Trattaro S (2019) 7q11. 23 Syndromes Reveal BAZ1B as a Master Regulator of the Modern Human Face and Validate the Self-Domestication Hypothesis. bioRxiv 570036

Zeder MA (2020) Straw foxes: domestication syndrome evaluation comes up short. Trends Ecol Evol

Zimen E (1971) Wölfe und Königspudel. Vergleichende Verhaltensbeobachtungen, München, Piper

WSC (Wolf Science Center) Ethogram for Pocket Observer

Behaviours	Behaviour type	Description
Social Interactions		
Affiliative		
Grooming (gro) *Modifier:* Animal ID	State event	To nip, lick or scratch the fur or skin occasionally the neck.
Play invitation (pi) *Modifier:* Animal ID	Point event	*Play bow* (one subject crouches down touching or almost touching forelimbs to the ground with rear end high in the air, orientation is directed towards play partner); *start-stop*: From a stand-still position, looking directly at play partner, the subject quickly lowers front end of body (though not as low as a play bow) and/or quickly bounces up in front of partner; *bounce-bow*: Similar to the start-stop, but involves repeated bounces as the actor appears to 'dance' around the play partner.
Play (py) *Modifier:* Animal ID *Modifier:* Animal ID	State event	Two or more subjects engage in motor patterns such as bite, chase, run around one another, kick, jump and maybe snap or bite without enough pressure to cause injury. May perform an invited chase sometimes with dodging behind obstacles, with frequent changes of direction around bushes, trees, rocks and the like, ultimately resulting in the subjects meeting face to face. These patterns can be typical of 'serious' functional contexts (e.g. agonistic, anti-predatory) but in a different manner. In fact, playful behaviours are often exaggerated, reordered, incomplete, brief, repeated, varied in sequence and inhibited. A play bout is considered over after 30 seconds of non-play.
Lie friendly (lf) *Modifier:* Animal ID	Point event	To lie on the back, tail-wagging, maybe kicking with the foreleg against another subject sometimes with open mouth.
Stand friendly (sf) *Modifier:* Animal ID	Point event	The subject stands with tail perpendicular to or below the plane of the back, wagging it, ears pointed forward, while another is approaching it or orienting/looking towards it.

(continued)

F. Range, S. Marshall-Pescini, *Wolves and Dogs*, Fascinating Life Sciences,
https://doi.org/10.1007/978-3-030-98411-3

Behaviours	Behaviour type	Description
Body contact (bc) *Modifier:* Animal ID, activity (rest, sleep, other)	State event	Two subjects stay (for at least 10 s) with at least a part of their bodies in contact and in a relaxed position. Please wait for 10 seconds before coding it, and wait another 10 seconds after it finishes to code the end of the behaviour.
Social sniff (sl) *Modifier:* Animal ID	Point event	To sniff another's body part except for its anogenital area.
Body rubbing (br) *Modifier:* Animal ID	Point event	To rub one's body against any part of the receiver's.
Agonistic Interactions		
Dominance Behaviour		
Stand tall (st) *Modifier:* Animal ID	Point event	Subject straightens up to full height, with a rigid posture and tail, may include raised hackles, ears erect and tail perpendicular or above the back.
Stand over (so) *Modifier:* Animal ID	Point event	To stand over another's body, with all four paws on the ground. The receiver may have either the whole body or just the forepaws under the actors' belly/side. With tail held high.
Paws on *Modifier:* Animal ID	Point event	Actor approaches recipient from the side and puts one forepaw or both over recipient's back; the actor's body posture being rigid. It refers to social interactions occurring outside both sexual and playful contexts.
Ride up (ru) *Modifier:* Animal ID	Point event	Actor places his forepaws around recipient's torso and he/she may or may not thrust his/her pelvis. It refers to social interactions occurring outside both sexual and playful contexts.
Head on (ho) *Modifier:* Animal ID	Point event	The subject approaches another's shoulder/back and puts its head on it. Most of the times formation looks like a capital 'T'. Actor can either place his/her head over recipient without making body contact or can push recipient's head downwards.
Muzzle bite (mz) *Modifier:* Animal ID	Point event	To grab the muzzle of another subject either softly or with enough pressure to make the other whimper.
Dominant-Affiliative Behaviour		
Dominant affiliation (daff) *Modifier:* Animal ID	Point event	A combination of any dominant and affiliative behaviours (see above of individual behaviours), excluding 'play' (e.g. stand tall while wagging the tail, or social sniff with a dominant posture). The individual behaviours shall not be registered separately (i.e. only 'dominant affiliation' should be coded, not the dominant and the affiliative behaviour).

(continued)

Behaviours	Behaviour type	Description
Aggressive Behaviour		
Low aggression		
Threat (th) *Modifier:* Animal ID	Point event	Subject orients towards another performing one or more of the following: Staring at, curling of the lips, baring of the canines, raising the hackles, snarling, growling, and barking, sometimes with the tail perpendicular or above the back.
Jaw spar (js) *Modifier:* Animal ID	Point event	Two subjects 'fencing' with open jaws.
Snapping (sp) *Modifier:* Animal ID	Point event	To snap teeth into the air, noisily. If snapping occurs with a crouched posture has to be considered as submissive aggression.
Lunge (lg) *Modifier:* Animal ID	Point event	Actor moves abruptly and rapidly towards recipient, for a distance of 1 or a few metres, while staring at him/her but without making physical contact.
High aggression		
Attack (at) *Modifier:* Animal ID	Point event	Running into or jumping onto another with tail, ears and sometimes hackles up, often with bites at the neck.
Knock down (kd) *Modifier:* Animal ID	Point event	To strike another subject sharply with the chest or shoulder so that the other falls to the ground.
Pin (pn) *Modifier:* Animal ID	Point event	To grab another at the neck or at the muzzle, forcing it down to the ground and holding it there.
Fight (fi) *Modifier:* Animal ID	Point event	The subject and the receiver engage in reciprocal biting and aggressive physical contact.
Chase (cs) *Modifier:* Animal ID	Point event	A subject runs after a conspecific, exhibiting threatening behaviours (see 'threat' above).
Joint aggression (ja) *Modifier:* Animal ID, other aggressors	Point event	The subject together with one or more others crowds around another one chasing, barking, biting, or wrestling with it in a harassing manner. The receiver is mostly submissive, for example lying back or trying to flee. If the focal individual is one of the aggressors, code the other aggressors as modifiers. If the focal individual is the victim, do not code any modifier and record all of the aggressors as actors.
Others		
Join victim (jv) *Modifier:* Animal ID	Point event	The supporter displays either dominant or aggressive behaviours towards the aggressor(s) of an ongoing agonistic interaction
Join aggressor (jag) *Modifier:* Animal ID	Point event	The supporter joins the actor displaying aggressive behaviours towards the victim. To be used whenever an individual joins an aggression or an ongoing joint aggression (i.e. if 'joint aggression' was already coded, additional aggressors can be added with this behaviour; aggressors in the original joint aggression have to be coded as modifiers and not by using this behaviour).

(continued)

Behaviours	Behaviour type	Description
Submissive-Aggressive Behaviour		
Submissive aggression (sa) *Modifier:* Animal ID, Submission cause	Point event	The subject is in a crouched posture, tail tucked in and sometimes with a paw up but shows teeth while having the corners of the mouth pushed forward, often accompanied by snapping, pushing the other away, and/or barking.
Submissive Behaviour		
Avert gaze (ag) *Modifier:* Animal ID, Submission cause	Point event	Actor turns his or her head away in response to the receiver either reducing the distance between, displaying dominant/aggressive behaviours, or orienting towards them. If occurring at the same time as 'head dip' and/or 'flattening ears', it has to be coded as 'submissive head posture'. Additionally, if it happens in conjunction with 'tail dip', it has to be coded as 'head +tail submission' instead of this behaviour.
Head dip (hd) *Modifier:* Animal ID, Submission cause	Point event	Actor lowers his/her head in response to the receiver either reducing the distance between, displaying dominant/aggressive behaviours, or orienting towards them. During the displaying of this behavioural pattern, the actor's muzzle may point downwards or forwards but the head and neck are leaning downwards. If occurring at the same time as 'avert gaze' and/or 'flattening ears', it has to be coded as 'submissive head posture'. Additionally, if it happens in conjunction with 'tail dip', it has to be coded as 'head+tail submission' instead of this behaviour.
Flattening ears (fe) *Modifier:* Animal ID, Submission cause	Point event	Actor retracts the ears into the neck or holds them backwards in response to the receiver either reducing the distance between, displaying dominant/aggressive behaviours, or orienting towards them. If occurring at the same time as 'head dip' and/or 'avert gaze', it has to be coded as 'submissive head posture'. Additionally, if it happens in conjunction with 'tail dip', it has to be coded as 'head+tail submission' instead of this behaviour.
Submissive head posture (shp) *Modifier:* Animal ID, Submission cause	Point event	Any combination of 'avert gaze', 'head dip', and 'flattening ears' occurring at the same time (within 5 seconds). These individual behaviours shall not be coded, as 'submissive head posture' covers all of them. Additionally, if it happens in conjunction with 'tail dip', it has to be coded as 'head+tail submission' instead of this behaviour.
Tail dip (td) *Modifier:* Animal ID, Submission cause	Point event	Actor lowers his/her tail in response to the receiver either reducing the distance between, displaying dominant/aggressive behaviours, or orienting towards them. The tail can be held near the hind legs, or tucked between them.

(continued)

Behaviours	Behaviour type	Description
Head + tail submission (hts) *Modifier:* Animal ID, Submission cause	Point event	Any combination of 'avert gaze', 'head dip', and 'flattening ears' with 'tail dip'. These individual behaviours shall not be coded, as 'head+tail submission' covers all of them.
Crouch (cr) *Modifier:* Animal ID, Submission cause	Point event	Lowered posture bending the legs, arching the back, lowering the tail between the hind legs, potentially including also gaze aversion head dip and ears flat.
Passive submission (ps) *Modifier:* Animal ID, Submission cause	Point event	To lie on the back showing the stomach potentially holding the tail between the legs. The ears are held back and close to the head and the subject can raise a hind leg for inguinal presentation.
Active submission (acs) *Modifier:* Animal ID, Submission cause	Point event	The subject has its tail tucked between the hind legs sometimes wagging it while he is in a crouched position (with hindquarters lowered) and may attempt to paw and lick the side of actors'/aggressor's muzzle. The behaviour may include urination.
Play submissive (pls) *Modifier:* Animal ID, Submission cause	State event	To play with the tail between the hind legs, often running away and snapping at the other.
Withdrawing (wd) *Modifier:* Animal ID, Submission cause	Point event	It occurs after the other individual approaches within one body length from the retreating one; or when a subject has been threatened, or attacked by another, or a fight has taken place. The subject withdraws from another moving away slowly in the opposite direction, displaying a submissive posture.
Flee (fl) *Modifier:* Animal ID, Submission cause	Point event	To run away from another with tail tucked between the legs and body ducked. It occurs when a subject has been threatened or attacked by another, or after a fight.
Avoidance (av) *Modifier:* Animal ID, Submission cause	Point event	In response to another reducing the distance towards it, the subject moves away displaying a submissive posture. The subject may also look at the individual he is trying to avoid.
Approach/Leave		
Neutral approach (na) *Modifier:* Animal ID	Point event	To approach another within one body length for at least 5 seconds, in a relaxed manner, without necessarily showing interest in the other.
Approach friendly (af) *Modifier:* Animal ID	Point event	To approach another subject within one body length remaining within that distance for at least 5 seconds. The approach is characterized by the subject holding the tail perpendicular to or below the plane of the back and wagging it.
Approach submissive (as) *Modifier:* Animal ID, Submission cause	Point event	To slowly approach another within one body length remaining within that distance for at least 5 seconds. The approach is characterized by a ducked posture, tail between the legs, and/or any of the behaviours that comprise 'submissive head posture'. Subject can also be moving in a wavy line and in a hesitant (stop-start) manner.

(continued)

Behaviours	Behaviour type	Description
Approach dominant (ad) *Modifier:* Animal ID	Point event	To approach another subject within one body length for at least 5 seconds, with the tail perpendicular or above the plane of the back and the ears erect and pointed forward.
Leave (le) *Modifier:* Animal ID	Point event	To move away from another subject, beyond one body length, in a neutral posture.
Submission cause	Modifier	It serves to record if the submissive behaviour was spontaneous (no aggres before) or it was caused by prior aggression (aggres before)
Others		
Genital sniffing (gs) *Modifier:* Focal animal-posture, animal ID, animal ID-posture (neutral, dominant, submissive).	Point event	To sniff the genital area of another subject.
Greeting (gre) *Modifier:* Animal ID	Point event	To interact in a friendly and relaxed manner holding the ears back, showing much tail wagging and licking of the other's mouth/muzzle. The subject however does not show crouching/lowered hindquarters nor is the tail tucked between the legs.
Stand Over tail low *Modifier:* Animal ID	Point event	To stand over another's body, with all four paws on the ground. The receiver may have either the whole body or just the forepaws under the actors' belly/side. With tail held neutral or low.
Not coded	State event	It must be coded when the focal animal does not display any other behaviours listed in the ethogram
Start	Point event	Has to be coded at the beginning of the observation.
Disturbance	Point event	Independent variable. Coded as 'yes' if a human-caused disturbance has taken place and 'no' otherwise. Coded within the last interval of the observation.
Weather	Point event	Independent variable. Describes the weather during the observation (several modifiers can be picked in case more than one applies). Coded after 'start'
Enclosure	Point event	Independent variable. Describes the enclosure in which the pack is. Coded after 'start'
Female in heat	Point event	Independent variable. Coded only in case there is a female in heat in the observed pack. Coded after 'start'
Self-directed Behaviours		
Autogrooming (au)	State event	To lick and nibble its own skin and fur.
Body shaking (bs)	Point event	To shake the body or the neck.
Yawning (ya)	Point event	To open the jaws wide without vocalizing
Lips licking (ll)	Point event	To extrude the tongue from the mouth and run it over the lips.

(continued)

www.ingramcontent.com/pod-product-compliance
Ingram Content Group UK Ltd.
Pitfield, Milton Keynes, MK11 3LW, UK
UKHW021421091026
15624UKWH00010B/366
9783030984106